CAMBRIDGE LIBRARY COLLECTION

Books of enduring scholarly value

Mathematical Sciences

From its pre-historic roots in simple counting to the algorithms powering modern desktop computers, from the genius of Archimedes to the genius of Einstein, advances in mathematical understanding and numerical techniques have been directly responsible for creating the modern world as we know it. This series will provide a library of the most influential publications and writers on mathematics in its broadest sense. As such, it will show not only the deep roots from which modern science and technology have grown, but also the astonishing breadth of application of mathematical techniques in the humanities and social sciences, and in everyday life.

The Scientific Papers of Sir George Darwin

Sir George Darwin (1845-1912) was the second son and fifth child of Charles Darwin. After studying mathematics at Cambridge he read for the Bar, but soon returned to science and to Cambridge, where in 1883 he was appointed Plumian Professor of Astronomy and Experimental Philosophy. His family home is now the location of Darwin College. His work was concerned primarily with the effect of the sun and moon on tidal forces on Earth, and with the theoretical cosmogony which evolved from practical observation: he formulated the fission theory of the formation of the moon (that the moon was formed from still-molten matter pulled away from the Earth by solar tides). He also developed a theory of evolution for the Sun–Earth–Moon system based on mathematical analysis in geophysical theory. This volume of his collected papers covers oceanic tides and lunar disturbances of gravity.

The Scientific Papers of Sir George Darwin

Volume 1: Oceanic Tides and Lunar Disturbance of Gravity

George Howard Darwin

CAMBRIDGE UNIVERSITY PRESS

Cambridge New York Melbourne Madrid Cape Town Singapore São Paolo Delhi

Published in the United States of America by Cambridge University Press, New York

www.cambridge.org
Information on this title: www.cambridge.org/9781108004428

This edition first published 1907
This digitally printed version 2009

ISBN 978-1-108-00442-8

SCIENTIFIC PAPERS

CAMBRIDGE UNIVERSITY PRESS WAREHOUSE,
C. F. CLAY, MANAGER.
London: FETTER LANE, E.C.
Glasgow: 50, WELLINGTON STREET.

Leipzig: F. A. BROCKHAUS.
New York: G. P. PUTNAM'S SONS.
Bombay and Calcutta: MACMILLAN AND CO., LTD.

SCIENTIFIC PAPERS

BY

SIR GEORGE HOWARD DARWIN,

K.C.B., F.R.S.

FELLOW OF TRINITY COLLEGE

PLUMIAN PROFESSOR IN THE UNIVERSITY OF CAMBRIDGE

VOLUME I

OCEANIC TIDES

AND

LUNAR DISTURBANCE OF GRAVITY

CAMBRIDGE:

AT THE UNIVERSITY PRESS

1907

Cambridge
PRINTED BY JOHN CLAY, M.A.
AT THE UNIVERSITY PRESS.

PREFACE.

WHEN the Syndics of the Cambridge University Press did me the honour of offering to publish a collection of my mathematical papers, I had to consider the method of arrangement which would be most convenient. A simple chronological order has been adopted in various collections of this kind, and this plan certainly has advantages; but an arrangement of papers according to subject may be more convenient. In the case of my own work the separation into well-defined groups of subjects was easy, and I have therefore adopted this latter method. I shall, however, give at the beginning of each volume a chronological list with a statement as to the volume in which each paper will be found.

This first volume contains papers on Oceanic Tides and on an attempt to measure the Lunar Disturbance of Gravity; the second will give my papers on Tidal Friction and on the astronomical speculations arising therefrom; the third will be devoted to papers on Figures of Equilibrium of Rotating Liquid and on cognate subjects; and the fourth will be on Periodic Orbits and on various miscellaneous subjects.

Throughout corrections and additions will be marked by inclusion in square parentheses.

The whole of my work on oceanic tides and the attempt made by my brother Horace and me to measure the attraction of the moon sprang from ideas initiated by Lord Kelvin, and I should wish to regard this present volume as being, in a special sense, a tribute to him.

Early in my scientific career it was my good fortune to be brought into close personal relationship with Lord Kelvin. Many visits to Glasgow and to Largs have taught me to look up to him as my master, and I cannot find words to express how much I owe to his friendship and to his inspiration.

The following statement gives in a few words an explanation of how the several investigations originated, and how they are connected together.

Part I is devoted to the consideration of OCEANIC TIDES. The advisability of applying to the tides a method of analysis similar to that used in the Lunar and Planetary Theories was first suggested by Lord Kelvin. Reports on the Harmonic Analysis of tidal observations were drawn up by him and presented to the British Association in 1868, 1870, 1871, 1872 and 1876. As the analysis employed in those reports needed coordination and revision, a new committee consisting of Professor Adams and myself was appointed in 1882. Professor Adams left the matter very much in my hands, although I enjoyed the great benefit of his advice from time to time. The first three papers below form the first, third and fourth reports of this Committee. The second report being merely formal is not reproduced. Some paragraphs of a merely temporary interest are omitted, and a few corrections and alterations have been made by the light of subsequent experience.

These papers explain the method of harmonic analysis, and its connection with the old method of hour-angles, declinations and parallaxes. The third paper shows how a tide-table may be computed from the harmonic tidal constants.

The fourth paper is an Article on the Tides written for the *Admiralty Scientific Manual,* and is reprinted by permission of the Admiralty. It explains the method of analysing a short series of observations and the computation of an approximate tide-table. A small mistake in the *Manual* has been corrected. The Article would have been too long for its place of publication if it had contained the analytical reasoning on which the several rules of procedure are based, but this analysis is contained in the three preceding papers.

In the fifth paper a method is devised for evaluating the harmonic constants from observations of high and low-water.

The sixth paper explains the use of a sort of abacus for the harmonic analysis of hourly observations.

In the seventh paper I return to the subject of the third paper, and show how a tide-table may be formed from the harmonic constants with any desirable degree of accuracy.

Next in logical order would have come two papers in which Colonel Baird and I began a systematic collection of tidal constants for several ports, but these papers are not reproduced. We did not continue our attempt, and fortunately Mr Rollin Harris has made a large collection of such results in his *Manual of Tides* published by the U.S. Coast and Geodetic Survey.

The rest of the papers in Part I are devoted to the theory of the tides.

In Thomson and Tait's *Natural Philosophy* Lord Kelvin discussed the correction required to make the equilibrium theory of tides true of an ocean interrupted by continents. The eighth paper gives the numerical evaluation of this correction by Professor H. H. Turner and me.

The ninth paper is an extract of paragraphs contributed to the second edition of Thomson and Tait's *Natural Philosophy*, in which an attempt is made to estimate the amount of tidal yielding to which the solid earth is subject.

Laplace was of opinion that the tidal oscillations of long period might be adequately represented by the equilibrium theory. But Lord Kelvin showed that a yielding of the solid earth would produce a calculable reduction of the oceanic tide according to the equilibrium theory. Accordingly it appeared that a measurement of the actual values of the tides of long period would afford a measure of the elastic yielding of the earth's mass. In this ninth paper an evaluation is made of the ranges of these tides at a number of ports; and thence an estimate is obtained of the tidal yielding of the earth.

The tenth paper is on the dynamical theory of the tides, being an extract from an unpublished article on Tides for a new edition of the *Encyclopædia Britannica*; it is reproduced by the kind permission of the proprietors of that work. My article in the old edition forms the basis of this new article, but Mr Hough's important addition to that theory is now included.

In the eleventh paper arguments are adduced controverting Laplace's opinion as to the adequacy of the equilibrium theory for discussing the tides of long period, and it is shown that the problem is in fact a dynamical one. Hence the evaluation of the amount of elastic yielding of the earth by means of the observed heights of these tides cannot be accepted as exact.

The last paper in the first part contains the reduction of the tidal observations made by the officers of the 'Discovery' in the course of their celebrated antarctic expedition. The volumes of scientific results of the 'Discovery' are not yet published.

Part II contains two papers on the LUNAR DISTURBANCE OF GRAVITY. They were reports of a Committee to the British Association on an attempt to make direct measurement of the attraction of the moon. The work was carried on at Cambridge in 1880 by my brother Horace and me, with a form of instrument of which the original idea is due to Lord Kelvin. Although the report is written in my name, it should be explained that my brother took the leading part in our experiments and designed all the mechanical appliances used.

The second of these papers, the fourteenth in the volume, is principally historical, but it contains an investigation of the results which may be expected to result from the varying elastic flexure of the soil under varying superincumbent weights.

The matters considered in these papers have been the subject of very many experimental investigations during the last twenty-five years, and a large bibliography of seismology would be required merely to enumerate all that has been written on them. I therefore make no attempt to furnish references, and leave the work in the form in which it appeared originally.

G. H. DARWIN.

September, 1907.

CONTENTS.

PAGE

Chronological List of Papers with References to the Volumes in which they probably will be contained xi

PART I.

OCEANIC TIDES.

ART.

1. The Harmonic Analysis of Tidal Observations 1

[Report of a Committee, consisting of Professors G. H. DARWIN and J. C. ADAMS, for the Harmonic Analysis of Tidal Observations. *British Association Report for* 1883, pp. 49—118.]

2. On the Periods Chosen for Harmonic Analysis, and a Comparison with the Older Methods by means of Hour-angles and Declinations 70

[Third Report of the Committee, consisting of Professors G. H. DARWIN and J. C. ADAMS, for the Harmonic Analysis of Tidal Observations. Drawn up by Professor G. H. DARWIN. *British Association Report for* 1885, pp. 35—60.]

3. Datum Levels: the Treatment of a Short Series of Tidal Observations and on Tidal Prediction 97

[Report of the Committee consisting of Professor G. H. DARWIN, Sir W. THOMSON, and Major BAIRD, for the purpose of preparing instructions for the practical work of Tidal Observation; and Fourth Report of the Committee consisting of Professors G. H. DARWIN and J. C. ADAMS, for the Harmonic Analysis of Tidal Observations. Drawn up by G. H. DARWIN. *British Association Report for* 1886, pp. 40—58.]

4. A General Article on the Tides 119

[Article "Tides," *Admiralty Scientific Manual* (1886), pp. 53—91.]

5. On the Harmonic Analysis of Tidal Observations of High and Low Water 157

[*Proceedings of the Royal Society*, XLVIII. (1890), pp. 278—340.]

6. On an Apparatus for Facilitating the Reduction of Tidal Observations 216

[*Proceedings of the Royal Society*, LII. (1892), pp. 345—389.]

7. On Tidal Prediction 258

[*Bakerian Lecture, Philosophical Transactions of the Royal Society of London*, CLXXXII. (1891), A, pp. 159—229.]

8. On the Correction to the Equilibrium Theory of Tides for the Continents. I. By G. H. DARWIN. II. By H. H. TURNER . 328

[*Proceedings of the Royal Society of London*, XL. (1886), pp. 303—315.]

ART. PAGE

9. Attempted Evaluation of the Rigidity of the Earth from the Tides of Long Period 340

[This is § 848 of the second edition of Thomson and Tait's *Natural Philosophy* (1883). There have been some changes of notation, so as to make the investigation consistent with the other papers in this volume. Some portions have been omitted, where omissions could be made without any interference with the main result.]

10. Dynamical Theory of the Tides 347

[This contains certain sections from the article "TIDES" written in 1906 for the new edition of the *Encyclopædia Britannica*, being based on the corresponding paragraphs in the original edition and on the article "TIDES" in the supplementary volumes. Reproduced by special permission of the Proprietors of the *Enc. Brit.*]

11. On the Dynamical Theory of the Tides of Long Period . . 366

[*Proceedings of the Royal Society of London*, XLI. (1886), pp. 337—342.]

12. On the Antarctic Tidal Observations of the 'Discovery' . 372

[To be published hereafter as a contribution to the scientific results of the voyage of the 'Discovery.']

PART II.

THE LUNAR DISTURBANCE OF GRAVITY.

13. On an Instrument for Detecting and Measuring Small Changes in the Direction of the Force of Gravity. By G. H. DARWIN and HORACE DARWIN 389

[Report of the Committee, consisting of Mr G. H. DARWIN, Professor Sir WILLIAM THOMSON, Professor TAIT, Professor GRANT, Dr SIEMENS, Professor PURSER, Professor G. FORBES, and Mr HORACE DARWIN, appointed for the Measurement of the Lunar Disturbance of Gravity. This Report is written in the name of G. H. DARWIN merely for the sake of verbal convenience. *British Association Report for* 1881, pp. 93—126.]

14. The Lunar Disturbance of Gravity; Variations in the Vertical due to Elasticity of the Earth's Surface 430

[Second Report of the Committee, consisting of Mr G. H. DARWIN, Professor Sir WILLIAM THOMSON, Professor TAIT, Professor GRANT, Dr SIEMENS, Professor PURSER, Professor G. FORBES, and Mr HORACE DARWIN, appointed for the Measurement of the Lunar Disturbance of Gravity. Written by Mr G. H. DARWIN. *British Association Report for* 1882, pp. 95—119.]

INDEX 461

PLATE. Abacus for the harmonic analysis of tidal observations *To face* 219

CHRONOLOGICAL LIST OF PAPERS WITH REFERENCES TO THE VOLUMES IN WHICH THEY WILL PROBABLY BE CONTAINED.

YEAR	TITLE AND REFERENCE	Probable volume in collected papers
1875	On two applications of Peaucellier's cells. London Math. Soc. Proc., 6, 1875, pp. 113, 114.	IV
1875	On some proposed forms of slide-rule. London Math. Soc. Proc., 6, 1875, p. 113.	IV
1875	The mechanical description of equipotential lines. London Math. Soc. Proc., 6, 1875, pp. 115—117.	IV
1875	On a mechanical representation of the second elliptic integral. Messenger of Math., 4, 1875, pp. 113—115.	IV
1875	On maps of the World. Phil. Mag., 50, 1875, pp. 431—444.	IV
1876	On graphical interpolation and integration. Brit. Assoc. Rep., 1876, p. 13.	IV
1876	On the influence of geological changes on the Earth's axis of rotation. Roy. Soc. Proc., 25, 1877, pp. 328--332; Phil. Trans., 167, 1877, pp. 271—312.	III
1876	On an oversight in the *Mécanique Céleste*, and on the internal densities of the planets. Astron. Soc. Month. Not., 37, 1877, pp. 77—89.	III
1877	A geometrical puzzle. Messenger of Math., 6, 1877, p. 87.	IV
1877	A geometrical illustration of the potential of a distant centre of force. Messenger of Math., 6, 1877, pp. 97—98.	IV
1877	Note on the ellipticity of the Earth's strata. Messenger of Math., 6, 1877, pp. 109, 110.	III
1877	On graphical interpolation and integration. Messenger of Math., 6, 1877, pp. 134—136.	IV
1877	On a theorem in spherical harmonic analysis. Messenger of Math., 6, 1877, pp. 165—168.	IV
1877	On a suggested explanation of the obliquity of planets to their orbits. Phil. Mag., 3, 1877, pp. 188—192.	II
1877	On fallible measures of variable quantities, and on the treatment of meteorological observations. Phil. Mag., 4, 1877, pp. 1—14.	IV
1878	On Professor Haughton's estimate of geological time. Roy. Soc. Proc., 27, 1878, pp. 179—183.	IV
1878	On the bodily tides of viscous and semi-elastic spheroids, and on the Ocean tides on a yielding nucleus. Roy. Soc. Proc., 27, 1878, pp. 419—424; Phil. Trans., 170, 1879, pp. 1—35.	II

YEAR	TITLE AND REFERENCE	Probable volume in collected papers
1878	On the precession of a viscous spheroid. Brit. Assoc. Rep., 1878, pp. 482—485.	II
1879	On the precession of a viscous spheroid, and on the remote history of the Earth. Roy. Soc. Proc., 28, 1879, pp. 184—194; Phil. Trans., 170, 1879, pp. 447—538.	II
1879	Problems connected with the tides of a viscous spheroid. Roy. Soc. Proc., 28, 1879, pp. 194—199; Phil. Trans., 170, 1879, pp. 539—593.	II
1879	Note on Thomson's theory of the tides of an elastic sphere. Messenger of Math., 8, 1879, pp. 23—26.	II
1879	The determination of the secular effects of tidal friction by a graphical method. Roy. Soc. Proc., 29, 1879, pp. 168—181.	II
1880	On the secular changes in the elements of the orbit of a satellite revolving about a tidally distorted planet. Roy. Soc. Proc., 30, 1880, pp. 1—10; Phil. Trans., 171, 1880, pp. 713—891.	II
1880	On the analytical expressions which give the history of a fluid planet of small viscosity, attended by a single satellite. Roy. Soc. Proc., 30, 1880, pp. 255—278.	II
1880	On the secular effects of tidal friction. Astr. Nachr., 96, 1880, col. 217—222.	omitted
1881	On the tidal friction of a planet attended by several satellites, and on the evolution of the solar system. Roy. Soc. Proc., 31, 1881, pp. 322—325; Phil. Trans., 172, 1881, pp. 491—535.	II
1881	On the stresses caused in the interior of the Earth by the weight of continents and mountains. Phil. Trans., 173, 1883, pp. 187—230; Amer. Journ. Sci., 24, 1882, pp. 256—269.	II
1881	(Together with Horace Darwin.) On an instrument for detecting and measuring small changes in the direction of the force of gravity. Brit. Assoc. Rep., 1881, pp. 93—126; Annal. Phys. Chem., Beibl. 6, 1882, pp. 59—62.	I
1882	On variations in the vertical due to elasticity of the Earth's surface. Brit. Assoc. Rep., 1882, pp. 106—119; Phil. Mag., 14, 1882, pp. 409—427.	I
1882	On the method of harmonic analysis used in deducing the numerical values of the tides of long period, and on a misprint in the Tidal Report for 1872. Brit. Assoc. Rep., 1882, pp. 319—327.	omitted
1882	A numerical estimate of the rigidity of the Earth. Brit. Assoc. Rep., 1882, pp. 472—474; § 848, Thomson and Tait's Nat. Phil. second edition.	I
1883	Report on the Harmonic analysis of tidal observations. Brit. Assoc. Rep., 1883, pp. 49—117.	I
1883	On the figure of equilibrium of a planet of heterogeneous density. Roy. Soc. Proc., 36, pp. 158—166.	III
1883	On the horizontal thrust of a mass of sand. Instit. Civ. Engin. Proc., 71, 1883, pp. 350—378.	IV
1884	On the formation of ripple-mark in sand. Roy. Soc. Proc., 36, 1884, pp. 18—43.	IV

YEAR	TITLE AND REFERENCE	Probable volume in collected papers
1884	Second Report of the Committee, consisting of Professors G. H. Darwin and J. C. Adams, for the harmonic analysis of tidal observations. Drawn up by Professor G. H. Darwin. Brit. Assoc. Rep., 1884, pp. 33—35.	omitted
1885	Note on a previous paper. Roy. Soc. Proc., 38, pp. 322—328.	II
1885	Results of the harmonic analysis of tidal observations. (Jointly with A. W. Baird.) Roy. Soc. Proc., 39, pp. 135—207.	omitted
1885	Third Report of the Committee, consisting of Professors G. H. Darwin and J. C. Adams, for the harmonic analysis of tidal observations. Drawn up by Professor G. H. Darwin. Brit. Assoc. Rep., 1885, pp. 35—60.	I
1886	Report of the Committee, consisting of Professor G. H. Darwin, Sir W. Thomson, and Major Baird, for preparing instructions for the practical work of tidal observation; and Fourth Report of the Committee, consisting of Professors G. H. Darwin and J. C. Adams, for the harmonic analysis of tidal observations. Drawn up by Professor G. H. Darwin. Brit. Assoc. Rep., 1886, pp. 40—58.	I
1886	Presidential Address. Section A, Mathematical and Physical Science. Brit. Assoc. Rep., 1886, pp. 511—518.	IV
1886	On the correction to the equilibrium theory of tides for the continents. I. By G. H. Darwin. II. By H. H. Turner. Roy. Soc. Proc., 40, pp. 303—315.	I
1886	On Jacobi's figure of equilibrium for a rotating mass of fluid. Roy. Soc. Proc., 41, pp. 319—336.	III
1886	On the dynamical theory of the tides of long period. Roy. Soc. Proc., 41, pp. 337—342.	I
1886	Article 'Tides.' (Admiralty) Manual of Scientific Inquiry.	I
1887	On figures of equilibrium of rotating masses of fluid. Roy. Soc. Proc., 42, pp. 359—362; Phil. Trans., 178A, pp. 379—428.	III
1887	Note on Mr Davison's Paper on the straining of the Earth's crust in cooling. Phil. Trans., 178A, pp. 242—249.	IV
1888	Article 'Tides.' Encyclopaedia Britannica.	Certain sections in I
1888	On the mechanical conditions of a swarm of meteorites, and on theories of cosmogony. Roy. Soc. Proc., 45, pp. 3—16; Phil. Trans., 180A, pp. 1—69.	IV
1889	Second series of results of the harmonic analysis of tidal observations. Roy. Soc. Proc., 45, pp. 556—611.	omitted
1889	Meteorites and the history of Stellar systems. Roy. Inst. Rep., Friday, Jan. 25, 1889.	omitted
1890	On the harmonic analysis of tidal observations of high and low water. Roy. Soc. Proc., 48, pp. 278—340.	I
1891	On tidal prediction. Bakerian Lecture. Roy. Soc. Proc., 49, pp. 130—133; Phil. Trans., 182A, pp. 159—229.	I

YEAR	TITLE AND REFERENCE	Probable volume in collected papers
1892	On an apparatus for facilitating the reduction of tidal observations. Roy. Soc. Proc., 52, pp. 345—389.	I
1896	On periodic orbits. Brit. Assoc. Rep., 1896, pp. 708, 709.	omitted
1897	Periodic orbits. Acta Mathematica, 21, pp. 101—242, also (with omission of certain tables of results) Mathem. Annalen, 51, pp. 523—583.	IV
	[by S. S. Hough. On certain discontinuities connected with periodic orbits. Acta Math., 24 (1901), pp. 257—288.]	IV
1899	The theory of the figure of the Earth carried to the second order of small quantities. Roy. Astron. Soc. Month. Not., 60, pp. 82—124.	III
1900	Address delivered by the President, Professor G. H. Darwin, on presenting the Gold Medal of the Society to M. H. Poincaré. Roy. Astron. Soc. Month. Not., 60, pp. 406—415.	IV
1901	Ellipsoidal harmonic analysis. Roy. Soc. Proc., 68, pp. 248—252; Phil. Trans., 197A, pp. 461—557.	III
1901	On the pear-shaped figure of equilibrium of a rotating mass of liquid. Roy. Soc. Proc., 69, pp. 147, 148; Phil. Trans., 198A, pp. 301—331.	III
1902	Article 'Tides.' Encyclopaedia Britannica, supplementary volumes.	Certain sections in I
1902	The stability of the pear-shaped figure of equilibrium of a rotating mass of liquid. Roy. Soc. Proc., 71, pp. 178—183; Phil. Trans., 200A, pp. 251—314.	III
1903	On the integrals of the squares of ellipsoidal surface harmonic functions. Roy. Soc. Proc., 72, p. 492; Phil. Trans., 203A, pp. 111—137.	III
1903	The approximate determination of the form of Maclaurin's spheroid. Trans. Amer. Math. Soc., 4, pp. 113—133.	III
1903	The Eulerian nutation of the Earth's axis. Bull. Acad. Roy. de Belgique (Sciences), pp. 147—161.	IV
1905	The analogy between Lesage's theory of gravitation and the repulsion of light. Roy. Soc. Proc., 76A, pp. 387—410.	IV
1905	Address by Professor G. H. Darwin, President. Brit. Assoc. Rep., 1905, pp. 3—32.	IV
1906	On the figure and stability of a liquid satellite. Roy. Soc. Proc., 77A, pp. 422—425; Phil. Trans., 206A, pp. 161—248.	III
	Unpublished Article 'Tides.' Encyclopaedia Britannica, new edition to be published hereafter (by permission of the proprietors).	Certain sections in I
	Unpublished Article 'Bewegung der Hydrosphäre' (The Tides). Encyklopädie der Mathematischen Wissenschaften, VI.	IV

PART I

OCEANIC TIDES

1.

THE HARMONIC ANALYSIS OF TIDAL OBSERVATIONS.

[Report of a Committee, consisting of Professors G. H. DARWIN and J. C. ADAMS, for the Harmonic Analysis of Tidal Observations. *British Association Report for* 1883, pp. 49—118.]

CONTENTS.

SECT.		PAGE
	PREFACE—Account of Operations	2
1.	Notation adopted in the Tidal Reports—[A] Schedule of Notation	4
2.	Development of the Equilibrium Theory of Tides, with reference to Tidal Observations — Tide-generating Potential — Evection — Variation — Equilibrium Tide—[B i.] [B ii.] [B iii.] Schedules of Lunar Tides—[C] Schedule of Solar Tides—[D] Schedule of Speeds in Degrees per m. s. hour—[E] Schedule of Theoretical Importance	6
3.	Tides depending on the Fourth Power of the Moon's Parallax	26
4.	Meteorological Tides, Over-tides, and Compound Tides—[F] Schedule of Over-tides—[G] Schedule of Speeds arising out of Combinations—[H] Schedule of Compound Tides	28
5.	The Method of Reduction of Tidal Observations—Definitions of A, B, R, ζ, f, H, $V+u$, κ—Treatment of Sidereal Diurnal, and Semi-diurnal Tides, K_1, K_2—The Tide L—The Tide M_1	34
6.	The Method of Computing the Arguments and Coefficients—Formulæ for Computing I, ν, ξ—Mean Values of the Coefficients in Schedules [B]—Formulæ for Computing f—Formulæ for s, p, h, p_1, N	40
7.	Summary of Initial Arguments and Factors of Reduction—Schedule [I]	45
8.	On the Reductions of the Published Results of Tidal Analysis—[Schedule [J] omitted]	48
9.	Description of the Numerical Harmonic Analysis for the Tides of Short Period—Incidence of Special Hours amongst m. s. hours—Schedule [K] Form for entry of Tidal Observations — Choice of Special Periods; [Schedule [L] omitted]—Augmenting Factors, Schedule [M]—Arrangement of Harmonic Analysis—Schedule [N], Form for Analysis—Schedule [O], Form for Evaluation of ζ, R, κ, H	48
10.	On the Harmonic Analysis for Tides of Long Period—Methods of taking Daily Means—Clearance from Effect of Tides of Short Period—Adams' use of Tide Predicter for this end—Ten final Equations for Components of Tides—Schedule [P] of the Coefficients in the Ten Equations—Clearance of Daily Means effected in the Final Equations, Schedule [Q] of Clearance Coefficients—Treatment for Gaps in the Series of Observations	57
11.	Method of Equivalent Multipliers for the Harmonic Analysis for the Tides of Long Period—Schedule [R], Form for Reduction	66
12.	Auxiliary Tables drawn up under the superintendence of Major A. W. Baird, R.E. [largely abridged, and replaced by formulæ]	68

Preface. Account of Operations.

A COMMITTEE appointed for the examination of the question of the Harmonic Analysis of Tidal Observations practically finds itself engaged in the question of the reduction of Indian Tidal Observations; since it is only in that country that any extensive system of observation with systematic publication of results* exists.

. . . .

[Early in 1883] I proceeded to draw up a considerable part of this Report, had it printed, and submitted it to Major Baird†. I was not at that time aware of the extent to which Mr Roberts, of the *Nautical Almanac* office, co-operated in England in the tidal operations, nor did I know that he was not unfrequently taking the advice of Professor Adams. It was not until Major Baird had read what I had written, and expressed his approval of the methods suggested, that these facts came to my knowledge; but it must be admitted that it was through my own carelessness that this was so. I then found that Professor Adams decidedly disapproved of the notation adopted, and would have preferred to throw over the notation of the old Reports‡ and take a new departure. The notation of the old Reports seems to me also to be unsatisfactory, but, seeing that Major Baird and his staff were already familiar with that notation, I considered that an entire change would be impolitic, and that it was better to allow the greater part of the existing notation to stand, but to introduce modifications. The fact that Major Baird, who was actually to work the method, approved of what had been written, and had already mastered it, went far to prejudge the question, and Professor Adams agreed, after discussion, that it would on the whole be best to allow the work to go on in the lines in which it had been started.

It has seemed proper to give this account of our operations in order that Professor Adams may be relieved from responsibility for the analytical methods and notation here adopted. I may state, however, that although the Report is drawn up in a form probably differing widely from that which it would have had if Professor Adams had been the author, yet he agrees with the correctness of the methods pursued. I have been in constant communication with him for the past eight months, and have received many valuable criticisms and suggestions.

Mr Roberts has been supervising the printing of a new edition of the computation forms; they have undergone some modification in accordance

* [This refers to the year 1883.]

† [Now Colonel Baird, R.E., F.R.S., at that time the officer in charge at Poona of the Tidal Department of the Survey of India.]

‡ [These are the Reports to the British Association for the years 1868, 1870, 1871, 1872, 1876.]

with this Report. He has also computed certain new coefficients [Schedule Q] which are required in the reductions.

. . . .

The general scope of this paper is to form a manual for the reduction of tidal observations by the Harmonic Analysis inaugurated by Sir William Thomson, and carried out by the previous Committee of the British Association*.

In the present Report the method of mathematical treatment differs considerably from that of Sir William Thomson†. In particular, he has followed, and extended to the diurnal tides, Laplace's method of referring each tide to the motion of an *astre fictif* in the heavens, and he considers that these fictitious satellites are helpful in forming a clear conception of the equilibrium theory of tides. As, however, I have found the fiction rather a hindrance than otherwise, I have ventured to depart from this method, and have connected each tide with an 'argument,' or an angle increasing uniformly with the time and giving by its hourly increase the 'speed' of the tide. In the method of the *astres fictifs*, the speed is the difference between the earth's angular velocity of rotation and the motion of the fictitious satellite amongst the stars. It is a consequence of the difference in the mode of treatment, and of the fact that the elliptic tides are here developed to a higher degree of approximation, that none of the present Report is quoted from the previous ones.

The Report of 1876 was not intended to be a final production, and it did not contain any complete explanation of a considerable portion of the numerical operations of the Harmonic Analysis. The present Report is intended to systematise the exposition of the theory of the harmonic analysis, to complete the methods of reduction, and to explain the whole process.

A careful survey of the methods hitherto in use has brought to light a good many minor points in which improvements may be introduced, but it has seemed desirable not to disturb the system, which is in working order, more than can be helped. It has also appeared that the published results have not been arranged in a form which lends itself to a satisfactory examination of the whole method. This defect will, we hope, now be remedied.

The first section refers to the notation, and contains a schedule of nomenclature by initials of the several tides under examination. The schedule is not, strictly speaking, in its proper position at the beginning, because it involves the results of subsequent analysis, but the advantage gained by having this list in a position of easy reference seems to outweigh the want of logic.

* See especially the Reports for 1872 and 1876.

† The present method of development is that pursued in a paper in the *Phil. Trans. R.S.*, Part II. 1880, p. 713. [To be included in Vol. II. of these collected papers.]

The forms for computation are privately printed for the India Office, and are therefore inaccessible to the public*. The type has been broken up, and very few copies remain, but we shall send copies to the Libraries of the following Societies, viz.: Royal Societies, London and Edinburgh; the Academies of Science of Dublin, Paris, Berlin, and Vienna, the Coast Survey of the United States at Washington, and the Cambridge Philosophical Society.

G. H. DARWIN.

§ 1. *The Notation adopted in the Tidal Reports.*

In considering the notation to be adopted, much weight should be given to the fact that a large mass of analysis and computation already exists in a certain form. We have not thus got a *tabula rasa* to work on, but had better accept a good deal that has grown up by a process of accretion. It is certainly unfortunate that a dual system should have been adopted, in which one set of letters are derived from the Greek and another from the English.

The letters γ, σ, η, ϖ are appropriated respectively to the earth's angular velocity of rotation, to the mean motions of the moon, sun, and lunar perigee. They form the initial letters of the words γῆ, σελήνη, ἥλιος, and perigee. There is also ω, derived from the obliquity of the ecliptic. In another category we have M, S, E, for the masses of the moon, sun, and earth. It is unfortunate that the letter S should thus be connected with the moon in σ; but it has not been thought advisable to change the notation in this matter. In this Report the already existing notation is adhered to, as far as might be without inconvenience; but it must be admitted that the notation is by no means satisfactory.

It is a matter of great practical utility to have a symbol for indicating special tides. In the endeavour to meet this want initial letters were assigned in the former Reports to each kind of tide; but, except in the case of M and S, for the principal 'moon' and 'sun' tides, the initials had no connection with the tide. Although a new system of initials might be devised which would have a direct connection with the tides to which they refer, yet it has appeared best to adhere to the old initials and to introduce certain new initials for the tides of long period and for some tides now considered for the first time.

* [Other methods of effecting the computations have been devised of which one is described in a subsequent paper in this volume "On an Apparatus, &c." The original method is still in use in India.]

[A.] *Schedule of Notation.*

Initials	Speed	Name of Tide
M_1	$\gamma-\sigma-\varpi$, and $\gamma-\sigma+\varpi$	Principal lunar series
M_2	$2(\gamma-\sigma)$	
M_3	$3(\gamma-\sigma)$	
&c.	&c.	
K_2	2γ	Luni-solar semi-diurnal
N	$2\gamma-3\sigma+\varpi$	Larger lunar elliptic
L	$2\gamma-\sigma-\varpi$ and $2\gamma-\sigma+\varpi$	Smaller lunar elliptic
	$2\gamma+\sigma-\varpi$	
2N	$2\gamma-4\sigma+2\varpi$	Lunar elliptic, second order
ν	$2\gamma-3\sigma-\varpi+2\eta$	Larger lunar evectional
λ	$2\gamma-\sigma+\varpi-2\eta$	Smaller lunar evectional
O	$\gamma-2\sigma$	Lunar diurnal
OO	$\gamma+2\sigma$	
K_1	γ	Luni-solar diurnal
Q	$\gamma-3\sigma+\varpi$	Larger lunar elliptic diurnal
	$\gamma-\sigma-\varpi$ included in M_1	Smaller lunar elliptic diurnal
J	$\gamma+\sigma-\varpi$	
	$\gamma-4\sigma+2\varpi$	Lunar elliptic diurnal, second order
	$\gamma-3\sigma-\varpi+2\eta$	Larger lunar evectional diurnal
S_1	$\gamma-\eta$	Principal solar series
S_2	$2(\gamma-\eta)$	
S_3	$3(\gamma-\eta)$	
&c.	&c.	
T	$2\gamma-3\eta$	Larger solar elliptic
R	$2\gamma-\eta$	Smaller solar elliptic
P	$\gamma-2\eta$	Solar diurnal
Mm	$\sigma-\varpi$	Lunar monthly
Mf	2σ	Lunar fortnightly
Sa	η	Solar annual
Ssa	2η	Solar semi-annual
MSf	$2(\sigma-\eta)$	Luni-solar synodic fortnightly
MS	$4\gamma-2\sigma-2\eta$	Compound tides
μ or 2MS	$2\gamma-4\sigma+2\eta$	
2SM	$2\gamma+2\sigma-4\eta$	
MK	$3\gamma-2\sigma$	
2MK	$3\gamma-4\sigma$	
MN	$4\gamma-5\sigma+\varpi$	

In the old notation the L tide was simply the tide of speed $2\gamma-\sigma-\varpi$. The values of this tide have probably been perturbed by another tide of speed $2\gamma-\sigma+\varpi$, and this tide is supposed also to be included in L.

Where it is necessary to refer to any other tides than those contained in this schedule, it will be best to use the scientific nomenclature simply by speed. For example, there may be a compound tide $3\gamma-2\eta$; and though this tide might be called SK, since $3\gamma-2\eta=2(\gamma-\eta)+\gamma$, yet reference to such a tide will be so infrequent as not to make the short notation desirable.

Both the old and the new initials are given in the preceding schedule [A].

§ 2. *Development of the Equilibrium Theory of Tides with reference to Tidal Observations.*

The first step is the formation of the tide-generating potential of the moon; that for the sun may then be written down by symmetry.

For the purpose we require to find certain spherical harmonic functions of the moon's coordinates, with reference to axes fixed in the earth.

Let A, B, C (Fig. 1) be such axes, C being the north pole and AB the equator.

Let X, Y, Z be a second set of axes, XY being the plane of the moon's orbit.

Let M be the projection of the moon in her orbit.

Let $I=$ ZC, the obliquity of the lunar orbit to the equator.

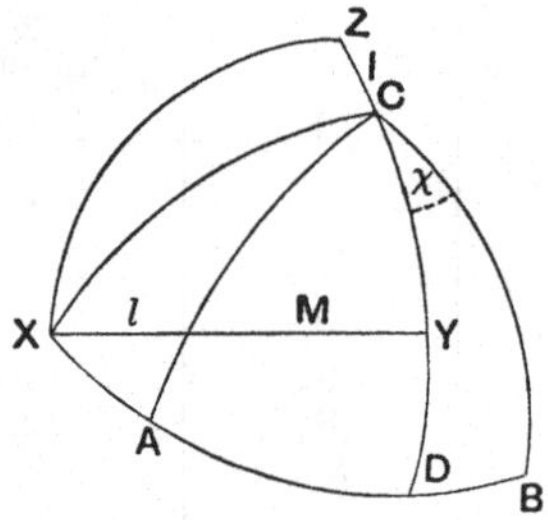

FIG. 1.

Let $\chi=$ AX $=$ BCY.

Let $l=$ MX, the moon's longitude in her orbit, measured from X.

$$\left.\begin{aligned} M_1&=\cos \mathrm{MA}\\ M_2&=\cos \mathrm{MB}\\ M_3&=\cos \mathrm{MC}\end{aligned}\right\}\ \text{the moon's direction-cosines with reference to ABC} \qquad \text{Let} \quad \ldots\ldots\ldots\ldots\ldots(1)$$

Then

$$\left.\begin{aligned} M_1&=\cos l\cos\chi+\sin l\sin\chi\cos I\\ M_2&=-\cos l\sin\chi+\sin l\cos\chi\cos I\\ M_3&=\sin l\sin I\end{aligned}\right\} \quad \ldots\ldots\ldots\ldots\ldots\ldots(2)$$

We may observe that M_2 is derivable from M_1 by putting $\chi+\frac{1}{2}\pi$ in place of χ.

Now for brevity let

$$p=\cos\tfrac{1}{2}I, \qquad q=\sin\tfrac{1}{2}I \quad \ldots\ldots\ldots\ldots\ldots\ldots(3)$$

Then (2) may be written

$$\left.\begin{aligned} M_1 &= \ \ p^2\cos(\chi - l) + q^2\cos(\chi + l) \\ M_2 &= -p^2\sin(\chi - l) - q^2\sin(\chi + l) \\ M_3 &= \ \ 2pq\sin l \end{aligned}\right\} \quad \text{.....................(4)}$$

Whence

$$\left.\begin{aligned} M_1^2 - M_2^2 &= \ \ p^4\cos 2(\chi - l) + 2p^2q^2\cos 2\chi + q^4\cos 2(\chi + l) \\ -2M_1M_2 &= \ \ \text{the same with sines in place of cosines} \\ M_2M_3 &= -p^3q\cos(\chi - 2l) + pq(p^2 - q^2)\cos\chi + pq^3\cos(\chi + 2l) \\ M_1M_3 &= \ \ \text{the same with sines in place of cosines} \\ \tfrac{1}{3} - M_3^2 &= \ \ \tfrac{1}{3}(p^4 - 4p^2q^2 + q^4) + 2p^2q^2\cos 2l \end{aligned}\right\} \text{......(5)}$$

These are the required spherical harmonic functions of M_1, M_2, M_3.

Let M denote the projection of the moon on the celestial sphere concentric with the earth, and P that of any other point.

Let r, ρ be the radius-vectors of the moon and of P respectively, and let ξ, η, ζ be the direction-cosines of P, with reference to the axes A, B, C.

Then $\rho\xi$, $\rho\eta$, $\rho\zeta$ are the coordinates of P, and rM_1, rM_2, rM_3 those of M.

If M be the moon's mass, and μ the attraction between unit masses at unit distance apart, then by the usual theory the tide-generating potential V, due to the moon, of the second order of harmonics, at the point P, is given by

$$V = \tfrac{3}{2}\frac{\mu M}{r^3}\rho^2(\cos^2 \mathrm{PM} - \tfrac{1}{3}) \quad \text{.......................(6)}$$

But since $\cos \mathrm{PM} = \xi M_1 + \eta M_2 + \zeta M_3$,

$$\cos^2 \mathrm{PM} - \tfrac{1}{3} = 2\xi\eta M_1M_2 + 2\frac{\xi^2 - \eta^2}{2}\frac{M_1^2 - M_2^2}{2} + 2\eta\zeta M_2M_3 + 2\xi\zeta M_1M_3$$
$$+ \tfrac{3}{2}\frac{\xi^2 + \eta^2 - 2\zeta^2}{3}\frac{M_1^2 + M_2^2 - 2M_3^2}{3} \quad \text{.....................(7)}$$

Now let c be the moon's mean distance, e the eccentricity of the moon's orbit, and let

$$\tau = \tfrac{3}{2}\frac{\mu M}{c^3} \quad \text{................................(8)}$$

Then putting

$$X = \left[\frac{c(1 - e^2)}{r}\right]^{\frac{3}{2}} M_1, \quad Y = \left[\frac{c(1 - e^2)}{r}\right]^{\frac{3}{2}} M_2, \quad Z = \left[\frac{c(1 - e^2)}{r}\right]^{\frac{3}{2}} M_3, \ \text{...(9)}$$

we have

$$V \div \frac{\tau}{(1 - e^2)^3}\rho^2 = 2\xi\eta XY + 2\frac{\xi^2 - \eta^2}{2}\frac{X^2 - Y^2}{2} + 2\eta\zeta YZ + 2\xi\zeta XZ$$
$$+ \tfrac{3}{2}\frac{\xi^2 + \eta^2 - 2\zeta^2}{3}\frac{X^2 + Y^2 - 2Z^2}{3} \quad \text{.................(10)}$$

A simple tide may be defined as a spherical harmonic deformation of the waters of the ocean which executes a simple harmonic motion in time. Corresponding to this definition the expression for each term of the tide-generating potential should consist of a solid spherical harmonic, multiplied by a simple time-harmonic.

In (10) $\rho^2\xi\eta$, $\rho^2(\xi^2-\eta^2)$, &c., are solid spherical harmonics, and in order to complete the expression for V it is necessary to develop the five functions of X, Y, Z in a series of simple time-harmonics.

It will be now convenient to introduce certain auxiliary functions, namely

$$\left.\begin{aligned} \Phi(\alpha) &= \left[\frac{c(1-e^2)}{r}\right]^3 \cos(2l+\alpha) \\ \Psi(\alpha) &= \left[\frac{c(1-e^2)}{r}\right]^3 \cos\alpha, \qquad \mathrm{R} = \left[\frac{c(1-e^2)}{r}\right]^3 \end{aligned}\right\} \quad \ldots\ldots(11)$$

Then from (5) and (9) we have

$$\left.\begin{aligned} X^2-Y^2 &= p^4\Phi(-2\chi)+2p^2q^2\Psi(2\chi)+q^4\Phi(2\chi) \\ 2XY &= \text{the same with } \chi+\tfrac{1}{4}\pi \text{ for } \chi \\ YZ &= -p^3q\Phi(-\chi)+pq(p^2-q^2)\Psi(\chi)+pq^3\Phi(\chi) \\ XZ &= \text{the same with } \chi-\tfrac{1}{2}\pi \text{ for } \chi \\ \tfrac{1}{3}(X^2+Y^2-2Z^2) &= \tfrac{1}{3}(p^4-4p^2q^2+q^4)\mathrm{R}+2p^2q^2\Phi(0) \end{aligned}\right\} \quad \ldots\ldots(12)$$

Thus when the functions Φ, Ψ, R are developed as a series of time-harmonics, the further development of the X-Y-Z functions consists in substitution in (12).

It will now be supposed that the moon moves in an elliptic orbit, undisturbed by the sun. The tides which arise from the lunar inequalities of the Evection and Variation will be the subject of separate treatment below.

The descending node of the equator on the lunar orbit will henceforth be called 'the Intersection.'

Let $\sigma_{,}$ be the moon's mean longitude measured in her orbit from the intersection, and $\varpi_{,}$ the longitude of the perigee measured in the same way. It has been already defined that l is the moon's longitude in her orbit measured from the intersection.

The equation of the ellipse described by the moon is

$$\frac{c(1-e^2)}{r} = 1+e\cos(l-\varpi_{,}) \quad \ldots\ldots\ldots\ldots\ldots\ldots\ldots(13)$$

Hence

$$\left.\begin{aligned} \mathrm{R} &= 1 + \tfrac{3}{2}e^2 + 3e\cos(l-\varpi_{,}) + \tfrac{9}{2}e^2\cos 2(l-\varpi_{,}) + \dots \\ \Phi(\alpha) &= \mathrm{R}\cos(2l+\alpha) \\ &= (1+\tfrac{3}{2}e^2)\cos(2l+\alpha) + \tfrac{3}{2}e\left[\cos(3l+\alpha-\varpi_{,}) + \cos(l+\alpha+\varpi_{,})\right] \\ &\qquad + \tfrac{9}{4}e^2\left[\cos(4l+\alpha-2\varpi_{,}) + \cos(\alpha+2\varpi_{,})\right] + \dots \\ \Psi(\alpha) &= \mathrm{R}\cos\alpha \end{aligned}\right\} \quad (14)$$

By the theory of elliptic motion

$$l = \sigma_{,} + 2e\sin(\sigma_{,}-\varpi_{,}) + \tfrac{5}{4}e^2\sin 2(\sigma_{,}-\varpi_{,}) + \dots \quad \dots\dots\dots(15)$$

In order to expand Φ, Ψ, R in terms of $\sigma_{,}$ (which increases uniformly with the time), we require $\cos(2l+\alpha)$ developed as far as e^2; $\cos(3l+\alpha-\varpi_{,})$, and $\cos(l+\alpha+\varpi_{,})$, as far as e; and only the first term of $\cos(4l+\alpha-2\varpi_{,})$.

Substituting for l its value (15) in terms of $\sigma_{,}$, it is easy to show that

$$\begin{aligned} \cos(2l+\alpha) &= (1-4e^2)\cos(2\sigma_{,}+\alpha) - 2e\cos(\sigma_{,}+\alpha+\varpi_{,}) + 2e\cos(3\sigma_{,}+\alpha-\varpi_{,}) \\ &\qquad + \tfrac{3}{4}e^2\cos(\alpha+2\varpi_{,}) + \tfrac{13}{4}e^2\cos(4\sigma_{,}+\alpha-2\varpi_{,}) + \dots \\ \cos(3l+\alpha-\varpi_{,}) &= \cos(3\sigma_{,}+\alpha-\varpi_{,}) \\ &\qquad - 3e\cos(2\sigma_{,}+\alpha) + 3e\cos(4\sigma_{,}+\alpha-2\varpi_{,}) + \dots \\ \cos(l+\alpha+\varpi_{,}) &= \cos(\sigma+\alpha+\varpi_{,}) + e\cos(2\sigma_{,}+\alpha) - e\cos(\alpha+2\varpi_{,}) + \dots \\ \cos(4l+\alpha-2\varpi_{,}) &= \cos(4\sigma_{,}+\alpha-2\varpi_{,}) + \dots \end{aligned}$$

Substituting these values in (14) we find,

$$\left.\begin{aligned} \Phi(\alpha) &= (1-\tfrac{11}{2}e^2)\cos(2\sigma_{,}+\alpha) - \tfrac{1}{2}e\cos(\sigma_{,}+\alpha+\varpi_{,}) \\ &\qquad + \tfrac{7}{2}e\cos(3\sigma_{,}+\alpha-\varpi_{,}) + \tfrac{17}{2}e^2\cos(4\sigma_{,}+\alpha-2\varpi_{,}) + \dots \\ \mathrm{R} &= (1-\tfrac{3}{2}e^2) + 3e\cos(\sigma_{,}-\varpi_{,}) + \tfrac{9}{2}e^2\cos 2(\sigma_{,}-\varpi_{,}) + \dots \\ \Psi(\alpha) &= (1-\tfrac{3}{2}e^2)\cos\alpha + \tfrac{3}{2}e\left[\cos(\sigma_{,}+\alpha-\varpi_{,}) + \cos(\sigma_{,}-\alpha-\varpi_{,})\right] \\ &\qquad + \tfrac{9}{4}e^2\left[\cos(2\sigma_{,}+\alpha-2\varpi_{,}) + \cos(2\sigma_{,}-\alpha-2\varpi_{,})\right] + \dots \end{aligned}\right\} \quad \dots\dots\dots(16)$$

Now substituting from (16) in (12), giving to α its appropriate value, we have

$$\left.\begin{aligned} X^2 - Y^2 &= (1-\tfrac{11}{2}e^2)\left[p^4\cos 2(\chi-\sigma_{,}) + q^4\cos 2(\chi+\sigma_{,})\right] \\ &\qquad + (1-\tfrac{3}{2}e^2)\,2p^2q^2\cos 2\chi \\ &+ \tfrac{7}{2}e\left[p^4\cos(2\chi-3\sigma_{,}+\varpi_{,}) + q^4\cos(2\chi+3\sigma_{,}-\varpi_{,})\right] \\ &- \tfrac{1}{2}e\left[p^4\cos(2\chi-\sigma_{,}-\varpi_{,}) + q^4\cos(2\chi+\sigma_{,}+\varpi_{,})\right] \\ &+ \tfrac{3}{2}e\,2p^2q^2\left[\cos(2\chi+\sigma_{,}-\varpi_{,}) + \cos(2\chi-\sigma_{,}+\varpi_{,})\right] \\ &+ \tfrac{17}{2}e^2\left[p^4\cos(2\chi-4\sigma_{,}+2\varpi_{,}) + q^4\cos(2\chi+4\sigma_{,}-2\varpi_{,})\right] \\ &+ \tfrac{9}{4}e^2\,2p^2q^2\left[\cos(2\chi+2\sigma_{,}-2\varpi_{,}) + \cos(2\chi-2\sigma_{,}+2\varpi_{,})\right] \end{aligned}\right\} \quad \dots\dots(17)$$

Clearly $-2XY$ is the same as (17) with sines in place of cosines. Also since YZ is the same as $X^2 - Y^2$ when χ replaces 2χ, $-p^3q$ replaces p^4,

$pq(p^2-q^2)$ replaces $2p^2q^2$, and pq^3 replaces q^4, and since XZ is the same as YZ with sines in place of cosines, we have from (17)

$$\left.\begin{aligned}XZ = -(1-\tfrac{11}{2}e^2)&[p^3q\sin(\chi-2\sigma_,)-pq^3\sin(\chi+2\sigma_,)]\\ &+(1-\tfrac{3}{2}e^2)pq(p^2-q^2)\sin\chi\\ -\tfrac{7}{2}e&[p^3q\sin(\chi-3\sigma_,+\varpi_,)-pq^3\sin(\chi+3\sigma_,-\varpi_,)]\\ +\tfrac{1}{2}e&[p^3q\sin(\chi-\sigma_,-\varpi_,)-pq^3\sin(\chi+\sigma_,+\varpi_,)]\\ +\tfrac{3}{2}e&pq(p^2-q^2)[\sin(\chi+\sigma_,-\varpi_,)+\sin(\chi-\sigma_,+\varpi_,)]\\ -\tfrac{17}{2}e^2&[p^3q\sin(\chi-4\sigma_,+2\varpi_,)-pq^3\sin(\chi+4\sigma_,-2\varpi_,)]\\ +\tfrac{9}{4}e^2&pq(p^2-q^2)[\sin(\chi+2\sigma_,-2\varpi_,)+\sin(\chi-2\sigma_,+2\varpi_,)]\end{aligned}\right\}\quad\ldots\ldots(18)$$

Lastly,

$$\begin{aligned}\tfrac{1}{3}(X^2+Y^2-2Z^2)=\tfrac{1}{3}(p^4-4p^2q^2+q^4)&[(1-\tfrac{3}{2}e^2)+3e\cos(\sigma_,-\varpi_,)\\ &+\tfrac{9}{2}e^2\cos2(\sigma_,-\varpi_,)]\\ +2p^2q^2[(1-\tfrac{11}{2}e^2)\cos2\sigma_,&+\tfrac{7}{2}e\cos(3\sigma_,-\varpi_,)-\tfrac{1}{2}e\cos(\sigma_,+\varpi_,)\\ &+\tfrac{17}{2}e^2\cos(4\sigma_,-2\varpi_,)]\ \ldots\ldots\ldots\ldots\ldots(19)\end{aligned}$$

Hitherto no approximation has been admitted with regard to I, the obliquity of the lunar orbit to the equator.

The obliquity of the ecliptic is 23° 27′·3, and I oscillates between 5° 8′·8 greater and 5° 8′·8 less than that value. The value of q or $\sin\frac{1}{2}I$, when I is 23° 27′·3, is ·203, and its square is ·041, and its cube ·0084. The eccentricity of the lunar orbit $e=\cdot0549$; hence q^2 is a little smaller than e.

The preceding developments have been carried as far as e^2, principally on account of the terms involving $\frac{17}{2}e^2$, which, as e is about $\frac{1}{18}$, have nearly the same magnitude as if the coefficient had been $\frac{1}{2}e$.

It is proposed, then, to regard q^2 and q^3 as of the same order as e, and to drop all terms of the order e^2, except in the case where the numerical factor is large. This rule will be neglected with regard to one term for a special reason, which appears below; and for another, because the numerical coefficient is just sufficiently large to make it worth retaining.

Adopting this approximation, we may write (17), (18), (19), thus,—

$$\left.\begin{aligned}X^2-Y^2=(1-\tfrac{11}{2}e^2)&p^4\cos2(\chi-\sigma_,)+(1-\tfrac{3}{2}e^2)2p^2q^2\cos2\chi\\ &+\tfrac{7}{2}ep^4\cos(2\chi-3\sigma_,+\varpi_,)\\ &-\tfrac{1}{2}ep^2[p^2\cos(2\chi-\sigma_,-\varpi_,)-6q^2\cos(2\chi-\sigma_,+\varpi_,)]\\ &+\tfrac{17}{2}e^2p^4\cos(2\chi-4\sigma_,+2\varpi_,)\\ XZ=-(1-\tfrac{11}{2}e^2)&[p^3q\sin(\chi-2\sigma_,)-pq^3\sin(\chi+2\sigma_,)]\\ &+(1-\tfrac{3}{2}e^2)pq(p^2-q^2)\sin\chi-\tfrac{7}{2}ep^3q\sin(\chi-3\sigma_,+\varpi_,)\\ &+\tfrac{1}{2}epq[p^2\sin(\chi-\sigma_,-\varpi_,)+3(p^2-q^2)\sin(\chi-\sigma_,+\varpi_,)]\\ &+\tfrac{3}{2}epq(p^2-q^2)\sin(\chi+\sigma_,-\varpi_,)-\tfrac{17}{2}e^2p^3q\sin(\chi-4\sigma_,+2\varpi_,)\\ \tfrac{1}{3}(X^2+Y^2-2Z^2)=\tfrac{1}{3}&(p^4-4p^2q^2+q^4)[(1-\tfrac{3}{2}e^2)+3e\cos(\sigma_,-\varpi_,)]\\ &+2p^2q^2[(1-\tfrac{11}{2}e^2)\cos2\sigma_,+\tfrac{7}{2}e\cos(3\sigma_,-\varpi_,)]\end{aligned}\right\}\quad(20)$$

The terms which have been retained in violation of the rule of approximation are that in $X^2 - Y^2$ with argument $2\chi - \sigma_{,} + \varpi_{,}$, and that in $\frac{1}{3}(X^2 + Y^2 - 2Z^2)$ with argument $3\sigma_{,} - \varpi_{,}$.

The only other term which could have any importance is

$$\tfrac{3}{2}e\, 2p^2q^2 \cos(2\chi + \sigma_{,} - \varpi_{,}) \quad \text{in} \quad X^2 - Y^2$$

Before proceeding to consider the tides due to lunar inequalities it will be well to consider two pairs of terms in the expressions (20).

First, in $X^2 - Y^2$ we have the terms

$$-\tfrac{1}{2}ep^2 [p^2 \cos(2\chi - \sigma_{,} - \varpi_{,}) - 6q^2 \cos(2\chi - \sigma_{,} + \varpi_{,})]$$

The expression within [ ] may be put in the form

$$(p^2 - 6q^2 \cos 2\varpi_{,}) \cos(2\chi - \sigma_{,} - \varpi_{,}) + 6q^2 \sin 2\varpi_{,} \sin(2\chi - \sigma_{,} - \varpi_{,})$$

If then we write $$\tan R = \frac{\sin 2\varpi_{,}}{\frac{1}{6}\cot^2 \frac{1}{2}I - \cos 2\varpi_{,}} \quad \text{.....................}(20^{\text{i}})$$

this pair of terms becomes

$$-\tfrac{1}{2}ep^4 \sqrt{\{1 - 12\tan^2 \tfrac{1}{2}I \cos 2\varpi_{,} + 36 \tan^4 \tfrac{1}{2}I\}} \cos(2\chi - \sigma_{,} - \varpi_{,} - R) \quad \text{...}(20^{\text{ii}})^*$$

[If we attribute to I its mean value 23° 27′·3 the square root becomes $\sqrt{\{1{\cdot}066 - 12\tan^2 \frac{1}{2}I \cos 2\varpi_{,}\}}$.]

Secondly, in XZ we have the terms

$$+\tfrac{1}{2}epq [p^2 \sin(\chi - \sigma_{,} - \varpi_{,}) + 3(p^2 - q^2) \sin(\chi - \sigma_{,} + \varpi_{,})]$$

[This is equal to

$$+\tfrac{1}{2}ep^3q [(4 - 3\tan^2 \tfrac{1}{2}I) \cos \varpi_{,} \sin(\chi - \sigma_{,}) + (2 - 3\tan^2 \tfrac{1}{2}I) \sin \varpi_{,} \cos(\chi - \sigma_{,})]$$

If then we write $$\tan Q = \frac{1}{2} \frac{1 - \frac{3}{2}\tan^2 \frac{1}{2}I}{1 - \frac{3}{4}\tan^2 \frac{1}{2}I} \tan \varpi_{,}$$

and if we attribute to I its mean value 23° 27′·3 the terms become

$$ep^3q \sqrt{\{2{\cdot}311 + 1{\cdot}436 \cos 2\varpi_{,}\}} \sin(\chi - \sigma_{,} + Q) \quad \text{.........}(20^{\text{iii}})$$

where $$\tan Q = {\cdot}483 \tan \varpi_{,} \quad \text{..........................}(20^{\text{iv}})$$

In the original paper $\tan^2 \frac{1}{2}I$ was neglected, and the numbers which appear here as 2·311, 1·436 and ·483 were taken respectively as $\frac{5}{2}$, $\frac{3}{2}$ and $\frac{1}{2}$.]

The object of the transformations (20^{ii}), (20^{iv}), which may seem theoretically undesirable, is as follows:—

The numerical harmonic analysis of the tides is made to extend over one year, and this period is not long enough to distinguish completely a tide whose argument is $2\chi - \sigma_{,} - \varpi_{,}$, from one whose argument is $2\chi - \sigma_{,} + \varpi_{,}$, nor one whose argument is $\chi - \sigma_{,} - \varpi_{,}$, from one whose

* [The term $36 \tan^4 \frac{1}{2}I$ ranges between $\frac{1}{7}$ and $\frac{1}{40}$, and it is neglected in the original paper.]

argument is $\chi - \sigma_{,} + \varpi_{,}$. In fact, the tide with argument $2\chi - \sigma_{,} + \varpi_{,}$ (for which no analysis has been as yet carried out) will only produce an irregularity in that of argument $2\chi - \sigma_{,} - \varpi_{,}$, called the smaller elliptic semidiurnal tide; such irregularity has in fact been noted, but no explanation has previously been given of it.

Again, the pair of terms with arguments $\chi - \sigma_{,} \pm \varpi_{,}$ will appear in the harmonic analysis with the single argument $\chi - \sigma_{,}$, and the resulting numbers will necessarily appear very irregular, unless compared with the theoretical expression (20^{iii}).

We will now consider the terms introduced by the two principal lunar inequalities due to the disturbing action of the sun.

The Evection.

Let θ be the moon's longitude in the ecliptic.

s the moon's mean longitude.

p the mean longitude of the perigee*.

h the sun's mean longitude.

m the ratio of the sun's to the moon's mean motion.

Then that inequality in longitude and radius vector is represented by

$$\theta = s + \tfrac{15}{4} me \sin (s - 2h + p) \quad \text{.....................(21)}$$

$$\frac{c(1-e^2)}{r} = 1 + \tfrac{15}{8} me \cos (s - 2h + p) \quad \text{.....................(22)}$$

If we neglect the distinction between longitudes in the orbit and in the ecliptic [which is in effect neglecting a term with coefficient $\sin^2(\frac{1}{2} \times 5^\circ 9')$], we have from (21),

$$l = \sigma_{,} + \tfrac{15}{4} me \sin (s - 2h + p)$$

whence

$$\cos(2l + \alpha) = \cos(2\sigma_{,} + \alpha) + \tfrac{15}{4} me [\cos (2\sigma_{,} + s - 2h + p + \alpha) - \cos (2\sigma_{,} - s + 2h - p + \alpha)]$$

And from (22) and the definitions of R, Ψ, Φ in (11),

$$\text{R} = \left[\frac{c(1-e^2)}{r}\right]^3 = 1 + \tfrac{45}{8} me \cos (s - 2h + p) \quad \text{..........................(23)}$$

$$\Psi(\alpha) = \cos \alpha + \tfrac{45}{16} me [\cos (s - 2h + p + \alpha) + \cos (s - 2h + p - \alpha)] \quad \text{......(24)}$$

$$\Phi(\alpha) = \cos(2\sigma_{,} + \alpha) + \tfrac{105}{16} me \cos (2\sigma_{,} + s - 2h + p + \alpha) - \tfrac{15}{16} me \cos (2\sigma_{,} - s + 2h - p + \alpha) \quad \text{......(25)}$$

* p in this sense will easily be distinguished from the p used to denote $\cos \frac{1}{2} I$ which latter will, moreover, be shortly discarded.

Then substituting from (23), (24), (25), in (12), and dropping the terms which are merely a reproduction of those already obtained, and neglecting terms in q^2 and q^3, we have

$$\left.\begin{aligned}
X^2 - Y^2 &= \tfrac{105}{16} mep^4 \cos(2\chi - 2\sigma_{,} - s + 2h - p) \\
&\qquad - \tfrac{15}{16} mep^4 \cos(2\chi - 2\sigma_{,} + s - 2h + p) \\
XZ &= -\tfrac{105}{16} mep^3 q \sin(\chi - 2\sigma_{,} - s + 2h - p) \\
&\qquad + \tfrac{15}{16} mep^3 q \sin(\chi - 2\sigma_{,} + s - 2h + p) \\
&\qquad + \tfrac{45}{16} mepq(p^2 - q^2)[\sin(\chi + s - 2h + p) + \sin(\chi - s + 2h - p)] \\
\tfrac{1}{3}(X^2 + Y^2 - 2Z^2) &= \tfrac{1}{3}(p^4 - 4p^2q^2 + q^4)\,\tfrac{45}{8} me \cos(s - 2h + p)
\end{aligned}\right\} \quad (25)$$

It must be noticed that $\frac{105}{16}me$ arises by the addition of the coefficient of the Evection in longitude to three halves of that in the reciprocal of the radius vector; that $\frac{15}{16}me$ is the difference of the same two quantities; and that $\frac{45}{8}me$ is three times the coefficient in the reciprocal of radius vector. When the development of the lunar theory is carried to higher orders these coefficients differ considerably from the amounts computed from the first term, which alone occurs in the above analysis. Hence, when these coefficients are computed, the full values of the coefficients in longitude and reciprocal of radius vector must be introduced. According to Professor Adams, the full values of the coefficients are, in longitude ·022233, and in c/r ·010022.

The ratio m of the mean motions is about $\frac{1}{13}$, and is therefore a little greater than e, hence me is somewhat greater than e^2. Thus we may abridge (25), and write the expressions thus:—

$$\left.\begin{aligned}
X^2 - Y^2 &= \tfrac{105}{16} mep^4 \cos(2\chi - 2\sigma_{,} - s + 2h - p) \\
&\qquad - \tfrac{15}{16} mep^4 \cos(2\chi - 2\sigma_{,} + s - 2h + p) \\
XZ &= -\tfrac{105}{16} mep^3 q \sin(\chi - 2\sigma_{,} - s + 2h - p) \\
\tfrac{1}{3}(X^2 + Y^2 - 2Z^2) &= \tfrac{1}{3}(p^4 - 4p^2q^2 + q^4)\,\tfrac{45}{8} me \cos(s - 2h + p)
\end{aligned}\right\} \quad \ldots(26)$$

The equations (26) contain the terms to be added to (20) on account of the Evection.

The Variation.

Treating this inequality in the same way as the Evection, we have

$$l = \sigma_{,} + \tfrac{11}{8} m^2 \sin 2(s - h)$$

$$\frac{c(1 - e^2)}{r} = 1 + m^2 \cos 2(s - h)$$

$$\mathrm{R} = 1 + 3m^2 \cos 2(s - h)$$

$$\Psi(\alpha) = \cos\alpha + \tfrac{3}{2} m^2 [\cos\{2(s - h) + \alpha\} + \cos\{2(s - h) - \alpha\}]$$

$$\Phi(\alpha) = \cos(2\sigma_{,} + \alpha) + \tfrac{23}{8} m^2 \cos(2\sigma_{,} + 2s - 2h + \alpha) + \tfrac{1}{8} m^2 \cos(2\sigma_{,} - 2s + 2h + \alpha)$$

Whence we have to a sufficient degree of approximation,

$$\left.\begin{aligned} X^2 - Y^2 &= \tfrac{23}{8} m^2 p^4 \cos(2\chi - 2\sigma_{,} - 2s + 2h), \quad XZ = 0 \\ \tfrac{1}{3}(X^2 + Y^2 - 2Z^2) &= \tfrac{1}{3}(p^4 - 4p^2q^2 + q^4)\, 3m^2 \cos(2s - 2h) \end{aligned}\right\} \quad \ldots\ldots(27)$$

In this case also the values of the coefficients are actually considerably greater than the amounts as computed from the first terms; and regard must be paid to this, as in the case of the Evection, when the values of the coefficients in the tidal expressions are computed. According to Professor Adams, the full values of the coefficients are, in longitude ·011489, and in c/r ·008249.

We have now obtained in (20), (26), (27), the complete expressions for the X-Y-Z functions in the shape of a series of simple time-harmonics; but they are not yet in a form in which the ordinary astronomical formulæ are applicable.

Further substitutions will now be made, and we shall pass from the potential to the height of tide generated by the forces corresponding to that potential.

The axes fixed in the earth may be taken to have their extremities as follows:

The axis A on the equator in the meridian of the place of observation of the tides; the axis B in the equator 90° east of A; the axis C at the north pole.

Now ξ, η, ζ are the direction-cosines of the place of observation, and if λ be the latitude of that place, we have

$$\xi = \cos\lambda, \qquad \eta = 0, \qquad \zeta = \sin\lambda$$

Thus

$$\xi^2 - \eta^2 = \cos^2\lambda, \quad \xi\eta = 0, \quad \eta\zeta = 0, \quad 2\xi\zeta = \sin 2\lambda, \quad \tfrac{1}{3}(\xi^2 + \eta^2 - 2\zeta^2) = \tfrac{1}{3} - \sin^2\lambda$$

Then writing a for the earth's radius, the expression (10) for V at the place of observation becomes

$$V = \frac{\tau a^2}{(1-e^2)^3}\left[\tfrac{1}{2}\cos^2\lambda (X^2 - Y^2) + \sin 2\lambda\, XZ + \tfrac{3}{2}(\tfrac{1}{3} - \sin^2\lambda)\,\tfrac{1}{3}(X^2 + Y^2 - 2Z^2)\right]$$

The X-Y-Z functions being simple time-harmonics, the principle of forced vibrations allows us to conclude that the forces corresponding to V will generate oscillations in the ocean of the same periods and types as the terms in V, but of unknown amplitudes and phases.

Now let $\mathfrak{X}^2 - \mathfrak{Y}^2$, $\mathfrak{X}\mathfrak{Z}$, $\tfrac{1}{3}(\mathfrak{X}^2 + \mathfrak{Y}^2 - 2\mathfrak{Z}^2)$ be three functions, having respectively similar forms to those of

$$\frac{X^2 - Y^2}{(1-e^2)^3}, \quad \frac{XZ}{(1-e^2)^3} \quad \text{and} \quad \tfrac{1}{3}\frac{(X^2 + Y^2 - 2Z^2)}{(1-e^2)^3}$$

but differing from them in that the argument of each of the simple time-harmonics has some angle subtracted from it, and that the term is multiplied by a numerical factor.

Then if g be gravity, and h the height of tide at the place of observation, we must have

$$h = \frac{\tau a^2}{g}\left[\tfrac{1}{2}\cos^2\lambda\,(\mathfrak{X}^2 - \mathfrak{Y}^2) + \sin 2\lambda\,\mathfrak{X}\mathfrak{Z} + \tfrac{3}{2}(\tfrac{1}{3} - \sin^2\lambda)\,\tfrac{1}{3}(\mathfrak{X}^2 + \mathfrak{Y}^2 - 2\mathfrak{Z}^2)\right] \qquad \text{.........(28)}$$

The factor $\frac{\tau a^2}{g}$ may be more conveniently written $\frac{3}{2}\frac{M}{E}\left(\frac{a}{c}\right)^3 a$, where E is the earth's mass. It has been so chosen that if the equilibrium theory of tides were fulfilled, with water covering the whole earth, the numerical factors in the $\mathfrak{X}$-$\mathfrak{Y}$-$\mathfrak{Z}$ functions would be each unity. The alterations of phase would also be zero, or, with land and sea as in reality, they might be computed by means of the five definite integrals involved in Sir William Thomson's amended equilibrium theory of tides*.

The actual results of tidal analysis at any place are intended (see below, § 5) to be presented in a series of terms of the form $\mathrm{fH}\cos(V + u - \kappa)$, where dV/dt or n, 'the speed,' is the rate of increase of the argument per unit time (say degrees per mean solar hour), and u is a constant. We require, therefore, to present all the terms of the $\mathfrak{X}$-$\mathfrak{Y}$-$\mathfrak{Z}$ functions as cosines with a positive sign. When, then, in these functions we meet with a negative cosine we must change its sign and add π to the argument; as the $\mathfrak{X}\mathfrak{Z}$ functions involve sines, we must add $\frac{1}{2}\pi$ to arguments of the negative sines, and subtract $\frac{1}{2}\pi$ from the arguments of the positive sines, and replace sines by cosines. The terms in the $\frac{1}{3}(\mathfrak{X}^2 + \mathfrak{Y}^2 - 2\mathfrak{Z}^2)$ function require special consideration. The function of the latitude being $\frac{1}{3} - \sin^2\lambda$, it follows that when in the northern hemisphere it is high-water north of a certain critical latitude, it is low water on the opposite side of that parallel; and the same is true of the southern hemisphere. The critical latitude is that in which $\sin^2\lambda = \frac{1}{3}$, or in Thomson's amended equilibrium* theory, where $\sin^2\lambda = \frac{1}{3}(1 + \mathfrak{E})$. An approximate evaluation of $\mathfrak{E}$, which depends on the distribution of land and sea, given in § 848 of the second edition of Thomson and Tait's *Natural Philosophy*, shows that the critical latitudes are 35° N. and S. It will be best to adopt a uniform system for the whole earth, and to regard high-tide and high-water as consentaneous in the equatorial belt, and of opposite meanings outside the critical latitudes. In this Report we conceive the function always to be written $\frac{1}{3} - \sin^2\lambda$, so that outside the critical latitudes high-tide is low-water. Accordingly we must add π to the arguments of the negative cosines (if any) which occur in the function $\frac{1}{3}(\mathfrak{X}^2 + \mathfrak{Y}^2 - 2\mathfrak{Z}^2)$.

* Thomson and Tait's *Nat. Phil.*, or the Report on Tides for 1876. [See also Paper 9 in this volume.]

In continuing the development, the $\mathfrak{X}$-$\mathfrak{Y}$-$\mathfrak{Z}$ functions will be written in the form appropriate to the equilibrium theory, with water covering the whole earth; for the actual case it is only necessary to multiply by the reducing factor, and to subtract the phase alteration κ. As these are unknown constants for each place, they would only occur in the development as symbols of quantities to be deduced from observation. It will be understood, therefore, that in the following schedules 'the argument' is that part of the argument which is derived from theory, the true complete argument being 'the argument' $-\kappa$, where κ is derived from observation.

Following the plan suggested, and collecting results from (20), (26), (27), we have

$$\begin{aligned}\mathfrak{X}^2 - \mathfrak{Y}^2 = {} & (1 - \tfrac{5}{2}e^2)\, p^4 \cos 2(\chi - \sigma_{,}) + (1 + \tfrac{3}{2}e^2)\, 2p^2q^2 \cos 2\chi \\ & + \tfrac{7}{2}ep^4 \cos(2\chi - 3\sigma_{,} + \varpi_{,}) \\ & + \tfrac{1}{2}ep^4 \sqrt{\{1{\cdot}066 - 12 \tan^2 \tfrac{1}{2}I \cos 2\varpi_{,}\}} \cos(2\chi - \sigma_{,} - \varpi_{,} - R + \pi) \\ & + \tfrac{17}{2}e^2p^4 \cos(2\chi - 4\sigma_{,} + 2\varpi_{,}) \\ & + \tfrac{105}{16}mep^4 \cos(2\chi - 2\sigma_{,} - s + 2h - p) \\ & + \tfrac{15}{16}mep^4 \cos(2\chi - 2\sigma_{,} + s - 2h + p + \pi) \\ & + \tfrac{23}{8}m^2p^4 \cos(2\chi - 2\sigma_{,} - 2s + 2h) \quad \ldots\ldots(29)\end{aligned}$$

$$\begin{aligned}\mathfrak{X}\mathfrak{Z} = {} & (1 - \tfrac{5}{2}e^2)\,[p^3q \cos(\chi - 2\sigma_{,} + \tfrac{1}{2}\pi) + pq^3 \cos(\chi + 2\sigma_{,} - \tfrac{1}{2}\pi)] \\ & + (1 + \tfrac{3}{2}e^2)\, pq\,(p^2 - q^2) \cos(\chi - \tfrac{1}{2}\pi) \\ & + \tfrac{7}{2}ep^3q \cos(\chi - 3\sigma_{,} + \varpi_{,} + \tfrac{1}{2}\pi) \\ & + ep^3q \sqrt{\{2{\cdot}31 + 1{\cdot}44 \cos 2\varpi_{,}\}} \cos(\chi - \sigma_{,} + Q - \tfrac{1}{2}\pi) \\ & + \tfrac{3}{2}epq\,(p^2 - q^2) \cos(\chi + \sigma_{,} - \varpi_{,} - \tfrac{1}{2}\pi) \\ & + \tfrac{17}{2}e^2p^3q \cos(\chi - 4\sigma_{,} + 2\varpi_{,} + \tfrac{1}{2}\pi) \\ & + \tfrac{105}{16}mep^3q \cos(\chi - 2\sigma_{,} - s + 2h - p + \tfrac{1}{2}\pi) \quad \ldots\ldots(30)\end{aligned}$$

$$\begin{aligned}\tfrac{1}{3}(\mathfrak{X}^2 + \mathfrak{Y}^2 - 2\mathfrak{Z}^2) = {} & \tfrac{1}{3}(p^4 - 4p^2q^2 + q^4)\,[1 + \tfrac{3}{2}e^2 + 3e \cos(\sigma_{,} - \varpi_{,}) \\ & + \tfrac{45}{8}me \cos(s - 2h + p) + 3m^2 \cos(2s - 2h)] \\ & + 2p^2q^2\,[(1 - \tfrac{5}{2}e^2) \cos 2\sigma_{,} + \tfrac{7}{2}e \cos(3\sigma_{,} - \varpi_{,})] \quad \ldots\ldots(31)\end{aligned}$$

In these expressions

$$\tan R = \frac{\sin 2\varpi_{,}}{\tfrac{1}{6}\cot^2 \tfrac{1}{2}I - \cos 2\varpi_{,}}, \quad \tan Q = {\cdot}483 \tan \varpi_{,}$$

The next step is to express the angles χ, $\sigma_{,}$, $\varpi_{,}$, each of which increases uniformly with the time, in terms of the sidereal hour-angle or of the local mean time, and of the mean longitudes of the moon, and of the perigee.

Let M be the moon in the orbit. A the extremity of the A-axis fixed in the earth.

g be the sidereal hour-angle.

N the longitude of the node ☊.

ν the right ascension of the intersection I.

ξ the longitude 'in the moon's orbit' of the intersection.

i the inclination of the moon's orbit to the ecliptic.

ω the obliquity of the ecliptic.

s the moon's mean longitude.

p the mean longitude of the perigee *.

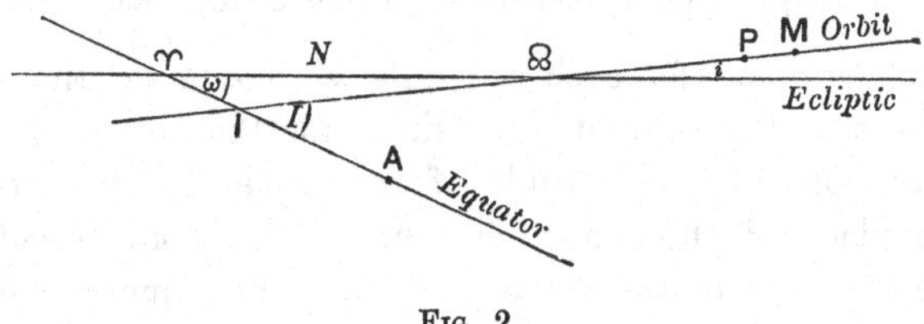

FIG. 2.

Then (Fig. 2) $\text{g} = \text{A♈}$, $\nu = \text{♈I}$, $\xi = \text{♈☊} - \text{☊I}$, $N = \text{♈☊}$

Now $\sigma_,$ and $\varpi_,$ have been defined above as the moon's mean longitude and the longitude of the perigee, both measured in the orbit from the intersection I.

Since $\sigma_, - \varpi_,$ is the moon's mean anomaly, we have

$$s - p = \sigma_, - \varpi_,$$

Let p' be the longitude of the perigee, measured from ♈ in the ecliptic.

If P in Fig. 2 be perigee, we have by the ordinary formula for reduction to the ecliptic,

$$\text{☊P} = p' - N + \tfrac{1}{4}\sin^2 i \sin 2(p' - N)$$

But
$$\varpi_, = \text{IP} = \text{I☊} + \text{☊P} = \text{♈☊} - \xi + \text{☊P}$$
$$= p' - \xi + \tfrac{1}{4}\sin^2 i \sin 2(p' - N)$$

Now
$$p = p' + \tfrac{1}{4}\sin^2 i \sin 2(p' - N)$$

and therefore
whence
$$\left.\begin{aligned}\varpi_, &= p - \xi\\ \sigma_, &= s - \xi\end{aligned}\right\} \quad \ldots\ldots\ldots\ldots\ldots\ldots\ldots\ldots\ldots\ldots(32)$$

Again
$$\chi = \text{IA} = \text{A♈} - \text{I♈} = \text{g} - \nu \quad \ldots\ldots\ldots\ldots\ldots\ldots(33)$$

In this formula we suppose g to increase uniformly from the time when the tidal observations begin.

* This p will easily be distinguished from the p used above to denote $\cos \tfrac{1}{2}I$.

Since in all the tidal observations local mean solar time is used, it will be better to substitute for g in terms of local mean solar time and the sun's mean longitude. Let t be local mean solar time reduced to angle, so that at noon $t = 0°$. Let h be the sun's mean longitude; hereafter we shall write p_1 for the longitude of the sun's perigee.

Then we have $$\chi = t + h - \nu \quad \text{.............................(34)}$$

We shall now substitute from (32) and (34) in the 𝔛-𝔜-ℨ functions (29), (30), (31); substitute from them in (28), and express the final result in the form of three schedules (pp. 20, 21, 22).

The schedules are arranged thus. First, there is the general coefficient $\frac{3}{2}\frac{M}{E}\left(\frac{a}{c}\right)^3 a$ which multiplies every term of all the schedules. Secondly, there are general coefficients one for each schedule, viz. $\cos^2\lambda$ for the semi-diurnal terms, $\sin 2\lambda$ for the diurnal, and $\frac{1}{2} - \frac{3}{2}\sin^2\lambda$ for the terms of long period. These three functions of the latitude of the place of observation are the values at that place of three surface spherical harmonic functions, which functions have the maximum value unity, at the equator for the semi-diurnal, in latitude 45° for the diurnal, and at the pole for the terms of long period.

First, in each schedule there is a column of coefficients, functions of I and e (and in two cases also of p).

In the second column is given the mean semi-range of the corresponding term. This is approximately the value of the coefficient in the first column when $I = \omega$. We forestall results given below so far as to state that the mean value is to be found by putting $I = \omega$ in the 'coefficient,' and when the function of I is $\cos^4 \frac{1}{2}I$, $\sin I \cos^2 \frac{1}{2}I$, $\sin I \sin^2 \frac{1}{2}I$, $\sin^2 I$ (in B, iii.) multiplying further by $\cos^4 \frac{1}{2}i$; and where the function of I is $\sin^2 I$ (in B, i.) $\sin I \cos I$, $1 - \frac{3}{2}\sin^2 I$ multiplying by $1 - \frac{3}{2}\sin^2 i$.

Thirdly, there is a column of arguments, linear functions of t, h, s, p, ν, ξ. In B, i. $2t + (2h - 2\nu)$, and in B, ii. $t + (h - \nu)$, are common to all the arguments, and they are written at the top of the column of arguments. The arguments are grouped in a manner convenient for subsequent computations.

Fourthly, there is a column of speeds, being the hourly increase of the arguments in the preceding column, the numerical values of which are added in a last column.

Every term is indicated by the initial letter (see § 1) adopted for the tide to which it corresponds, except in the case of certain unimportant terms to which no initials have been appropriated.

To write down any term: take the general coefficient; the coefficient for the class of tides; the special coefficient, and multiply by the cosine of the argument. The result is a term in the equilibrium tide, with water covering the whole earth. The transition to the actual case by the introduction of a factor and a delay of phase (to be derived from observation) has been already explained.

The solar tides.

The expression for the tides depending on the sun may be written down at once by symmetry. The eccentricity of the solar orbit is so small, being ·01679, that the elliptic tides may be omitted, excepting the larger elliptic semi-diurnal tide.

The lunar schedule is to be transformed by putting $s=h$, $p=p_1$, $\xi=\nu=0$, $\sigma=\eta$, $I=\omega$, $e=e_1$, $\varpi=\varpi_1$. In order that the comparison of the importance of the solar tides with the lunar may be complete, the same general coefficient $\frac{3}{2}\frac{M}{E}\left(\frac{a}{c}\right)^3 a$ will be retained, and the special coefficient for each term will be made to involve the factor τ_1/τ. Here $\tau_1=\frac{3}{2}\frac{\mu S}{c_1^3}$, S being the sun's mass.

With $E/M=81{\cdot}5$, $\qquad \frac{\tau_1}{\tau}=\cdot46035=\frac{1}{2{\cdot}17226}$

The schedule [C] of solar tides is given on page 23.

The subsequent schedules [D] and [E] give all the tides of purely astronomical origin contained in the previous developments, arranged first in order of speed, and secondly in order of the magnitude of the coefficient. In schedule [E] the tides K_1, K_2 originate both from the moon and sun, but the lunar and solar parts are also entered separately.

The coefficients of the evectional and variational tides are computed from the full values to those inequalities.

In the schedule [E] the tides are marked which occur in the 'Tide-predicter' of the Indian Government in its present condition.

Schedule of Lunar Tides [B, i.].

$$\text{General Coefficient} = \tfrac{3}{2}\frac{M}{E}\left(\frac{a}{c}\right)^3 a.$$

Semi-diurnal Tides; General Coefficient $= \cos^2 \lambda$.

Initial	Coefficient	Mean Value of Coefficient	Argument $2t+2(h-\nu)$	Speed	Speed in degrees per m. s. hour
M_2	$\frac{1}{2}(1-\frac{5}{2}e^2)\cos^4\frac{1}{2}I$	·45426	$-2(s-\xi)$	$2(\gamma-\sigma)$	28°·9841042
K_2	$\frac{1}{2}(1+\frac{3}{2}e^2)\frac{1}{2}\sin^2 I$	·03929	—	2γ	30°·0821372
N	$\frac{1}{2}\cdot\frac{7}{2}e\cos^4\frac{1}{2}I$	·08796	$-2(s-\xi)-(s-p)$	$2\gamma-3\sigma+\varpi$	28°·4397296
L	$\frac{1}{2}\cdot\frac{1}{2}e\cos^4\frac{1}{2}I$ $\times\sqrt{\{1{\cdot}066-12\tan^2\frac{1}{2}I\cos 2(p-\xi)\}}$	·01257	$-2(s-\xi)+(s-p)-R+\pi$ where $\tan R=\dfrac{6\sin 2(p-\xi)}{\cot^2\frac{1}{2}I-6\cos 2(p-\xi)}$	$2\gamma-\sigma-\varpi$	29°·5284788
2N	$\frac{1}{2}\cdot\frac{17}{2}e^2\cos^4\frac{1}{2}I$	·01173	$-2(s-\xi)-2(s-p)$	$2\gamma-4\sigma+2\varpi$	27°·8953548
ν	$\frac{1}{2}\cdot\frac{105}{16}me\cos^4\frac{1}{2}I$	·01234† ·01706	$-2(s-\xi)+(s-p)+2h-2s$	$2\gamma-3\sigma-\varpi+2\eta$	28°·5125830
λ	$\frac{1}{2}\cdot\frac{15}{16}me\cos^4\frac{1}{2}I$	·00176† ·00330	$-2(s-\xi)-(s-p)-2h+2s+\pi$	$2\gamma-\sigma+\varpi-2\eta$	29°·4556254
μ*	$\frac{1}{2}\cdot\frac{23}{8}m^2\cos^4\frac{1}{2}I$	·00736† ·01094	$-2(s-\xi)+2h-2s$	$2\gamma-4\sigma+2\eta$	27°·9682084

* Indicated by 2MS as a compound tide.

† In these three entries the lower of the two numbers gives the value when the coefficients of the Evection and Variation have their full values as derived from Lunar Theory.

[B, ii.]

Diurnal Tides; General Coefficient = sin 2λ.

Initial	Coefficient	Mean Value of Coefficient	Argument $t+(h-\nu)$	Speed	Speed in degrees per m. s. hour
O	$(1-\frac{5}{2}e^2)\frac{1}{2}\sin I\cos^2\frac{1}{2}I$	·18856	$-2(s-\xi)+\frac{1}{2}\pi$	$\gamma-2\sigma$	13°·9430356
OO	$(1-\frac{5}{2}e^2)\frac{1}{2}\sin I\sin^2\frac{1}{2}I$	·00812	$+2(s-\xi)-\frac{1}{2}\pi$	$\gamma+2\sigma$	16°·1391016
K_1	$(1+\frac{3}{2}e^2)\frac{1}{2}\sin I\cos I$	·18115	$-\frac{1}{2}\pi$	γ	15°·0410686
Q	$\frac{7}{2}e\,\frac{1}{2}\sin I\cos^2\frac{1}{2}I$	·03651	$-2(s-\xi)-(s-p)+\frac{1}{2}\pi$	$\gamma-3\sigma+\varpi$	13°·3986609
M_1	$e\,\frac{1}{2}\sin I\cos^2\frac{1}{2}I$ $\times\sqrt{\{2{\cdot}31+1{\cdot}44\cos 2(p-\xi)\}}$	·00522* ·01585	$-(s-\xi)+Q-\frac{1}{2}\pi$ where $\tan Q = {\cdot}483\tan(p-\xi)$	$\gamma-\sigma$	14°·4920521
J	$\frac{3}{2}e\,\frac{1}{2}\sin I\cos I$	·01485	$+(s-p)-\frac{1}{2}\pi$	$\gamma+\sigma-\varpi$	15°·5854433
*	$\frac{17}{2}e^2\,\frac{1}{2}\sin I\cos^2\frac{1}{2}I$	·00487	$-2(s-\xi)-2(s-p)+\frac{1}{2}\pi$	$\gamma-4\sigma+2\varpi$	12°·8542862
*	$\frac{105}{16}me\,\frac{1}{2}\sin I\cos^2\frac{1}{2}I$	·00512† ·00708	$-2(s-\xi)+(s-p)+2h-2s+\frac{1}{2}\pi$	$\gamma-3\sigma-\varpi+2\eta$	13°·4715144

* The first of these two numbers is the mean value of the coefficient of the tide $\gamma-\sigma-\varpi$; the second applies to the tide M_1 compounded from $\gamma-\sigma-\varpi$ and $\gamma-\sigma+\varpi$.

† The lower of these two figures gives the value when the coefficients in the Evection have the full value as derived from the Lunar Theory.

[B, iii.]

Long Period Tides; General Coefficient $= \frac{1}{2} - \frac{3}{2}\sin^2\lambda$.

Name or Initial	Coefficient	Mean Value of Coefficient	Argument	Speed	Speed in degrees per m. s. hour
	$(1+\frac{3}{2}e^2)\frac{1}{3}(1-\frac{3}{2}\sin^2 I)$	·25224 †	Of variable part is N, the long. of node	$\frac{dN}{dt}$	19°·34 per annum
Mm	$3e \cdot \frac{1}{3}(1-\frac{3}{2}\sin^2 I)$	·04136	$s-p$	$\sigma-\varpi$	0°·5443747
Evect. monthly	$\frac{45}{8}me \cdot \frac{1}{3}(1-\frac{3}{2}\sin^2 I)$	·00580 ‡ ·00755	$-(s-p)+2s-2h$	$\sigma-2\eta+\varpi$	0°·4715211
*	$3m^2 \cdot \frac{1}{3}(1-\frac{3}{2}\sin^2 I)$	·00422 ‡ ·00621	$2(s-h)$	$2(\sigma-\eta)$	1°·0158958
Mf	$(1-\frac{5}{2}e^2)\frac{1}{2}\sin^2 I$	·07827	$2(s-\xi)$	2σ	1°·0980330
Termensual	$\frac{7}{2}e \cdot \frac{1}{2}\sin^2 I$	·01516	$(s-p)+2(s-\xi)$	$3\sigma-\varpi$	1°·6424077

* Indicated by MSf as a compound tide.

† The mean value of this coefficient is $\frac{1}{3}(1+\frac{3}{2}e^2)(1-\frac{3}{2}\sin^2 i)(1-\frac{3}{2}\sin^2\omega) = $ ·25224, and the variable part is approximately

$$-(1+\tfrac{3}{2}e^2)\sin i\cos i\sin\omega\cos\omega\cos N = -\cdot 0328\cos N$$

‡ The lower of these figures give the value when the coefficients in the Evection and Variation have their full values as derived from the Lunar Theory.

[C.]

Schedule of Solar Tides.

Solar Tides; General Coefficient $= \frac{3}{2}\frac{M}{E}\left(\frac{a}{c}\right)^3 a$.

Initial	Coefficient	Value of Coefficient	Argument	Speed	Speed in degrees per m. s. hour
[i.] *Semi-diurnal Tides; General Coefficient* $= \cos^2 \lambda$.					
S_2	$\frac{\tau_1}{\tau}(1-\frac{5}{2}e_1^2)\frac{1}{2}\cos^4\frac{1}{2}\omega$	·21137	$2t$	$2(\gamma-\eta)$	30°·000000
K_2	$\frac{\tau_1}{\tau}(1+\frac{3}{2}e_1^2)\frac{1}{4}\sin^2\omega$	·01823	$2t+2h$	2γ	30°·0821372
T	$\frac{\tau_1}{\tau}\cdot\frac{1}{2}\cdot\frac{7}{2}e_1\cos^4\frac{1}{2}\omega$	·01243	$2t-(h-p_1)$	$2\gamma-3\eta$	29°·9589314
[ii.] *Diurnal Tides; General Coefficient* $= \sin 2\lambda$.					
P	$\frac{\tau_1}{\tau}(1-\frac{5}{2}e_1^2)\frac{1}{2}\sin\omega\cos^2\frac{1}{2}\omega$	·08775	$t-h+\frac{1}{2}\pi$	$\gamma-2\eta$	14°·9589314
K_1	$\frac{\tau_1}{\tau}(1+\frac{3}{2}e_1^2)\frac{1}{2}\sin\omega\cos\omega$	·08407	$t+h-\frac{1}{2}\pi$	γ	15°·0410686
[iii.] *Long Period Tides; General Coefficient* $= \frac{1}{2}-\frac{3}{2}\sin^2\lambda$.					
Ssa	$\frac{\tau_1}{\tau}(1-\frac{5}{2}e_1^2)\frac{1}{2}\sin^2\omega$	·03643	$2h$	2η	0°·0821372

[D.]

Schedule of Speeds in Degrees per Mean Solar Hour.

Semi-diurnal Tides		Diurnal Tides		Long-period Tides	
Initial	Speed	Initial	Speed	Initial	Speed
K_2	30°·0821372	OO	16°·1391016	$3\sigma-\varpi$	1°·6424077
S_2	30°·0000000	J	15°·5854433	Mf	1°·0980330
T	29°·9589314	K_1	15°·0410686	$2(\sigma-\eta)$	1°·0158958
L	29°·5284788	P	14°·9589314	Mm	0°·5443747
λ	29°·4556254	M_1	14°·4920521	$\sigma-2\eta+\varpi$	0°·4715211
M_2	28°·9841042	O	13°·9430356	Ssa	0°·0821372
ν	28°·5125830	$\gamma-3\sigma-\varpi+2\eta$	13°·4715144		
N	28°·4397296	Q	13°·3986609		
μ	27°·9682084	$\gamma-4\sigma+2\varpi$	12°·8542862		
2N	27°·8953548				

Besides these there is the tide initialled R, with speed $2\gamma-\eta=30°·0410686$, which may be neglected.

The variational tides μ and $2(\sigma-\eta)$ have the same speeds respectively as the compound tides 2MS and MSf, and they will occur again when we come to those tides in § 4.

[E.]

*Schedule of Theoretical Importance.**

Initial	Indian Predicter	Coefficient	Coefficient in terms of $M_2=1$	Initial	Indian Predicter	Coefficient	Coefficient in terms of $M_2=1$
M_2	M_2	·45426	1·00000	ν	ν	·01706	·03756
K_1	K_1	·26522	·58385	M_1	—	·01649	·03630
S_2	S_2	·21137	·46531	J	J	·01485	·03269
O	O	·18856	·41509	L	L	·01257	·02767
[lunar K_1	. . . .	·18115	·39878]	T	—	·01243	·02736
N	N	·08796	·19363	2N	—	·01173	·02582
P	P	·08775	·19317	variational μ	—	·01094	·02408
[solar K_1	. . . .	·08407	·18507]	OO	—	·00812	·01788
K_2	K_2	·05752	·12662	$3\sigma-\varpi$	—	·00758	·01669
[lunar K_2	. . . .	·03929	·08649]	$\gamma-3\sigma-\varpi+2\eta$	—	·00708	·01559
Mf	—	·03914	·08616	variational $2(\sigma-\eta)$	—	·00621	·01367
Q	Q	·03651	·08037	$\gamma-4\sigma+2\varpi$	—	·00487	·01072
Mm	—	·02068	·04552	$\sigma-2\eta+\varpi$	—	·00378	·00832
[solar K_2	. . . .	·01823	·04013]	λ	λ	·00330	·00726
Ssa	semi-annual	·01822	·04011				

* [See Vol. xvi., *Great Trig. Survey of India*, "Tidal Observations" by J. Eccles, p. 48.]

A tide of greater importance than some of those retained here is that referred to where the approximation with regard to I was introduced, viz. with speed $2\gamma + \sigma - \varpi$; the value of its coefficient is ·00323. There is also the larger variational diurnal tide, which has been omitted: it would have a coefficient ·00450; also an evectional termensual tide, $\frac{105}{16} me \frac{1}{2} \sin^2 I \cos(3s - 2h + p)$, with coefficient of magnitude ·00292. All other tides in a complete development as far as the second order of small quantities, without any approximation as to the obliquity of the lunar orbit, would have smaller coefficients than those comprised in the above list. Such a development has been made by Professor J. C. Adams, and the values of all the coefficients computed therefrom, in comparison with the above.

Besides the tides above enumerated, the predicter of the India Office also has the over-tides M_4 and M_6, of speeds $4(\gamma - \sigma)$, $6(\gamma - \sigma)$, and the compound tides 2MS, 2SM, MS, of speeds $2\gamma - 4\sigma + 2\eta$, $2\gamma + 2\sigma - 4\eta$, $4\gamma - 2\sigma - 2\eta$, and the meteorological tides S_1, Sa, of speeds $\gamma - \eta$, η.

If this schedule is worth anything, it seems probable that the India Office predicter would do better with some other term substituted for λ.

If further examination of the tidal records should show that the tide M_1 is in reality regular, it should be introduced.

§ 3. *Tides Depending on the Fourth Power of the Moon's Parallax.*

The potential corresponding to these tides is

$$V = \frac{\mu M}{r^4} \rho^3 \left(\tfrac{5}{2} \cos^3 \text{PM} - \tfrac{3}{2} \cos \text{PM}\right)$$

We may obviously neglect the eccentricity of the lunar orbit, and it will appear below, when the principal terms are evaluated, that the declinational tides may be safely omitted.

By these approximations we may put $r = c$, and $M_3 = 0$, and neglect the terms in M_1, M_2 which involve q^2. Following the same plan as in the previous development of § 2, we have, when $M_3 = 0$,

$$\begin{aligned} V \div \frac{\mu M}{c^4} \rho^3 = {} & \tfrac{5}{8} (\xi^3 - 3\xi\eta^2)(M_1^3 - 3M_1 M_2^2) + \tfrac{5}{8} (\eta^2 - 3\xi^2\eta)(M_2^3 - 3M_1^2 M_2) \\ & + \tfrac{3}{8} (\xi^3 + \xi\eta^2 - 4\xi\zeta^2)(M_1^3 + M_1 M_2^2) \\ & + \tfrac{3}{8} (\xi^2\eta + \eta^3 - 4\eta\zeta^2)(M_1^2 M_2 + M_2^3) \end{aligned}$$

The four functions of ξ, η, ζ, in this expression are surface spherical harmonics of the third order, and therefore, corresponding to these four terms, there will be four tides of the types determined by those functions.

Now, we have approximately

$$M_1 = p^2 \cos(\chi - l), \quad M_2 = -p^2 \sin(\chi - l)$$

From which we have

$$M_1^3 - 3M_1M_2^2 = p^6 \cos 3(\chi - l)$$

$$M_1^3 + \ M_1M_2^2 = p^6 \cos \ (\chi - l)$$

When $\eta = 0$; $\xi^3 - 3\xi\eta^2 = \cos^3\lambda$, $\xi^3 + \xi\eta^2 - 4\xi\zeta^2 = \cos\lambda(1 - 5\sin^2\lambda)$

Then, following the same procedure as before, we have for the height of tide

$$h = \tfrac{3}{2}\frac{M}{E}\left(\frac{a}{c}\right)^3 \frac{a^2}{c}\left[\tfrac{5}{12}\cos^3\lambda \,.\, p^6 \cos 3(\chi - l) + \tfrac{1}{4}\cos\lambda(1 - 5\sin^2\lambda) \,.\, p^6\cos(\chi - l)\right] \quad \text{.........(35)}$$

Now, $\cos\lambda(5\sin^2\lambda - 1)$ has its maximum value $\dfrac{16}{3\sqrt{15}}$ when $\cos\lambda = \tfrac{2}{15}\sqrt{15}$: that is to say, when $\lambda = 58° 54'$; thus we may write (35)

$$h = \tfrac{3}{2}\frac{M}{E}\left(\frac{a}{c}\right)^3 a \left[\cos^3\lambda \,.\, \tfrac{5}{12}\left(\frac{a}{c}\right)\cos^6\tfrac{1}{2}I \,.\, \cos[3t + 3(h - \nu) - 3(s - \xi)] \right.$$
$$\left. + \frac{3\,.\,15^{\frac{1}{2}}}{16}\cos\lambda(1 - 5\sin^2\lambda)\frac{4\,.\,15^{\frac{1}{2}}}{45}\left(\frac{a}{c}\right)\cos^6\tfrac{1}{2}I\cos[t + (h - \nu) - (s - \xi)]\right] \quad \text{.........(36)}$$

In this expression observe that there is the same 'general coefficient' outside [ ] as in the previous development; that the spherical harmonics $\cos^3\lambda$, $\dfrac{3\,.\,15^{\frac{1}{2}}}{16}\cos\lambda(5\sin^2\lambda - 1)$ have the maximum values unity, the first at the equator and the second in latitude 58° 54′. The 'speeds' of these two tides are respectively $3(\gamma - \sigma)$ or 43°·4761563 per mean solar hour, and $\gamma - \sigma$, or 14°·4920521 per mean solar hour.

The coefficient of the tide $3(\gamma - \sigma)$, which is comparable with those in the previous schedules [B], [C], [E], is

$$\tfrac{5}{12}\left(\frac{a}{c}\right)\cos^6\tfrac{1}{2}I$$

and the mean value of this function multiplied by $\cos 3(\nu - \xi)$ is ·00599;

also the coefficient of the tide $(\gamma-\sigma)$, likewise comparable with previous coefficients, is

$$\frac{4.15^{\frac{1}{2}}}{45}\left(\frac{a}{c}\right)\cos^6 \tfrac{1}{2}I$$

and the mean value of this function multiplied by $\cos(\nu-\xi)$ is ·0050.

The expression for the tides is written in the form applicable to the equatorial belt bounded by latitudes 26° 34′ N. and S. (viz. where $\sin l = \frac{1}{5}\sqrt{5}$). Outside this belt, what may be called high tide, will correspond with low water. The distribution of land on the earth may, however, perhaps somewhat disturb the latitude of evanescent tide.

It must be noticed that the $\gamma-\sigma$ tide is comparatively small in the equatorial belt, having at the equator only about $\frac{3}{4}$ of its value in latitude 58° 54′.

Referring to the schedule [E] of theoretical importance, we see that the ter-diurnal tide M_3 would come in last but four on the list, and the diurnal tide M_1 (with *rigorous* speed $\gamma-\sigma$) would only be about a half as great again as the synodic fortnightly variational tide.

It thus appears that the ter-diurnal tide is smaller than some of the tides not included in our approximation, and that the diurnal tide should certainly be negligeable.

The value of the M_3 tide, however, is found with scarcely any trouble, from the numerical analysis of the tidal observations, and therefore it is proposed that it should still be evaluated.

§ 4. *Meteorological Tides, Over-tides, and Compound Tides.*

Meteorological Tides.

A rise and fall of water due to regular day and night breezes, prevalent winds, rainfall and evaporation, is called a meteorological tide.

All tides whose period is an exact multiple or sub-multiple of a mean solar day, or of a tropical year, are affected by meteorological conditions. Thus all the tides of the principal solar astronomical series S, with speeds $\gamma-\eta$, $2(\gamma-\eta)$, $3(\gamma-\eta)$, &c., are subject to more or less meteorological perturbation. Although the diurnal elliptic tide, S_1 or $\gamma-\eta$, the semi-annual and annual tides of speeds 2η and η, are all probably quite insensible as arising from astronomical causes, yet they have been found of sufficient importance to be included on the tide-predicter.

The annual and semi-annual tides are of enormous importance in some rivers; in such cases the ter-annual tide (3η) is probably also important, although no harmonic analysis has been as yet made for it.

In the reduction of these tides the arguments of the S series are t, $2t$, $3t$, &c., and of the annual, semi-annual, ter-annual tides are h, $2h$, $3h$. As far as can be foreseen, the magnitudes of these tides will be constant from year to year.

Over-tides.

When a wave runs into shallow water its form undergoes a progressive change as it advances; the front slope generally becomes steeper and the back slope less steep. The most striking example of such a change is when the tide runs up a river in the form of a 'bore.'

A wave which in deep water presented an approximately simple harmonic contour departs largely from that form when it has run into shallow water. Thus in rivers the rise and fall of the water is not even approximately a simple harmonic motion. From the nature of harmonic analysis we are, however, only able to represent the motion by simple harmonic oscillations, and thus to give the non-harmonic rise and fall of tide in shallow water it is necessary to introduce a series of over-tides whose speeds are double, triple, quadruple the speed of the fundamental astronomical tide.

The only tides, in which it has hitherto been thought necessary to represent this change of form in shallow water, belong to the principal lunar and principal solar series. Thus, besides the fundamental astronomical tides M_2 and S_2, the over-tides M_4, M_6, M_8, and S_4, S_6 have been deduced by harmonic analysis.

The height of the fundamental tide M_2 varies from year to year, according to the variation in the obliquity of the lunar orbit, and this variability is represented by the coefficient $\cos^4 \frac{1}{2}I$. It is probable that the variability of M_4, M_6, M_8, will be represented by the square, cube and fourth power of that coefficient.

The law connecting the phase of an over-tide with the height of the fundamental tide is unknown, and under these circumstances it is only possible to make the argument of the over-tide a multiple of the argument of the fundamental, with a constant subtracted. If that constant is found to be the same from year to year, then it will be known that the phase of an over-tide is independent of the height of the fundamental tide.

The following schedule gives the over-tides which must be taken into consideration, the notation being the same as before:—

[F.]

Schedule of Over-tides.

Tide	Coefficient	Argument	Speed	Speed in degrees per m. s. hour
M_4	$(\cos^4 \frac{1}{2}I)^2$	$4t+4(h-\nu)-4(s-\xi)$	$4\gamma-4\sigma$	57°·9682084
M_6	$(\cos^4 \frac{1}{2}I)^3$	$6t+6(h-\nu)-6(s-\xi)$	$6\gamma-6\sigma$	86°·9523126
M_8	$(\cos^4 \frac{1}{2}I)^4$	$8t+8(h-\nu)-8(s-\xi)$	$8\gamma-8\sigma$	115°·9364168
S_4	1	$4t$	$4\gamma-4\eta$	60°·0000000
S_6	1	$6t$	$6\gamma-6\eta$	90°·0000000

It will be understood that here, as elsewhere, the column of arguments only gives that part of the argument which is derived from theory, and the constant to be subtracted from the argument is derivable from observation. It is necessary to have recourse also to observation to determine whether the suggested law of variability in the magnitude of the M over-tides holds good.

Compound Tides.

When two waves of different speeds are propagated in the same water the vertical displacement at the surface is generally determined with sufficient accuracy by summing the displacements due to each wave separately. If, however, the height of the waves is not a small fraction of the depth of the water, the principle of superposition leads to inaccuracy, and it becomes necessary to take into consideration the squares and products of the displacements.

It may be shown that the result of the interaction of two waves is represented by introducing two simple harmonic waves, whose speeds are the sum and the difference of those of the interacting waves. When the interacting waves are tidal these two resultant waves may be called compound tides. They are found to be of considerable importance in estuaries.

A compound tide being derived from the consideration of the product of displacements, we may form an index number, indicative of the probable importance of each compound tide, by multiplying together the semi-ranges of the component tides.

Probably the best way of searching at any station for the compound tides, which are likely to be important, would be to take the semi-ranges of the five or six largest tides at that station and to form index numbers of importance by multiplying the semi-ranges together two and two. Since

[G.]

Schedule of Speeds arising out of Combinations.

	K_1	S_2	O	N	M_4	S_4
M_2	$3\gamma-2\sigma$ $\gamma-2\sigma$ (O)	$4\gamma-2\sigma-2\eta$ $2\sigma-2\eta$	$3\gamma-4\sigma$ γ (K_1)	$4\gamma-5\sigma+\varpi$ $\sigma-\varpi$ (Mm)	—	$6\gamma-2\sigma-4\eta$ $2\gamma+2\sigma-4\eta$
K_1	—	$3\gamma-2\eta$ $\gamma-2\eta$ (P)	$2\gamma-2\sigma$ (M_2) 2σ (Mf)	$3\gamma-3\sigma+\varpi$ $\gamma-3\sigma+\varpi$ (Q)	$5\gamma-4\sigma$ $3\gamma-4\sigma$ (bis)	$5\gamma-4\eta$ $3\gamma-4\eta$
S_2	—	—	$3\gamma-2\sigma-2\eta$ $\gamma+2\sigma-2\eta$	$4\gamma-3\sigma+\varpi-2\eta$ $3\sigma-\varpi-2\eta$	$6\gamma-4\sigma-2\eta$ $2\gamma-4\sigma+2\eta$	—
O	—	—	—	$3\gamma-5\sigma+\varpi$ $\gamma-\sigma+\varpi$ (M_1)	$5\gamma-6\sigma$ $3\gamma-2\sigma$ (bis)	$5\gamma-2\sigma-4\eta$ $3\gamma+2\sigma-4\eta$
N	—	—	—	—	$6\gamma-7\sigma+\varpi$ $2\gamma-\sigma-\varpi$ (L)	$6\gamma-3\sigma+\varpi-4\eta$ $2\gamma+3\sigma-\varpi-4\eta$

these index numbers have no absolute magnitudes, we may omit the decimal point in forming them. Having selected as many of these combinations, in order of importance as may be thought expedient, the arguments of the compound tides are to be found by adding and subtracting the arguments of the components taken in pairs.

[Theory would, however, seem to indicate (see *Encyc. Brit.* Art. Tides) that a better index of importance would be the product of the theoretical importance of the combining tides multiplied by the sum or difference, as the case may be, of their respective speeds.]

In the general case it is only possible to take the tides which the previous schedules have shown usually to be large, and to form a list of compound speeds, with index numbers derived from the multiplication together of the mean values of the coefficients of the astronomical tides. The tides selected here will be M_2, K_1, S_2, O, and N; but to these we shall add M_4 and S_4, although it will not be possible to affix index numbers to combinations involving them.

The schedule [G] gives the speeds of the compound tides. In many cases it will be observed that the compound tide has itself a speed identical with that of an astronomical or meteorological tide. These cases are indicated by the addition of the initial after the speed in question. We thus learn that the tides O, K_1, Mm, P, M_2, Mf, Q, M_1, L will be liable to perturbation in shallow water.

The schedule [G] contains 36 speeds of compound tides: 9 of these fall into the category of astronomical or meteorological tides, 2 are repeated twice, and of the remaining 25 we need only consider, say, the twelve most important.

If either or both the component tides are of lunar origin, the height of the compound tide will change from year to year, and will probably vary proportionally to the product of the coefficients of the component tides.

For the purpose of properly reducing the numerical value of the compound tides, we require not merely the speed, but also the argument.

The following schedule [H] gives the index of importance, argument and speed of the compound tides. The index of importance is the product of the coefficients of the two tides to be compounded.

[But the numbers given in square brackets immediately after these coefficients are proportional to the product of the semi-ranges of the component tides, multiplied by the sum or difference of their speeds. In accordance with the considerations adduced above as to the theoretical height of a compound tide, these numbers ought to give a better index of importance than the simple product of the semi-ranges.]

As in the case of the over-tides, the law of variability of the amplitudes of compound tides in various years is only to be tested by observation.

[H.]

Schedule of Compound Tides.

Index of Importance	Initials	Arguments combined	Speed	Speed in degrees per m. s. hour
1205 [53] —	MK	M_2+K_1 M_4-O	$\left\{ 3\gamma-2\sigma \right\}$	44°·0251728
960 [57]	MS	M_2+S_2	$4\gamma-2\sigma-2\eta$	58°·9841042
960 [1]	MSf	S_2-M_2	$2\sigma-2\eta$	1°·0158958
857 [37] —	2MK	M_2+O M_4-K_1	$\left\{ 3\gamma-4\sigma \right\}$	42°·9271398
561 [25]	—	S_2+K_1	$3\gamma-2\eta$	45°·0410686
400 [23]	MN	M_2+N	$4\gamma-5\sigma+\varpi$	57°·4238338
399 [18]	—	S_2+O	$3\gamma-2\sigma-2\eta$	43°·9430356
399 [6]	—	S_2-O	$\gamma+2\sigma-2\eta$	16°·0569644
—	2SM	S_4-M_2	$2\gamma+2\sigma-4\eta$	31°·0158958
—	—	M_2+S_4	$6\gamma-2\sigma-4\eta$	88°·9841042
—	2MS	M_4-S_2	$2\gamma-4\sigma+2\eta$	27°·9682084
—	—	M_4+S_2	$6\gamma-4\sigma-2\eta$	87°·9682084

It will be noticed that in two cases an over-tide of one speed arises in more than one way, and accordingly different parts of it have different arguments and coefficients. In these cases the utilisation of the results of one year for prediction in future years can only be made by dividing up the compound tide into several parts, according to its theoretical origin. In order to do this it is necessary that the law should be known which connects the heights of a summation and a difference compound tide. A like difficulty arises from the fact that MSf and 2SM are also variational tides.

In practice, however, the compound tide will generally be so small that we may probably treat it as though it arose entirely in one way: and accordingly it is proposed to treat the tides $3\gamma-2\sigma$ or MK, and $3\gamma-4\sigma$ or 2MK, as though they arose entirely from M_2+K_1, M_4-K_1 respectively, and MSf and 2SM as though they were entirely compound tides.

§ 5. *The Method of Reduction of Tidal Observations.*

The printed tabular forms on which the numerical harmonic analysis of the tides is carried out are arranged so that the series of observations to be analysed is supposed to begin at noon, or 0^h, of the first day, and to extend for a year from that time. It has not been found practicable to arrange that the first day shall be the same at all the ports of observation.

Supposing n to be the speed of any tide in degrees per mean solar hour, and t to be mean solar time elapsing since 0^h of the first day; then the immediate result of the harmonic analysis is to obtain A and B, two heights (estimated in feet and tenths) such that the height of this tide at the time t is given by

$$A \cos nt + B \sin nt$$

The question then arises as to what further reductions it will be convenient to make, in order to present the results in the most convenient form.

First, let us put $R = \sqrt{(A^2 + B^2)}$, and $\tan \zeta = \frac{B}{A}$, then the tide is represented by

$$R \cos (nt - \zeta)$$

In this form R is the semi-range of the tide in British feet, and ζ is an angle such that ζ/n is the time elapsing after 0^h of the first day until it is high-water of this particular tide.

It is obvious that ζ may have any value from 0° to 360°, and that the results of the analysis of successive years of observation will not be comparable with one another, when presented in this form.

Secondly, let us suppose that the results of the analysis are to be presented in a number of terms of the form

$$fH \cos (V + u - \kappa)$$

Here V is a linear function of the moon's and sun's mean longitudes, the mean longitude of the moon's and sun's perigees, and the local mean solar time at the place of observation, reduced to angle at 15° per hour. V increases uniformly with the time, and its rate of increase per mean solar hour is the n of the first method, and is called the 'speed' of the tide.

It is supposed that u stands for a certain function of the longitude of the node of the lunar orbit at an epoch half a year later than 0^h of the first day. Strictly speaking, u should be taken as the same function of the longitude of the moon's node, varying as the node moves; but as the variation is but small in the course of a year, u may be treated as a constant and put equal to an average value for the year, which average value is taken as the true value of u at exactly mid-year. Together $V+u$ constitutes that function which has been tabulated as 'the argument' in the schedules B, C, F, H.

Since $V+u$ are together the whole argument according to the equilibrium theory of tides, with sea covering the whole earth, it follows that κ/n is the lagging of the tide which arises from kinetic action, friction of the water, imperfect elasticity of the earth, and the distribution of land.

It is supposed that H is the mean value in British feet of the semi-range of the particular tide in question.

f is a numerical factor of augmentation or diminution, due to the variability of the obliquity of the lunar orbit. The value of f is the ratio of 'the coefficient' in the column of coefficients of the preceding schedules to the mean value of the same term. For example, for all the solar tides f is unity, and for the principal lunar tide M_2, f is equal to $\cos^4 \frac{1}{2}I/\cos^4 \frac{1}{2}\omega \cos^4 \frac{1}{2}i$; for as we shall see below, the mean value of this term has a coefficient $\cos^4 \frac{1}{2}\omega \cos^4 \frac{1}{2}i$.

It is obvious, then, that, if the tidal observations are consistent from year to year, H and κ should come out the same from each year's reductions. It is only when the results are presented in such a form as this that it will be possible to judge whether the harmonic analysis is presenting us with satisfactory results. This mode of giving the tidal results is also essential for the use of the tide-predicting machine.

We must now show how to determine H and κ from R and ζ.

It is clear that $H=R/f$, and the mode of determination of f from the schedules has been explained above, although the proof has been deferred.

If V_0 be the value of V at 0^h of the first day, then clearly

$$-\zeta = V_0 + u - \kappa$$

So that

$$\kappa = \zeta + V_0 + u$$

Thus the rule for the determination of κ is: *Add to the value of* ζ *the value of the argument at* 0^h *of the first day.*

It is suggested that it will henceforth be advisable to tabulate R and ζ, so as to give the results of harmonic analysis in the form $R\cos(nt-\zeta)$; and also H and κ, so as to give it in the form $fH\cos(V+u-\kappa)$, when the results will be comparable from year to year.

A third method of presenting tidal results will be very valuable for the discussion of the theory of tidal oscillations, although it is doubtful whether it will at present be worth while to tabulate the results in this proposed form. This method is to substitute for the H of the second method FK, where F is the mean value of the coefficient as tabulated in the column of coefficients in the schedules—for example, in the case of M_2 we should

3—2

have $F = \frac{1}{2}(1 - \frac{5}{2}e^2)\cos^4 \frac{1}{2}\omega \cos^4 \frac{1}{2}i$, and in the case of S_2 we should have $F = \frac{\tau_1}{\tau} \cdot \frac{1}{2}\cos^4 \frac{1}{2}\omega$. When this process is carried out it will enable us to compare together the several K's corresponding to each of the three classes of tides, but not the several classes *inter se*.

It might perhaps be advisable to proceed still further and to purify K of the coefficient $\frac{3}{2}\frac{M}{E}\left(\frac{a}{c}\right)^3 a$, and of the function of the latitude, viz. $\cos^2\lambda$, $\sin 2\lambda$, $\frac{1}{2} - \frac{3}{2}\sin^2\lambda$, as the case may be. Then we should simply be left with a numerical factor as a residuum, which would represent the augmentation above or diminution below the equilibrium value of the tide. This further reduction may, however, be left out of consideration for the present, since it is superfluous for the proper presentation of the results of harmonic analysis.

For the purpose of using the tide-predicting machine the process of determining H and κ from R and ζ has simply to be reversed, with the difference that the instant of time to which the argument is to refer is 0^h of the first day of the new year, and we must take note of the different value of u and f for the new year. Thus supposing V_1 to be the value of V at 0^h of the first day of the year to which the predictions are to apply, and u_1, f_1, the values of u and f half a year after that 0^h, we have

$$R = f_1 H$$

$$\zeta = \kappa - (V_1 + u_1)$$

This value of R will give the proper throw of the crank of the tide-predicter, and ζ will give the angle at which the crank is to be set. Mr Roberts states, however, that the subtraction, in the predicter of the India Office, of $V_1 + u_1$ from κ is actually performed on the machine, one index being set at κ and the other at $V_1 + u_1$.

We learn also from him that one portion of the term u_1 has been systematically neglected up to the present time: namely, that part which arises in the form $\nu - \xi$ or its multiples. If in the schedules above we were to write $\xi = \nu$ throughout we should arrive at the rule by which the tide-predicter has hitherto been used.

The above statement of procedure is applicable to nearly all the tides, but there are certain tides, viz. K_1, K_2, which have their origins jointly in the tide-generating forces of the moon and sun; also the tides L and M_1 which are rendered complex from the fact that the tidal analysis only extends over a year.

Treatment of the Sidereal Diurnal and Semi-diurnal Tides K_1, K_2.—The expression for the whole K_1 tide of luni-solar origin must, as we see from the schedules B and C, § 3, be of the form

$$M \cos(t + h - \tfrac{1}{2}\pi - \nu - \kappa) + S \cos(t + h - \tfrac{1}{2}\pi - \kappa) \quad \text{......(39)}$$

If now we put

$$\left. \begin{aligned} R &= M \left\{1 + \left(\frac{S}{M}\right)^2 + 2\frac{S}{M}\cos\nu\right\}^{\frac{1}{2}} \\ \tan\nu' &= \frac{\sin\nu}{\cos\nu + S/M} \end{aligned} \right\} \quad \text{..................(40)}$$

these two terms may be written

$$R \cos(t + h - \tfrac{1}{2}\pi - \nu' - \kappa)$$

If h_0 be the sun's mean longitude at 0^h of the first day, $t + h - h_0$ is equal to γt, where t is now mean solar time measured from that 0^h and *not* reduced to angle.

Hence if we write $\zeta = \kappa + \frac{1}{2}\pi - h_0 + \nu'$ ....(41)

the two terms become $R \cos(\gamma t - \zeta)$

But this is the form in which the results of harmonic analysis for the total K_1 tide is expressed in the first method.

From (41) we have

$$\kappa = \zeta + (h_0 - \tfrac{1}{2}\pi) - \nu' \quad \text{.......................(42)}$$

In this formula $h_0 - \frac{1}{2}\pi$ is V_0 for the solar K_1 tide, and ν' is a complex function of the longitude of the moon's node, to be computed (as explained below) from the second of (40).

We must now consider the coefficient f.

If M_0 be the mean value of the lunar K_1 tide, then we know that its ratio to M should according to theory be given by

$$\frac{M}{M_0} = \frac{\sin I \cos I}{\sin\omega\cos\omega\,(1 - \frac{3}{2}\sin^2 i)}$$

The ratio of M to S should also according to theory be given by

$$\frac{M}{S} = \frac{\tau(1 + \frac{3}{2}e^2)\sin I\cos I}{\tau_1(1 + \frac{3}{2}e_1^2)\sin\omega\cos\omega}$$

We must therefore put the coefficient

$$\left. \begin{aligned} f &= \frac{M}{M_0}\,\frac{\left\{1 + \left(\frac{S}{M}\right)^2 + 2\frac{S}{M}\cos\nu\right\}^{\frac{1}{2}}}{1 + \frac{S_0}{M_0}} \\ \text{where}\quad \frac{S_0}{M_0} &= \frac{\tau_1(1 + \frac{3}{2}e_1^2)}{\tau(1 + \frac{3}{2}e^2)} \cdot \frac{1}{(1 - \frac{3}{2}\sin^2 i)} \\ \frac{S}{M} &= \frac{S_0}{M_0}\,\frac{\sin\omega\cos\omega\,(1 - \frac{3}{2}\sin^2 i)}{\sin I\cos I} \end{aligned} \right\} \quad \text{......... ...(43)}$$

f is clearly a complex function of the longitude of the moon's node to be computed as shown below.

The reversal of the process of reduction for the use of the instrument for prediction is obvious.

In the case of the K_2 semi-diurnal tide, if we follow exactly the same process, and put

$$\left.\begin{aligned} \tan 2\nu'' &= \frac{\sin 2\nu}{\cos 2\nu + S/M} \\ f &= \frac{M}{M_0} \frac{\left\{1 + \left(\frac{S}{M}\right)^2 + 2\frac{S}{M}\cos 2\nu\right\}^{\frac{1}{2}}}{1 + \frac{S_0}{M_0}} \\ \text{where}\quad \frac{S_0}{M_0} &= \frac{\tau_1 (1 + \frac{3}{2} e_1^2)}{\tau (1 + \frac{3}{2} e^2)} \cdot \frac{1}{(1 - \frac{3}{2}\sin^2 i)} \\ \frac{S}{M} &= \frac{S_0}{M_0} \frac{\sin^2 \omega (1 - \frac{3}{2}\sin^2 i)}{\sin^2 I} \end{aligned}\right\} \quad \text{..............(44)}$$

the argument of the K_2 tide is $2t + 2h - 2\nu''$, and f is the factor for reduction.

The numerical value of $\frac{S_0}{M_0}$ both for K_1 and K_2 is ·46407.

. . . .

The Tide L.

Reference to the theoretical development in § 3 shows that this tide requires special treatment.

In schedule B (i.) it appears that it must be proportional to

$$\cos^4 \tfrac{1}{2} I \sqrt{1 \cdot 066 - 12 \tan^2 \tfrac{1}{2} I \cos 2(p - \xi)}$$
$$\times \cos[2t + 2(h - \nu) - 2(s - \xi) + (s - p) - R + \pi] \quad \text{......(51)}$$

where $$\tan R = \frac{\sin 2(p - \xi)}{\frac{1}{6}\cot^2 \frac{1}{2} I - \cos 2(p - \xi)}$$

In this expression we must deem R to form a part of the function u, for which a mean value is to be taken. This is, it must be admitted, not very satisfactory, since p increases by nearly 41° per annum.

Suppose, then, that P be the longitude of the perigee at mid-year, measured from the intersection, and that we compute R from the formula

$$\tan R = \frac{\sin 2P}{\frac{1}{6}\cot^2 \frac{1}{2} I - \cos 2P} \quad \text{....................(52)}$$

Then the treatment will be the same as in all the other cases, if the argument $V + u$ be taken as $2t + 2(h - \nu) - 2(s - \xi) + (s - p) - R + \pi$.

The factor f in this case is equal to

$$\frac{\cos^4 \frac{1}{2} I}{\cos^4 \frac{1}{2}\omega \cos^4 \frac{1}{2} i} \sqrt{\{1{\cdot}066 - 12 \tan^2 \tfrac{1}{2} I \cos 2P\}}$$

The Tide M_1.

Reference to schedule B (ii.) shows that this tide must be proportional to

$$e \tfrac{1}{2} \sin I \cos^2 \tfrac{1}{2} I \sqrt{\{2{\cdot}307 + 1{\cdot}435 \cos 2(p - \xi)\}} \times \cos [t + (h - \nu) - (s - \xi) + Q - \tfrac{1}{2}\pi] \ldots\ldots (52')$$

where $\tan Q = {\cdot}483 \tan (p - \xi)$.

We must here deem Q to form a part of the function u, for which a mean value is to be taken; but as in the case of the L tide, this course is not very satisfactory.

If P as before denotes the longitude of the perigee at mid-year, measured from the intersection, and Q be computed from

$$\tan Q = {\cdot}483 \tan P \qquad \ldots\ldots\ldots\ldots\ldots\ldots\ldots (52'')$$

then the argument $V + u$ will be

$$t + (h - \nu) - (s - \xi) + Q - \tfrac{1}{2}\pi$$

And the factor f is

$$\frac{\sin I \cos^2 \frac{1}{2} I}{\sin \omega \cos^2 \frac{1}{2}\omega \cos^4 \frac{1}{2} i} \frac{\sqrt{\{2{\cdot}307 + 1{\cdot}435 \cos 2P\}}}{\sqrt{2{\cdot}307}} \qquad \ldots\ldots\ldots (52''')$$

[In the computation forms the ·483 of (52″) has been taken as $\frac{1}{2}$, and this value of f has been taken as being

$$\frac{\sin I \cos^2 \frac{1}{2} I}{\sin \omega \sin^2 \frac{1}{2}\omega \cos^4 \frac{1}{2} i} \sqrt{(\tfrac{5}{2} + \tfrac{3}{2} \cos 2P)}$$

because $\sqrt{(\frac{5}{2})}$ which should have been in the denominator was accidentally omitted, and $\tan^2 \frac{1}{2} I$ was treated as zero, instead of having a mean value. It is clearly a matter of indifference for practical work whether or not this factor $\sqrt{(\frac{5}{2})}$ stands in the denominator or not. All the Indian work has been carried out in the form just shown, and the practice may well be adhered to. In so small a tide it cannot be of much importance whether or not $\tan^2 \frac{1}{2} I$ is treated as zero*.]

It has been shown that the tide M_1, in as far as it depends on the fourth power of the moon's parallax, is too small to be worth including in the numerical analysis.

* [See Eccles, Vol. XVI., *Great Trig. Survey of India*, p. 57.]

§ 6. *On the Method of Computing the Arguments and Coefficients.*

In performing the reductions of the preceding sections a number of numerical quantities are required, which are to be derived from the position of the heavenly bodies.

Formulæ for Computing I, ν, ξ.

From Fig. 2, § 3, we see that

$$\left.\begin{aligned} \cot(N-\xi)\sin N &= \cos N\cos i + \sin i\cot\omega \\ \cot\nu\sin N &= \cos N\cos\omega + \sin\omega\cot i \\ \cos I &= \cos i\cos\omega - \sin i\sin\omega\cos N \end{aligned}\right\} \quad \ldots\ldots(53)$$

If β be an auxiliary angle defined by

$$\tan\beta = \tan i\cos N \quad \ldots\ldots\ldots\ldots\ldots\ldots\ldots\ldots(54)$$

then

$$\left.\begin{aligned} \cos I &= \cos i\sec\beta\cos(\omega+\beta) \\ \sin\nu &= \sin i\,\mathrm{cosec}\, I\sin N \\ \sin(N-\xi) &= \sin\omega\,\mathrm{cosec}\, I\sin N \end{aligned}\right\} \quad \ldots\ldots\ldots\ldots\ldots\ldots(55)$$

The formulæ (53) also lead to the rigorous formulæ

$$\left.\begin{aligned} \tan\xi &= \frac{\sin i\cot\omega\sin N(1-\tan\tfrac{1}{2}i\tan\omega\cos N)}{\cos^2\tfrac{1}{2}i + \sin i\cot\omega\cos N - \sin^2\tfrac{1}{2}i\cos 2N} \\ \tan\nu &= \frac{\tan i\,\mathrm{cosec}\,\omega\sin N}{1+\tan i\cot\omega\cos N} \end{aligned}\right\} \quad \ldots\ldots(53')$$

But, if we treat i as small, (53′) may be reduced to

$$\left.\begin{aligned} \tan\xi &= i\cot\omega\sin N - \tfrac{1}{2}i^2\sin 2N\,\frac{1-\tfrac{1}{2}\sin^2\omega}{\sin^2\omega} \\ \tan\nu &= i\,\mathrm{cosec}\,\omega\sin N - \tfrac{1}{2}i^2\sin 2N\,\frac{\cos\omega}{\sin^2\omega} \\ \cos I &= (1-\tfrac{1}{2}i^2)\cos\omega - i\sin\omega\cos N \end{aligned}\right\} \quad \ldots\ldots\ldots(53'')$$

A table of values of ξ, ν, I, for different values of N, with $\omega = 23°\ 27'\cdot3$, $i = 5°\ 8'\cdot8$, may be computed either directly from (53) or from (55).

The approximate formulæ (53″) will be of service hereafter.

On the Mean Values of the Coefficients in Schedules [B].

In the three schedules [B] of lunar tides, 'the coefficients' are certain functions of I, and there are certain terms in the arguments which are functions of ν and ξ. We may typify all the terms by $J\cos(T+u)$, where J is a function of I, and u of ν and ξ. If we substitute for J and u in terms of ω, i, N, and develop the result, we shall obtain a series of terms of which the one independent of N is, say, $J_1 \cos T$. Then J_1 is the mean value of the semi-range of the tide in question. Such a development may be carried out rigorously, but it involves a good deal of analysis to do so; we shall therefore confine ourselves to an approximate treatment of the question, using the formulæ (53″) for ξ and ν.

It may be proved that in no case does J involve a term with a sine of an odd multiple of N, and the formulæ (54) or (55) show that in every term of $\sin u$ there will occur a sine of an odd multiple of N; whence it follows that $J \sin u$ has mean value zero, and J_1 is the term independent of N in $J \cos u$.

It may also be proved that in no case does $\cos u$ involve a term in $\cos N$, and that the terms in $\cos 2N$ are all of order i^2; also it appears that J always involves a term in $\cos N$, and also terms in $\cos 2N$ of order i^2.

Hence to the degree of approximation adopted, J_1 is equal to $J_0 \cos u_0$, where J_0 is the mean value of J, and $\cos u_0$ the mean value of $\cos u$.

In evaluating $\cos u_0$ from the formulæ (53″), we may observe that wherever $\sin^2 N$ occurs it may be replaced by $\frac{1}{2}$; for $\sin^2 N = \frac{1}{2} - \frac{1}{2}\cos 2N$, and the $\cos 2N$ has mean value zero.

The following are the values of $\cos u_0$ thus determined from (53″):—

$$(\alpha) \quad \cos 2(\nu - \xi)_0 = 1 - i^2 \left(\frac{1 - \cos\omega}{\sin\omega}\right)^2$$

$$(\beta) \quad \cos 2\nu_0 = 1 - i^2 \frac{1}{\sin^2\omega}$$

$$(\gamma) \quad \cos(2\xi - \nu)_0 = 1 - \tfrac{1}{4} i^2 \left(\frac{1 - 2\cos\omega}{\sin\omega}\right)^2$$

$$(\delta) \quad \cos(2\xi + \nu)_0 = 1 - \tfrac{1}{4} i^2 \left(\frac{1 + 2\cos\omega}{\sin\omega}\right)^2$$

$$(\epsilon) \quad \cos\nu_0 = 1 - \tfrac{1}{4} i^2 \frac{1}{\sin^2\omega}$$

$$(\zeta) \quad \cos 2\xi_0 = 1 - i^2 \cot^2\omega$$

The suffix $_0$ indicating the mean value.

Similarly the following are the J_0's or mean values of J:—

$$(\alpha') \quad \cos^4 \tfrac{1}{2} I_0 = \cos^4 \tfrac{1}{2}\omega \left[1 + \tfrac{1}{2} i^2 \frac{\sin^2 \frac{1}{2}\omega - \cos\omega}{\cos^2 \frac{1}{2}\omega}\right]$$

$$(\beta') \;\&\; (\zeta') \quad \sin^2 I_0 = \sin^2\omega \left[1 + i^2 \frac{1 - \frac{3}{2}\sin^2\omega}{\sin^2\omega}\right]$$

$$(\gamma') \quad \sin I_0 \cos^2 \tfrac{1}{2} I_0 = \sin\omega \cos^2 \tfrac{1}{2}\omega \left[1 + \tfrac{1}{4} i^2 \left(\frac{\cos 2\omega}{\sin^2\omega} - \frac{2\cos\omega}{\cos^2 \frac{1}{2}\omega}\right)\right]$$

$$(\delta') \quad \sin I_0 \sin^2 \tfrac{1}{2} I_0 = \sin \omega \sin^2 \tfrac{1}{2}\omega \left[1 + \tfrac{1}{4} i^2 \left(\frac{\cos 2\omega}{\sin^2 \omega} + \frac{2\cos \omega}{\sin^2 \frac{1}{2}\omega}\right)\right]$$

$$(\epsilon') \quad \sin I_0 \cos I_0 = \sin \omega \cos \omega \left[1 + \tfrac{1}{4} i^2 (\cot^2 \omega - 5)\right]$$

On referring to schedules [B], it appears that (α) multiplied by (α') is the mean value of the $\cos^4 \frac{1}{2} I \cos 2(\nu - \xi)$ which occurs in the semidiurnal terms; and so on with the other letters, two and two. Performing these multiplications, and putting $1 - \frac{1}{2} i^2$ in the results as equal to $\cos^4 \frac{1}{2} i$, and $1 - \frac{3}{2} i^2$ as equal to $1 - \frac{3}{2}\sin^2 i$, we find that the mean values are all unity for the following functions, viz.:

$$\frac{\cos^4 \frac{1}{2} I \cos 2(\nu - \xi)}{\cos^4 \frac{1}{2}\omega \cos^4 \frac{1}{2} i}, \quad \frac{\sin^2 I \cos 2\nu}{\sin^2 \omega (1 - \frac{3}{2}\sin^2 i)}, \quad \frac{\sin I \cos^2 \frac{1}{2} I \cos(2\xi - \nu)}{\sin \omega \cos^2 \frac{1}{2}\omega \cos^4 \frac{1}{2} i},$$

$$\frac{\sin I \sin^2 \frac{1}{2} I \cos(2\xi + \nu)}{\sin \omega \sin^2 \frac{1}{2}\omega \cos^4 \frac{1}{2} i}, \quad \frac{\sin I \cos I \cos \nu}{\sin \omega \cos \omega (1 - \frac{3}{2}\sin^2 i)}, \quad \frac{\sin^2 I \cos 2\xi}{\sin^2 \omega \cos^4 \frac{1}{2} i}$$

Lastly, it is easy to show rigorously that the mean value of

$$\frac{1 - \frac{3}{2}\sin^2 I}{(1 - \frac{3}{2}\sin^2 \omega)(1 - \frac{3}{2}\sin^2 i)}$$

is also unity.

If we write
$$\varpi = \cos \tfrac{1}{2}\omega \cos \tfrac{1}{2} i - \sin \tfrac{1}{2}\omega \sin \tfrac{1}{2} i \, e^{N\iota}$$
$$\kappa = \sin \tfrac{1}{2}\omega \cos \tfrac{1}{2} i + \cos \tfrac{1}{2}\omega \sin \tfrac{1}{2} i \, e^{N\iota}$$

where ι stands for $\sqrt{-1}$; and let ϖ_1, κ_1 denote the same functions with the sign of N changed, then it may be proved rigorously that

$$\cos^4 \tfrac{1}{2} I \cos 2(\nu - \xi) = \tfrac{1}{2}(\varpi^4 + \varpi_1^4)$$
$$\sin^2 I \cos 2\nu = 2(\varpi^2 \kappa_1^2 + \varpi_1^2 \kappa^2)$$
$$\sin I \cos^2 \tfrac{1}{2} I \cos(2\xi - \nu) = \varpi^3 \kappa + \varpi_1^3 \kappa_1$$
$$\sin I \sin^2 \tfrac{1}{2} I \cos(2\xi + \nu) = \varpi \kappa^3 + \varpi_1 \kappa_1^3$$
$$\sin I \cos I \cos \nu = (\varpi \kappa_1 + \varpi_1 \kappa)(\varpi \varpi_1 - \kappa \kappa_1)$$
$$\sin^2 I \cos 2\xi = 2(\varpi^2 \kappa^2 + \varpi_1^2 \kappa_1^2)$$
$$1 - \tfrac{3}{2}\sin^2 I = \varpi^2 \varpi_1^2 - 4\varpi \varpi_1 \kappa \kappa_1 + \kappa^2 \kappa_1^2$$

The proof of these formulæ, and the subsequent development of the functions of the ϖ's and κ's, constitute the rigorous proof of the formulæ, of which the approximate proof has been indicated above. The analogy between the ϖ's and κ's, and the p, q of the earlier developments of this Report, is that if i vanishes $\varpi = \varpi_1 = p$, $\kappa = \kappa_1 = q$.

(See a paper in the *Phil. Trans. R. S.*, Part II. 1880, p. 713; to be reproduced also in Vol. II. of the present work.)

This investigation justifies the statements preceding the schedules [B] as to the mean values of the coefficients.

Formulæ for computing f.

In the original reduction of tidal observations we want 1/f; in the use of the tide-predicter f is required.

On looking through the schedules [B], we see that the following values of 1/f are required.

$$(1)\ \frac{\cos^4\frac{1}{2}\omega\cos^4\frac{1}{2}i}{\cos^4\frac{1}{2}I},\quad (2)\ \frac{\sin^2\omega\,(1-\frac{3}{2}\sin^2 i)}{\sin^2 I},\quad (3)\ \frac{\sin\omega\cos^2\frac{1}{2}\omega\cos^4\frac{1}{2}i}{\sin I\cos^2\frac{1}{2}I},$$

$$(4)\ \frac{\sin\omega\sin^2\frac{1}{2}\omega\cos^4\frac{1}{2}i}{\sin I\sin^2\frac{1}{2}I},\quad (5)\ \frac{\sin\omega\cos\omega\,(1-\frac{3}{2}\sin^2 i)}{\sin I\cos I},$$

$$(6)\ \frac{\sin^2\omega\cos^4\frac{1}{2}i}{\sin^2 I},\quad (7)\ \frac{(1-\frac{3}{2}\sin^2\omega)(1-\frac{3}{2}\sin^2 i)}{(1-\frac{3}{2}\sin^2 I)}$$

And in the case of the over-tides and compound tides (schedules [F], [H]), powers and products of these quantities.

A table of values of these functions for various values of I is given in § 12.

The functions (2) and (5) are required for computing f for the K_1 and K_2 tides.

In this list of functions let us call that numbered (2) k_2, and that numbered (5) k_1; k_2 and k_1 being the values of the reciprocal of f which would have to be applied in the cases of the K_2 and K_1 tides, if the sun did not exist.

On referring back to the paragraph in § 5 in which the treatment of the K_2 and K_1 tides is explained we see that for K_2

$$\frac{S}{M} = \cdot 46407 \times k_2$$

and therefore from (44) we see that for K_2

$$\left.\begin{aligned}\frac{1}{f} &= \frac{1\cdot 46407 \times k_2}{\{1+(0\cdot 46407\times k_2)^2+0\cdot 92814 k_2\cos 2\nu\}^{\frac{1}{2}}}\\ \tan 2\nu'' &= \frac{\sin 2\nu}{\cos 2\nu + \cdot 46407 k_2}\end{aligned}\right\}\quad\text{.........(56)}$$

And for K_1 the similar formulæ hold with k_1 in place of k_2, and ν in place of 2ν*.

Tables of 1/f and ν', $2\nu''$ for the K_1 and K_2 tides have been formed from (56), and are given in Col. Baird's *Manual of Tidal Observations*.

The angle I ranges from 18° 18′·5, when it is $\omega - i$, to 28° 36′·1, when it is $\omega + i$.

* This method of treating these tides is due to Professor Adams. I had proposed to divide the K tides into their lunar and solar parts.—G. H. D.

Then in using these tables, we first extract I for any value of N, and afterwards find the coefficients from the subsequent tables.

The coefficients for the over-tides and compound tides may be found from tables of squares and cubes and by multiplication.

[Algebraic formulæ for the several f's are given at the end of the present paper; they are so simple that the auxiliary tables may be dispensed with, without much loss.]

Formulæ for s, p, h, p_1, N.

The numerical values may be deduced from the formulæ given in Hansen's *Tables de la Lune*. The following are reduced to a more convenient epoch, and to forms appropriate to the present investigation.

$$\left.\begin{aligned} s &= 150^\circ{\cdot}0419 + [13 \times 360^\circ + 132^\circ{\cdot}67900]\, T + 13^\circ{\cdot}1764\, D \\ &\qquad + 0^\circ{\cdot}5490165\, H \\ p &= 240^\circ{\cdot}6322 + 40^\circ{\cdot}69035\, T + 0^\circ{\cdot}1114\, D + 0^\circ{\cdot}0046418\, H \\ h &= 280^\circ{\cdot}5287 + 360^\circ{\cdot}00769\, T + 0^\circ{\cdot}9856\, D + 0^\circ{\cdot}0410686\, H \\ p_1 &= 280^\circ{\cdot}8748 + 0^\circ{\cdot}01711\, T + 0^\circ{\cdot}000047\, D \\ N &= 285^\circ{\cdot}9569 - 19^\circ{\cdot}34146\, T - 0^\circ{\cdot}0529540\, D \end{aligned}\right\} \quad \ldots\ldots(57)$$

Where

T is the number of Julian years of $365\frac{1}{4}$ mean solar days,

D the number of mean solar days,

H the number of mean solar hours,

after 0^h Greenwich mean time, January 1, 1880.

From the coefficients of H we see that

$$\sigma = 0^\circ{\cdot}5490165, \qquad \varpi = 0^\circ{\cdot}0046418, \qquad \eta = 0^\circ{\cdot}0410686, \quad \ldots\ldots(58)$$

whence $\gamma = 15^\circ{\cdot}0410686$.

For the purposes of using the forms for harmonic analysis of the tidal observations, these formulæ may be reduced to more convenient and simpler forms.

The mean values of N and p_1 are required, and for the treatment of the L and M_1 tides the mean value of $p - \xi$, denoted by P. For determining these three quantities, we may therefore add half the coefficient of T once for all, and write

$$\left.\begin{aligned} N &= 276^\circ{\cdot}2861 - 0^\circ{\cdot}05295\, D - 19^\circ{\cdot}34146\, T \\ p_1 &= 280^\circ{\cdot}8833 + 0^\circ{\cdot}00005\, D + 0^\circ{\cdot}01711\, T \\ P + \xi &= 261^\circ{\cdot}0 \qquad + 0^\circ{\cdot}111\, D \quad + 40^\circ{\cdot}69\, T \end{aligned}\right\} \quad \ldots\ldots\ldots(59)$$

where T is simply the number of years, whether there be leap-years or not amongst them, since 1880, and D the number of days from Jan. 1, numbered as zero up to the first day of the year to be analysed.

Now, suppose d to denote the number of quarter days either one, two, or three in excess of the Julian years which have elapsed since 0^h Jan. 1, 1880, up to 0^h Jan. 1 of the year in question; let D denote the same as before; and let L be the East Longitude of the place of observation in hours and decimals of hours.

Then for s_0, p_0, h_0, the values of s, p, h at 0^h of the first day, we have

$$\left.\begin{aligned} s_0 &= 150^\circ{\cdot}0419 + 132^\circ{\cdot}67900\,T + 3^\circ{\cdot}29410\,d + 13^\circ{\cdot}1764\,D - 0^\circ{\cdot}54902\,L \\ p_0 &= 240^\circ{\cdot}6322 + 40^\circ{\cdot}69035\,T + 0^\circ{\cdot}02785\,d + 0^\circ{\cdot}1114\,D - 0^\circ{\cdot}00464\,L \\ h_0 &= 280^\circ{\cdot}5287 + 0^\circ{\cdot}00769\,T + 0^\circ{\cdot}24641\,d + 0^\circ{\cdot}9856\,D - 0^\circ{\cdot}04107\,L \end{aligned}\right\} \quad (60)$$

In these formulæ T is an integer, being the excess of the year in question above 1880, and d is to be determined thus:—if the excess of the year above 1880 divided by 4 leaves remainder 3, d is 1; if remainder 2, it is 2; if remainder 1, it is 3; and if remainder zero, it is zero. For example for 1895, $T=15$, $d=1$; because from 0^h Jan. 1, 1880 to 0^h Jan. 1, 1895, is 15 Julian years and a quarter day.

For all dates after Feb. 28, 1900, *one day's motion must be subtracted from* s_0, p_0, h_0, p_1, $P+\xi$, *and one day's motion added to* N.

The terms in L may be described as corrections for longitude.

The $13 \times 360^\circ$ and 360° which occurred in the previous formulæ for s and h are now omitted, because T is essentially an integer.

If it be preferred, the values of s_0 and N may be extracted from the *Nautical Almanac*, and h_0 is (neglecting nutation) the sidereal time reduced to angle. We may take p_0 from a formula given by Hansen at p. 300 of the *Tables de la Lune*. This latter course is that which is followed in the forms for computation.

§ 7. *Summary of Initial Arguments and Factors of Reduction.*

The results for the various kinds of tide are scattered in various parts of the above, and it will therefore be convenient to collect them together. In order to present the results in a form convenient for computation, each argument is given by reference to any previous argument which contains the same element. In the following schedule Arg. M_2 and Fac. M_2 (for example) mean the argument and factor computed for the tide M_2.

[I.]

Schedule of Arguments at 0^h of the first day, and Factors for Ensuing Year.

	Initial Arguments. V_0+u	Factors for Reduction. $\frac{1}{f}$
S_1, S_2, S_3, S_4	zero	unity
P	$-h_0+\frac{1}{2}\pi$	unity
T	$-(h_0-p_1)$	unity
M_1	$(h_0-\nu)-(s_0-\xi)+Q-\frac{1}{2}\pi$ where $\tan Q = \cdot483 \tan P$	Fac. O $\div\sqrt{\{\frac{5}{2}+\frac{3}{2}\cos 2P\}}$ *
M_2	$2(h_0-\nu)-2(s_0-\xi)$	$\left(\frac{\cos\frac{1}{2}\omega\cos\frac{1}{2}i}{\cos\frac{1}{2}I}\right)^4$
M_3	$\frac{3}{2}$ Arg. M_2	(Fac. $M_2)^{\frac{3}{2}}$
M_4	2 Arg. M_2	(Fac. $M_2)^2$
M_6	3 Arg. M_2	(Fac. $M_2)^3$
M_8	4 Arg. M_2	(Fac. $M_2)^4$
K_2	$2h_0-2\nu''$ where $\tan 2\nu''=\frac{\sin 2\nu}{\cos 2\nu+\cdot464\times k}$	$\frac{1\cdot46407k}{\sqrt{\{1+(\cdot464\times k)^2+\cdot928k\cos 2\nu\}}}$ where $k=\frac{\sin^2\omega(1-\frac{3}{2}\sin^2 i)}{\sin^2 I}$
K_1	$h_0-\nu'-\frac{1}{2}\pi$ where $\tan\nu'=\frac{\sin\nu}{\cos\nu+\cdot464\times k}$	$\frac{1\cdot46407k}{\sqrt{\{1+(\cdot464\times k)^2+\cdot928k\cos\nu\}}}$ where $k=\frac{\sin 2\omega(1-\frac{3}{2}\sin^2 i)}{\sin 2I}$
N	Arg. $M_2-(s_0-p_0)$	Fac. M_2
2N	Arg. N $-(s_0-p_0)$	Fac. M_2
L	Arg. $M_2+(s_0-p_0)-R+\pi$ where $\tan R=\frac{\sin 2P}{\frac{1}{6}\cot^2\frac{1}{2}I-\cos 2P}$	Fac. $M_2\div\sqrt{1\cdot066-12\tan^2\frac{1}{2}I\cos 2P}$
ν	Arg. $M_2+(s_0-p_0)+2h_0-2s_0$	Fac. M_2
O	$(h_0-\nu)-2(s_0-\xi)+\frac{1}{2}\pi$	$\frac{\sin\omega\cos^2\frac{1}{2}\omega\cos^4\frac{1}{2}i}{\sin I\cos^2\frac{1}{2}I}$
OO	$(h_0-\nu)+2(s_0-\xi)-\frac{1}{2}\pi$	$\frac{\sin\omega\sin^2\frac{1}{2}\omega\cos^4\frac{1}{2}i}{\sin I\sin^2\frac{1}{2}I}$
Q	Arg. O $-(s_0-p_0)$	Fac. O
J	$(h_0-\nu)+(s_0-p_0)-\frac{1}{2}\pi$	$\frac{\sin 2\omega(1-\frac{3}{2}\sin^2 i)}{\sin 2I}$

* [See § 5 above as to a more correct value of f in this case, and the reasons which have led to the use of the value here given.]

Schedule [I.] *continued.*

	Initial Arguments. V_0+u	Factors for Reduction. $\frac{1}{f}$
MS	Arg. M_2	Fac. M_2
2MS	Arg. M_4	Fac. M_4
2SM	2π − Arg. M_2	Fac. M_2
MK	Arg. M_2+Arg. K_1	Fac. M_2 × Fac. K_1
2MK	Arg. M_4−Arg. K_1	Fac. M_4 × Fac. K_1
MN	Arg. M_2+Arg. N	Fac. M_2 × Fac. N
MSf	2π − Arg. M_2	Fac. M_2
Mm	(s_0-p_0)	$\dfrac{(1-\frac{3}{2}\sin^2\omega)(1-\frac{3}{2}\sin^2 i)}{1-\frac{3}{2}\sin^2 I}$
Mf	$2(s_0-\xi)$	$\dfrac{\sin^2\omega\cos^4\frac{1}{2}i}{\sin^2 I}$
Sa	h_0	unity
Ssa	$2h_0$	unity

There are two tables, numbered I. and II., given at pp. 304 and 305 of the Report for 1876 of the Committee of the British Association on Tidal Observations. The columns headed ϵ give functions which, when their signs are reversed, are the arguments at the epoch. To show the identity of these expressions with those in the above schedule [I], we must put

$$f=-h_0,\quad g=-h_0,\quad ☽=s_0+\nu-\xi,\quad \odot=h_0,\quad \varpi'=p_0+\nu-\xi,\quad \varpi=p_1$$

For the sake of symmetry these tables contain several entries which we have omitted from our schedule, because of the smallness of the tides to which they refer. The entries of the tides of long period, Nos. 3 and 4, are given with the opposite sign from that here adopted*; thus those entries require alteration by 180° to bring them into accordance with our schedule.

The following corrections have to be made in Table II.: No. 8, for 2ν read 3ν; No. 15, add 4ν; Nos. 17 and 19, add $2(\nu-\xi)$; Nos. 18 and 20, subtract $2(\nu-\xi)$.

The K_1, K_2 tides, Nos. 9 and 16 of both tables, are entered separately as to their lunar and solar parts. The two parts of the M_1 tide, Nos. 7 and 11, are entered separately. Also No. 14 only gives one part of the tide here entered as L.

The reader is warned that the definition of ϵ on p. 293 is incomplete, and incorrect for proper reference to the equilibrium theory of tides. The definition of ϖ' on p. 302 is incorrect.

* See the passage in § 2 between equations (28) and (29).

§ 8. *On the Reductions of the Published Results of Tidal Analysis.*

In the Tide Tables published by the Indian Government, it is stated that each tide is expressed in the form $R \cos(nt - \epsilon)$, where R is the semi-range in feet, n the speed of the tide, and ϵ/n is the time in mean solar hours which elapses, after an epoch appropriate to the tide, until the next high-water of that tide. Tables are then given for R and ϵ at each station for each year.

The mode of tabulation is the same as that followed in the Tidal Reports of the British Association for 1872 and 1876.

It is advisable that all the results should be reduced according to one system, such that the observations of the several years and the values for the several speeds of tide may be comparable *inter se.*

In § 5 it has been proposed that the tide should be recorded in the form

$$fH \cos(V + u - \kappa)$$

It appears from the statements in the Reports for 1872 and 1876 and from an examination of the reductions of the published results that the ϵ of the tables is equal to $\kappa - u$, and that the R of the tables is equal to fH. Thus in order to reduce the published results to proper forms, comparable *inter se*, it is necessary to add to ϵ the appropriate u, and to divide R by the proper f. Following this process we obtain certain corrections to the ϵ's to obtain the κ's. The values of 1/f by which the R's are to be multiplied to obtain the H's, are those given in the preceding schedule [I]. [But it does not seem worth while to reproduce the schedule of instructions for correcting these old results.]

§ 9. *Description of the Numerical Harmonic Analysis for the Tides of Short Period.*

It forms no part of the plan of this Report to give an account of the instruments with which the tidal observations are made, or of the tide-predicting instrument. A description of the tide-gauge, which is now in general use in India and elsewhere, and of the tide-predicter, which is at the India Store Department in Lambeth, and of designs for modifications of those instruments, has been given in a paper by Sir William Thomson, read before the Institution of Civil Engineers on March 1, 1881 *, and to this paper we refer the reader. Our present object is to place on record the manner in which the observations have been or are to be henceforth treated, and to give the requisite information for the subsequent use of the tide-predicting instrument.

* "The Tide Gauge, Tidal Harmonic Analyser, and Tide Predicter," *Proc. Inst. C. E.*, Vol. 45, Part III.

The tide-gauge furnishes us with a continuous graphical record of the height of the water above some known datum mark for every instant of time.

It is probable that at some future time the Harmonic Analyser of Professors James and Sir William Thomson may be applied to the tide-curves. The instrument is nearly completed, and now lies in the Physical Laboratory of the University of Glasgow, but it has not yet been put into use*. The treatment of the observations which we shall describe is the numerical process used at the office of the Indian Survey at Poona, under the immediate superintendence of Major A. W. Baird, R.E. The printed forms for computation were admirably drawn up by Mr Edward Roberts, of the 'Nautical Almanac' Office; but they have now undergone certain small modifications in accordance with this Report. The work of computation is to a great extent carried out by native Indian computers. The results of the harmonic analysis are afterwards sent to Mr Roberts, who works out the instrumental tide-predictions for the several ports for the ensuing year. The use of that instrument requires great skill and care. The results of the tidal reductions have hitherto been presented in a somewhat chaotic form, and we believe that it is only due to Mr Roberts' knowledge of the manner in which the tidal results have been treated that they have been correctly used for prediction. It may be hoped that the use of the methods recommended in the present Report will remove some of the factitious difficulties in the use of the instrument†.

The first operation performed on the tidal record is the measurement in feet and decimals of the height of water above the datum at every mean solar hour. The period chosen for analysis is about one year, and the first measurement corresponds to noon. It has been found impracticable to make the initial noon belong to the same day at the several ports. It would seem, at first sight, preferable to take the measurements at every mean lunar hour; but the whole of the actual process in use is based on measurements taken at the mean solar hours, and a change to lunar time would involve a great deal of fresh labour and expense.

If T be the period of any one of the diurnal tides, or twice the period of any one of the semi-diurnal tides, it approximates more or less nearly to 24 m. s. hours, and if we divide it into 24 equal parts, we may speak of each as a T-hour. We shall for brevity refer to mean solar time as S-time.

Suppose, now, that we have two clocks, each marked with 360°, or 24 hours, and that the hand of the first, or S-clock, goes round once in 24

* See Appendix, Thomson and Tait's *Nat. Phil.*, 2nd ed. 1883. [It has not been found expedient to use this interesting instrument, and it is deposited in the Museum at South Kensington.]

† [The tide-predicter was transferred to the National Physical Laboratory in 1904.

S-hours, and that of the second, or T-clock, goes round once in 24 T-hours, and suppose that the two clocks are started at 0° or 0^h at noon of the initial day. For the sake of distinctness, let us imagine that a T-hour is longer than an S-hour, so that the T-clock goes slower than the S-clock. The measurements of the tide-curve give us the height of water exactly at each S-hour; and it is required from these data to determine the height of water at each T-hour.

For this end we are, in fact, instructed to count T-time, but are only allowed to do so by reference to S-time, and, moreover, the time is always to be specified as an integral number of hours.

Beginning, then, with 0^h of the first day, we shall begin counting 0, 1, 2, &c., as the T-hand comes up to its hour-marks. But as the S-hand gains on the T-hand, there will come a time when the T-hand, being exactly at the p hour-mark, the S-hand is nearly as far as $p+\frac{1}{2}$. When, however, the T-hand has advanced to the $p+1$ hour-mark, the S-hand will be a little beyond $p+1+\frac{1}{2}$: that is to say, a little less than half an hour before $p+2$. Counting, then, in T-time by reference to S-time, we shall jump from p to $p+2$. The counting will go on continuously for a number of hours nearly equal to $2p$, and then another number will be dropped, and so on throughout the whole year. If it had been the T-hand which went faster than the S-hand, it is obvious that one number would be repeated at two successive hours instead of one being dropped. We may describe each such process as a 'change.'

Now, if we have a sheet marked for entry of heights of water according to T-hours from results measured at S-hours, we must enter the S-measurements continuously up to p, and we then come to a 'change,' and dropping one of the S-series, we go on again continuously until another 'change,' when another is dropped, and so on.

Since a 'change' occurs at the time when a T-hour falls almost exactly half way between two S-hours, it will be more accurate at a 'change' to insert the two S-entries which fall on each side of the truth. If this be done the whole of the S-series of measurements is entered on the T-sheet. Similarly, if it be the T-hand which goes faster than the S-hand, we may leave a gap in the T-series instead of duplicating an entry. For the analysis of the T-tide there is therefore prepared a sheet arranged in rows and columns; each row corresponds to one T-day, and the columns are marked $0^h, 1^h, \dots 23^h$; the 0^h's may be called T-noons. A dot is put in each space for entry, and where there is a change two dots are put if there is to be a double entry, and a bar if there is to be no entry. Black vertical lines mark the end of each S-day. These black lines will of course fall into slightly irregular diagonal lines across the page, and such lines are steeper and steeper the more nearly T-time approaches to S-time. They slope downwards

Schedule [K].

Form for Entry of Tidal Observations.

Series M.

	0^h	1^h	2^h	3^h	4^h	5^h	6^h	7^h	8^h	9^h	10^h	11^h	12^h	13^h	14^h	15^h	16^h	17^h	18^h	19^h	20^h	21^h	22^h	23^h	S-hour
1	.	.	.	.	.	.	.	.	.	.	.	.	.	.	:	.	.	.	.	.	.	.	.	▌ .	2^d 0^h
2	.	.	.	.	.	.	.	.	.	.	.	.	.	.	.	.	.	.	:	.	.	.	▌ .	.	3 1
3	.	.	.	.	.	.	.	.	.	.	.	.	.	.	.	.	.	.	.	.	.	.	▌ .	:	4 2
4	.	.	.	.	.	.	.	.	.	.	.	.	.	.	.	.	.	.	.	.	.	▌ .	.	.	5 2
5	.	.	.	:	.	.	.	.	.	.	.	.	.	.	.	.	.	.	.	.	▌ .	.	.	.	6 3

&c. &c. &c.

	0^h	1^h	2^h	3^h	4^h	5^h	6^h	7^h	8^h	9^h	10^h	11^h	12^h	13^h	14^h	15^h	16^h	17^h	18^h	19^h	20^h	21^h	22^h	23^h	S-hour
67	.	.	.	.	.	.	.	.	.	.	.	.	.	.	.	.	▌ .	.	.	.	.	.	.	.	70 7
68	.	.	.	:	.	.	.	.	.	.	.	.	.	.	.	▌ .	.	.	.	.	.	.	.	.	71 8
69	.	.	.	.	.	.	.	.	:	.	.	.	.	.	▌ .	.	.	.	.	.	.	.	.	.	72 9
70	.	.	.	.	.	.	.	.	.	.	.	.	.	:	▌ .	.	.	.	.	.	.	.	.	.	73 9
71	.	.	.	.	.	.	.	.	.	.	.	.	.	▌ .	.	.	.	:	.	.	.	.	.	.	74 11
Sum No.	73	73	74	74	73	73	73	73	75	73	73	73	74	74	74	73	73	75	74	73	73	74	73	74	No. of Days 73^d 12^h

from right to left if the T-hour is longer than the S-hour, and the other way in the opposite case. The 'changes' also run diagonally, with a slope in the opposite direction to that of the black lines.

We annex a diminished sample of a part of a page drawn up for the entry of the M-series of tides, in which T-time is mean lunar time.

The incidence of the hours in the computation forms for the several series was determined by Mr Roberts.

Since the first day is numbered 1, and the first hour 0^h, it follows that the hourly observation numbered $74^d\ 11^h$ is the observation which completes a period of $73^d\ 12^h$ of mean solar time since the beginning; in fact, to find the period elapsed since 0^h of the first day we must subtract 1 from the number of the day and add one to the number of the hour. The $73^d\ 12^h$ of m. s. time, inserted at the foot of the form, is very nearly equal to 71 days of mean lunar or M-time. For each class of tide there are five pages, giving in all about 370 values for the height of the water at each of the 24 special hours; the number of values for each hour varies slightly according as more or less 'changes' fall into each column.

The numbers entered in each column are summed on each of the five pages; the five sets of results being summed, the results are then divided each by the proper divisor for its column, and thus is obtained the mean value for that column. In this way 24 numbers are found which give the mean height of water at each of the 24 special hours.

It is obvious that if this process were continued over a very long time we should in the end extract the tide under analysis from amongst all the others, but as the process only extends over about a year, the elimination of the others is not quite complete.

. . . .

[The choice of appropriate periods is considered in the next paper in this volume, and I therefore omit the consideration of the forms which were in use in India in 1883.]

Let us now return to our general notation, and consider the 24 mean values, each pertaining to the 24 T-hours. We suppose that all the tides excepting the T-tide are adequately eliminated, and, in fact, a computation of the necessary corrections for the absence of complete elimination, which is given in the Tidal Report of 1872, shows that this is the case.

It is obvious that any one of the 24 values does not give the true height of the T-tide at that T-hour, but gives the average height of the water, as due to the T-tide, estimated over half a T-hour before and half a T-hour after that hour. We must now consider the correction necessary on this account.

Suppose we have a function

$$h = A_1 \cos\theta + B_1 \sin\theta + A_2 \cos 2\theta + B_2 \sin 2\theta + \ldots + A_r \cos r\theta + B_r \sin r\theta + \ldots$$

Then we see by integration that the function

$$h' = A_1' \cos\theta + B_1' \sin\theta + A_2' \cos 2\theta + B_2' \sin 2\theta + \ldots + A_r' \cos r\theta + B_r' \sin r\theta + \ldots$$

where

$$\frac{A_1'}{A_1} = \frac{B_1'}{B_1} = \frac{\sin\frac{1}{2}\alpha}{\frac{1}{2}\alpha}; \quad \frac{A_2'}{A_2} = \frac{B_2'}{B_2} = \frac{\sin\frac{1}{2}2\alpha}{\frac{1}{2}2\alpha}; \ \ldots \ \frac{A_r'}{A_r} = \frac{B_r'}{B_r} = \frac{\sin\frac{1}{2}r\alpha}{\frac{1}{2}r\alpha}; \ \ldots$$

is derivable from h by substituting for the h, corresponding to any value of θ, the mean value of h estimated over the interval from $\theta + \frac{1}{2}\alpha$ to $\theta - \frac{1}{2}\alpha$.

Thus when harmonic analysis is applied to the 24 T-hourly values, the coefficients which express that oscillation which goes through its period r times in the 24 T-hours must be augmented by the factor $\frac{1}{2}r\alpha/\sin\frac{1}{2}r\alpha$. Thus we get the following expressions for the augmenting factors for the diurnal, semi-diurnal, ter-diurnal oscillations, &c., viz. :—

$$\frac{7{\cdot}5\pi}{180}\Big/\sin 7^\circ 30'; \ \frac{15\pi}{180}\Big/\sin 15^\circ; \ \frac{22{\cdot}5\pi}{180}\Big/\sin 22^\circ 30', \text{\&c.} \ \ldots\ldots(63)$$

Computing from these we find the following augmenting factors.

[M.]

Augmenting Factors.

For A_1, B_1	1·00286
A_2, B_2	1·01152
A_3, B_3	1·02617
A_4, B_4	1·04720
A_6, B_6	1·11072
A_8, B_8	1·20920

In the reduction of the S-series of tides, the numbers treated are the actual heights of the water exactly at the S-hours, and therefore no augmenting factor is requisite.

We must now explain how the harmonic analysis, which the use of these factors presupposes, is carried out.

If t denotes T-time expressed in hours, and n is 15°, we express the height h, as given by the averaging process above explained, by the formula

$$h = A_0 + A_1 \cos nt + B_1 \sin nt + A_2 \cos 2nt + B_2 \sin 2nt + \ldots$$

where t is 0, 1, 2 … 23.

[N.]

Copy of Harmonic Analysis for M-*Series, Karachi* 1880–81, *abridged by omission of some of the decimals.*

h	I.	h	II.	III. I.+II.	IV. I.−II.	V. Lower half of IV. reversed	VI. IV.+V.	M	VII. M×VI.	M	VIII. M×VII.	IX. IV.−V.	M	X.	M	XI. M×IX.
0	7·1	12	6·9	14·0	+·19	. . .	+·19	0	·00	0	·00	+·19	1	+ ·19	1	+ ·19
1	8·3	13	8·1	16·4	+·20	−·14	+·05	S_1	+ ·01	S_3	+ ·04	+·34	S_5	+ ·33	S_3	+ ·24
2	9·3	14	9·1	18·4	+·16	−·09	+·08	S_2	+ ·04	1	+ ·08	+·25	S_4	+ ·22	0	+ ·00
3	9·8	15	9·7	19·5	+·09	−·02	+·07	S_3	+ ·05	S_3	+ ·05	+·11	S_3	+ ·08	$-S_3$	− ·08
4	9·6	16	9·6	19·3	+·01	+·04	+·06	S_4	+ ·05	0	·00	−·03	S_2	− ·01	−1	+ ·03
5	8·8	17	8·8	17·6	−·02	+·03	+·01	S_5	+ ·01	$-S_3$	− ·01	−·05	S_1	− ·01	$-S_3$	+ ·04
6	7·6	18	7·6	15·1	−·02	. . .	−·02	1	− ·02	−1	+ ·02	−·02	0	− ·00	0	+ ·00
7	6·3	19	6·3	12·6	+·03			12	+ ·14	12	+ ·18		12	+ ·78	12	+ ·42
8	5·3	20	5·3	10·6	+·04											
9	4·7	21	4·7	9·5	−·02			$B_1=$	+·012	$B_3=$	+·015		$A_1=$	+·065	$A_3=$	+·035
10	4·9	22	5·0	9·8	−·09											
11	5·7	23	5·9	11·6	−·14											

XII. First half of III.	XIII. Second half of III.	XIV. XII. + XIII.	XV. XII. – XIII.	M	XVI. M × XV.	M	XVII. M × XV.	M	XVIII. M × XV.	M	XIX. M × XV.
14·0	15·1	29·1	– 1·1	0	·00	1	– 1·1	0	·0	1	– 1·1
16·4	12·6	29·0	+ 3·8	S_2	+ 1·9	S_4	+ 3·3	1	+ 3·8	0	·0
18·4	10·6	29·0	+ 7·8	S_4	+ 6·8	S_2	+ 3·9	0	·0	– 1	– 7·8
19·5	9·5	29·0	+ 10·0	1	+ 10·0	0	0	– 1	– 10·0	0	·0
19·3	9·8	29·1	+ 9·4	S_4	+ 8·2	$-S_2$	– 4·7	0	·0	1	+ 9·4
17·6	11·6	29·2	+ 6·0	S_2	+ 3·0	$-S_4$	– 5·2	1	+ 6·0	0	·0
				12	+ 29·8	12	– 3·8	12	– ·2	12	+ ·5
				$B_2 =$	+ 2·49	$A_2 =$	– ·32	$B_6 =$	– ·02	$A_6 =$	+ ·04

XX. First half of XIV.	XXI. Sec. half of XIV.	XXII. XX. – XXI.	M	XXIII. M × XXII.	M	XXIV. M × XXII.	XXV. XX. + XXI.	M	XXVI. M × XXV.	M	XXVII. M × XXV.	Values of Multipliers (M)
29·1	29·0	+ ·18	0	·00	1	+ ·18	58·1	0	0	1	+ 58·1	0 = 0 $S_1 =$ ·25882
29·0	29·1	– ·03	S_4	– ·02	S_2	– ·01	58·1	S_4	+ 50·3	$-S_2$	– 29·1	$S_2 =$ ·50000
29·0	29·2	– ·16	S_4	– ·14	$-S_2$	+ ·08	58·1	$-S_4$	– 50·4	$-S_2$	– 29·1	$S_3 =$ ·70711
			12	– ·17	12	+ ·25		12	– ·02	12	+ ·00	$S_4 =$ ·86603 $S_5 =$ ·96593
			$B_4 =$	– ·014	$A_4 =$	+ ·21		$B_8 =$	– ·002	$A_8 =$	·00	1 = 1·00000

Then if Σ denotes summation of the series of 24 terms found by attributing to t its 24 values, it is obvious that

$$A_0 = \tfrac{1}{24}\Sigma h;\ A_1 = \tfrac{1}{12}\Sigma h \cos nt;\ B_1 = \tfrac{1}{12}\Sigma h \sin nt;$$

$$A_2 = \tfrac{1}{12}\Sigma h \cos 2nt;\ B_2 = \tfrac{1}{12}\Sigma h \sin 2nt;\ \&c., \&c.$$

Since n is 15° and t is an integer, it follows that all the cosines and sines involved in these series are equal to one of the following: viz.: $0, \pm\sin 15°, \pm\sin 30°, \pm\sin 45°, \pm\sin 60°, \pm\sin 75°, \pm 1$. It is found convenient to denote these sines, as $0, \pm S_1, \pm S_2, \pm S_3, \pm S_4, \pm S_5, \pm 1$. The multiplication of the 24 h's by the various S's, and the subsequent additions may be arranged in a very neat tabular form.

We give on the last two pages the form for the reduction of the M-tides, filled in for Karachi 1880–81, but abridged by the omission of some of the decimals. The columns marked M are the multipliers appropriate for each series.

The columns I. and II. contain the 24 hourly values to be submitted to analysis. The subsequent operations are sufficiently indicated by the headings to the columns, and it will be found on examination that the results are in reality the sums of the several series indicated above. We believe that this mode of arranging the harmonic analysis is due to Archibald Smith, who gives it in the Admiralty manual on the Compass. The arrangement seems to be very nearly the same as that adopted by Everett (*Trans. Roy. Soc. Edin.* 1860) in his reductions of observations on underground temperature.

In most cases it is not necessary to deduce more than the tide of the speed indicated by astronomical theory, but we give the full form by which the over-tides are deducible. If we want only a diurnal tide, then the only columns necessary are I. to VII. and IX. and X.; if only a semi-diurnal tide, the columns to be retained are I., II., III., XII., XIII., XV., XVI., XVII.

The A's and B's having been thus deduced, we have $R = \sqrt{(A^2 + B^2)}$. R must then be multiplied by the augmenting factors which we have already evaluated (Schedule [M]). We thus have the augmented R. Next the angle whose tangent is B/A gives ζ. The addition to ζ of the appropriate $V_0 + u$ (see Schedule [I]) gives κ, and the multiplication of R by the appropriate 1/f (see Schedule [I]) gives H. The reduction is then complete.

The following is a sample of the form used.

[O.]

Form for Evaluation of ζ, R, κ, H.

log B = ·
log A = ·
log tan ζ = ·

°
ζ = ·
$V_0 + u$ = ·
κ = ·

B^2 = ·
A^2 = ·
R^2 = ·
R = ·
Augtn. = ·
Augd. R = ·
1/f = ·
H = ·

A form similar to [O] serves for the same purpose in the treatment of the tides of long period, to the consideration of which we now pass; it will be seen, however, that for these tides there is no augmenting factor, and that the increase of n for $11\frac{1}{2}$ hours has to be added to ζ.

§ 10. *On the Harmonic Analysis for the Tides of Long Period.*

For the purpose of determining these tides we have to eliminate the oscillations of water-level arising from the tides of short period. As the quickest of these tides has a period of many days, the height of mean water at one instant for each day gives sufficient data. Thus there will in a year's observations be 365 heights to be submitted to harmonic analysis. In leap-years the last day's observation must be dropped, because the treatment is adapted for analysing 365 values.

To find the daily mean for any day it has hitherto been usual to take the arithmetic mean of 24 consecutive hourly values, beginning with the height at noon. This height will then apply to the middle instant of the period from 0^h to 23^h: that is to say, to $11^h\ 30^m$ at night. We shall propose some new modes of treating the observations, and in the first of them it will probably be more convenient that the mean for the day should apply to midnight instead of to $11^h\ 30^m$. For finding a mean applicable to midnight we take the 25 consecutive heights for 0^h to 24^h, and add the half of the first value to the 23 intermediate and to the half of the last and divide by

24. It would probably be sufficiently accurate if we took $\frac{1}{25}$ of the sum of the 25 consecutive values, if it is found that the division of every 24th hourly value into two halves materially increases the labour of computing the daily means. The three plans for finding the daily mean are then

$$\left.\begin{array}{lr} \frac{1}{24}(h_0+h_1+\ldots+h_{23}) \quad \ldots \quad \ldots & \text{(i)} \\ \frac{1}{24}(\frac{1}{2}h_0+h_1+\ldots+h_{23}+\frac{1}{2}h_{24}) \quad \ldots & \text{(ii)} \\ \frac{1}{25}(h_0+h_1+\ldots+h_{23}+h_{24}) \ldots \quad \ldots & \text{(iii)} \end{array}\right\} \quad \ldots\ldots\ldots\ldots(64)$$

And they will be denoted as methods (i), (ii), (iii) respectively. It does not, however, seem very desirable to use the third method. Major Baird considers that the use of method (i) is most convenient for the computers.

The formation of a daily mean does not obliterate the tidal oscillations of short period, because none of the tides, excepting those of the principal solar series, have commensurable periods in mean solar time.

A correction, or 'clearance of the daily mean,' has therefore to be applied for all the important tides of short period, excepting for the solar tides.

Let $R\cos(nt-\zeta)$ be the expression for one of the tides of short period as evaluated by the harmonic analysis for the same year, and let α be the value of $nt-\zeta$ at any noon. Then the 25 consecutive hourly heights of water, beginning with that noon, are—

$$R\cos\alpha,\ R\cos(n+\alpha),\ R\cos(2n+\alpha)\ldots R\cos(23n+\alpha),\ R\cos(24n+\alpha)$$

In the method (i) of taking the daily mean it is obvious that the 'clearance' is

$$-\frac{1}{24}R\frac{\sin 12n}{\sin\frac{1}{2}n}\cos(\alpha+11\tfrac{1}{2}n)$$

In the method (ii) it is easily proved to be

$$-\frac{1}{24}R\frac{\sin 12n}{\tan\frac{1}{2}n}\cos(\alpha+12n) \quad \ldots\ldots\ldots\ldots\ldots\ldots\ldots(65)$$

and in method (iii) it is

$$-\frac{1}{25}R\frac{\sin\frac{25}{2}n}{\sin\frac{1}{2}n}\cos(\alpha+12n)$$

The clearance, as written here, is additive.

It was found practically in the computation for these tides that only three tides of short period exercise an appreciable effect, so that clearances for them have to be applied. These tides are the M_2, N, O tides. It was usual to compute these three clearances for every day in the year, and to correct the daily values accordingly. But in following this plan a great deal of unnecessary labour has been incurred, and when a simpler plan is followed it may perhaps be worth while to include more of the short-period tides in the clearances.

Professor J. C. Adams suggests the use of the tide-predicting machine for the evaluation of the sum of the clearances, and if this plan is not found to inconveniently delay operations in India, it may perhaps be tried.

In explaining the process we will suppose that method (i) has been followed; if either of the other plans be adopted it will be easy to change the formulæ accordingly.

It is clear that $R\cos(\alpha + 11\frac{1}{2}n)$ is the height of the tide n at $11^h\ 30^m$; and the same is true for each such tide. Hence if we use the tide-predicter to run off a year of fictitious tides with the semi-range of each tide equal to $\frac{1}{24}\sin 12n/\sin\frac{1}{2}n$ of its true semi-range, and with all the solar series and the annual and semi-annual tides put at zero, the height given at each $11^h\ 30^m$ in the year is the sum for each day of all the clearances to be *subtracted*. The scale to which the ranges are set may of course be chosen so as to give the clearances to a high degree of accuracy.

In the other process of clearance, which will be explained below, a single correction for each short-period tide is applied to each of the final equations, instead of to each daily mean.

We next take the 365 daily means, and find their mean value. This gives the mean height of water for the year. If the daily means be uncleared, the result cannot be sensibly vitiated.

We next subtract the mean height from each of the 365 values, and find 365 quantities δh giving the daily height of water above the mean height.

These quantities are to be the subject of the harmonic analysis; and the tides chosen for evaluation are those which have been denoted above as Mm, Mf, MSf, Sa, Ssa.

Let
$$\left.\begin{aligned}\delta h = \ & A\cos(\sigma-\varpi)t + B\sin(\sigma-\varpi)t\\ &+ C\cos 2\sigma t \quad\ + D\sin 2\sigma t\\ &+ C'\cos 2(\sigma-\eta)t + D'\sin 2(\sigma-\eta)t\\ &+ E\cos\eta t \quad\ + F\sin\eta t\\ &+ G\cos 2\eta t \quad + H\sin 2\eta t\end{aligned}\right\} \text{..............(66)}$$
where t is time measured from the first $11^h\ 30^m$.

Now suppose l_1, l_2 are the increments in 24 m. s. hours of any two of the five arguments $(\sigma-\varpi)t$, $2\sigma t$, $2(\sigma-\eta)t$, ηt, $2\eta t$, and that A_1, B_1; A_2, B_2, are the corresponding coefficients of the cosine and sine in the expression for δh.

Then if δh_i be the value of δh at the $(i+1)^{th}$ $11^h\ 30^m$ in the year, we may write
$$\delta h_i = A_1\cos l_1 i + B_1\sin l_1 i + A_2\cos l_2 i + B_2\sin l_2 i + \dots \quad \text{......(67)}$$

And therefore

$$\delta h_i \cos l_1 i = \tfrac{1}{2}A_2 \{\cos (l_1 + l_2) i + \cos (l_1 - l_2) i\} + \tfrac{1}{2}B_2 \{\sin (l_1 + l_2) i - \sin (l_1 - l_2) i\} + \ldots$$

$$\delta h_i \sin l_1 i = \tfrac{1}{2}A_2 \{\sin (l_1 + l_2) i + \sin (l_1 - l_2) i\} + \tfrac{1}{2}B_2 \{-\cos (l_1 + l_2) i + \cos (l_1 - l_2) i\} + \ldots$$

Now let
$$\phi (x) = \tfrac{1}{2} \frac{\sin \frac{365}{2} x}{\sin \frac{1}{2} x}$$

so that
$$\phi (l_1 \pm l_2) = \tfrac{1}{2} \frac{\sin \frac{365}{2} (l_1 \pm l_2)}{\sin \frac{1}{2} (l_1 \pm l_2)}$$

We may observe that

$$\phi (x) = \phi (-x), \text{ and } \phi (0) = 182\tfrac{1}{2}$$

If therefore Σ denotes summation for the 365 values from $i=0$ to $i=364$, we have

$$\left.\begin{aligned}\Sigma\delta h \cos l_1 i &= [\phi (l_1 + l_2) \cos 182 (l_1 + l_2) + \phi (l_1 - l_2) \cos 182 (l_1 - l_2)]\, A_2 \\ &\quad + [\phi (l_1 + l_2) \sin 182 (l_1 + l_2) - \phi (l_1 - l_2) \sin 182 (l_1 - l_2)]\, B_2 + \ldots \\ \Sigma\delta h \sin l_1 i &= [\phi (l_1 + l_2) \sin 182 (l_1 + l_2) + \phi (l_1 - l_2) \sin 182 (l_1 - l_2)]\, A_2 \\ &\quad + [-\phi (l_1 + l_2) \cos 182 (l_1 + l_2) + \phi (l_1 - l_2) \cos 182 (l_1 - l_2)]\, B_2 + \ldots\end{aligned}\right\} \quad \ldots\ldots\ldots(68)$$

In these equations there is always one pair of terms in which l_2 is identical with l_1, and since $\phi (l_1 - l_1) = 182\tfrac{1}{2}$, and $\cos 182 (l_1 - l_1) = 1$, it follows that there is one term in each equation in which there is a coefficient nearly equal to 182·5. In the cosine series it will be a coefficient of an A; in the sine series, of a B.

The following are the equations (copied from the Report for 1872*) with the coefficients inserted, as computed from these formulæ, or their equivalents:—

* [Some small corrections have been introduced.]

[P.]

Final Equations for Tides of Long Period.

	Coefft. of A	Coefft. of B	Coefft. of C	Coefft. of D	Coefft. of C′	Coefft. of D′	Coefft. of E	Coefft. of F	Coefft. of G	Coefft. of H
$\Sigma\delta h\times\cos(\sigma-\varpi)t=$	+183·05	+ 2·14	+ 0·72	+ 4·29	+ 0·76	+ 5·04	+ 4·88	− 0·34	+ 4·96	− 0·70
$\times\sin(\sigma-\varpi)t=$	+ 2·14	+181·95	− 4·15	+ 1·01	− 4·90	+ 1·06	+ 3·80	+ 0·34	+ 3·88	+ 0·68
$\times\cos 2\sigma t=$	+ 0·72	− 4·15	+183·17	+ 0·88	+ 0·56	+ 0·92	− 1·50	− 0·09	− 1·51	− 0·18
$\times\sin 2\sigma t=$	+ 4·29	+ 1·01	+ 0·88	+181·83	+ 0·92	− 0·80	+ 3·05	− 0·08	+ 3·06	− 0·17
$\times\cos 2(\sigma-\eta)t=$	+ 0·76	− 4·90	+ 0·56	+ 0·92	+183·19	+ 0·97	− 1·68	− 0·11	− 1·70	− 0·21
$\times\sin 2(\sigma-\eta)t=$	+ 5·04	+ 1·06	+ 0·92	− 0·80	+ 0·97	+181·81	+ 3·24	− 0·10	+ 3·25	− 0·20
$\times\cos\eta t=$	+ 4·88	+ 3·80	− 1·50	+ 3·05	− 1·68	+ 3·24	+182·38	+ 0·00	− 0·24	+ 0·01
$\times\sin\eta t=$	− 0·34	+ 0·34	− 0·09	− 0·08	− 0·11	− 0·10	+ 0·00	+182·62	+ 0·00	+ 0·00
$\times\cos 2\eta t=$	+ 4·96	+ 3·88	− 1·51	+ 3·06	− 1·70	+ 3·25	− 0·24	+ 0·00	+182·38	+ 0·00
$\times\sin 2\eta t=$	− 0·70	+ 0·68	− 0·18	− 0·17	− 0·21	− 0·20	+ 0·01	+ 0·00	+ 0·00	+182·62

If the daily means have been cleared by the use of the tide-predicter as above described, these ten equations are to be solved by successive approximation, and we are then furnished with the two component semi-amplitudes, say A_1, B_1 of the five long-period tides. But the initial instant of time is the first $11^h\ 30^m$ in the year instead of the first noon. Hence if as before we put $R^2 = A_1^2 + B_1^2$, and $\tan \zeta_1 = B_1/A_1$, we must, in order to reduce the results to the normal form in which noon of the first day is the initial instant of time, add to ζ_1 the increment of the corresponding argument for $11^h\ 30^m$, according to method (i), or for 12 hours according to methods (ii) or (iii).

If, however, the daily means have not been cleared, then before solution of the final equations corrections for clearance will have to be applied, which we shall now proceed to evaluate.

For this process we still suppose method (i) to be adopted.

Let n be the speed of a short-period tide in degrees per m. s. hour, and let $\psi(n) = \frac{1}{24}\frac{\sin 12n}{\sin \frac{1}{2}n}$. Then we have already seen that the clearance to δh_i, the mean height of water at $11^h\ 30^m$ of the $(i+1)^{th}$ day, will be

$$-\psi(n)\,R\cos[n\{24i + 11\tfrac{1}{2}\} - \zeta]$$

If we write $m = 24n$ (so that m is the daily increase of argument of the tide of short period), and $\beta = n \times 11\frac{1}{2} - \zeta$, this becomes

$$-\psi(n)\,R\cos(mi + \beta)$$

Hence the clearance for $\delta h_i \cos li$ is

$$-\tfrac{1}{2}\psi(n)\,R\,\{\cos[(m+l)\,i + \beta] + \cos[(m-l)\,i + \beta]\}$$

and for $\delta h_i \sin li$ is

$$-\tfrac{1}{2}\psi(n)\,R\,\{\sin[(m+l)\,i + \beta] - \sin[(m-l)\,i + \beta]\}$$

Summing the series of 365 terms we find that the additive clearance for $\Sigma\delta h \cos li$ is

$$-R\psi(n)\,\{\phi(m+l)\cos[182(m+l) + \beta] + \phi(m-l)\cos[182(m-l) + \beta]\}$$

where as before $$\phi(x) = \tfrac{1}{2}\frac{\sin \frac{365}{2}x}{\sin \frac{1}{2}x} \quad \text{..........................(69)}$$

If Δn denotes the increase of the argument nt in $182^d\ 11^h\ 30^m$, this may now be written

$$-R\psi(n)\,\{\phi(m+l)\cos[\Delta n + 182l - \zeta] + \phi(m-l)\cos[\Delta n - 182l - \zeta]\}$$

If therefore $R\cos\zeta = A$, $R\sin\zeta = B$, so that A and B are the component semi-ranges of the tide n as immediately deduced from the harmonic analysis for the tides of short period, we have for the clearance to $\Sigma\delta h \cos li$

$$-[\psi(n)\,\phi(m+l)\cos(\Delta n + 182l) + \psi(n)\,\phi(m-l)\cos(\Delta n - 182l)]\,A$$
$$-[\psi(n)\,\phi(m+l)\sin(\Delta n + 182l) + \psi(n)\,\phi(m-l)\sin(\Delta n - 182l)]\,B$$

In precisely the same manner we find the clearance for $\Sigma\delta h \sin li$ to be

$$-[\psi(n)\,\phi(m+l)\sin(\Delta n+182l)-\psi(n)\,\phi(m-l)\sin(\Delta n-182l)]\,A$$
$$+[\psi(n)\,\phi(m+l)\cos(\Delta n+182l)-\psi(n)\,\phi(m-l)\cos(\Delta n-182l)]\,B$$

These coefficients may be written in a form more convenient for computation. For

$$\phi(m\pm l)=\frac{\sin\frac{365}{2}(m\pm l)}{2\sin\frac{1}{2}(m\pm l)}$$
$$=\tfrac{1}{2}\cos 182\,(m\pm l)+\tfrac{1}{2}\sin 182\,(m\pm l)\cot\tfrac{1}{2}(m\pm l)\;\ldots\ldots\ldots(70)$$

Then let

$$\left.\begin{aligned}K(n,l)&=\phi(m+l)+\phi(m-l)\\ Z(n,l)&=\phi(m+l)-\phi(m-l)\end{aligned}\right\}\;\ldots\ldots\ldots\ldots\ldots\ldots(71)$$

Also let

$$\left.\begin{aligned}\psi(n)\cos\Delta n&=\frac{1}{24}\frac{\sin 12n}{\sin\frac{1}{2}n}\cos\Delta n=C(n)\\ \psi(n)\sin\Delta n&=S(n)\end{aligned}\right\}\;\ldots\ldots\ldots\ldots(72)$$

The functions $K(n,l)$, $Z(n,l)$, $C(n)$, $S(n)$ may be easily computed from (70), (71), (72).

Then if we denote the additive clearance for $\Sigma\delta h\cos li$ by

$$[A,n,l,\cos]\,A+[B,n,l,\cos]\,B$$

and that for $\Sigma\delta h\sin li$ by

$$[A,n,l,\sin]\,A+[B,n,l,\sin]\,B$$

We have

$$\left.\begin{aligned}[A,n,l,\cos]&=-C(n)\,K(n,l)\cos 182l+S(n)\,Z(n,l)\sin 182l\\ [B,n,l,\cos]&=-S(n)\,K(n,l)\cos 182l-C(n)\,Z(n,l)\sin 182l\\ [A,n,l,\sin]&=-S(n)\,Z(n,l)\cos 182l-C(n)\,K(n,l)\sin 182l\\ [B,n,l,\sin]&=C(n)\,Z(n,l)\cos 182l-S(n)\,K(n,l)\sin 182l\end{aligned}\right\}\;\ldots\ldots(73)$$

We must remark that if $\frac{1}{2}(m+l)=360°$, $\phi(m+l)$ is equal to 182·5.

This case arises when l is the tide MSf of speed $2(\sigma-\eta)$, and m the tide M_2 of speed $2(\gamma-\sigma)$, for $m+l$ is then $24\times 2(\gamma-\eta)=720°$.

The clearance of the long-period tide l from the effects of the short-period tide n requires the computation of these four coefficients. For the clearance of the five long-period tides from the effects of the three tides M_2, N, O, it will be necessary to compute 60 coefficients.

If it shall be found convenient to make the initial instant or epoch for the tides of long period different from that chosen in the reductions of those of short period, it will, of course, be necessary to compute the values which A and B would have had if the two epochs had been identical. A and B are, of course, the component semi-ranges of the tide of short period at the epoch chosen for the tides of long period; to determine them it is necessary to multiply R by the cosine and sine of $V+u-\kappa$ at the epoch.

[Q.]

Schedule of Coefficients for Clearance of Daily Means in the Final Equations.*

l =	$\sigma-\varpi$	2σ	$2(\sigma-\eta)$	η	2η
$(M_2)\ n=2(\gamma-\sigma)$					
[A, n, l, cos]	−0·05557	+0·00302	+5·7393	−0·10410	−0·10465
[B, n, l, cos]	−0·17036	−0·03773	−2·9228	−0·07525	−0·07546
[A, n, l, sin]	−0·17075	+0·04170	−2·8400	−0·00176	−0·00353
[B, n, l, sin]	+0·04410	+0·01052	−5·7271	+0·00476	+0·00958
(N) $n=2\gamma-3\sigma+\varpi$					
[A, n, l, cos]	−0·05884	+0·03680	+0·02938	−0·01760	−0·01760
[B, n, l, cos]	−0·07758	−0·22357	−0·19384	+0·00254	+0·00254
[A, n, l, sin]	−0·02059	−0·15257	−0·12210	+0·00020	+0·00041
[B, n, l, sin]	+0·11381	−0·08544	−0·08081	+0·00007	+0·00015
(O) $n=\gamma-2\sigma$					
[A, n, l, cos]	−0·06485	+0·01662	+0·01571	−0·19240	−0·19340
[B, n, l, cos]	−0·34765	−0·07775	−0·08158	−0·18260	−0·18311
[A, n, l, sin]	−0·34523	+0·08411	+0·08754	−0·00460	−0·00926
[B, n, l, sin]	+0·04052	+0·03384	+0·03306	+0·00897	+0·01802

It may happen from time to time that the tide-gauge breaks down for a few days, from the stoppage of the clock, the choking of the tube, or some other such accident. In this case there will be a hiatus in the values of δh. Now, the whole process employed depends on the existence of 365 continuous values of δh. Unless, therefore, the year's observations are to be sacrificed, this hiatus must be filled. If not more than three or four days' observations are wanting, it will be best to plot out the values of δh graphically on each side of the hiatus, and filling in the gap with a curve drawn by hand, use the values of δh given by the conjectural curve. If the gap is somewhat longer, several plans may be suggested, and judgment must be used as to which of them is to be adopted.

* [A few small corrections have been introduced.]

If there is another station of observation in the neighbourhood, the values of δh for that station may be inserted.

The values of δh for another part of the year, in which the moon's and sun's declinations are as nearly as may be the same as they were during the gap, may be used.

It may be, however, that the hiatus is of considerable length, so that the preceding methods are inapplicable: as when in 1882 the tidal record for Vizagapatam is wanting for 67 days. The following method of treatment will then be applicable:—

We find approximate values of the tidal constituents of long period, and fill in the hiatus, so as to complete the 365 values, with the computed height of the tide during the hiatus.

To find these approximate values we form $\Sigma\delta h \cos lt$ and $\Sigma\delta h \sin lt$ for the days of observation; next, in the ten final equations of Schedule P we neglect all the terms with small coefficients, and in the terms whose coefficients are approximately 182·5, we substitute a coefficient equal to 182·5 diminished by half the number of days of hiatus. For example, for Vizagapatam in 1882 we have $182{\cdot}5 - \frac{1}{2} \times 67 = 149$, and, e.g., $\Sigma\delta h \cos(\sigma - \varpi)t = 149 A$ approximately. After the approximate values of A, B, C, D, &c., have been found, it is easy to find the approximate height of tide for the days of the hiatus. This plan will also apply where the hiatus is of short duration.

It may be pursued whether or not we are working with cleared daily means; for if the daily means are uncleared, as will henceforth be the case, we import with the numbers by which the hiatus is filled exactly those fictitious tides of long period which are cleared away by the use of the "clearance coefficients," in preparing the ten final equations for solution.

Other methods of treating a stoppage of the record may be devised. If the stoppage be near the beginning of the year, or near the end, we may neglect the observations before or after the gap, and compute afresh the 100 coefficients of Schedule P, and the clearance coefficients of Schedule Q for the number of days remaining. If the gap is in the middle we might compute the values of the coefficients of Schedules P and Q as though the days of hiatus were days of observation, bearing in mind that the formulæ are to be altered by the consideration that time is to be measured from the initial $11^h\,30^m$ of the year, instead of from the initial $11^h\,30^m$ of the days of hiatus.

The so computed coefficients are then to be subtracted from the values given in Schedules P and Q, and the amended final equations and amended clearance coefficients to be used.

It must remain a matter of judgment as to which of these various methods is to be adopted in each case.

§ 11. *Method of Equivalent Multipliers for the Harmonic Analysis for the Tides of Long Period.*

Up to the present time (1883) the harmonic analysis for these tides has been conducted on a plan which seems to involve a great deal of unnecessary labour. If l be the speed of any one of the five tides for which the analysis has been carried out, in degrees per m. s. day, the values of $\cos lt$ and $\sin lt$ have been computed for $t = 0, 1, 2 \dots 364$, so that there are 730 values for each of the five tides. These 730 values have then been multiplied by the 365 δh's corresponding to each value of t, and the summations gave $\Sigma\delta h \cos lt$ and $\Sigma\delta h \sin lt$, the numerical results being the left-hand sides of one pair of the ten final equations explained in § 10. Now, it appears that this labour may be largely abridged, without any substantial loss of accuracy.

The plan proposed by Professor Adams is that of equivalent multipliers. The values of $\cos lt$ may be divided into eleven groups, according as they fall nearest to 1·0, ·9, ·8, ·7...·2, ·1, 0. Then, as all the values of δh are to be multiplied by some value of $\cos lt$, and that value of $\cos lt$ must fall into one of these groups, we collect together all the values of δh which belong to one of these groups, sum them, and multiply the sum by the corresponding multiplier, 1·0, ·9, ·8, &c., as the case may be. Since there are as many values of $\cos lt$ which are negative as positive, we must change the sign of half of the δh's. This changing of sign may be effected mechanically as follows:—In the spaces for entry of the δh's, those δh's whose sign is to be unchanged are to be entered on the left side of the space if positive, and to the right if negative; when the sign is to be altered this order of entry is to be reversed. Thus in the column corresponding to each multiplier we shall have two sub-columns, on the left all the δh's which, when the signs are appropriately altered, are +, and on the right those which are −. The sub-columns are to be separately summed, and their difference gives the total of the column, which is to be multiplied by the multiplier appropriate to the column. The treatment for the formation of $\Sigma\delta h \sin lt$ is precisely similar.

The annexed form [Schedule R] is designed for entry for determination of $\Sigma\delta h \cos(\sigma - \eta) t$.

The entries of δh are to be made continuously in the marked squares from left to right, and back again from right to left. The numbers in the squares, which in the computation forms are to be printed small and put in the corner, indicate the days of observation. The rows are arranged in sets of four corresponding to each complete period of $2(\sigma - \eta)$. In the middle pair for each period the + values of δh are to be written on the right, and in the rest on the left. The word 'change' opposite half the rows is to show the computer that he is to change the mode of entry. Each column, excepting that for zero, is to be summed at the foot of the page, and multiplied by the multiplier corresponding to its column. A pair of forms is required for each tide of long period; they are very easily prepared from the existing forms, in which the values of the multipliers are already computed.

[In the form as finally prepared for the Indian Government the rows marked 'change' are treated in a schedule by themselves separated from the others.]

[R.]

Form for Reduction of the Tide MSf.

		+ − 1·0	+ − ·9	+ − ·8	+ − ·7	+ − ·6	+ − ·5	+ − ·4	+ − ·3	+ − ·2	+ − ·1	No entries	
1	→	0	1		2				3				
	←	7		6			5				4		change
	→	8		9				10				11	change
	←	14			13			12					
2	→	15	16			17				18			
	←		21			20				19			change
	→	22	23		24			25					change
	←	29		28			27					26	
3	→	30		31			32				33		
	←		36		35				34				change
	→	37	38			39			40				change
	←	44	43			42				41			
4	→	45		46				47					
	←	51		50				49				48	change
	→	52	53			54				55			change
	←		58			57			56				
5	→	59	60		61				62				
	←	66		65			64				63		change
	→	67		68			69				70		change
	←	74	73		72			71					
			&c.				&c.				&c.		
Total + . Total − .												No entries	
Total . . Multiply .		× 1·0	× ·9	× ·8	× ·7	× ·6	× ·5	× ·4	× ·3	× ·2	× ·1	× ·0	
Results .													

Sum laterally . . . Sum of + = . . Sum of − = .

$\Sigma \delta h \cos 2(\sigma - \eta) t =$ . .

§ 12. Auxiliary Tables drawn up under the superintendence of Major Baird, R.E.

[Largely abridged.]

Values of N (Long. Moon's Ascending Node) for 0^h *Jan.* 1, *G.M.T.*

Value at 0^h *G.M.T. Jan.* 1, 1880 = 285°·956863
Motion per Julian year in 1880 = 19°·34146248
Motion for 365 *days* = 19°·32822387, *and for* 1 *day* = 0°·052954

Year	N	Year	N
1900	259°·1276	1910	65°·7395
1	239·7994	1	46·4112
2	220·4712	2	27·0830
3	201·1429	3	7·7018
4	181·8147	4	348·3736
1905	162·4335	1915	329·0454
6	143·1053	6	309·7172
7	123·7771	7	290·3360
8	104·4489	8	271·0078
9	85·0677	9	251·6795

Decrement of N since 0^h *Jan.* 1 *up to midnight of certain days of the year.*

[Omitted.]

Values of p_1 *(Mean Long. of Solar Perigee) for* 0^h *Jan.* 1.

[Abridged.]

Value at 0^h *Jan.* 1, 1880 = 280°·874802
Motion per Julian year = 0°·01710693
Motion for 365 *days* = 0°·01709295
Motion for 1 *day* = 0°·00004683

Year	p_1	Year	p_1
1900	281°·2169	1910	281°·3879
1	·2340	1	·4050
2	·2511	2	·4221
3	·2682	3	·4393
4	·2853	4	·4564
1905	·3024	5	·4735
6	·3195	6	·4906
7	·3366	7	·5078
8	·3537	8	·5249
9	·3708	9	·5420

[Tables, computed by Colonel Baird, for I, ν, ξ and the factors f in terms of N the longitude of the node are here replaced by algebraic formulæ, which are sufficiently accurate for all practical purposes. The formulæ are derived from the tables in Baird's Manual by the method of special values.

The angles ν, ξ, ν', $2\nu''$ are expressible in the form

$$A_1 \sin N + A_2 \sin 2N + A_3 \sin 3N$$

The values of the A's for the several angles are given in the following schedule:—

The angle	A_1	A_2	A_3
ν	12°·94	−1°·34	+0°·19
ξ	11°·87	−1°·34	+0°·19
ν'	8°·86	−0°·68	+0°·07
$2\nu''$	17°·74	−0°·68	+0°·04

The factors f are expressible in the form

$$f = B_0 + B_1 \cos N + B_2 \cos 2N + B_3 \cos 3N$$

Tides for which f is applicable	B_0	B_1	B_2	B_3
M_2, N, 2N, ν, MS, 2SM, MSf	1·00035	−·03733	+·00017	+·00001
K_2	1·0241	+·2863	+·0083	−·0015
K_1	1·0060	+·1150	−·0088	+·0006
O, Q	1·0089	+·1871	−·0147	+·0014
OO	1·1027	+·6504	+·0317	−·0014
J	1·0129	+·1676	−·0170	+·0016
Mf	1·0429	+·4135	−·0040	0000
Mm	1·0000	−·1300	+·0013	0000

May 1906.]

2.

ON THE PERIODS CHOSEN FOR HARMONIC ANALYSIS, AND A COMPARISON WITH THE OLDER METHODS BY MEANS OF HOUR-ANGLES AND DECLINATIONS.

[Third Report of the Committee, consisting of Professors G. H. DARWIN and J. C. ADAMS, for the Harmonic Analysis of Tidal Observations. Drawn up by Professor G. H. DARWIN. *British Association Report for* 1885, pp. 35—60.]

I. RECORD OF WORK DURING THE PAST YEAR.

. . . .

A LARGE number of tidal results have been obtained by the United States Coast Survey, and reduced under the superintendence of Professor Ferrel. Although the method pursued by him has been slightly different from that of the British Association, it appears that the American results should be comparable with those at the Indian and European ports. Professor Ferrel has given an assurance that this is the case; nevertheless, there appears to be strong internal evidence that, at some of the ports, some of the phases should be altered by 180°.

. . . .

II. CERTAIN FACTORS AND ANGLES USED IN THE REDUCTION OF TIDAL OBSERVATIONS.

[These are given at the end of the last Paper.]

III. On the Periods chosen for Harmonic Analysis in the Computation Forms.

Before proceeding to the subject of this section, it may be remarked that it is unfortunate that the days of the year in the computation forms should have been numbered from unity upwards, instead of from zero, as in the case of the hours. It would have been preferable that the first entry should have been numbered Day 0, Hour 0, instead of Day 1, Hour 0. This may be rectified with advantage if ever a new issue of the forms is required, but the existing notation is adhered to in this section.

The computation form for each tide consists of pages for entry of the hourly tide-heights, in which the entries are grouped according to rules appropriate to that tide. The forms terminate with a broken number of hours. This, as we shall now show, is erroneous, although this error may not be of much practical importance.

In § 9 of the Report for 1883 the following passage [omitted in the preceding paper] occurs:—

'The elimination of the effects of the other tides may be improved by choosing the period for analysis not exactly equal to one year. For suppose that the expression for the height of water is

$$\mathrm{A}_1 \cos n_1 t + \mathrm{B}_1 \sin n_1 t + \mathrm{A}_2 \cos n_2 t + \mathrm{B}_2 \sin n_2 t \quad \ldots\ldots\ldots\ldots(61)$$

'where n_2 is nearly equal to n_1, and that we wish to eliminate the n_2-tide, so as to be left only with the n_1-tide.

'Now, this expression is equal to

$$\left.\begin{array}{l}\{\mathrm{A}_1 + \mathrm{A}_2 \cos (n_1 - n_2) t - \mathrm{B}_2 \sin (n_1 - n_2) t\} \cos n_1 t \\ + \{\mathrm{B}_1 + \mathrm{A}_2 \sin (n_1 - n_2) t + \mathrm{B}_2 \cos (n_1 - n_2) t\} \sin n_1 t\end{array}\right\} \quad \ldots\ldots\ldots(62)$$

'That is to say, we may regard the tide as oscillating with a speed n_1, but with slowly varying range.'

Although this is thus far correct, yet the subsequent justification of the plan according to which the computation forms have been compiled is wrong.

In the column appertaining to any hour in the form we have $n_1 t$ a multiple of 15°, if n_1 be a diurnal, and of 30°, if n_1 be a semidiurnal tide.

Consider the column headed 'p-hours'; then $n_1 t = 15° p$ for diurnals, and $30° p$ for semidiurnals.

Hence (62), quoted above, shows us that, for diurnal tides, the sum of all the entries (of which suppose there are q) in the column numbered p-hours, is

$$\cos 15° p \left\{ A_1 q + A_2 \left[\cos (n_1 - n_2) \frac{15p}{n_1} + \cos \left[(n_1 - n_2) \left(\frac{2\pi}{n_1} + \frac{15p}{n_1} \right) \right] \right.\right.$$
$$\left.\left. + \cos \left[(n_1 - n_2) \left(2 \frac{2\pi}{n_1} + \frac{15p}{n_1} \right) \right] + \ldots \right] + B_2 [\&\text{c.}] \right\} + \sin 15° p \{\&\text{c.}\} \ldots (a)$$

And for semidiurnal tides the arguments of all the circular functions in (a) are to be doubled.

Now, we want to choose such a number of terms that the series by which A_2 and B_2 are multiplied may vanish. This is the case if the series is exactly re-entrant, and is nearly the case if nearly re-entrant.

The condition is exactly satisfied for diurnal tides, if

$$(n_1 - n_2)\, q \frac{2\pi}{n_1} = 2\pi r$$

where r is either a positive or negative integer. And for semidiurnal tides, if

$$(n_1 - n_2)\, q \frac{4\pi}{n_1} = 2\pi r$$

That is to say, $(n_1 - n_2)\, q = n_1 r$, for diurnal tides

or $(n_1 - n_2)\, q = \frac{1}{2} n_1 r$, for semidiurnal tides

It is not worth while attempting to eliminate the effect of the semidiurnal tides on the diurnal tides, and *vice versâ*, because we cannot be more than a fraction of a day out, and on account of the incommensurability of the speeds we cannot help being wrong to that amount.

S *Series*.

Now suppose we are analysing for the S_2 tide, and wish to minimise the effect of the M_2 tide.

Then $$n_1 = 2(\gamma - \eta) = 2 \times 15^\circ \text{ per hour}$$

$$n_2 = 2(\gamma - \sigma)$$

$$n_1 - n_2 = 2(\sigma - \eta) = 1^\circ{\cdot}0158958 \text{ per hour}$$

The equation is $1{\cdot}0158958 q = 15 r$

If $r = 25$, $q = 369{\cdot}13$

Thus 25 periods of $2(\sigma - \eta)$ is 369·13 mean solar days. It follows, therefore, that we must sum the series over 369 days in order to be as near right as possible.

Now this is equally true of all the columns, and each should have 369 entries.

Hence, in order to have 369 entries in each column, the S_2 computation form (as used in India) should be corrected accordingly.

M *Series*.

Now consider that we are analysing for M_2, and wish to minimise the effect of the S_2 tide. Hence

$$n_1 = 2(\gamma - \sigma) = 2 \times 14^\circ{\cdot}4920521 \text{ per hour}$$

$$n_2 = 2(\gamma - \eta)$$

$$n_1 - n_2 = -1^\circ{\cdot}0158958 \text{ per hour}$$

Hence, taking r negative, the equation is

$$1{\cdot}0158958q = 14{\cdot}4920521r$$

If $r = 25$, $q = 356{\cdot}63$

Thus 25 periods of $2(\sigma - \eta)$ is 356·63 of mean lunar time.

It follows, therefore, that we must have 357 entries in each column.

The M_2 computation form in use should be also corrected by adding 9 entries amongst which there are no 'changes.'

K *Series.*

To minimise the effect of M_2 on K_2, we have

$$n_1 = 2\gamma = 2 \times 15^{\circ}{\cdot}0410686 \text{ per hour}$$

$$n_2 = 2(\gamma - \sigma)$$

$$n_1 - n_2 = 2\sigma = 1^{\circ}{\cdot}0980330 \text{ per hour}$$

$$1{\cdot}0980330q = 15{\cdot}0410686r$$

If $r = 27$, $q = 369{\cdot}85$

Hence we should complete the row numbered 370; and correct the form accordingly.

To minimise the effect of O on K_1, we have

$$n_1 = \gamma = 15^{\circ}{\cdot}0410686 \text{ per hour}$$

$$n_2 = \gamma - 2\sigma$$

$$n_1 - n_2 = 2\sigma = 1^{\circ}{\cdot}0980330 \text{ per hour}$$

$$1{\cdot}0980330q = 15{\cdot}0410686r$$

This is the same equation again, and it confirms the conclusion that the row numbered 370 should be completed.

The N *Series.*

Here $\quad n_1 = 2\gamma - 3\sigma + \varpi = 2 \times 14^{\circ}{\cdot}2198648$ per hour

To minimise the effect of M_2,

$$n_2 = 2\gamma - 2\sigma$$

$$n_1 - n_2 = -(\sigma - \varpi) = -0^{\circ}{\cdot}5443747 \text{ per hour}$$

$$0{\cdot}5443747q = 14{\cdot}2198648r$$

If $r = 13$, $q = 339{\cdot}58$

Hence in the computation form we should complete the row numbered 340.

There is no justification for the alternative offered in the computation forms of continuing the entries up to $369^{d}\, 3^{h}$ of mean solar time.

The L *Series.*

Here $n_1 = 2\gamma - \sigma - \varpi = 2 \times 14°\cdot7642394$ per hour

To minimise the effect of M_2,

$$n_2 = 2\gamma - 2\sigma$$

$$n_1 - n_2 = \sigma - \varpi = 0°\cdot5443747 \text{ per hour}$$

$$0\cdot5443747q = 14\cdot7642394r$$

If $r = 13$, $q = 352\cdot58$

Hence we should complete the row numbered 353.

There is no justification for the alternative offered in the computation forms of continuing the entries up to $369^d\ 3^h$ of mean solar time.

The ν *Series.*

Here $n_1 = 2\gamma - 3\sigma - \varpi + 2\eta = 2 \times 14°\cdot2562915$ per hour

To minimise the effect of M_2,

$$n_2 = 2\gamma - 2\sigma$$

$$n_1 - n_2 = -\sigma - \varpi + 2\eta = -0°\cdot4715211 \text{ per hour}$$

$$0\cdot4715211q = 14\cdot2562915r$$

If $r = 11$, $q = 332\cdot6$

Hence we should complete the row numbered 333.

There is no justification for the alternative offered in the computation forms of continuing the entries up to $369^d\ 3^h$ of mean solar time.

The λ *Series.*

Here $n_1 = 2\gamma - \sigma + \varpi - 2\eta = 2 \times 14°\cdot7278127$ per hour

To minimise the effect of M_2,

$$n_2 = 2\gamma - 2\sigma$$

$$n_1 - n_2 = \sigma + \varpi - 2\eta = 0°\cdot4715211 \text{ per hour}$$

$$0\cdot4715211q = 14\cdot7278127r$$

If $r = 11$, $q = 343\cdot58$

Hence we should complete the row numbered 344.

There is no justification for the alternative offered in the computation forms of continuing the entries up to $369^d\ 3^h$ of mean solar time.

The 2N *Series.*

Here $n_1 = 2\gamma - 4\sigma + 2\varpi = 2 \times 13°\cdot9476774$ per hour

To minimise the effect of M_2,

$$n_2 = 2\gamma - 2\sigma$$

$$n_1 - n_2 = -2(\sigma - \varpi) = 1^\circ\!\cdot\!0887494 \text{ per hour}$$

$$1\cdot0887494q = 13\cdot9476774r$$

If $r = 26$, $q = 333\cdot08$

Hence we must in the computation form complete the row numbered 333.

The T *Series.*

Here $$n_1 = 2\gamma - 3\eta = 2 \times 14^\circ\!\cdot\!9794657 \text{ per hour}$$

To minimise the effect of M_2,

$$n_2 = 2\gamma - 2\sigma$$

$$n_1 - n_2 = 2\sigma - 3\eta = 0^\circ\!\cdot\!9748272 \text{ per hour}$$

$$0\cdot9748272q = 14\cdot9794657r$$

If $r = 24$, $q = 368\cdot79$

Hence we must in the form complete the row numbered 369.

The R *Series.*

Here $$n_1 = 2\gamma - \eta = 2 \times 15^\circ\!\cdot\!0205343 \text{ per hour}$$

To minimise the effect of M_2,

$$n_2 = 2\gamma - 2\sigma$$

$$n_1 - n_2 = 2\sigma - \eta = 1^\circ\!\cdot\!0569644 \text{ per hour}$$

$$1\cdot0569644q = 15\cdot0205343r$$

If $r = 25$, $q = 355\cdot28$, and $r = 26$, $q = 369\cdot49$

Hence we should either complete the row numbered 355 or that numbered 369.

The 2MS *Series.*

Here $$n_1 = 2\gamma - 4\sigma + 2\eta = 2 \times 13^\circ\!\cdot\!9841042 \text{ per hour}$$

To minimise the effect of M_2,

$$n_2 = 2\gamma - 2\sigma$$

$$n_1 - n_2 = -2(\sigma - \eta) = -1^\circ\!\cdot\!0158958 \text{ per hour}$$

$$1\cdot0158958q = 13\cdot9841042r$$

If $r = 24$, $q = 330\cdot37$, and $r = 25$, $q = 344\cdot13$

Hence we should either complete the row numbered 330 or that numbered 344.

The 2SM *Series.*

Here $n_1 = 2\gamma + 2\sigma - 4\eta = 2 \times 15^\circ{\cdot}5079479$ per hour

To minimise the effect of M_2,

$$n_2 = 2\gamma - 2\sigma$$

$$n_1 - n_2 = 4(\sigma - \eta) = 2^\circ{\cdot}0317916 \text{ per hour}$$

$$2{\cdot}0317916q = 15{\cdot}5079479r$$

If $r = 48$, $q = 366{\cdot}37$

Hence we should complete the row numbered 366.

The O *Series.*

Here $n_1 = \gamma - 2\sigma = 13^\circ{\cdot}9430356$ per hour

To minimise the effect of K_1,

$$n_2 = \gamma$$

$$n_1 - n_2 = -2\sigma = -1^\circ{\cdot}0980330 \text{ per hour}$$

$$1{\cdot}0980330q = 13{\cdot}9430356r$$

If $r = 27$, $q = 342{\cdot}85$

Hence we should complete the row numbered 343, cutting off the last three entries in the forms as in use.

The P *Series.*

Here $n_1 = \gamma - 2\eta = 14^\circ{\cdot}9589314$ per hour

It is open to question whether it is best to minimise the effect of K_1 or of O.

For K_1 take $n_2 = \gamma$

$$n_1 - n_2 = -2\eta = -0^\circ{\cdot}0821372 \text{ per hour}$$

$$0{\cdot}0821372q = 14{\cdot}9589314r$$

If $r = 2$, $q = 364{\cdot}24$

Hence we should complete the row numbered 364.

For O, take $n_2 = \gamma - 2\sigma$

$$n_1 - n_2 = 2(\sigma - \eta) = 1^\circ{\cdot}0158958 \text{ per hour}$$

$$1{\cdot}0158958q = 14{\cdot}9589314r$$

If $r = 25$, $q = 368{\cdot}12$

Hence we should complete the row numbered 368.

It is better to abide by this, for in the former case $n_1 - n_2$ varies very slowly; and we may be satisfied that on stopping with row 368 the effects of O and K_1 will both be adequately eliminated.

The J *Series.*

Here $$n_1 = \gamma + \sigma - \varpi = 15^\circ{\cdot}5854433 \text{ per hour}$$

To minimise the effect of K_1,

$$n_2 = \gamma$$

$$n_1 - n_2 = \sigma - \varpi = 0^\circ{\cdot}5443747 \text{ per hour}$$

$$0{\cdot}5443747q = 15{\cdot}5854433r$$

If $r = 12$, $q = 343{\cdot}56$, and $r = 13$, $q = 372{\cdot}19$

To minimise the effect of O,

$$n_2 = \gamma - 2\sigma$$

$$n_1 - n_2 = 3\sigma - \varpi = 1^\circ{\cdot}6424077 \text{ per hour}$$

$$1{\cdot}6424077q = 15{\cdot}5854433r$$

If $r = 36$, $q = 341{\cdot}6$, and $r = 39$, $q = 370{\cdot}09$

Since in the latter case $n_1 - n_2$ varies three times as fast as in the former, it will be better to abide by this, and stop either with the row numbered 342 or that numbered 370.

The Q *Series.*

Here $$n_1 = \gamma - 3\sigma + \varpi = 13^\circ{\cdot}3986609 \text{ per hour}$$

To minimise the effect of K_1,

$$n_2 = \gamma$$

$$n_1 - n_2 = -(3\sigma - \varpi) = -1^\circ{\cdot}6424077 \text{ per hour}$$

$$1{\cdot}6424077q = 13{\cdot}3986609r$$

If $r = 38$, $q = 310{\cdot}00$

To minimise the effect of O,

$$n_2 = \gamma - 2\sigma$$

$$n_1 - n_2 = -(\sigma - \varpi) = -0^\circ{\cdot}5443747 \text{ per hour}$$

$$0{\cdot}5443747q = 13{\cdot}3986609r$$

If $r = 12$, $q = 307{\cdot}36$

Since in the former case $n_1 - n_2$ varies about three times as fast as in the latter, it will be better to abide by the former, and stop with the row numbered 310.

With regard to the quaterdiurnal and terdiurnal tides, it does not signify where we stop; but it seems more reasonable to stop with the exact year of 365 mean solar days. These tides are called MS, MN, MK, 2MK.

Schedule II.

Periods over which the Harmonic Analysis should extend.

Initial of series	Number of day and hour of last entry in special time	Period elapsing from 0^h of special day 1 to 23^h of last special day in mean solar hours
S	369^d 23^h	368^d 23^h
M	357 23	369 11
K	370 23	368 23
N	340 23	358 15
L	353 23	358 14
ν	333 23	350 8
λ	344 23	350 8
2N	333 23	358 2
T	369 23	369 11
R	355 23 or 370 23	354 11 or 369 11
2MS	330 23 or 344 23	353 22 or 368 23
2SM	366 23	353 23
O	343 23	368 23
P	368 23	368 23
J	342 23 or 370 23	329 3 or 356 1
Q	310 23	347 0

In the second column the numbers are given to the nearest mean solar hour.

IV. A Comparison of the Harmonic Treatment of Tidal Observations with the Older Methods.

§ 1. *On the Method of Computing Tide-tables.*

There is nothing in the harmonic reduction of tidal observations which necessitates recourse to mechanical prediction of the tides. It may happen that it is desirable to produce a tide-table by arithmetical processes, and that the computers prefer to use the older methods of corrections, or it may be desired to obtain the tidal constants in the harmonic notation from older observations. For either of these purposes it is necessary to show how the harmonically expressed results may be converted into the older form, so that the constants for the fortnightly inequality in time and height, and the corrections for parallax and declination may be obtained from those of the harmonic analysis, and conversely.

In the following sections I propose, therefore, first to reduce the harmonic presentment of the resultant tide into the synthetic form, where we have a single harmonic term depending on the local mean solar time of moon's transit, and on corrections depending on the R.A., declination, and parallax of the perturbing bodies. Subsequently it will be shown how a synthesis may be carried out more simply by retaining the mean longitudes and elements of the orbits.

§ 2. *Notation for Mean Heights and Retardations derived from the Harmonic Method.*

The notation of the Report of 1883 [Paper 1] is adopted; and I shall carry the approximation to about the same degree as has been adopted by the older writers. Closer approximation may, of course, be easily obtained.

In the Report of 1883 the mean height* of a tide is denoted by H, and the retardation or lag by κ. In the present note it will be necessary to refer to several of the H's and κ's at the same time, and therefore it is expedient to introduce the following notation:—

Schedule III.

Initial of tide	Mean height (H)	Retardation (κ)	Initial of tide	Mean height (H)	Retardation (κ)
M_2	M	2μ	L	L	2λ
S_2	S	2ζ	T	T	2ζ
Lunar K_2	K''	2κ	R	R	2ζ
Solar K_2	$K_{\prime}''$	2κ	O	M'	μ'
K_2	K_2	2κ	P	S'	ζ'
N	N	2ν	K_1	K_1	κ_1

In this schedule we assume T and R (of speeds $2\gamma - 3\eta$ and $2\gamma - \eta$) to have the same lag as S_2; and we use ν in a new sense, the old ν, the R.A. of the intersection of the equator with the lunar orbit, being denoted by ν_0. The initials of each tide are used to denote its height at any time.

* I use height to denote semi-range. All references to this Report will simply be by the date 1883.

§ 3. *Introduction of Hour-angles, Declinations, and Parallaxes.*

We must now get rid of the elements of the orbit and of the mean longitudes, and introduce hour-angles, declinations, and parallaxes.

At the time t let α, δ, ψ be ☽'s R.A., and declination, and hour-angle,

and $\alpha_{,}$, $\delta_{,}$, $\psi_{,}$, ☉'s R.A., and declination, and hour-angle.

Let l be ☽'s longitude in her orbit measured from 'the intersection,' and $\alpha - \nu_0$ (ν_0 being the ν of 1883) be ☽'s R.A. measured from the intersection.

The annexed figure exhibits the relation of the several angles to one another.

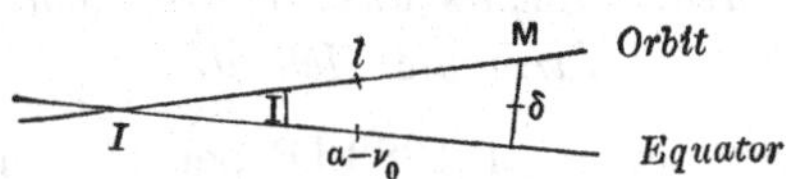

The spherical triangle affords the relations

$$\tan(\alpha - \nu_0) = \cos I \tan l, \quad \sin\delta = \sin I \sin l \quad \ldots\ldots\ldots\ldots(1)$$

From the first of (1) we have, approximately,

$$\alpha = l + \nu_0 - \tan^2 \tfrac{1}{2} I \sin 2l \quad \ldots\ldots\ldots\ldots(2)$$

Now, $s - \xi$ is the moon's mean longitude measured from I, and $s - p$ is the mean anomaly. Hence, approximately,

$$l = s - \xi + 2e \sin(s - p) \quad \ldots\ldots\ldots\ldots(3)$$

And therefore, approximately,

$$\alpha = s + \nu_0 - \xi + 2e\sin(s - p) - \tan^2 \tfrac{1}{2} I \sin 2(s - \xi) \quad \ldots\ldots\ldots(4)$$

Now, $t + h$ being the sidereal hour-angle,

$$\psi = t + h - \alpha \quad \ldots\ldots\ldots\ldots(5)$$

Therefore, from (4) and (5),

$$t + h - s - (\nu_0 - \xi) = \psi + 2e\sin(s - p) - \tan^2 \tfrac{1}{2} I \sin 2(s - \xi) \quad \ldots(6)$$

By the second of (1) we have, approximately,

$$\cos^2\delta = 1 - \tfrac{1}{2}\sin^2 I + \tfrac{1}{2}\sin^2 I \cos 2(s - \xi) \quad \ldots\ldots\ldots\ldots(7)$$

Hence, if Δ be such a declination that $\cos^2\Delta$ is the mean value of $\cos^2\delta$, we have

$$\left.\begin{aligned} \cos^2\Delta &= 1 - \tfrac{1}{2}\sin^2 I \\ \text{and} \qquad \cos^2\Delta_{,} &= 1 - \tfrac{1}{2}\sin^2\omega \end{aligned}\right\} \quad \ldots\ldots\ldots\ldots(8)$$

From this we have (neglecting terms in $\sin^4\Delta$) the following relations:—

$$\cos^4 \tfrac{1}{2} I = \cos^2\Delta, \quad \sin I \cos^2 \tfrac{1}{2} I = \sqrt{2}\sin\Delta\cos\Delta, \quad \sin^2 I = 2\sin^2\Delta$$

$$\cos^4 \tfrac{1}{2}\omega = \cos^2\Delta_{,}, \quad \sin\omega\cos^2\tfrac{1}{2}\omega = \sqrt{2}\sin\omega\cos\omega, \quad \sin^2\omega = 2\sin^2\Delta_{,}$$

Thus we may put

$$\left.\begin{array}{ll}\dfrac{\cos^4 \frac{1}{2}I}{\cos^4 \frac{1}{2}\omega \cos^4 \frac{1}{2}i} = \dfrac{\cos^2 \Delta}{\cos^2 \Delta_{,}}, & \dfrac{\sin I \cos^2 \frac{1}{2}I}{\sin \omega \cos^2 \frac{1}{2}\omega \cos^4 \frac{1}{2}i} = \dfrac{\sin 2\Delta}{\sin 2\Delta_{,}} \\ \dfrac{\sin^2 I}{\sin^2 \omega (1 - \frac{3}{2}\sin^2 i)} = \dfrac{\sin^2 \Delta}{\sin^2 \Delta_{,}}, & \tan^2 \frac{1}{2}I = \frac{1}{2}\tan^2 \Delta\end{array}\right\} \quad \ldots(9)$$

An approximate formula for Δ and the value of $\Delta_{,}$ are

$$\Delta = 16^\circ{\cdot}51 + 3^\circ{\cdot}44 \cos \mathrm{N} - 0^\circ{\cdot}19 \cos 2\mathrm{N}, \quad \Delta_{,} = 16^\circ{\cdot}36 \quad \ldots\ldots(10)$$

The introduction of Δ and $\Delta_{,}$ in place of I and ω entails a loss of accuracy, and it is only here made because former writers have followed that plan. It may easily be dispensed with.

Now let us write

$$\left.\begin{array}{ll} D = \cos 2(s - \xi), & D' = \sin 2(s - \xi) \\ \Pi = \cos(s - p), & \Pi' = \sin(s - p)\end{array}\right\} \quad \ldots\ldots\ldots\ldots\ldots(11)$$

From (7) and (8),

$$D = \frac{\cos^2 \delta - \cos^2 \Delta}{\sin^2 \Delta}, \qquad D' = \frac{\sin \delta \cos \delta}{\sigma \sin^2 \Delta}\frac{d\delta}{dt} \quad \ldots\ldots\ldots\ldots(12)$$

Then, if we write for the ratio of the moon's parallax to her mean parallax P, we have

$$P - 1 = e \cos(s - p)$$

and

$$\Pi = \frac{1}{e}(P - 1), \qquad \Pi' = -\frac{1}{e(\sigma - \varpi)}\frac{dP}{dt} \quad \ldots\ldots\ldots\ldots\ldots(13)$$

Hence D, D', Π, Π' are functions of declinations and parallaxes. The similar symbols with subscript accents are to apply to the sun.

Now (6) may be written by aid of (9) and (11),

$$2[t + h - s - (\nu_0 - \xi)] = 2\psi + 4e\Pi' - D' \tan^2 \Delta \quad \ldots\ldots\ldots(14)$$

The left-hand side of (14) is the argument of M_2 (see Sched. B. i. 1883), and from (9) the factor of M_2 is $\cos^2 \Delta / \cos^2 \Delta_{,}$. Hence, subtracting the retardation 2μ from (14) we have

$$(\mathrm{M}_2) = \frac{\cos^2 \Delta}{\cos^2 \Delta_{,}} M \cos[(2\psi + 4e\Pi' - D' \tan^2 \Delta) - 2\mu]$$

expanding approximately,

$$\begin{aligned}(\mathrm{M}_2) = {} & \frac{\cos^2 \Delta}{\cos^2 \Delta_{,}} M \cos 2(\psi - \mu) \\ & - \frac{\cos^2 \Delta}{\cos^2 \Delta_{,}} \Pi' \, 4Me \sin 2(\psi - \mu) \\ & + \frac{\sin^2 \Delta}{\cos^2 \Delta_{,}} D' M \sin 2(\psi - \mu) \quad \ldots\ldots\ldots\ldots\ldots\ldots(15)\end{aligned}$$

We shall see later that the two latter terms of (15) are nearly annulled by terms arising from other tides, and as in the case of the sun the rates of change of parallax and declination are small, we may write by symmetry,

$$(S_2) = S \cos 2(\psi_{,} - \zeta) \quad \text{.......................(16)}$$

In all the smaller tides we may write

$$t + h - s - (\nu_0 - \xi) = \psi$$

A general formula of transformation will be required below. Thus, if $\cos 2x = X$, $\sin 2x = X'$,

$$\cos 2(\psi \mp x - \alpha) = \{X \mp \tan 2(\alpha - \mu) X'\} \cos 2(\psi - \alpha)$$
$$\pm \frac{X'}{\cos 2(\alpha - \mu)} \sin 2(\psi - \mu) \quad \text{......(17)}$$

The lunar K_2 *tide.*

From Sched. B. i., 1883, we have

$$\text{Lunar } K_2 = \frac{\sin^2 I}{\sin^2 \omega (1 - \frac{3}{2}\sin^2 i)} K'' \cos 2[t + h - \nu_0 - \kappa]$$
$$= \frac{\sin^2 \Delta}{\sin^2 \Delta_{,}} K'' \cos 2[\psi + (s - \xi) - \kappa]$$

Applying (17) with $X = D$, $X' = D'$, $\alpha = \kappa$, and taking the lower sign,

$$\text{Lunar } K_2 = \frac{\sin^2 \Delta}{\sin^2 \Delta_{,}} K'' \left[\{D + \tan 2(\kappa - \mu) D'\} \cos 2(\psi - \kappa) - \frac{D'}{\cos 2(\kappa - \mu)} \sin 2(\psi - \mu) \right] \quad \text{...(18)}$$

In the case of the sun we neglect the terms in D', for the same reasons as were assigned for the similar neglect in (16), and have

$$\text{Solar } K_2 = K_{,}'' D_{,} \cos 2(\psi_{,} - \kappa) \quad \text{.................(19)}$$

The tide N.

From Schedule B. i., Report 1883,

$$(N) = \frac{\cos^4 \frac{1}{2} I}{\cos^4 \frac{1}{2}\omega \cos^4 \frac{1}{2} i} N \cos [2(t + h - s - \nu_0 + \xi) - (s - p) - 2\nu]$$

Then
$$(N) = \frac{\cos^2 \Delta}{\cos^2 \Delta_{,}} N \cos 2[\psi - \nu - \tfrac{1}{2}(s - p)]$$

Then applying (17) with $X = \Pi$, $X' = \Pi'$, $\alpha = \nu$, and taking the upper sign, but writing $\mu - \nu$ instead of $\nu - \mu$, because this tide being slower than M_2 suffers less retardation,

$$(N) = \frac{\cos^2 \Delta}{\cos^2 \Delta_{,}} N \left[\{\Pi + \tan 2(\mu - \nu) \Pi'\} \cos 2(\psi - \nu) + \frac{\Pi'}{\cos 2(\mu - \nu)} \sin 2(\psi - \mu) \right] \quad \text{...(20)}$$

The tide L.

We shall here omit the small tide of speed $2\gamma - \sigma + \varpi$, by which the true elliptic tide is perturbed. Thus the R in the column of arguments in Sched. B. i., 1883 is neglected, and we have

$$(\mathrm{L}) = -\frac{\cos^4 \frac{1}{2} I}{\cos^4 \frac{1}{2}\omega \cos^4 \frac{1}{2} i} \cdot L \cos [2(t+h-s-\nu+\xi) + (s-p) - 2\lambda]$$

$$= -\frac{\cos^2 \Delta}{\cos^2 \Delta_,} L \cos 2 [\psi - \lambda + \tfrac{1}{2}(s-p)]$$

Applying (17) with $X = \Pi$, $X' = \Pi'$, $\alpha = \lambda$, and taking the lower sign, and changing the sign of the whole, because of the initial negative sign,

$$(\mathrm{L}) = \frac{\cos^2 \Delta}{\cos^2 \Delta_,} L \left[\{-\Pi - \tan 2(\lambda - \mu)\Pi'\} \cos 2(\psi - \lambda) + \frac{\Pi'}{\cos 2(\lambda - \mu)} \sin 2(\psi - \mu) \right] \quad \ldots(21)$$

The sum of N *and* L.

In order to fuse these terms an approximation will be adopted. The L tide is just as much faster than M_2 as N is slower, but the N tide should be nearly 7 times as great as the L tide; hence the $\tan 2(\lambda - \mu)$ in (21) will be put equal to $\tan 2(\mu - \nu)$. We then have

$$(\mathrm{N}) + (\mathrm{L}) = \frac{\cos^2 \Delta}{\cos^2 \Delta_,} [\{\Pi + \tan 2(\mu - \nu)\Pi'\} \{N \cos 2(\psi - \nu) - L \cos 2(\psi - \lambda)\} + \Pi' \{N \sec 2(\mu - \nu) + L \sec 2(\lambda - \mu)\} \sin 2(\psi - \mu)]$$

But

$$N \cos 2(\psi - \nu) - L \cos 2(\psi - \lambda) = \cos 2\psi (N \cos 2\nu - L \cos 2\lambda) + \sin 2\psi (N \sin 2\nu - L \sin 2\lambda)$$

Then writing

$$\tan 2\epsilon = \frac{N \sin 2\nu - L \sin 2\lambda}{N \cos 2\nu - L \cos 2\lambda} \quad \ldots\ldots\ldots\ldots(22)$$

so that ϵ is nearly equal to ν, we have

$$(\mathrm{N}) + (\mathrm{L}) = \frac{\cos^2 \Delta}{\cos^2 \Delta_,} \frac{N \cos 2\nu - L \cos 2\lambda}{\cos 2\epsilon} [\{\Pi + \tan 2(\mu - \nu)\Pi'\} \cos 2(\psi - \epsilon)]$$

$$+ \frac{\cos^2 \Delta}{\cos^2 \Delta_,} [\{N \sec 2(\mu - \nu) + L \sec 2(\lambda - \mu)\} \sin 2(\psi - \mu)] \quad \ldots(23)$$

In the symmetrical term for the sun, with approximation as in (16), we get

$$(\mathrm{T}) + (\mathrm{R}) = (T - R)\Pi_, \cos 2(\psi_, - \zeta) \quad \ldots\ldots\ldots\ldots(24)$$

This terminates the semidiurnal tides which we are considering; but before proceeding to collect the results some further transformations must be exhibited.

Let us consider the function $D + xD'$, where x is small. From (12) we see that

$$D + xD' = \frac{\cos^2 \delta - \cos^2 \Delta}{\sin^2 \Delta} + \tfrac{1}{2} x \frac{2 \sin \delta \cos \delta}{\sigma \sin^2 \Delta} \frac{d\delta}{dt}$$

Hence, if δ' be the moon's declination at a time earlier than the time of observation by $x/2\sigma$, then

$$D + xD' = \frac{\cos^2 \delta' - \cos^2 \Delta}{\sin^2 \Delta}$$

Hence, in (17),

$$D + \tan 2(\kappa - \mu) D' = \frac{\cos^2 \delta' - \cos^2 \Delta}{\sin^2 \Delta} \quad \text{..............(25)}$$

when δ' is the moon's declination at time $\frac{1}{15}t - 57^\circ\!\cdot\!3 \tan 2(\kappa - \mu)/2\sigma$. The period $57^\circ\!\cdot\!3 \tan 2(\kappa - \mu)/2\sigma$ may be called 'the age of the declinational inequality.'

Again, $$\Pi + x\Pi' = \frac{1}{e}\left\{P - 1 - \frac{x}{\sigma - \varpi}\frac{dP}{dt}\right\}$$

Hence, if $(P' - 1)/e$ denotes the value of $(P - 1)/e$ at a time $x/(\sigma - \varpi)$ earlier than that of observation, then

$$\Pi + x\Pi' = \frac{1}{e}(P' - 1)$$

Hence, in (23), $$\Pi + \tan 2(\mu - \nu)\Pi' = \frac{1}{e}(P' - 1) \quad \text{.................(26)}$$

where P' is the ratio of the moon's parallax to her mean parallax at a time $\frac{1}{15}t - 57^\circ\!\cdot\!3 \tan 2(\mu - \nu)/(\sigma - \varpi)$. The period $57^\circ\!\cdot\!3 \tan 2(\mu - \nu)/(\sigma - \varpi)$ may be called 'the age of the parallactic inequality.'

In collecting results we shall write the sum

$$\mathrm{M}_2 + \mathrm{S}_2 + \mathrm{K}_2 + \mathrm{N} + \mathrm{L} + \mathrm{R} + \mathrm{T} = h_2$$

For reasons explained below we omit terms depending on the rate of change of solar parallax and declination.

Then, from (15), (16), (18), (19), (23), (24), (25), (26), we have

$$\begin{aligned} h_2 = {} & \frac{\cos^2 \Delta}{\cos^2 \Delta_,} M \cos 2(\psi - \mu) + S \cos 2(\psi_, - \zeta) \\ & + \frac{\cos^2 \delta' - \cos^2 \Delta}{\sin^2 \Delta_,} K'' \cos 2(\psi - \kappa) + \frac{\cos^2 \delta_, - \cos^2 \Delta_,}{\sin^2 \Delta_,} K_,'' \cos 2(\psi_, - \kappa) \\ & - \frac{\sin \delta \cos \delta}{\sigma \sin^2 \Delta_,} \frac{d\delta}{dt} \left(\frac{K''}{\cos 2(\kappa - \mu)} - M \tan^2 \Delta_,\right) \sin 2(\psi - \mu) \\ & + \frac{\cos^2 \Delta}{\cos^2 \Delta_,} (P' - 1) \frac{N \cos 2\nu - L \cos 2\lambda}{e \cos 2\epsilon} \cos 2(\psi - \epsilon) \\ & + (P_, - 1) \frac{(T - R)}{e_,} \cos 2(\psi_, - \zeta) \\ & + \frac{\cos^2 \Delta}{\cos^2 \Delta_,} \frac{1}{\sigma - \varpi} \frac{dP}{dt} \left(4M - \frac{N \sec 2(\mu - \nu) + L \sec 2(\lambda - \mu)}{e}\right) \sin 2(\psi - \mu) \end{aligned}$$

.........(27)

It may easily be shown, from Schedule B. i., 1883, that in the equilibrium theory $K'' - M\tan^2\Delta_{,} = 0$, and $4M - (N+L)/e = 0$; hence the terms depending on rates of change of declination and parallax are small. This also shows that we were justified in neglecting the corresponding terms in the case of the sun. Also, since the faster tides are more augmented by kinetic action that the slow ones, the two functions, written above, which vanish in the equilibrium theory are normally actually positive. The formula (27) gives the complete expression for the semidiurnal tide in terms of hour-angles, declinations, and parallaxes, with the constants of the harmonic analysis.

We shall now show that with rougher approximation (27) is reducible to a much simpler form.

The retardation of each tide should be approximately a constant, plus a term varying with the speed. Hence all the retardations may be expressed in terms of ζ and μ, and

$$\kappa = \mu + \frac{\zeta - \mu}{\sigma - \eta}\sigma$$

$$\nu = \mu - \frac{\zeta - \mu}{\sigma - \eta}\tfrac{1}{2}(\sigma - \varpi)$$

$$\lambda = \mu + \frac{\zeta - \mu}{\sigma - \eta}\tfrac{1}{2}(\sigma - \varpi)$$

It is clear that κ differs very little from ζ, and that

$$\frac{\kappa - \mu}{\sigma} = \frac{2(\mu - \nu)}{\sigma - \varpi} = \frac{\zeta - \mu}{\sigma - \eta}$$

The time $(\zeta - \mu)/(\sigma - \eta)$ is called 'the age of the tide,' for reasons explained below, and $\kappa - \mu$, $\mu - \nu$, not being large angles, do not differ much from their tangents. Hence the ages of the declinational and parallactic inequalities are both approximately equal to the age of the tide.

Let $\text{\ae}$, then, denote $(\zeta - \mu)/(\sigma - \eta)$, the age of the tide.

Now, as an approximation, we may suppose that heights of the lunar K_2 tide, the N and L tides bear the same ratio to the M_2 tide as in the equilibrium theory; and that the solar K_2, the T and R tides bear the same ratio to the S_2 tide as in that theory. Then reverting to the notation with I, ω, i in place of Δ, $\Delta_{,}$, and writing

$$\left(\frac{\cos\frac{1}{2}I}{\cos\frac{1}{2}\omega\cos\frac{1}{2}i}\right)^4 = \mathrm{f}$$

we have

$$\frac{\sin^2\Delta}{\sin^2\Delta_{,}}K'' = \frac{\frac{1}{2}\sin^2 I}{\cos^4\frac{1}{2}I}\mathrm{f}M, \quad \frac{\cos^2\Delta}{\cos^2\Delta_{,}}N = \tfrac{7}{2}e\mathrm{f}M, \quad \frac{\cos^2\Delta}{\cos^2\Delta_{,}}L = \tfrac{1}{2}e\mathrm{f}M$$

$$K_{,}'' = \frac{\frac{1}{2}\sin^2\omega}{\cos^4\frac{1}{2}\omega}S, \quad T = \tfrac{7}{2}e_{,}S, \quad R = \tfrac{1}{2}e_{,}S$$

Also, since (22) may be written

$$\tan(2\mu - 2\epsilon) = \frac{N \sin 2(\mu - \nu) + L \sin 2(\lambda - \mu)}{N \cos 2(\mu - \nu) - L \cos 2(\lambda - \mu)}$$

we have, treating $\mu - \nu$, $\lambda - \mu$, $\mu - \epsilon$ as small, approximately,

$$\epsilon = \mu - \tfrac{2}{3}\text{æ}(\sigma - \varpi) = \mu - \tfrac{2}{3}(\lambda - \nu)$$

Also $$\frac{\cos^2 \Delta}{\cos^2 \Delta_{,}} \frac{N \cos 2\nu - L \cos 2\lambda}{\cos 2\epsilon} = 3e\text{f}M$$

Then reverting to mean longitudes, and substituting the age of tide where required, we find, on neglecting the difference between κ and ζ,

For the lunar declinational term,

$$2 \tan^2 \tfrac{1}{2} I \,\text{f}M \cos 2[s - \text{æ}\sigma - \xi] \cos 2(\psi - \zeta)$$

For the solar declinational term,

$$2 \tan^2 \tfrac{1}{2}\omega \; S \cos 2h \cos 2(\psi_{,} - \zeta)$$

For the lunar parallactic term,

$$3e\text{f}M \cos[s - p - \text{æ}(\sigma - \varpi)] \cos 2[\psi - \mu + \tfrac{2}{3}\text{æ}(\sigma - \varpi)]$$

For the solar parallactic term,

$$3e_{,}S \cos(h - p_{,}) \cos 2[\psi_{,} - \zeta]$$

Then omitting the terms depending on changes of declination and parallax, we have as an approximation,

$$\begin{aligned} h_2 = \text{f}M \{&\cos 2(\psi - \mu) + 2 \tan^2 \tfrac{1}{2} I \cos 2[s - \text{æ}\sigma - \xi] \cos 2(\psi - \zeta) \\ &+ 3e \cos[s - p - \text{æ}(\sigma - \varpi)] \cos 2[\psi - \mu + \tfrac{2}{3}\text{æ}(\sigma - \varpi)]\} \\ &+ S[1 + 2\tan^2 \tfrac{1}{2}\omega \cos 2h + 3e_{,} \cos(h - p_{,})] \cos 2(\psi_{,} - \zeta) \quad \text{..................(28)} \end{aligned}$$

In the equilibrium theory we have the lunar semidiurnal tide depending on $r^{-3} \cos^2 \delta \cos 2\psi$. Now it is obvious that $\cos^2 \delta$ introduces a factor $1 + 2\tan^2 \frac{1}{2} I \cos 2(s - \xi)$, and r^{-3} a factor $1 + 3e \cos(s - p)$. Thus, if we could have foreseen the exact disturbance introduced by friction and other causes in the various angles, the formula (28) might have been established at once; but it seems to have been necessary to have recourse to the complete development in order to find how the age of the tide will enter.

§ 4. *Reference to Time of Moon's Transit.*

It has been usual to refer the tide to the time of moon's transit, and we shall now proceed to the transformations necessary to do so.

$\cos^2\Delta/\cos^2\Delta_{,}$ goes through its oscillation about the value unity in 19 years; it is therefore convenient to write for, say, a whole year,

and similarly,
$$\left.\begin{aligned} M_0 &= \frac{\cos^2\Delta}{\cos^2\Delta_{,}}M \\ N_0 &= \frac{\cos^2\Delta}{\cos^2\Delta_{,}}N \\ L_0 &= \frac{\cos^2\Delta}{\cos^2\Delta_{,}}L \end{aligned}\right\} \quad\ldots\ldots\ldots\ldots\ldots\ldots\ldots\ldots\ldots(29)$$

We also observe that K'' and $K_{,}''$, being the lunar and solar parts of the mean K_2 tide, and their ratio being ·464 (Report, 1883),

$$K'' = \cdot 68303K_2, \quad K_{,}'' = \cdot 31697K_2 \quad\ldots\ldots\ldots\ldots(30)$$

It will also be seen that in all the terms arising from the sun, excepting that in $K_{,}''$, the argument of the cosine is $2(\psi_{,} - \zeta)$. It will be convenient, and sufficiently accurate for all practical purposes, to replace κ by ζ in this solar declinational term $K_{,}''$.

We shall now proceed to refer the tide to the moon's transit at the place of observation.

Let α_0, h_0 be ☽'s R.A. and ☉'s mean longitude at ☽'s transit—say upper transit, for distinctness. Then the local time of transit is given by the vanishing of ψ, and since $\psi = t + h - \alpha$, it follows that the time-angle of ☽'s transit (at 15° to the hour) is $\alpha_0 - h_0$.

Now let τ (mean solar hours) be the interval after transit to which the time-angle t refers; then, since

$$\frac{dh}{dt} = \eta, \quad \frac{d\alpha}{dt} = \sigma + \left(\frac{d\alpha}{dt} - \sigma\right), \quad \gamma - \eta = 15^\circ, \quad \gamma - \sigma = 14^\circ{\cdot}49$$

$$\psi = t + h - \alpha$$

$$= [(\gamma - \eta)\tau + \alpha_0 - h_0] + [h_0 + \eta\tau] - \left[\alpha_0 + \sigma\tau + \left(\frac{d\alpha}{dt} - \sigma\right)\tau\right]$$

$$\psi = (\gamma - \sigma)\tau - \left(\frac{d\alpha}{dt} - \sigma\right)\tau$$

For the sake of brevity, put

$$\mathrm{T} = (\gamma - \sigma)\tau$$

so that T is τ converted to angle at the rate of 14°·49 per hour. Then we have

$$\psi = \mathrm{T} - \left(\frac{d\alpha}{dt} - \sigma\right)\tau \quad\ldots\ldots\ldots\ldots\ldots\ldots\ldots(31)$$

Similarly putting $\alpha_,$ for ☉'s R.A. at ☽'s transit, we have

$$\psi_, = t + h - \alpha_,$$

$$= [(\gamma - \eta)\tau + \alpha_0 - h_0] + [h_0 + \eta\tau] - \left[\alpha_, + \sigma\tau - \left(\sigma - \frac{d\alpha_,}{dt}\right)\tau\right]$$

so that
$$\psi_, = (\gamma - \sigma)\tau + \alpha_0 - \alpha_, + \left(\sigma - \frac{d\alpha_,}{dt}\right)\tau$$

Then let
$$A = \alpha_0 - \alpha_, \quad \text{.............................(32)*}$$

So that A is the apparent time of ☽'s transit, reduced to angle at 15° per hour, and we have

$$\psi_, = \mathrm{T} + A + \left(\sigma - \frac{d\alpha_,}{dt}\right)\tau \quad \text{........................(33)}$$

It is only in the two principal tides that we need regard the changes of R.A. since ☽'s transit, and in all the smaller terms we may simply put

$$\psi = \mathrm{T}, \quad \psi_, = \mathrm{T} + A$$

The first pair of terms of (27) now become

$$M_0 \cos 2\left[\mathrm{T} - \left(\frac{d\alpha}{dt} - \sigma\right)\tau - \mu\right] + S \cos 2\left[\mathrm{T} + A + \left(\sigma - \frac{d\alpha_,}{dt}\right)\tau - \zeta\right]$$

and these are equal to

$$M_0 \cos 2(\mathrm{T} - \mu) + S \cos 2(\mathrm{T} + A - \zeta)$$

$$+ \frac{2\pi}{180}\left(\frac{d\alpha}{dt} - \sigma\right)\tau M_0 \sin 2(\mathrm{T} - \mu) - \frac{2\pi}{180}\left(\sigma - \frac{d\alpha_,}{dt}\right)\tau S \sin 2[\mathrm{T} + A - \zeta]$$

.........(34)

We may now collect together all the results, and write them in the form of a schedule [Schedule IV.].

Definition of symbols:—

α, δ, $\alpha_,$, $\delta_,$ ☽'s and ☉'s R.A. and declination at moon's transit; $A = \alpha - \alpha_,$, apparent time of ☽'s transit at the port.

δ' ☽'s decl. at the time (generally earlier than transit) $\tau - \frac{57°\cdot3}{2\sigma}\tan 2(\kappa - \mu)$.

P, $P_,$ the ratio of ☽'s and ☉'s parallax to mean parallaxes.

P' the ratio for ☽ at the time (generally earlier than transit)

$$\tau - \frac{57°\cdot3}{\sigma - \varpi}\tan 2(\mu - \nu)$$

* It would be better to put

$$A = \alpha_0 - \alpha_, + \frac{\sigma - \eta}{\gamma - \sigma}\mu$$

If this be used the correction (40) for ☉'s change of R.A. becomes small.

Semidiurnal Tides.

Schedule IV.

Description of Term	Coefficient	Periodic Factor
Principal lunar	M_0	$\times \cos 2(\mathrm{T}-\mu)$
Principal solar	$+S$	$\times \cos 2(\mathrm{T}+A-\zeta)$
☽ Change of R.A.	$+\frac{2\pi}{180}\left(\frac{da}{dt}-\sigma\right)\tau M_0$	$\times \sin 2(\mathrm{T}-\mu)$
☉ Change of R.A.	$-\frac{2\pi}{180}\left(\sigma-\frac{da_{,}}{dt}\right)\tau S$	$\times \sin 2(\mathrm{T}+A-\zeta)$
☽ Declination	$+\frac{\cos^2\delta'-\cos^2\Delta}{\sin^2\Delta_{,}}\cdot 683\, K_2$	$\times \cos 2(\mathrm{T}-\kappa)$
☉ Declination	$+\frac{\cos^2\delta_{,}-\cos^2\Delta_{,}}{\sin^2\Delta_{,}}\cdot 317\, K_2$	$\times \cos 2(\mathrm{T}+A-\zeta)$
☽ Change of declination	$-\frac{\sin\delta\cos\delta}{\sigma\sin^2\Delta_{,}}\frac{d\delta}{dt}\left(\frac{\cdot 683\, K_2}{\cos 2(\kappa-\mu)}-M\tan^2\Delta_{,}\right)$	$\times \sin 2(\mathrm{T}-\mu)$
☽ Parallax	$+(P'-1)\frac{N_0\cos 2\nu-L_0\cos 2\lambda}{e\cos 2\epsilon}$	$\times \cos 2(\mathrm{T}-\epsilon)$
☉ Parallax	$+(P_{,}-1)\frac{T-R}{e_{,}}$	$\times \cos 2(\mathrm{T}+A-\zeta)$
☽ Change of parallax	$+\frac{1}{\sigma-\varpi}\frac{dP}{dt}\left(4M_0-\frac{N_0\sec 2(\mu-\nu)+L_0\sec 2(\lambda-\mu)}{e}\right)$	$\times \sin 2(\mathrm{T}-\mu)$

τ the time elapsed since ☽'s transit in m. s. hours; T the same time reduced to angle at 14°·49 per hour.

Δ such a declination that $\cos^2\Delta$ is the mean value of $\cos^2\delta$; Δ has a 19-yearly period.

$\Delta_{,}$ such a declination that $\cos^2\Delta_{,}$ is the mean value of $\cos^2\delta_{,}$.

e, $e_{,}$ eccentricities of lunar and solar orbits; σ the ☽'s mean motion; ϖ the mean motion of the ☽'s perigee.

$$\frac{M_0}{M} = \frac{N_0}{N} = \frac{L_0}{L} = \frac{\cos^2\Delta}{\cos^2\Delta_{,}}$$

M, S, K_2, N, L, T, R the mean semi-ranges H of the tides of those denominations in the harmonic method. The retardations found by harmonic analysis are 2μ for M_2, 2ζ for S_2, 2κ for K_2, 2ν for N, 2λ for L, and 2ζ for T and R.

Lastly $\tan 2\epsilon = \dfrac{N \sin 2\nu - L \sin 2\lambda}{N \cos 2\nu - L \cos 2\lambda}$, 2ϵ to be taken in the same quadrant as 2ν.

§ 5. *Synthesis of the Several Terms.*

Consider the two principal terms in Schedule IV.,

$$M_0 \cos 2(\mathrm{T} - \mu) + S \cos 2(\mathrm{T} + A - \zeta)$$

They may be written in the form

$$H \cos 2(\mathrm{T} - \phi)$$

where

$$H \cos 2(\mu - \phi) = M_0 + S \cos 2(A - \zeta + \mu)$$
$$H \sin 2(\mu - \phi) = S \sin 2(A - \zeta + \mu)$$

If we compute ϕ corresponding to the time of moon's transit from the formula

$$\tan 2(\mu - \phi) = \frac{S \sin 2(A - \zeta + \mu)}{M_0 + S \cos 2(A - \zeta + \mu)}$$

then ϕ reduced to time at the rate of 14°·49 per hour is the interval after moon's transit to high water, to a first approximation. The angle $\phi \pm 90°$, similarly reduced, gives the low waters before and after the high water, and $\phi \pm 180°$ gives another high water. The high waters and low waters are to be referred to the nearest transit of the moon.

The height or depression is given to a first approximation by

$$H = \sqrt{\{M_0^2 + S^2 + 2M_0 S \cos 2(\mu - \phi)\}}$$

This variability in the time and height of high water, due to variability of ϕ, is called the fortnightly or semi-menstrual inequality in the height and interval. The period $(\zeta - \mu)/(\sigma \dot{-} \eta)$ is called 'the age of the tide,' because this is the mean period after new and full moon before the occurrence of spring tide.

§ 6. *Corrections.*

The smaller terms in Schedule IV. may be regarded as inequalities in the principal terms. They are of several types. Consider a term $B\cos 2(\mathrm{T}-\beta)$.

Then

$$B\cos 2(\mathrm{T}-\beta) = B\cos 2(\beta-\phi)\cos 2(\mathrm{T}-\phi) + B\sin 2(\beta-\phi)\sin 2(\mathrm{T}-\phi)$$

Hence the addition of such a term to $H\cos 2(\mathrm{T}-\phi)$ gives us

$$(H+\delta H)\cos 2(\mathrm{T}-\phi-\delta\phi)$$

where $$\delta H = B\cos 2(\beta-\phi),\quad 2H\delta\phi = B\sin 2(\beta-\phi) \quad\text{.........(35)}$$

Next consider a term $C\sin 2(\mathrm{T}-\mu)$. Putting $\beta=\mu+\frac{1}{4}\pi$, we have

$$\delta H = -C\sin 2(\mu-\phi),\quad 2H\delta\phi = C\cos 2(\mu-\phi) \quad\text{.........(36)}$$

Next consider a term $E\cos 2(\mathrm{T}+A-\zeta)$. Putting $\beta=\zeta-A$, we have

$$\delta H = E\cos 2(A-\zeta+\phi),\quad 2H\delta\phi = -E\sin 2(A-\zeta+\phi) \quad\text{...(37)}$$

Lastly, consider a term $F\sin 2(\mathrm{T}+A-\zeta)$. Putting $\beta=\zeta-A+\frac{1}{4}\pi$, we have

$$\delta H = F\sin 2(A-\zeta+\phi),\quad 2H\delta\phi = F\cos 2(A-\zeta+\phi) \quad\text{......(38)}$$

In writing down the corrections we substitute $14{\cdot}49\delta t$ for $\delta\phi$, and introduce a factor so that the times may be given in mean solar hours and the angular velocities in degrees per hour.

Change of Moon's R.A., Sched. IV.

This is of type (36), and gives

$$\left.\begin{aligned}\delta H &= -\frac{2\pi}{180}\left(\frac{d\alpha}{dt}-\sigma\right)\tau M_0\sin 2(\mu-\phi)\\ \delta t &= 1^{\mathrm{h}}{\cdot}977\,\frac{2\pi}{180}\left(\frac{d\alpha}{dt}-\sigma\right)\tau\frac{M_0}{H}\cos 2(\mu-\phi)\end{aligned}\right\}\quad\text{...........(39)}$$

This correction to the height is very small.

Change of Sun's R.A., Sched. IV.*

This is of type (38), and gives

$$\left.\begin{aligned}\delta H &= -\frac{2\pi}{180}\left(\sigma-\frac{d\alpha_{\prime}}{dt}\right)\tau S\sin 2(A-\zeta+\phi)\\ \delta t &= -1^{\mathrm{h}}{\cdot}977\,\frac{2\pi}{180}\left(\sigma-\frac{d\alpha_{\prime}}{dt}\right)\tau\frac{S}{H}\cos 2(A-\zeta+\phi)\end{aligned}\right\}\quad\text{......(40)}$$

* With the value of A suggested in footnote to (32)

$$(\sigma-d\alpha_{\prime}/dt)\,\tau \text{ becomes } [(\phi-\mu)\,\sigma-(\phi d\alpha_{\prime}/dt-\mu\eta)]/(\gamma-\sigma)$$

at high water. This is obviously very small.

Moon's Declination, Sched. IV.

This is of type (35), and gives

$$\left.\begin{aligned}\delta H &= \frac{\cos^2 \delta' - \cos^2 \Delta}{\sin^2 \Delta_{,}} \cdot 683\, K_2 \cos 2(\kappa - \phi) \\ \delta t &= 1^{\text{h}}{\cdot}977 \frac{\cos^2 \delta' - \cos^2 \Delta}{\sin^2 \Delta_{,}} \cdot 683 \frac{K_2}{H} \sin 2(\kappa - \phi)\end{aligned}\right\} \quad \text{.........(41)}$$

Sun's Declination, Sched. IV.

This is of type (37), and gives

$$\left.\begin{aligned}\delta H &= \frac{\cos^2 \delta_{,} - \cos^2 \Delta_{,}}{\sin^2 \Delta_{,}} \cdot 317\, K_2 \cos 2(A - \zeta + \phi) \\ \delta t &= -1^{\text{h}}{\cdot}977 \frac{\cos^2 \delta_{,} - \cos^2 \Delta_{,}}{\sin^2 \Delta_{,}} \cdot 317 \frac{K_2}{H} \sin 2(A - \zeta + \phi)\end{aligned}\right\} \quad \text{......(42)}$$

Change of Moon's Declination, Sched. IV.

This is of type (36), and gives

$$\left.\begin{aligned}\delta H &= \frac{\sin \delta \cos \delta}{\sigma \sin^2 \Delta_{,}} \frac{d\delta}{dt} \left(\frac{\cdot 683 K_2}{\cos 2(\kappa - \mu)} - M \tan^2 \Delta_{,} \right) \sin 2(\mu - \phi) \\ \delta t &= -1^{\text{h}}{\cdot}977 \frac{\sin \delta \cos \delta}{\sigma H \sin^2 \Delta_{,}} \frac{d\delta}{dt} \left(\frac{\cdot 683 K_2}{\cos 2(\kappa - \mu)} - M \tan^2 \Delta_{,} \right) \cos 2(\mu - \phi)\end{aligned}\right\} \quad \text{...(43)}$$

Moon's Parallax, Sched. IV.

This is of type (35), and gives

$$\left.\begin{aligned}\delta H &= (P' - 1) \frac{N_0 \cos 2\nu - L_0 \cos 2\lambda}{e \cos 2\epsilon} \cos 2(\epsilon - \phi) \\ \delta t &= 1^{\text{h}}{\cdot}977 (P' - 1) \frac{N_0 \cos 2\nu - L_0 \cos 2\lambda}{He \cos 2\epsilon} \sin 2(\epsilon - \phi)\end{aligned}\right\} \quad \text{.........(44)}$$

Sun's Parallax, Sched. IV.

This is of type (37), and gives

$$\left.\begin{aligned}\delta H &= (P_{,} - 1) \frac{T - R}{e_{,}} \cos 2(A - \zeta + \phi) \\ \delta t &= -1^{\text{h}}{\cdot}977 (P_{,} - 1) \frac{T - R}{e_{,} H} \sin 2(A - \zeta + \phi)\end{aligned}\right\} \quad \text{.........(45)}$$

Change of Moon's Parallax, Sched. IV.

This is of type (36), and gives

$$\left.\begin{aligned}\delta H &= \frac{-1}{\sigma - \varpi} \frac{dP}{dt} \left(4M_0 - \frac{N_0 \sec 2(\mu - \nu) + L_0 \sec 2(\lambda - \mu)}{e} \right) \sin 2(\mu - \phi) \\ \delta t &= \frac{1^{\text{h}}{\cdot}977}{(\sigma - \varpi) H} \frac{dP}{dt} \left(4M_0 - \frac{N_0 \sec 2(\mu - \nu) + L_0 \sec 2(\lambda - \mu)}{e} \right) \cos 2(\mu - \phi)\end{aligned}\right\}$$

.........(46)

The lunar corrections involving sines are small compared with those involving cosines.

To evaluate these corrections we must compute τ from ϕ reduced to time at 14°·49 per hour.

In the right ascensional terms, $d\alpha/dt$ and σ are to be expressed in degrees per hour. $d\alpha/dt$ is the hourly change of ☽'s R.A. at time of ☽'s transit, and $d\alpha_{,}/dt$ is the hourly change of ☉'s R.A. at time of ☽'s transit.

Similarly, $d\delta/dt$ is to be expressed in degrees, if σ be in degrees.

δ', P' can be found for the antecedent moments, $57°\cdot3 \tan 2(\kappa - \mu)/2\sigma$, and $57°\cdot3 \tan 2(\mu - \nu)/(\sigma - \varpi)$, before the time τ.

§ 7. *The Diurnal Tides.*

I shall not consider these tides so completely as the semidiurnal ones, although the method indicated would serve for an accurate discussion, if it be desired to make one.

The important diurnal tides are K_1, O, P.

From Schedule B. ii., 1883, we have

$$(O) = \frac{\sin I \cos^2 \frac{1}{2}I}{\sin \omega \cos^2 \frac{1}{2}\omega \cos^4 \frac{1}{2}i} . M' \cos [t + h - \nu_0 - 2(s - \xi) + \tfrac{1}{2}\pi - \mu']$$

By (9) the coefficient is $\sin 2\Delta / \sin 2\Delta_{,}$, and we shall put, as in the case of the semidiurnal tides,

$$M_0' = \frac{\sin 2\Delta}{\sin 2\Delta_{,}} M'$$

Then, since $t + h = \psi + \alpha$

$$(O) = M_0' \cos [\psi + (\alpha + \nu_0) - 2(s - \xi) + \tfrac{1}{2}\pi - \mu']$$
$$= M_0' \cos \Omega, \text{ for brevity} \quad \ldots\ldots\ldots\ldots\ldots\ldots\ldots\ldots\ldots\ldots (47)$$

Again, from Schedule C., 1883,

$$(P) = S' \cos [t - h + \tfrac{1}{2}\pi - \zeta']$$

Then let $$\chi = 2(s - h) + \nu_0 - 2\xi - \zeta' + \mu'$$

and we have $$(P) = S' \cos (\Omega + \chi) \quad \ldots\ldots\ldots\ldots\ldots\ldots\ldots\ldots (48)$$

Whence $$(O) + (P) = [M_0' + S' \cos \chi] \cos \Omega - S' \sin \chi \sin \Omega$$

If we put $$H' \cos (\mu' - \phi') = M_0' + S' \cos \chi$$
$$H' \sin (\mu' - \phi') = S' \sin \chi$$

$$(O) + (P) = H' \cos (\Omega + \mu' - \phi')$$
$$= H' \cos [\psi + (\alpha - \nu_0) - 2(s - \xi) + \tfrac{1}{2}\pi - \phi'] \quad \ldots\ldots (49)$$

Where $$H' = \sqrt{\{M_0'^2 + S'^2 + 2M_0'S' \cos\chi\}}$$

and $$\tan(\mu' - \phi') = \frac{S' \sin\chi}{M_0' + S' \cos\chi} \qquad \text{.................(50)}$$

The rate of increase of the angle χ is twice the difference of the mean motions of the moon and sun, but it would be more correct to substitute for s and h the true longitudes of the bodies. It follows from (50) that ϕ' has a fortnightly inequality like that of ϕ.

ψ is very nearly equal to T, and where the diurnal tide is not very large we may with sufficient approximation put

$$(\alpha - \nu_0) - 2(s - \xi) = -(s - \xi)$$

So that with fair approximation

$$(\mathrm{O}) + (\mathrm{P}) = H' \cos[\mathrm{T} - (s - \xi) + \tfrac{1}{2}\pi - \phi'] \qquad \text{...........(51)}$$

The synthesis of the two parts of the K_1 tide has been performed in the harmonic method (Report, 1883), and we have

$$(\mathrm{K}_1) = \mathrm{f}_1 K_1 \cos(t + h - \nu' - \tfrac{1}{2}\pi - \kappa_1)$$

Then, writing $\mathrm{f}_1 K_1 = K_0$, we have

$$(\mathrm{K}_1) = K_0 \cos(\mathrm{T} + \alpha - \nu' - \tfrac{1}{2}\pi - \kappa_1) \qquad \text{.................(52)}$$

We have next to consider what corrections to the time and height of high and low water are necessary on account of these diurnal tides.

If we have a function

$$h = B + H\cos 2(\mathrm{T} - \phi) + H_1 \cos(n\mathrm{T} - \beta)$$

where n is nearly equal to unity, and H_1 is small compared with H; its maxima and minima are determined by

$$\sin 2(\mathrm{T} - \phi) = -\frac{H_1 n}{2H} \sin(n\mathrm{T} - \beta)$$

If $\mathrm{T} = \mathrm{T}_0$ be the approximate time of maximum, and $\mathrm{T}_0 + \delta\mathrm{T}_0$ the true time, then, since the mean lunar day is 24·84 hours, and the quotient when this is divided by 8π is $0^{\mathrm{h}}{\cdot}988$, we have in mean solar hours,

$$\delta\mathrm{T}_0 = -0^{\mathrm{h}}{\cdot}988 \frac{H_1 n}{H} \sin(n\mathrm{T}_0 - \beta)$$

And the correction to the maximum is

$$\delta H = H_1 \cos(n\mathrm{T}_0 - \beta) \qquad \text{.................(53)}$$

Again if $\mathrm{T} = \mathrm{T}_1$ be the approximate time of minimum, and $\mathrm{T}_1 + \delta\mathrm{T}_1$ the true time, then

$$\delta\mathrm{T}_1 = 0^{\mathrm{h}}{\cdot}988 \frac{H_1 n}{H} \sin(n\mathrm{T}_1 - \beta)$$

And the correction to the minimum is

$$\delta H = H_1 \cos(n\mathrm{T}_1 - \beta) \qquad \text{.................(54)}$$

In the case of the correction due to (O) + (P), n is approximately $1-\dfrac{\sigma}{\gamma-\sigma}$, and for the correction due to K_1, n is approximately $1+\dfrac{\sigma}{\gamma-\sigma}$.

§ 8. *Direct Synthesis of the Harmonic Expression for the Tide.*

The scope of the preceding investigation is the establishment of the nature of the connection between the older treatment of tidal observation and the harmonic method. It appears, however, that if the results of harmonic analysis are to be applied to the numerical computation of a tide-table, then a direct synthesis of the harmonic form may be preferable to a transformation to moon's transit, declinations, and parallaxes.

Semidiurnal Tides.

We shall now suppose that M_0 is the height of the M_2 tide, augmented or diminished by the factor for the particular year of observation, according to the longitude of the moon's node, and similarly K_0 generically for the augmented or diminished height of any of the smaller tides. As before, let 2μ, 2ζ be the lags of M_2, S_2; and 2κ, generically, the lag of the K tide.

Let
$$\theta = t + h - s - \nu_0 + \xi$$

Then θ might be defined as the mean moon's hour-angle, the mean moon coinciding with the true, not at Aries, but at the intersection.

Let the argument of the K tide be written generically $2[\theta + u - \kappa]$.

Then

$$h_2 = M_0 \cos 2(\theta-\mu) + S\cos 2[\theta + s - h + \nu_0 - \xi - \zeta] + K_0 \cos 2[\theta + u - \kappa] \quad \text{.........(55)}$$

If we write
$$\zeta_0 = \zeta - \nu_0 + \xi$$

and
$$H\cos 2(\mu-\phi) = M_0 + S\cos 2[s - h - \zeta_0 + \mu]$$
$$H\sin 2(\mu-\phi) = S\sin 2[s - h - \zeta_0 + \mu]$$

the first two terms of (55) are united into

$$H\cos 2(\theta-\phi) \quad \text{..............................(56)}$$

with fortnightly inequality of time and height defined by

$$\left.\begin{aligned} \tan 2(\mu-\phi) &= \frac{S\sin 2(s-h-\zeta_0+\mu)}{M_0 + S\cos 2(s-h-\zeta_0+\mu)} \\ H &= \sqrt{[M_0^2 + S^2 + 2MS\cos 2(s-h-\zeta_0+\mu)]} \end{aligned}\right\} \quad \text{.........(57)}$$

The amount of the fortnightly inequality depends to a small extent on the longitude of the moon's node, since ζ_0 and M_0 are both functions of that longitude.

For the K tide we have

$$K_0 \cos 2(\theta + u - \kappa) = K_0 \cos 2(u - \kappa + \phi) \cos 2(\theta - \phi) - K_0 \sin 2(u - \kappa + \phi) \sin 2(\theta - \phi)$$

Hence

$$\left.\begin{aligned} \delta H &= K_0 \cos 2(u - \kappa + \phi) \\ \delta\phi &= -\frac{K_0}{2H} \sin 2(u - \kappa + \phi) \end{aligned}\right\} \quad \text{.....................(58)}$$

It is easy to find from the *Nautical Almanac* the exact time of mean moon's transit on any day, and then the successive additions of $12^h{\cdot}420601$ or $12^h\ 25^m\ 14^s{\cdot}16$ give the successive upper and lower transits. The successive values of $2(s-h)$ may be easily found by successively adding $12^\circ{\cdot}618036$ to the initial value at the time of the first transit of the mean moon, and ϕ may be obtained from the table of the fortnightly inequality for each value of $2(s-h)$.

The function u is slowly varying, e.g., for the K_2 tide

$$2u = 2(s - \xi) + 2(\nu_0 - \nu'')$$

and the increment of argument for each $12^h{\cdot}420601$ may be easily computed once for all, and added to the initial value.

In the case of the diurnal tides it will probably be most convenient to apply corrections for each independently, following the same lines as those sketched out in § 5.

The corrections for the over-tides M_4, S_4, &c., and for the terdiurnal and quaterdiurnal compound tides, would also require special treatment, which may easily be devised.

At ports, where the diurnal tide is nearly as large or larger than the semidiurnal, special methods will be necessary.

Although the treatment in terms of mean longitudes makes the corrections larger than in the other method, yet it appears that the computation of a tide-table may thus be made easier, with less reference to ephemerides, and with amply sufficient accuracy.

[This subject is considered hereafter in the paper (7) on 'Tidal Prediction.']

3.

DATUM LEVELS; THE TREATMENT OF A SHORT SERIES OF TIDAL OBSERVATIONS AND ON TIDAL PREDICTION.

[Report of the Committee consisting of Professor G. H. DARWIN, Sir W. THOMSON, and Major BAIRD, for the purpose of preparing instructions for the practical work of Tidal Observation; and Fourth Report of the Committee consisting of Professors G. H. DARWIN and J. C. ADAMS, for the Harmonic Analysis of Tidal Observations. Drawn up by G. H. DARWIN. *British Association Report for* 1886, pp. 40—58.]

I. RECORD OF WORK DURING THE PAST YEAR. DATUM LEVELS.

. . . .

IN the course of the Indian tidal operations a discussion has arisen as to the determination of a datum level for tide-tables. The custom of the Admiralty is to refer the tides to 'the mean low-water mark of ordinary spring tides.' This datum has not a precise scientific meaning, but, at ports where there are but few observations, has been derived from a mean of the spring-tides available. At some of the Indian ports this datum has been found by taking the mean of all spring-tides on the tide diagram for a year, with the exception of those which occur when the moon is near perigee. The diurnal tides enter into the determination of the datum in an undefined manner. It follows that two determinations of this datum level, both equally defensible, might differ sensibly from one another.

A datum level should be sufficiently low to obviate the frequent occurrence of negative entries in a tide-table, and it should be rigorously determinable from tidal theory. It is now proposed to adopt as the datum level at any new ports in India, for which tide-tables are to be issued, a datum to be called 'the Indian spring low-water mark,' and which is to be below mean sea-level by the sum of the mean semi-ranges of the tides M_2, S_2, K_1, O; or, in the notation used below,

$$H_m + H_s + H' + H_o$$

below mean water mark.

This datum is found to agree pretty nearly with the Admiralty datum, but is usually a few inches lower. The definition is not founded on any precise theoretical considerations, but it satisfies the conditions of a good datum, and is precisely referable to tidal theory.

If, when further observations are made, it is found that the values of the several H's require correction, it is not proposed that the datum level shall be altered accordingly, but when once fixed it is to be always adhered to.

II. On the Treatment of a Short Series of Tidal Observations and on Tidal Prediction.

§ 1. *Harmonic Analysis.*

Having been asked to write an article on the tides in a new edition of the *Admiralty Scientific Manual* [see Paper 4 below], now in the press, I thought it would be useful to show how harmonic analysis might be applied to the reduction of a short series of tidal observations, such as might be made when a ship lies for a fortnight or a month in a port.

The process of harmonic analysis, as applicable to a year of continuous observation, needs some modification for a short series, and as it was not possible to explain the reasons for the rules laid down within the limits of the article, it seems desirable to place on record an explanation of the instructions given.

The observations to be treated are supposed to consist of hourly observations extending over a fortnight or a month. In the reduction of a long series of observations the various tides are disentangled from one another by means of an appropriate grouping of the hourly observations. When, however, the series is short, the method of grouping is not sufficient in all cases.

With the amount of observation supposed to be available, a determination of the elliptic tides was not possible, and it was therefore proposed to consider only the tides M_2, S_2, K_2, K_1, O, P—that is to say, the principal lunar, solar, and luni-solar semidiurnal tides, and the luni-solar, lunar, and solar diurnal tides. The luni-solar and solar semidiurnal tides have, however, so nearly the same speed that we cannot hope for a direct separation of them by the grouping of the hourly values, and we must have recourse to theory for completing the process; and the like is true of the luni-solar and solar diurnal tides.

Also, the tides K_1 and P have very nearly half the speed of S_2; hence the diurnal tides K_1 and P will appear together as the diurnal constituent, whilst S_2 and K_2 will appear as the semidiurnal constituent, from the harmonic analysis of the same table of entries.

It thus appears that three different harmonic analyses will suffice to determine the six tides, viz.:—

First, an analysis for M_2; second, an analysis for O; third, an analysis for S_2, K_2, K_1, P.

The rules therefore begin with instructions for drawing up three schedules, to be called M, O, S, for the entry of hourly tide-heights. Each schedule consists of twenty-four hour columns, and a number of rows for the successive days. In M and O certain squares are marked, in which two successive hourly entries are to be put. The instructions for drawing up the schedules are simply rules for preparing part of the first page of the series M, O, S of the computation forms for a year of observation.

In order to minimise the vitiation of the results derived from the M sheet by the S_2 tide, and *vice-versâ*, and similarly to minimise the vitiation of the results from the O sheet by the K_1 tide, it is important to choose the proper number of entries in each of the three sheets.

It was shown in Section III. of the Tidal Report to the British Association for 1885 [Paper 2] how these periods were to be determined. The equation by which we find how many rows to take to minimise the effect of the S_2 tide on the M_2 tide is there shown to be

$$1{\cdot}0158958q = 14{\cdot}4920521r$$

If $r = 1$, $q = 14{\cdot}26$; and if $r = 2$, $q = 28{\cdot}5$.

For a reason similar to that given in 1885 we conclude that, in analysing about a fortnight of observation we must have 14 rows of values on the M sheet, and for a month's observation 29 rows of values.

Similarly, to minimise the effect of the M_2 tide on the S_2 tide the equation is

$$1{\cdot}0158958q = 15r$$

If $r = 1$, $q = 14{\cdot}76$; and if $r = 2$, $q = 29{\cdot}5$.

Whence we must have 15 rows of values on the S sheet for a fortnight's observation, and 30 rows of values for a month's observation.

These two rules are simply a statement that on the M and S sheets we are to take a period equal to the interval from spring-tide to spring-tide, or twice that period.

Similarly, to minimise the effect of the K_1 tide on the O tide, the equation is

$$1\cdot0980330q = 13\cdot9430356r$$

If $r = 1$, $q = 12\cdot69$; and if $r = 2$, $q = 25\cdot38$.

Whence we must have 13 rows of values on the O sheet for a fortnight's observation, and 25 rows for a month's observation.

Lastly, to minimise the effect of the O tide on the K_1 tide, the equation is

$$1\cdot0980330q = 15\cdot0410686r$$

If $r = 1$, $q = 13\cdot70$; and if $r = 2$, $q = 26\cdot4$.

Hence, in using the numbers on the S sheet for determining the diurnal tides, we must use 14 rows of values for a fortnight's observation, and 26 rows for a month's observation.

Thus, on the S sheet we use more rows for the semidiurnal tides than for the diurnal—namely, one more for a fortnight and three more for a month.

The rules for drawing up the computation forms then specify, in accordance with the above results, where the entries are to stop on the three sheets, and give directions for the dual use of the S sheet, according as it is for finding semidiurnal or diurnal tides.

When the entries have been made, the twenty-four columns on each sheet are summed, and each is divided by the number of entries in the column. On the S sheet there are two sets of sums and divisions, one with and the other without the additional row or rows.

The three sheets thus provide us with four sets of twenty-four mean hourly values; the M sheet corresponds with mean lunar time, the hour being $15 \div 14\cdot49$ of a mean solar hour; both the means on the S sheet correspond with mean solar time; and the O sheet corresponds with a special time, in which the hour is $15 \div 13\cdot94$ of a mean solar hour.

The four sets of means are then submitted to harmonic analysis: the semidiurnal components are only evaluated on the M sheet; the diurnal components are evaluated from the shorter series on S, and the semidiurnal from the longer series; and the diurnal components from the O sheet. We may also evaluate the quaterdiurnal components from the M and S sheets.

It might, perhaps, be useful to evaluate the diurnal component on the M sheet, for if it does not come out small it is certain that the amount of observations analysed is not sufficient to give satisfactory results.

In the article the harmonic analysis is arranged according to a rule devised by General [Sir Richard] Strachey*, which is less laborious than that usually employed, and which is sufficiently accurate for the purpose.

§ 2. *On the Notation employed.*

It will be convenient to collect together the definitions of the principal symbols employed in this paper.

The mean semi-range and angle of lagging of each of the harmonic constituent tides have, in the Tidal Report for 1883, been denoted generically by H, κ; but when several of the H's and κ's occur in the same algebraic expression it is necessary to distinguish between them. The tides to which we shall refer are M_2, S_2, N, L, T, R, O, P, and K_2, K_1; the H and κ for the first eight of these will be distinguished by writing the suffix letters $_m$, $_s$, $_n$, &c., *e.g.*, H_m, κ_m for the M_2 tide. With regard to the K tides, we may put H'', κ'', and H', κ'.

Again, the factors of augmentation f (functions of longitude of moon's node), as applicable to the several tides, will be denoted thus:—for M_2, N, L, simply f; for K_2, K_1, f'', f' respectively; for O, f_o.

The K_2, K_1 tides take their origin jointly from the moon and sun, and it will be necessary in computing the tide-table to separate the lunar from the solar portion of K_2. Now, the ratio of the lunar to the solar tide-generating force is such that ·683H'' is the lunar portion and ·317H'' is the solar portion of H''.

In the Report of 1885 [Paper 2] a slightly different notation was employed for the H's and κ's, but it is easy to see how the results of that Report are to be transformed into the present notation.

As in the Report of 1883 [Paper 1] we write t, h, s for local mean solar hour-angle, sun's and moon's mean longitude, and ν, ξ, ν', $2\nu''$ for functions of the longitude of moon's node depending on the intersection of the equator with the lunar orbit; also $\gamma - \eta$, η, σ, ϖ are the hourly increments of t, h, s and longitude of moon's perigee, and e, $e_{,}$ the eccentricities of lunar and solar orbits.

Let p, $p_{,}$ denote the cubes of the ratios of the moon's and sun's parallaxes to their mean parallaxes; δ, $\delta_{,}$ the moon's and sun's declinations; p' the value of p at a time $\tan(\kappa_m - \kappa_n)/(\sigma - \varpi)$, or $105^{h}{\cdot}3 \tan(\kappa_m - \kappa_n)$ earlier than t; δ' the moon's declination at a time $\tan(\kappa'' - \kappa_m)/2\sigma$ or $52^{h}{\cdot}2 \tan(\kappa'' - \kappa_m)$ earlier than t.

Let P, $P_{,}$, P' be the cube roots of p, $p_{,}$, p'.

* *Proc. Roy. Soc.* Vol. XLII. (1887), p. 61.

Let Δ, $\Delta_{\prime}$ be declinations such that $\cos^2\Delta$, $\cos^2\Delta_{\prime}$ are respectively the mean values of $\cos^2\delta$, $\cos^2\delta_{\prime}$: obviously Δ has a small inequality with the longitude of the moon's node.

Let ϵ be an auxiliary angle defined by

$$\tan\epsilon = \frac{H_n \sin\kappa_n - H_l \sin\kappa_l}{H_n \cos\kappa_n - H_l \cos\kappa_l}$$

Lastly, let ψ, $\psi_{\prime}$ be the moon's and sun's local hour angles.

§ 3. *The Reduction of the Results of Harmonic Analysis**.

We now suppose the harmonic analysis of the hourly means on the three sheets M, O, S completed.

The deduction of H_m, κ_m and H_o, κ_o from the M and O sheets follows exactly the same rules as in a long series of observations, and the reader is referred to the Report of 1883 [Paper 1] for an explanation.

With regard to the S sheet, the results of the harmonic analysis do not separate the S_2 tide from the K_2 tide, nor the K_1 tide from the P tide, and we have to employ theoretical considerations for effecting the separation.

The semidiurnal tides will be taken first.

The solar tide, as derived from a short series of observations, is of course affected by the sun's parallax, and as the sun changes his parallax slowly, the solar tide will follow the equilibrium law and vary as the cube of the sun's parallax. Thus the height of the purely solar semidiurnal tide as derived from our short series of observations will be $p_{\prime}H_s$ instead of H_s, and this will be fused with the luni-solar tide K_2.

The schedules of the Report of 1883 thus show that we shall have as the expression for this tide, compounded of S_2 (with parallactic inequality) and K_2,

$$h_2 = p_{\prime}H_s \cos(2t - \kappa_s) + f''H'' \cos(2t + 2h - 2\nu'' - \kappa'') \quad\text{.........(1)}$$

The theoretical ratio of H'' to H_s is (see Schedule E, 1883) that of ·12662 to ·46531, or 1 to 3·67; and the tides having nearly the same speed, we may assume $\kappa'' = \kappa_s$.

Hence:

$$h_2 = H_s \left[p_{\prime} \cos(2t - \kappa_s) + \frac{f''}{3{\cdot}67} \cos(2t + 2h - 2\nu'' - \kappa_s) \right]$$

Again, the schedules of the Report of 1883 show that we shall have as the expression for the tide which is compounded of K_1 and P

$$h_1 = f'H' \cos(t + h - \nu' - \tfrac{1}{2}\pi - \kappa') + H_p \cos(t - h + \tfrac{1}{2}\pi - \kappa_p) \quad\text{......(5)}$$

* [Certain mistakes in this section have been corrected. This has involved the omission of certain equations originally numbered (2), (3), (4). The portion of this section which has been rewritten is enclosed in square brackets on pp. 103—4.]

The theoretical ratio of H_p to H' is (see Sched. E, 1883) that of ·19317 to ·58385, or 1 to 3, and the tides having nearly the same speed, we may assume $\kappa_p = \kappa_s$. Hence:

$$h_1 = H' \{f' \cos(t + h - \nu' - \tfrac{1}{2}\pi - \kappa') - \tfrac{1}{3}\cos[t + h - \nu' - \tfrac{1}{2}\pi - \kappa' - (2h - \nu')]\} \quad \ldots\ldots\ldots\ldots(6)$$

[The processes employed in the harmonic analysis are nearly the same for the semidiurnal and for the diurnal tides. We find the mean height of water at each of the 24 hours of mean solar time, as estimated over a succession of days, and then submit the 24 mean heights to harmonic analysis. The algebraic details of the process will be found in Paper 6, 'On an apparatus for facilitating the reduction of tidal observations,' § 5; and I shall here only give the results. It must, however, be remarked that the present case differs slightly from the case treated in the paper referred to, because we now suppose the analysis for the diurnal tides to embrace 27 days or 14 days, instead of 30 or 15, which are the periods adopted for the semidiurnal tides.

For both classes of tide the sun's mean longitude at 0^h of the first day of observation will be denoted by h_0.

The formulæ for semidiurnal tides for 30 or for 15 days may be written down together by means of a simple alternative notation; and the formulæ for diurnal tides for 27 or for 14 days may be written in a similar manner.

For semidiurnal tides the harmonic analysis of the 24 hourly means gives us the two components A_2 and B_2, and we may arrange the result in terms of R_s and ζ_s, where

$$A_2 = R_s \cos\zeta_s, \quad B_2 = R_s \sin\zeta_s$$

We now put
$$\tan\psi = \frac{f'' \sin(2h_0 - 2\nu'' + \alpha)}{3{\cdot}67 p_{\prime} \mathfrak{F}_2 + f'' \cos(2h_0 - 2\nu'' + \alpha)}$$

where, for $\left.\begin{matrix}30\\15\end{matrix}\right\}$ days,

$$\alpha = \left.\begin{matrix}29^\circ{\cdot}53\\14^\circ{\cdot}76\end{matrix}\right\} \text{ and } \log \mathfrak{F}_2 = \left.\begin{matrix}{\cdot}01945\\{\cdot}00483\end{matrix}\right\}; \text{ also } 3{\cdot}67\mathfrak{F}_2 = \left.\begin{matrix}3{\cdot}84\\3{\cdot}71\end{matrix}\right\}$$

The investigation then shows that

$$\kappa_s = \zeta_s + \psi, \quad \kappa'' = \kappa_s$$

$$H_s = \frac{3{\cdot}67 \mathfrak{F}_2 R_s \cos\psi}{3{\cdot}67 p_{\prime} \mathfrak{F}_2 + f'' \cos(2h_0 - 2\nu'' + \alpha)}, \quad H'' = \frac{1}{3{\cdot}67} H_s$$

Turning now to the diurnal tides, the harmonic analysis gives us

$$A_1 = R' \cos\zeta', \quad B_1 = R' \sin\zeta'$$

We then put $$\tan\phi = \frac{\sin(2h_0 - \nu' + \beta)}{3f' - \cos(2h_0 - \nu' + \beta)}$$

where, for $\left.\begin{matrix}27\\14\end{matrix}\right\}$ days, $$\beta = \left.\begin{matrix}26^\circ\cdot57\\13^\circ\cdot76\end{matrix}\right\}$$

The investigation then shows that

$$\kappa' = \zeta' + (h_0 - \tfrac{1}{2}\pi - \nu') + \phi + \left.\begin{matrix}13^\circ\cdot29\\6^\circ\cdot88\end{matrix}\right\}, \quad \kappa_p = \kappa'$$

and $$\mathrm{H}' = \frac{3\mathfrak{F}_1 \mathrm{R}' \cos\phi}{3f' - \cos(2h_0 - \nu' + \beta)}, \quad \mathrm{H}_p = \tfrac{1}{3}\mathrm{H}' \quad \ldots\ldots\ldots\ldots\ldots(7)$$

where, for $\left.\begin{matrix}27\\14\end{matrix}\right\}$ days, $\log\mathfrak{F}_1 = \left.\begin{matrix}\cdot00391\\\cdot00105\end{matrix}\right\}$ and $3\mathfrak{F}_1 = \left.\begin{matrix}3\cdot027\\3\cdot007\end{matrix}\right\}$

These are the formulæ which should have been used in the *Admiralty Scientific Manual*, and are used in Paper 4, as given in the present volume. There was a mistake in the *Manual* as published, and although pains have been taken to insert errata in as many copies as possible, it is certain that several uncorrected copies must be in circulation. It is fortunate that the mistake was such as not to make a large difference in the result.]

§ 4. *Computation of a Tide-table. Semidiurnal Tides.*

The computation of a tide-table from tidal constants which do not contain the elliptic tides N and L presents some difficulty, because the total neglect of these tides would make the results very considerably in error. On this account it was found necessary to use the moon's hour-angle, declination, and parallax in making the computations.

We shall begin by considering only the semidiurnal tide.

In the Tidal Report of 1885 [Paper 2] it was shown how the expression for this tide in the harmonic notation may be transformed so as to involve hour-angles, declinations and parallaxes, instead of mean longitudes and eccentricities of orbits.

The formula (27) of the Report of 1885 for the total semidiurnal tide, when written in the notation of § 2, is

$$h_2 = \frac{\cos^2\Delta}{\cos^2\Delta_{,}}\mathrm{H}_m\cos(2\psi - \kappa_m) + \mathrm{H}_s\cos(2\psi_{,} - \kappa_s)$$

$$+ \frac{\cos^2\delta' - \cos^2\Delta}{\sin^2\Delta_{,}}\cdot683\mathrm{H}''\cos(2\psi - \kappa'')$$

$$+ \frac{\cos^2\delta_{,} - \cos^2\Delta_{,}}{\sin^2\Delta_{,}}\cdot317\mathrm{H}''\cos(2\psi_{,} - \kappa'')$$

$$-\frac{\sin\delta\cos\delta}{\sigma\sin^2\Delta_,}\frac{d\delta}{dt}\left[\frac{\cdot 683\mathrm{H}''}{\cos(\kappa''-\kappa_m)}-\mathrm{H}_m\tan^2\Delta_,\right]\sin(2\psi-\kappa_m)$$

$$+\frac{\cos^2\Delta}{\cos^2\Delta_,}(\mathrm{P}'-1)\frac{\mathrm{H}_n\cos\kappa_n-\mathrm{H}_l\cos\kappa_l}{e\cos\epsilon}\cos(2\psi-\epsilon)$$

$$+(\mathrm{P}_,-1)\frac{\mathrm{H}_t-\mathrm{H}_r}{e_,}\cos(2\psi_,-\kappa_s)$$

$$+\frac{\cos^2\Delta}{\cos^2\Delta_,}\frac{d\mathrm{P}/dt}{\sigma-\varpi}\left[4\mathrm{H}_m-\frac{\mathrm{H}_n\sec(\kappa_m-\kappa_n)+\mathrm{H}_l\sec(\kappa_l-\kappa_m)}{e}\right]\sin(2\psi-\kappa_m)$$

$$\ldots\ldots\ldots\ldots(8)$$

where $$\tan\epsilon=\frac{\mathrm{H}_n\sin\kappa_n-\mathrm{H}_l\sin\kappa_l}{\mathrm{H}_n\cos\kappa_n-\mathrm{H}_l\cos\kappa_l}$$

We shall now proceed to simplify this.

In the first place, the terms depending on $d\delta/dt$ and $d\mathrm{P}/dt$ are certainly small, and may be neglected.

Then let

$$\mathrm{M}=\frac{\cos^2\Delta}{\cos^2\Delta_,}\mathrm{H}_m+\frac{\cos^2\delta'-\cos^2\Delta}{\sin^2\Delta_,}\cdot 683\mathrm{H}''\cos(\kappa''-\kappa_m)$$

$$+\frac{\cos^2\Delta}{\cos^2\Delta_,}(\mathrm{P}'-1)\frac{\mathrm{H}_n\cos\kappa_n-\mathrm{H}_l\cos\kappa_l}{e\cos\epsilon}\cos(\epsilon-\kappa_m)$$

$$\mu=\kappa_m+\frac{\cos^2\delta'-\cos^2\Delta}{\sin^2\Delta_,}\cdot 683\frac{\mathrm{H}''}{\mathrm{H}_m}\sin(\kappa''-\kappa_m)$$

$$+\frac{\cos^2\Delta}{\cos^2\Delta_,}(\mathrm{P}'-1)\frac{\mathrm{H}_n\cos\kappa_n-\mathrm{H}_l\cos\kappa_l}{\mathrm{H}_m e\cos\epsilon}\sin(\epsilon-\kappa_m)$$

$$\mathrm{M}_,=\mathrm{H}_s+\frac{\cos^2\delta_,-\cos^2\Delta_,}{\sin^2\Delta_,}\cdot 317\mathrm{H}''+(\mathrm{P}_,-1)\frac{\mathrm{H}_t-\mathrm{H}_r}{e_,}$$

$$\mu_,=\kappa_s\quad\ldots\ldots\ldots\ldots(9)$$

Now observation and theory agree in showing that κ'' is very nearly equal to κ_s; hence we are justified in substituting κ_s for κ'' in the small solar declinational term of (8) involving $\cdot 317\mathrm{H}''$.

This being so, (8) becomes

$$h_2=\mathrm{M}\cos(2\psi-\mu)+\mathrm{M}_,\cos(2\psi_,-\mu_,)\quad\ldots\ldots\ldots\ldots(10)$$

In the equilibrium theory each H is proportional to the corresponding term in the harmonically developed potential. This proportionality holds nearly between tides of nearly the same speed; hence in the solar tides we may assume (see Sched. B, 1883, and note that $\cot^2\Delta_,=\frac{1}{2}\cot^2\frac{1}{2}\omega$) that,

$$\frac{\cos^2\Delta_,}{\sin^2\Delta_,}\cdot 317\mathrm{H}''=\frac{1}{3e_,}(\mathrm{H}_t-\mathrm{H}_r)=\mathrm{H}_s$$

and $M_,$ reduces to

$$M_, = \frac{\cos^2 \delta_,}{\cos^2 \Delta_,} H_s + 3(P_, - 1) H_s = \frac{\cos^2 \delta_,}{\cos^2 \Delta_,} H_s [1 + 3(P_, - 1)] \text{ nearly}$$

$$= P_,^3 \frac{\cos^2 \delta_,}{\cos^2 \Delta_,} H_s \quad \ldots\ldots\ldots\ldots(11)$$

Now, $\Delta_, = 16^\circ\!\cdot\!36 = 16^\circ\ 22'$, $\sec^2 \Delta_, = 1\!\cdot\!086$, also $P_,^3 = p_,$, and therefore

$$M_, = 1\!\cdot\!086 p_, \cos^2 \delta_, H_s \quad \ldots\ldots\ldots\ldots(12)$$

In a similar way, according to the equilibrium theory, we should have

$$\frac{1}{3e}(H_n - H_l) = H_m$$

Although this proportionality is probably not actually very exact, yet in our supposed ignorance of the lunar elliptic tides we have to assume its truth. Also, we must assume that the two elliptic tides N and L suffer the same retardation, and therefore $\kappa_n = \kappa_l = \epsilon$.

With these assumptions,

$$H_m + (P' - 1) \frac{H_n \cos \kappa_n - H_l \cos \kappa_l}{e \cos \epsilon} \cos (\epsilon - \kappa_m) = H_m [1 + 3(P' - 1)] = H_m P'^3$$

Then, since $\dfrac{\cos^2 \Delta}{\cos^2 \Delta_,} = \mathrm{f}$, and $P'^3 = p'$

we have $$M = \mathrm{f} p' H_m + \frac{\cos^2 \delta' - \cos^2 \Delta}{\sin^2 \Delta_,} \cdot 683 H'' \cos (\kappa'' - \kappa_m)$$

$$\mu = \kappa_m + \frac{\cos^2 \delta' - \cos^2 \Delta}{\sin^2 \Delta_,} \cdot 683 \frac{H''}{H_m} \sin (\kappa'' - \kappa_m) \quad \ldots\ldots\ldots\ldots(13)$$

If we put $$C_1 = \frac{\cdot 683}{2 \sin^2 \Delta_,}, \quad C_2 = \frac{\cdot 683}{2 \sin^2 \Delta_,} \times 57^\circ\!\cdot\!3$$

then $$\log C_1 = \cdot 6344, \quad \log C_2 = 2\!\cdot\!3925$$

and C_1, C_2 are absolute constants for all times and places.

Next, if we put

$$\alpha = C_1 H'' \cos (\kappa'' - \kappa_m), \qquad \beta = C_2 \frac{H''}{H_m} \sin (\kappa'' - \kappa_m)$$

$$A = \alpha \cos 2\Delta, \qquad B = \beta \cos 2\Delta \quad \ldots\ldots\ldots\ldots(14)$$

then obviously α, β are absolute constants for the port, and A and B are nearly constant, for their small variability only depends on the longitude of the moon's node entering through Δ.

Thus we have, from (9), (12), (13), (14),

$$M = \mathrm{f} H_m + (p' - 1) \mathrm{f} H_m + (\alpha \cos 2\delta' - A)$$

$$\mu = \kappa_m + (\beta \cos 2\delta' - B), \text{ expressed in degrees}$$

$$M_, = 1\!\cdot\!086 p_, \cos^2 \delta_, H_s$$

$$\mu_, = \kappa_s \quad \ldots\ldots\ldots\ldots(15)$$

where p′, δ' are the values of p and δ at a time earlier than that corresponding to ψ by 'the age' $52^{\text{h}}\cdot 2 \tan(\kappa'' - \kappa_m)$.

In the article [Paper 4 in this volume] fH_m is called R_m; $(p'-1) fH_m$, the parallactic correction, is called $\delta_1 R_m$; $(\alpha \cos 2\delta' - A)$, the declinational correction, is called $\delta_2 R_m$. Similarly, $\beta \cos 2\delta' - B$, the declinational correction to κ_m, is called $\delta_2 \kappa_m$. Also, $M_,$ is called S.

Thus, with this notation the whole semidiurnal tide is

$$h_2 = (R_m + \delta_1 R_m + \delta_2 R_m) \cos(2\psi - \kappa_m - \delta_2 \kappa_m) + S \cos(2\psi_, - \kappa_s) \ldots(16)$$

The mean rate of increase of ψ is $\gamma - \sigma$, or $14°\cdot 49$ per hour; hence the interval from moon's transit to lunar high water is approximately $\frac{1}{29}(\kappa_m + \delta_2 \kappa_m)$ hours, when κ_m is expressed in degrees. If i be the mean interval, and $\delta_2 i$ its declinational correction,

$$i + \delta_2 i = \tfrac{1}{29}\kappa_m + \tfrac{1}{29}\delta_2 \kappa_m \quad \ldots\ldots\ldots\ldots\ldots\ldots\ldots\ldots(17)$$

Now, let A be twice the apparent time of moon's transit reduced to angle at 15° per hour, or the apparent time reduced at 30° per hour.

Then the excess of the moon's over the sun's R.A. at lunar high water is $\frac{1}{2}A$ plus the increase of the difference of R.A.'s in the interval i. This increase is approximately $\frac{1}{2}\frac{\sigma - \eta}{\gamma - \sigma}\kappa_m$, and at lunar high water the sun's hour-angle is given by

$$2\psi_, = 2\psi + A + \frac{\sigma - \eta}{\gamma - \sigma}\kappa_m \quad \ldots\ldots\ldots\ldots\ldots\ldots\ldots\ldots(18)$$

Since the difference of time between lunar high water and actual high water never exceeds about an hour and a half, if we neglect the separation of the moon from the sun in that time, this relationship also holds at actual luni-solar high water.

Now, let

$$H \cos(\mu - \phi) = M + S \cos\left[A + \frac{\sigma - \eta}{\gamma - \sigma}\kappa_m - \kappa_s + \kappa_m + \delta_2 \kappa_m\right]$$

$$= M + S \cos(A - \kappa_s + \tfrac{30}{29}\kappa_m + \delta_2 \kappa_m)$$

$$H \sin(\mu - \phi) = \qquad S \sin(A - \kappa_s + \tfrac{30}{29}\kappa_m + \delta_2 \kappa_m) \quad \ldots\ldots\ldots\ldots(19)$$

and we have for the whole luni-solar semidiurnal tide

$$h_2 = H \cos(2\psi - \phi) \quad \ldots\ldots\ldots\ldots\ldots\ldots\ldots\ldots\ldots(20)$$

If we put

$$\gamma + \delta_2 \gamma = \kappa_s - \tfrac{30}{29}\kappa_m + \delta_2 \kappa_m$$

$$x = A - (\gamma + \delta_2 \gamma)$$

we have, from (19),

$$\left.\begin{aligned} \tan(\mu - \phi) &= \frac{S \sin x}{M + S \cos x} \\ H^2 &= M^2 + S^2 + 2MS \cos x \end{aligned}\right\} \quad \ldots\ldots\ldots\ldots\ldots\ldots\ldots(21)$$

High water occurs approximately $\dfrac{\phi}{2(\gamma - \sigma)}$ or $\frac{1}{29}\phi$ after moon's transit.

The determination of ϕ and H may be conveniently carried out by a graphical construction. If we take O as a fixed centre, OS as an initial line, and S a point in it such that OS = S, and set off the angle AOM equal to x, and OM equal to M; then OMS is the angle $\mu - \phi$, and SM is the height H.

The angle x increases by 360° from spring-tide to spring-tide, and therefore one revolution in the figure corresponds to 15 days.

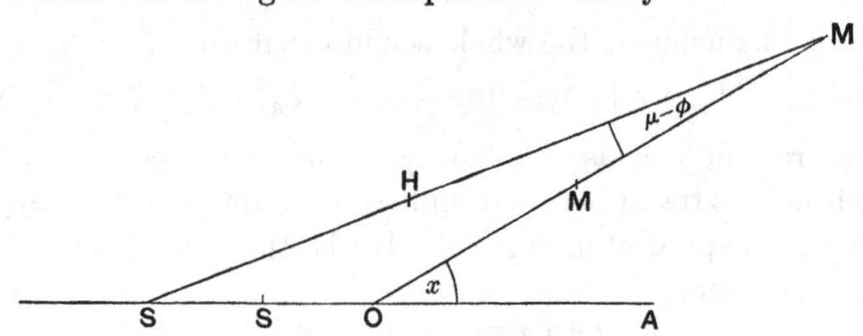

As a very rough approximation, M lies on a circle, but the parallactic and declinational corrections $\delta_1 R_m$ and $\delta_2 R_m$ cause a considerable departure from the circle.

The angle ϕ and the height H are also easily computed numerically.

If cos x is positive, let θ be an auxiliary angle determined by

$$\tan^2 \theta = \frac{S}{M} \cos x$$

and we have

$$\tan(\mu - \phi) = \sin^2 \theta \tan x, \qquad H = S \operatorname{cosec}(\mu - \phi) \sin x$$

If cos x is negative, let θ be an auxiliary angle determined by

$$\sin^2 \theta = -\frac{S}{M} \cos x$$

and we have

$$\tan(\mu - \phi) = \tan^2 \theta \tan x, \qquad H = S \operatorname{cosec}(\mu - \phi) \sin x$$

These formulæ are adapted for logarithmic computation.

§ 5. *Correction for Diurnal Tides.*

The tide-table has to be corrected for the effect of three diurnal tides, designated O, K_1, P.

If we write

$$V_o = t + h - 2s - \nu + 2\xi + \tfrac{1}{2}\pi$$
$$V' = t + h - \nu' - \tfrac{1}{2}\pi$$

then, in accordance with Schedules B of the Report of 1883, the expressions for the three tides are

$$\begin{aligned} O &= f_o H_o \cos(V_o - \kappa_o) \\ K_1 &= f' H' \cos(V' - \kappa') \\ P &= -H_p \cos[V' - \kappa' - (2h - \nu') + (\kappa' - \kappa_p)] \end{aligned} \qquad \text{.........(22)}$$

[The tides K_1 and P change their phases relatively to one another in half a year, and since P is considerably smaller than K_1, we may without serious error attribute to $2h$ a mean value for a short period such as a month. Also we may assume $\kappa_p = \kappa'$.

Hence if $\odot$ denotes the sun's mean longitude at the middle of the month (or other short period), we may with rough approximation write the expression for the P tide in the form

$$P = -H_p \cos [V' - \kappa' - (2\odot - \nu')]$$

Therefore if we put

$$\tan \phi = \frac{\sin(2\odot - \nu')}{3f' - \cos(2\odot - \nu')}, \quad R' = \frac{3f' - \cos(2\odot - \nu')}{3\cos\phi} H'$$

and if further we write $f_o H_o = R_o$ the diurnal tides, reduced to two, are]

$$O = R_o \cos(V_o - \kappa_o)$$

$$K_1 + P = R' \cos(V' - \kappa' - \phi) \quad \text{.....................(23)}$$

ϕ and R′, having a semi-annual inequality, must be recomputed for each month.

Now, suppose that we compute V_o and V′ at the epoch, that is, at the initial noon of the period during which we wish to predict the tides, and with these values put

$$\zeta_o = \kappa_o - V_o \text{ at epoch}$$

$$\zeta' = \kappa' - \phi - V' \text{ at epoch}$$

then the speed of V_o is $\gamma - 2\sigma$, or 13°·94 per hour, or 360° − 25°·37 per day; and the speed of V′ is γ, or 15°·04 per hour, or 360°·986 per day. Hence, if t be the mean solar time in hours on the $(n+1)$th day since the epoch,

$$V_o - \kappa_o = 360°n + 13°{\cdot}94t - \zeta_o - 25°{\cdot}37n$$

$$V' + \phi - \kappa' = 360°n + 15°{\cdot}04t - \zeta' + 0°{\cdot}986n$$

Therefore the diurnal tide at the time t hours on the $(n+1)$th day is given approximately by

$$O = R_o \cos [14°t - \zeta_o - 25\tfrac{1}{3}° \times n]$$

$$K_1 + P = R' \cos [15°t - \zeta' + 1° \times n] \quad \text{.......................(24)}$$

If we substitute for t the time of high or low water as computed simply from the semidiurnal tide, it is clear that the sum of these two expressions will give us the diurnal correction for height of tide at high or low water.

If we consider the maximum of a function,

$$A \cos 2n(\mathrm{t}-\alpha) + B \cos n'(\mathrm{t}-\beta)$$

where n is nearly equal to n', we see that the time of maximum is given approximately by $\mathrm{t}=\alpha$, with a correction $\delta\mathrm{t}$ determined from

$$-2An \sin(2n\delta\mathrm{t}) - n'B \sin n'(\mathrm{t}-\beta) = 0$$

or

$$\delta\mathrm{t} = -\frac{180}{4\pi n}\frac{n'B}{nA}\sin n(\mathrm{t}-\beta)$$

In this way we find the corrections to the time of high water from O and $\mathrm{K}_1+\mathrm{P}$; and since $n=\gamma-\sigma$, and $\frac{180}{4\pi n} = 0^{\mathrm{h}}{\cdot}988$, and $\frac{n'}{n} = 1 - \frac{\sigma}{\gamma-\sigma}$ for O, and $1+\frac{\sigma}{\gamma-\sigma}$ for K_1, we have

$$\delta\mathrm{t}_o = -0^{\mathrm{h}}{\cdot}988\left(1-\frac{\sigma}{\gamma-\sigma}\right)\frac{\mathrm{R}_o}{\mathrm{H}}\sin[14^\circ\mathrm{t} - \zeta_o - 25\tfrac{1}{3}^\circ \times n]$$

$$\delta\mathrm{t}' = -0^{\mathrm{h}}{\cdot}988\left(1+\frac{\sigma}{\gamma-\sigma}\right)\frac{\mathrm{R}'}{\mathrm{H}}\sin[15^\circ\mathrm{t} - \zeta' + 1^\circ \times n] \ldots\ldots\ldots (25)$$

where H is the height of the semidiurnal high water.

With sufficient approximation we may write these corrections:

$$\delta\mathrm{t}_o = -1^{\mathrm{h}} \times \frac{\mathrm{R}_o}{\mathrm{H}}\sin[14^\circ\mathrm{t} - \zeta_o - 25\tfrac{1}{3}^\circ \times n]$$

$$\delta\mathrm{t}' = -1^{\mathrm{h}} \times \frac{\mathrm{R}'}{\mathrm{H}}\sin[15^\circ\mathrm{t} - \zeta' + 1^\circ \times n] \ldots\ldots\ldots\ldots\ldots (26)$$

The computations are easily carried out, although the arithmetic is necessarily tedious. Since two places of decimals are generally sufficient for R_o and R', the multiplications by the sines and cosines are very easily made with a Traverse Table.

The successive high and low waters follow one another on the average at $6^{\mathrm{h}}\ 12^{\mathrm{m}}$; now, $14^\circ \times 6{\cdot}2 = 87^\circ$, and $15^\circ \times 6{\cdot}2 = 93^\circ$. Hence, if we compute $14^\circ\mathrm{t} - \zeta_o - 25\tfrac{1}{3}^\circ \times n$ for the first tide on any day, the remaining values are found with sufficient approximation by adding once, twice, thrice 87°; and similarly, in the case of $15^\circ\mathrm{t} - \zeta' + 1^\circ \times n$ we add once, twice, thrice 93°.

[If the diurnal tide is the predominant one, as occurs at some places, this method of correction would of course be insufficient.]

§ 6. *Certain Details in the Computation of the Tide-table.*

It will be well to give some explanatory details concerning the manner of carrying out the computations.

The angle Δ is given by $16^\circ{\cdot}51 + 3^\circ{\cdot}44 \cos ☊ - 0^\circ{\cdot}19 \cos 2☊$, where ☊ is the longitude of the moon's node. It is clear that Δ varies so slowly that it may be regarded as constant for many months, and the same is true of the factors f, f″, f′, f_o, and the small angles ν, ξ, ν', $2\nu''$. Approximate formulæ for these quantities in terms of ☊ were given at the end of the first paper in this volume, and are used in the article in the *Manual* [Paper 4].

To find the cube of the ratio of the sun's parallax to his mean parallax, the following rule is given: Subtract the mean parallax from the parallax, multiply the difference by $19\frac{1}{3}$, read as degrees instead of seconds, look out the sine, and add 1. This rule is founded on the fact that a mean parallax $8''{\cdot}85$ multiplied by $19\frac{1}{3}$ gives $3 \times 57''$, and 57° is the unit angle or radian, whilst the sine of a small angle is equal to the angle in radians*. Similarly, the cube of the ratio of the moon's parallax to her mean parallax is

$$1 + 3 \sin [60 (\text{parx} - \text{mean parx})]$$

That is to say, for the moon: Subtract the mean parallax from the parallax, read as degrees instead of minutes, look out the sine, multiply by 3, and add 1. This rule depends on the fact that the moon's mean parallax in radians is $\frac{1}{60}$.

For the purpose of applying the corrections $\delta_1 R_m$, $\delta_2 R_m$, $\delta_2 \kappa_m$, $\delta_2 i$, $\delta_2 \gamma$, it is most convenient to compute auxiliary tables for each degree of declination of the moon and minute of her parallax, and then the actual corrections are easily applied by interpolation.

These tables serve for the port as long as the longitude of the moon's node is nearly constant, or with rougher approximation for all time.

The declinational and parallactic corrections to high water depend on the moon's declination and parallax at a time anterior to high water by 'the age.' Hence, in order to find these corrections we have to know the time of high water in round numbers. Each high water follows a moon's transit at the port approximately by the interval i. The Greenwich time of the moon's transit at the port is the G.M.T. of moon's transit at Greenwich, less 2 minutes for each hour of E. longitude, less the E. longitude in hours. Then, if we subtract from this 'the age' and add the interval i, we find the G.M.T.'s at which we want the moon's declination and parallax.

* [The mean solar parallax is now taken, in the *Nautical Almanac*, as $8''{\cdot}80$, but the rule still remains sufficiently exact.]

$$\left.\begin{array}{l}\text{Thus, at Port Blair}\\ \text{the G.M.T. at which we}\\ \text{want parx. and decl.}\end{array}\right\} = \left\{\begin{array}{l}\text{G.M.T. of ☽'s}\\ \text{transit at Gr.}\end{array}\right\} - \text{long. corr. for transit } (0^{h}\cdot 2)$$

$$- \text{E. long. of port } (6^{h}\cdot 2) - \text{age of tide } (32^{h}\cdot 6)$$

$$+ \text{mean interval } (9^{h}\cdot 6)$$

$$= \text{G.M.T. of ☽'s tr. at Gr.} - 29^{h}\cdot 4.$$

Thus at Greenwich, on Feb. 1st, 1885, the moon's lower transit was at 2^h, and hence, corresponding to the lower transit at Port Blair of Feb. 1, we require the moon's parallax and declination at 21^h Jan. 30, G.M.T. The parallax at the nearest Greenwich noon or midnight is sufficiently near the truth, and therefore we take the parallax at 0^h Jan. 31, which is $60'\cdot 0$, and the excess above the mean is $3'\cdot 0$, and $1 + 3 \sin 3°$ is 1·157, which is the factor p′. Actually, however, we read off the correction $\delta_1 R_m$ and the other corrections $\delta_2 R_m$, $\delta_2 i$, $\delta_2 \gamma$ straight from the auxiliary tables.

§ 7. *On Tide-tables Computed by the above Method.*

A great deal of arithmetical work was necessary in making trial of the rules devised above and in various modifications of them, and I must record my thanks to Mr Allnutt, who has been indefatigable in working out tide-tables for various ports, and in comparing them with official tables. The whole of the results, to which I now refer, are due to him. The following table exhibits the amount of agreement between a computed table and one obtained by the tide-predicting instrument. It must be borne in mind that the instrument is rigorous in principle, and makes use of far more ample data than are supposed to be available in our computations. The columns headed 'Indian tables' are taken from the official Indian tide-tables. The datum level, however, in those tables is 3·13 ft. below mean water mark, whereas 'Indian spring low-water mark' is 3·55 ft. below the mean. Thus, to convert the heights given in the Indian tables to our datum 0·42 ft. or 5 ins. have been added to all the heights in the official table.

A tide-table was computed for Aden for a fortnight, and the results were found to be somewhat less satisfactory than those in the following table. It must be remarked, however, that the sum of the semi-ranges of the three diurnal tides K_1, O, P is 2·340 ft. and is actually greater than the sum of the semi-ranges of the tides M_2 and S_2, which is 2·265 ft. Thus, at some parts of some lunations the semidiurnal tide is obliterated by the diurnal tide, and there is only one high water and one low water in the day. In this case it is obvious that the approximation, by which we determine semidiurnal high and low water and apply a correction for the diurnal tides, becomes inapplicable. In the greater part of our computed table the concordance is fairly good; but the tide-predicting instrument shows that on each of the days,

7th and 8th February, 1885, there was only one high and low water, whereas our table, of course, gives a double tide as usual. Again, on the 9th February there is an error of 68 minutes in a high water. These discrepancies are to be expected, since the approximate method is here pushed beyond its due limits; and for such a port as Aden special methods of numerical approximation would have to be devised.

TIDE-TABLE FOR PORT BLAIR, 1885.

	Calculated Times	Indian tables Times	Calculated Heights	Indian tables Heights
	h. m.	h. m.	ft.	ft. in.
Feb. 1, H.W. .	11 3 p.m.	11 4 p.m.	7·4	7 2
Feb. 2, L.W. .	5 21 a.m.	5 18 a.m.	0·0	−0 2
H.W. .	11 26 a.m.	11 31 a.m.	6·6	6 5
L.W. .	5 28 p.m.	5 25 p.m.	0·4	0 0
H.W. .	11 39 p.m.	11 43 p.m.	7·1	6 11
Feb. 3, L.W. .	5 56 a.m.	5 56 a.m.	0·2	0 1
H.W. .	0 3 p.m.	0 9 p.m.	6·4	6 3
L.W. .	6 4 p.m.	6 5 p.m.	0·7	0 7
Feb. 4, H.W. .	0 14 a.m.	0 20 a.m.	6·7	6 6
L.W. .	6 31 a.m.	6 33 a.m.	0·5	0 5
H.W. .	0 40 p.m.	0 48 p.m.	6·1	6 0
L.W. .	6 42 p.m.	6 44 p.m.	1·2	1 0
Feb. 5, H.W. .	0 48 a.m.	0 56 a.m.	6·1	5 11
L.W. .	7 5 a.m.	7 9 a.m.	1·0	0 10
H.W. .	1 18 p.m.	1 28 p.m.	5·7	5 7
L.W. .	7 20 p.m.	7 25 p.m.	1·7	1 7
Feb. 6, H.W. .	1 24 a.m.	1 33 a.m.	5·5	5 4
L.W. .	7 41 a.m.	7 45 a.m.	1·5	1 4
H.W. .	2 1 p.m.	2 10 p.m.	5·3	5 2
L.W. .	8 6 p.m.	8 12 p.m.	2·2	2 1
Feb. 7, H.W. .	2 4 a.m.	2 13 a.m.	4·9	4 9
L.W. .	8 23 a.m.	8 25 a.m.	1·9	1 10
H.W. .	2 53 p.m.	2 57 p.m.	4·9	4 10
L.W. .	9 7 p.m.	9 8 p.m.	2·7	2 6
Feb. 8, H.W. .	2 58 a.m.	3 8 a.m.	4·4	4 3
L.W. .	9 20 a.m.	9 24 a.m.	2·4	2 2
H.W. .	4 10 p.m.	4 14 p.m.	4·7	4 7
L.W. .	10 42 p.m.	10 40 p.m.	3·0	2 4
Feb. 9, H.W. .	4 29 a.m.	4 40 a.m.	4·0	3 10
L.W. .	10 46 a.m.	10 57 a.m.	2·6	2 6
H.W. .	5 47 p.m.	5 48 p.m.	4·7	4 7

In a table computed for Amherst the agreement is not quite so good as was to be hoped; the error in heights amounts in two cases in fifteen days to nearly a foot, and in two other cases to three-quarters of an hour in time. It may be remarked, however, that the tides are large at Amherst, having a spring range of 20 ft. and a neap range of 6 ft., that the diurnal tide is considerable, and that the sum of the semi-ranges of the over-tides M_4, S_4 (which we neglect entirely) amounts to 6 inches. It appears also that the tidal constants are somewhat abnormal, for $H'' = \frac{1}{2\cdot5} H_s$ instead of $H'' = \frac{1}{3\cdot67} H_s$, and further $H_p = \frac{1}{3\cdot9} H'$ instead of $H_p = \frac{1}{3} H'$. Under these circumstances it is perhaps not surprising that the discrepancies are as great as they are.

Tables were also computed for Liverpool and West Hartlepool, but no correction was here applied for the diurnal tides. The results were compared with the Admiralty tide-tables for Liverpool and Sunderland. In the case of Liverpool there were four tides in a fortnight in which there was a discrepancy in the times amounting to 12 minutes, and four other tides in which there was a discrepancy of a foot, and one with a discrepancy of 1 ft. 2 ins. It was obvious, however, that the agreement would have been better if the correction for the diurnal tides had been applied. The spring rise of tide at Liverpool is 26 ft.

In the case of Sunderland there were in a fortnight two discrepancies of 15 m., two of 14 m., two of 13 m., two of 12 m., &c. in the times, and in the heights one discrepancy of 3 ins., and four of 2 ins., &c. The spring rise at West Hartlepool is 14 ft.

These two tables are quite as satisfactory as could be expected considering the approximate nature of the methods employed.

Finally, in order to test the methods both of reduction and of prediction, Mr Allnutt took the harmonic constants derived from our analysis of a fortnight of hourly observation at Port Blair, from April 19 to May 2, 1880, and computed therefrom a tide-table for that same fortnight. He then, by interpolation in the observed hourly heights, determined the actual high waters and low waters during that period.

The results of the comparison are exhibited in the table on next page.

If our method had been perfect, of course, the errors should be everywhere zero.

It must be admitted that the agreement is less perfect than might have been hoped. If, however, the calculated and observed tide curves are plotted down graphically side by side, it will be seen that the errors are inconsiderable fractions of the whole intervals of time and heights under consideration.

When we consider the extreme complication of tidal phenomena, together with meteorological perturbation, it is, perhaps, not reasonable to expect any better results from an admittedly approximate method, adapted for all ports, and making use of a very limited number of tidal constants. In devising these rules for reduction and prediction I could find no model to work from,

COMPARISON OF A TIDE-TABLE COMPUTED FROM A FORTNIGHT'S OBSERVATION WITH ACTUALITY.

High Water							Low Water						
Astr. Date 1880	Observed Time	Calc. Time	C – O	Observed Height	Calc. Height	C – O	Astr. Date 1880	Observed Time	Calc. Time	C – O	Observed Height	Calc. Height	C – O
	h.	h.	h.	ft.	ft.	ft.		h.	h.	h.	ft.	ft.	ft.
April 19	4·76	4·76	·00	5·83	6·24	+·41	April 19	11·60	11·48	–·12	3·28	3·07	–·21
	18·29	17·86	–·43	5·59	5·94	+·35		23·92	23·78	–·14	3·71	3·62	–·09
„ 20	6·20	5·91	–·29	6·22	6·35	+·13	„ 20	12·58	12·41	–·17	2·84	2·76	–·08
	19·07	18·82	–·25	6·26	6·46	+·20	„ 21	1·13	·85	–·28	3·31	3·22	–·09
„ 21	7·16	6·88	–·28	6·60	6·61	+·01		13·40	13·20	–·20	2·33	2·38	+·05
	19·68	19·59	–·09	6·93	7·03	+·10	„ 22	1·82	1·72	–·10	2·74	2·74	·00
„ 22	7·80	7·72	–·08	6·97	6·90	–·07		14·12	13·91	–·21	1·76	2·01	+·25
	20·41	20·27	–·14	7·60	7·59	–·01	„ 23	2·73	2·47	–·26	2·27	2·28	+·01
„ 23	8·67	8·46	–·21	7·39	7·20	–·19		14·77	14·61	–·16	1·53	1·69	+·16
	21·06	20·97	–·09	8·22	8·06	–·16	„ 24	3·28	3·22	–·06	1·71	1·89	+·18
„ 24	9·30	9·22	–·08	7·50	7·40	–·10		15·26	15·32	+·06	1·25	1·46	+·21
	21·76	21·68	–·08	8·46	8·40	–·06	„ 25	3·90	3·98	+·08	1·50	1·62	+·12
„ 25	9·99	9·98	–·01	7·41	7·50	+·09		16·03	16·04	+·01	1·21	1·39	+·18
	22·37	22·41	+·04	8·40	8·57	+·17	„ 26	4·68	4·75	+·07	1·37	1·49	+·12
„ 26	10·80	10·76	–·04	7·31	7·44	+·13		16·78	16·80	+·02	1·36	1·49	+·13
	23·20	23·17	–·03	8·18	8·55	+·37	„ 27	5·30	5·57	+·27	1·48	1·55	+·07
„ 27	11·55	11·60	+·05	7·00	7·22	+·22		17·37	17·60	+·23	1·64	·80	+·16
	23·98	23·98	·00	7·89	8·29	+·40	„ 28	6·22	6·44	+·22	1·90	·78	–·12
„ 28	12·36	12·50	+·14	6·50	6·89	+·39		18·23	18·46	+·23	2·24	2·22	–·02
„ 29	0·70	·83	+·13	7·53	7·90	+·37	„ 29	7·20	7·38	+·18	2·43	2·08	–·35
	13·36	13·47	+·11	6·14	6·53	+·39		18·88	19·39	+·51	2·90	2·70	–·20
„ 30	1·64	1·76	+·12	7·29	7·42	+·13	„ 30	8·28	8·38	+·10	2·88	2·41	–·47
	14·17	14·53	+·36	5·86	6·22	+·36		20·00	20·43	+·43	3·52	3·14	–·38
May 1	2·46	2·78	+·32	6·74	6·98	+·24	May 1	9·43	9·49	+·06	3·15	2·65	–·50
	15·50	15·76	+·26	5·63	6·08	+·45		21·66	21·65	–·01	3·89	3·42	–·47
„ 2	3·81	3·92	+·11	6·36	6·65	+·29	„ 2	10·80	10·60	–·20	3·23	2·72	–·41
	17·24	16·98	–·26	5·84	6·17	+·33		23·20	22·94	–·26	3·95	3·45	–·50
„ 3	5·17	5·11	–·06	6·40	6·51	+·11							

and it seems probable that advantageous modifications may be introduced. I spared, however, no pains to reduce the labour of computation. Nearly half the work in forming a short tide-table is preparatory, and would serve for a systematic computation of tables for all time.

III. An Attempt to Detect the 19-Yearly Tide.

If M, E be the moon's and earth's masses; a the earth's mean radius; c the moon's mean distance; ω the obliquity of the ecliptic; i the inclination of the lunar orbit; e the eccentricity of the lunar orbit; ☊ the longitude of the moon's node; and λ the latitude of the port of observation; then the term in the equilibrium tidal theory which is independent of the moon's longitude (see Schedule B, iii., Paper 1, p. 22) is

$$\tfrac{3}{2}\frac{M}{E}\left(\frac{a}{c}\right)^3 a\left(\tfrac{1}{2}-\tfrac{3}{2}\sin^2\lambda\right)\left(1+\tfrac{3}{2}e^2\right)\sin i\cos i\sin\omega\cos\omega$$
$$\left[-\cos ☊ + \tfrac{1}{4}\tan i\tan\omega\cos 2☊\right]$$

Since $\frac{1}{4}\tan i \tan\omega = \cdot 00975$, the second term is negligeable compared with the first.

If we take

$$\frac{M}{E}=\frac{1}{81\cdot 5},\quad \frac{a}{c}=\frac{1}{60\cdot 27},\quad a=21\times 10^6 \text{ feet},\quad i=5^\circ\ 8',\quad \omega=23^\circ\ 28'$$

the expression for this tide is, in British feet,

$$-0\cdot 0579\left(\tfrac{1}{2}-\tfrac{3}{2}\sin^2\lambda\right)\cos ☊$$

Thus, at the poles this tide gives an oscillation of sea-level of 0·695 of an inch, or a total range of $1\frac{2}{5}$ of an inch, and at the equator it is half as great.

In the *Mécanique Céleste* Laplace argues that all the tides of long period (such as the fortnightly tide) must conform nearly to the equilibrium law. I shall adduce arguments elsewhere* which seem to invalidate his conclusion, and to show that in these tides inertia still plays the principal part, so that the oscillations must take place nearly as though the sea were a frictionless fluid.

With a tide, however, of as long a period as nineteen years Laplace's argument must hold good, and hence the equilibrium tide of which the above is the expression must represent an actual oscillation of sea-level, provided that the earth is absolutely rigid. The actual observation of the 19-yearly tide would therefore be a result of the greatest interest for determining the elasticity of the earth's mass.

* See Paper 11 below "On the Tides of Long Period."

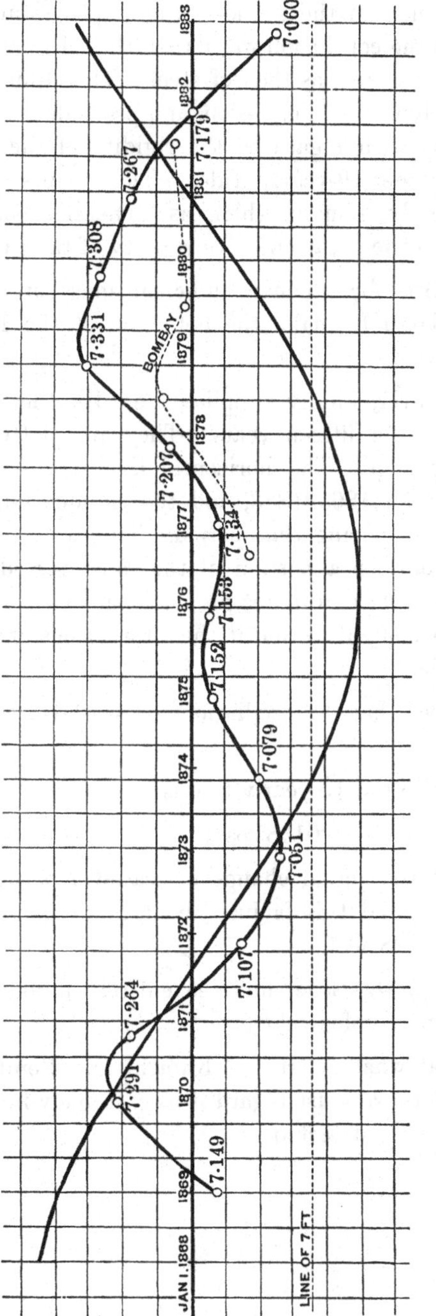

The ordinates of the curve of sines represent $-0{\cdot}25$ ft. cos ☊. The sinuous curve presents the variations of mean water at Karachi. The dotted line is the level of 7 ft. above the zero of the tide-gauge, and the numbers on the curve are the mean depths evaluated for each year. The side of each square represents horizontally, in time, 5 months, and vertically, in depth of sea, 0·05 ft., or three-fifths of an inch.

A reduction of the observed tides of long period at a number of ports was carried out in Thomson and Tait's *Natural Philosophy*, Part II., 1883 [Paper 9 below], in the belief in the soundness of Laplace's argument with regard to those tides, and the conclusion was drawn that the earth must have an effective rigidity about as great as that of steel. The failure of Laplace's argument, however, condemns this conclusion, and precludes us from making any numerical conclusions with regard to the rigidity of the earth's mass, excepting by means of the 19-yearly tide. The results given in the *Natural Philosophy* merely remain, then, as generally confirmatory of Thomson's conclusion as to the great effective rigidity of the earth's mass.

There are but few ports for which a sufficient mass of accurate tidal observations are accumulated to make the detection of the 19-yearly tide a possibility.

Major Baird has, however, kindly supplied me with the values of the mean sea-level at Karachi for fifteen years. They are plotted out in the figure on the preceding page. The horizontal line represents the mean sea-level for the period from 1869–1883, and the sinuous curve gives the variations of mean sea-level during that period. The dotted sinuous curve gives the annual variations for a portion of the same period for Bombay. The full-line sweeping curve has ordinates proportional to $-\cos \Omega$, and shows the kind of curve which we ought to find if the alternations of sea-level were due to the 19-yearly tide.

It is obvious at a glance that the oscillations of sea-level are not due to astronomical causes.

At Karachi (lat. 24° 47′) the 19-yearly tide is

$$-0^{\text{ft.}}\cdot 0138 \cos \Omega$$

The figure shows that the actual change of sea-level between 1870 and 1873 was nearly 0·25 feet, and this is just about nine times the range of the 19-yearly tide, viz., 0·028 feet.

It is thus obvious that this tide must be entirely masked by changes of sea-level arising from meteorological causes.

It seems unlikely that what is true of Karachi and Bombay is untrue at other ports, and therefore we must regard it as extremely improbable that the 19-yearly tide will ever be detected.

4.

A GENERAL ARTICLE ON THE TIDES.

[Article 'Tides,' *Admiralty Scientific Manual* (1886), pp. 53—91.]

I. Introduction.

The object of the present article is to show how the best use may be made, for scientific purposes, of a short visit to any port.

We refer to the article "Hydrography" [*Admiralty Scientific Manual*] for an account of the method of observing the tides, and shall here assume that the height of the water above some zero mark may be measured, in feet and decimals of a foot, at any time, and that the zero of the tide gauge may be referred by levelling to a bench-mark ashore.

Something of the law of the tide might be discovered from hourly or half-hourly observations even through a single day and night, but to discover the law at all adequately it is necessary that the observations should embrace at least one spring tide and one neap tide. For the full use of the methods given below, the observations should be taken each hour for 360 hours, or 720 hours. A longer series must be regarded as a new set of observations, and the means must be taken of the results of the several sets.

It has been usual to recommend observations of the times and heights of high and low water, but hourly observations are far preferable, the hours being reckoned according to mean time of the port.

We shall, however, begin by a sketch of the treatment of observations of high and low water, and shall then give more detailed instructions for hourly observations and the formation of a tide table.

The height of the water is subject to considerable perturbation from the weather, and the most perfect tide table is one which gives the height of the water, when abstraction is made of the disturbing causes. Such a table can only be made from observations of such extent as to eliminate irregularities by averages.

No general rule can be given for wind disturbance, but it is often considerable in bays and estuaries. The water stands higher with low, and lower with high, barometer; the amount of the effect appears to be very uncertain, the estimates varying from 7 inches to 20 inches rise of water for an inch fall of the mercury. It appears probable that the rule differs in different ports, and even in the same port with different winds. To make the most, however, of a short series of observations, it might perhaps be best to reduce each hourly tide height to a standard height of barometer at the rate of a foot of water to an inch of mercury, before undertaking the tidal reductions.

In order to discover the general run of the tide in any part of the world, observations should be taken at several stations separated by 50 to 100 miles; and this is the more important if some of the stations have to be chosen in estuaries, since the tide wave takes a considerable time to run up from the open sea and changes its form in doing so.

In estuaries and rivers it is important not to confuse flood and ebb with high and low water, for the water often still runs up-stream for long after the tide has turned and when the water-level is falling; and the converse is true of ebb and low water. We refer to "Hydrography" for remarks on tidal currents and streams.

II. Tidal Observations of High and Low Water*.

The immediate object is to connect the times and heights of high and low water (H.W. and L.W.) with the time of the moon's transit. About high and low tide the water often rises and falls irregularly, and the critical moment cannot be found from a single observation. Observations are, therefore, to be taken every 5 or 10 minutes for half-an-hour or an hour about H.W. and L.W. The time and height of H.W. or L.W. are then to be found by graphical interpolation, *i.e.*, take a straight line to represent time, and at the points corresponding to the observations erect perpendiculars or ordinates corresponding to the observed heights, draw a sweeping curve nearly through the tops of the ordinates, so as to obliterate minor irregularities and measure the height of the maximum or minimum ordinate, and note its incidence in the time scale†.

Dr Whewell recommends that the observation should begin with half-hourly observations during 24 hours, for if there should be found to be double H.W. or L.W., or only a single tide in the 24 hours, this method will fail; he also advises that tidal observations be referred to the moon's transit

* Founded on Dr Whewell's article in a former edition of the *Admiralty Scientific Manual.*

† A similar but less elaborate process would render hourly observations more perfect. The readings might be every 2½ minutes, from five minutes before to five minutes after the hour.

during their course, in order to detect irregularities in "the interval" from transit to H.W., which might cause the observations to prove useless.

The object of the observations is to find "the establishment," or time of high water, on days of full and change of moon, the heights of tide at spring and neap, and "the fortnightly or semi-mensual irregularity" in the time and height. The reference of the tide to the "establishment" is not, however, scientifically desirable, and it is better to determine the mean or corrected establishment, being the average interval from moon's transit to H.W. at spring tide, and "the age of the tide," being the average interval from full and change to spring tide. For these purposes the observations are conveniently treated graphically*.

An equally divided horizontal scale is taken to represent the 12 hours of the clock of civil time, regulated to the time of the port—or more accurately arranged always to show apparent time by being fast or slow by the equation of time; this time scale represents the time of the clock of the moon's transit, either upper or lower. The scale is perhaps most conveniently arranged in the order V, VI, ..., XII, I, ..., IIII. Then each "interval" of time from transit to H.W. is set off as an ordinate above the corresponding time-of-clock of moon's transit. A sweeping curve is then drawn so as to pass nearly through the tops of the ordinates, cutting off minor irregularities. Next along the same ordinates are set off lengths corresponding to the height of water at each H.W.

A second similar figure may also be made for the interval and height at L.W.

In the curve of H.W. intervals the ordinate corresponding to XII is the vulgar "establishment," since it gives the time of H.W. at full and change of moon. That ordinate of H.W. intervals which is coincident with the greatest ordinate of H.W. heights gives the "mean establishment†."

Since the moon's transit falls about 50 minutes later on each day, in setting off a fortnight's observation there will be about five days for every four hours-of-clock of moon's upper transit. Hence in these figures we may regard each division of the time scale I to II, II to III, &c. as representing 25 hours instead of one hour. Then the distance from the maximum ordinate of H.W. heights to XII, each division being estimated as 25 hours, is called "the age of the tide."

From these two figures the times and heights of H.W. and L.W. may in general be predicted with fair approximation; we find the time-of-clock of moon's upper or lower transit on the day, correct by the equation of time,

* For numerical treatment, see *Directions for reducing Tidal Observations*. By Staff Commander John Burdwood, R.N. London, 1876. J. D. Potter. Price 6*d*.

† See Section IV. for the numerical computation of mean from vulgar establishment.

read off the corresponding heights of H.W. and L.W. from the figures; and the intervals to H.W. and L.W. being also read off are added to the time of moon's transit, and give the times of H.W. and L.W.

We shall show below how a tide table may be otherwise computed from establishment and spring rise and neap rise.

At all ports, however, there is an irregularity of intervals and heights between successive tides, and in consequence of this our curves will present more or less of a zig-zag appearance. Where the zig-zag is perceptible to the eye, the curves must be smoothed by drawing them so as to bisect the zig-zags, because these "diurnal inequalities" will not present themselves similarly in the future. When, as in many equatorial ports, the diurnal tides are large, this method of tidal prediction fails, but we shall show below how the observations may then be treated scientifically.

III. Instructions for the Reduction of Hourly Tidal Observations, with an Example.

We now suppose that the observations of the tides are taken at each hour. If the observations are only taken every two hours, or if there are gaps in the series, the hourly numbers must be filled in as indicated below.

All the measurements should be positive, and if the zero of the tide gauge has been fixed too high it will be well to refer the measurement to an ideal zero 10 feet lower. The following instructions for reduction should be read along with the example.

Computation Forms.

Mark three large sheets of paper with the letters M, O, S; divide them into squares, with 24 columns; head the columns, 0^h, 1^h, 2^h...23^h for the several hours. On the left margin write the numbers of the days, 0^d, 1^d, 2^d, &c. Each square may be specified by its day and hour.

M *Sheet.*

Place dots in the squares of each row, as follows: 0^d, 14^h; 1^d, 18^h; 2^d, 23^h; 3^d, none; 4^d, 3^h; 5^d, 8^h; 6^d, 12^h; 7^d, 17^h; 8^d, 21^h; 9^d, none; 10^d, 2^h; 11^d, 7^h; 12^d, 11^h; 13^d, 16^h; (13^d is the last row required for a fortnight's observation); 14^d, 20^h; 15^d, none; 16^d, 1^h; 17^d, 5^h; 18^d, 10^h; 19^d, 14^h; 20^d, 19^h; 21^d, 23^h; 22^d, none; 23^d, 4^h; 24^d, 8^h; 25^d, 13^h; 26^d, 17^h; 27^d, 22^h. (27^d is the last row required for a month's observation.)

O *Sheet.*

Place dots in the following squares: 0^d, 6^h and 19^h; 1^d, 8^h and 22^h; 2^d, 11^h 3^d, 0^h and 13^h; 4^d, 2^h and 16^h; 5^d, 5^h and 18^h; 6^d, 7^h and 20^h; 7^d, 10^h

and 23^h; 8^d, 12^h; 9^d, 1^h and 14^h; 10^d, 4^h and 17^h; 11^d, 6^h and 19^h; 12^d, 8^h and 22^h (12^d is the last row required for a fortnight's observation); 13^d, 11^h; 14^d, 0^h and 13^h; 15^d, 2^h and 15^h; 16^d, 5^h and 18^h; 17^d, 7^h and 20^h; 18^d, 9^h and 23^h; 19^d, 12^h; 20^d, 1^h and 14^h; 21^d, 3^h and 17^h; 22^d, 6^h and 19^h; 23^d, 8^h and 21^h; 24^d, 11^h. (24^d is the last row required for a month's observation.)

S *Sheet.*

There are no dots. For a fortnight's observation, let row 13^d be the last; then leave three rows blank, and add another row numbered 14^d.

For a month's observation, let row 26^d be the last; then leave three rows blank, and add three more rows numbered 27^d, 28^d, 29^d.

The shorter series (13^d or 26^d) is for diurnal tides, the longer (14^d or 29^d) for semi-diurnal tides.

Entry in S.

The first hourly observation is supposed to be at noon or 0^d, 0^h; enter it in that square; enter the second in 0^d, 1^h; the third in 0^d, 2^h, and so on.

The 0^h's of the rows are the noons of each day; write the day of the month opposite each row.

If any hourly observations are wanting or obviously vitiated through accident or weather, fill in the blanks thus. Consider a column in which a blank occurs; take a number of equidistant points along a line, and let each point correspond to one of the days before and after the blank; then draw lines perpendicular to the first line proportional to the height of water on each of the days. Draw a sweeping curve through the extremities of the lines, and measure off the height of the curve where it passes above the blank. The resulting number is to be used for filling in the blank.

If the observations at night are taken at rarer intervals, say every two or three hours, the same process of filling in blanks must be employed, by considering the hourly observations before and after the blank.

Every deficiency of data of course weakens the strength of the result.

The entry of observations is to stop as indicated in the instructions for forming the S sheet.

Entry in M.

Copy the numbers on the S sheet; begin entering as in S, but when a dot is reached, enter two successive hourly values, the first above, the second below; then continue copying from S (the entries falling of course in a different hour column) until a second dot is reached, where again make a double entry; and so on.

The entries stop as indicated in the instructions for forming the M sheet.

For a fortnight's observation the last entry (which is in square $13^d\ 23^h$) is that which is copied from square $14^d\ 11^h$ of S.

For a month's observation the last entry (which is in square $27^d\ 23^h$) is that which is copied from square $28^d\ 23^h$ of S.

Entry in O.

Follow the same rules for entry as in M.

For a fortnight's observation the last entry (which is in square $12^d\ 23^h$) is that which is copied from square $13^d\ 23^h$ of S.

For a month's observation the last entry (which is in square $24^d\ 23^h$) is that which is copied from square $26^d\ 20^h$ of S.

Rules of reduction on all three Sheets.

Add up the numbers in each column (in S there will be two sums, one without rows 14^d or 27^d, 28^d, 29^d, as the case may be, and the other with those additional rows). Divide the sum in each column by the number of entries of which it is the sum. In S the divisors for each of the two sets of sums are all the same, since there are not duplicate entries.

The results are 24 hourly mean values, and there are four of them, viz. two from S, one from M, one from O.

The set from M, and the second set (longer series) on S are to be harmonically analysed for semi-diurnal inequality; the first set (shorter series) on S, and the set from O are to be harmonically analysed for diurnal inequality.

Thus where it is known that the diurnal tides are small, the O sheet and the shorter series on S may be omitted for rough results.

The height of mean water with reference to the zero of the tide gauge has also to be determined from the second set on S.

Harmonic Analysis.

If we have any quantity which is variable during the day and night, such as temperature or the height of the barometer, it is often desirable to express it by the formula

$$A_0 + A_1 \cos\theta + B_1 \sin\theta + A_2 \cos 2\theta + B_2 \sin 2\theta + A_3 \cos 3\theta + B_3 \sin 3\theta + A_4 \cos 4\theta + B_4 \sin 4\theta$$

where θ is an angle which increases at the rate of 15° per hour, and is zero at noon.

If we put

$$\tan \zeta_1 = \frac{B_1}{A_1}, \quad \tan \zeta_2 = \frac{B_2}{A_2}, \quad \tan \zeta_3 = \frac{B_3}{A_3}, \quad \tan \zeta_4 = \frac{B_4}{A_4}$$

and $$R_1 = A_1 \sec \zeta_1 = B_1 \operatorname{cosec} \zeta_1, \quad R_2 = A_2 \sec \zeta_2 = B_2 \operatorname{cosec} \zeta_2$$

and so on, the formula may be written

$$A_0 + R_1 \cos (\theta - \zeta_1) + R_2 \cos (2\theta - \zeta_2) + R_3 \cos (3\theta - \zeta_3) + R_4 \cos (4\theta - \zeta_4)$$

The term in R_1 is diurnal, that in R_2 semi-diurnal, that in R_3 ter-diurnal, that in R_4 quater-diurnal; that is to say, they go through their changes once, twice, thrice, and four times a day.

The term A_0 gives the mean value for the day.

The A's and B's are the numbers which are derived from harmonic analysis, as explained below.

The same process is applicable to the tides, but with the difference that there are several kinds of days, viz.:—first, the ordinary or mean solar or S day; second, the mean lunar or M day; and a third kind—the O day—for which there is no name.

Thus in application to the tides there are to be four harmonic analyses, two performed on the means on the S sheet, one on the means on the M sheet, and one on the means on the O sheet. The matter is simplified, however, by the fact that from the first means on S we only want A_1, B_1; from the second means on S we only want A_2, B_2, and A_0; from the means on M we only want A_2, B_2; and from the means on O we only want A_1, B_1.

The following schedule gives General [Sir Richard] Strachey's rules for harmonic analysis. Columns I. and II. contain the 24 hourly values to be analysed, and the headings to each successive column give the rules for its derivation from the preceding ones.

If we only want A_1, B_1 the columns I. to VIII. inclusive are required; if we only want A_2, B_2 the columns I., II., and IX. to XIV. inclusive are required, and for A_0 we require also column XV.

A comparison of this complete schedule with the numerical example below will render the process intelligible.

Form for Harmonic Analysis.

Hours	I.	Hours	II.	III.	IV.	V.	VI.	VII.	VIII.	IX.	X.	XI.	XII.	XIII.	XIV.	XV.	XVI.	XVII.
	Hourly Values, 0 to 11		Hourly Values, 12 to 23	I. − II.	Middle 4 of III. with opp. sign	Last 4 of III.	III. + IV. + V.	See III. and VI.	See III. and VI.	I. + II.	Second half of IX.	IX. − X.	Second half of XI.	XI. − XII. omitting top entry of XII.	XI. + XII. omitting top entry of XI.	IX. + X.	Second half of XV.	XV. − XVI.
0	.	12	.	(a) .	.	.	(c) .	M .	M .	.	.	.	.	.	.	.	.	(g) .
1	.	13	.	.	.	.	(d) .	−N .	N .	.	.	.	.	.	.	.	.	(h) .
2	.	14	.	.	.	.	(e) .	(a) .	(b) .	.	.	.	.	.	.	.	.	(i) .
3	.	15	.	sum .	.	.	(f) .	$\frac{1}{3}a$.	$-\frac{1}{3}\beta$.	.	.	.				.		
4	.	16	.	.	$a=(c)+(d)-(f)$			Sum .	Sum .	.	.	.				.	$(g)+(h)-(i)$	
5	.	17	.	.	= .			× ·0658	× ·0658	.	.	.				.	= .	
				M = .	× ·0690			A_1 = .	B_1 = .		Sum of XIII. = .				Sum = .		÷ 16	
6	.	18	.	(b) .	A_3 =					.	× ·067				÷ 24		A_4 = .	
7	.	19	.	.						.	A_2 = .				A_0 = .			
8	.	20	.	.						.								
9	.	21	.	sum .	$\beta=(d)+(e)+(f)$					.	Sum of XIV. = .						$(h)+(i)$ = .	
10	.	22	.	.	= .					.	× ·067						× ·072	
11	.	23	.	.	× ·0690					.	B_2 = .						B_4 = .	
				N = .	B_3 =													

General Rule for the Determination of ζ *and* R *from* A *and* B.

If A is + and B is +, ζ is less than 90°, or in 1st quadrant.

If A is − and B is +, ζ lies between 90° and 180°, in 2nd quadrant.

If A is − and B is −, ζ lies between 180° and 270°, in 3rd quadrant.

If A is + and B is −, ζ lies between 270° and 360°, in 4th quadrant.

If $\tan\zeta$ is numerically less than 1, compute from $R = A \sec\zeta$; if greater than 1, compute from $R = B \operatorname{cosec}\zeta$.

In certain cases mentioned below, we shall have also to augment the result R by a factor which is nearly equal to unity, as there explained.

General Rule as to Angles.

All angles are to be written as positive angles less than 360°; if an angle is greater than 360°, subtract 360°.

Certain small angles, however, determined below are to be estimated as either positive or negative (*e.g.*, 355° will in this case be written as − 5°), and do not fall under this rule. The occurrence of the exception will always be noted at the time.

It will often be convenient to write angles in degrees and decimals of a degree.

Harmonic analysis of M.

Analyse the hourly means for A_2, B_2.

Find ζ_m from $\tan\zeta_m = \dfrac{B_2}{A_2}$.

Find R_m from $R_m = A_2 \sec\zeta_m \times 1{\cdot}0115$ or $B_2 \operatorname{cosec}\zeta_m \times 1{\cdot}0115$, according as $\tan\zeta_m$ is numerically less or greater than 1.

N.B. $\operatorname{Log} 1{\cdot}0115 = {\cdot}0050$.

Harmonic analysis of S.

Analyse the second hourly means (the longer series) for A_2, B_2, and for A_0.

Find ζ_s from $\tan\zeta_s = \dfrac{B_2}{A_2}$.

Find R_s from $R_s = A_2 \sec\zeta_s$ or $B_2 \operatorname{cosec}\zeta_s$, according as $\tan\zeta_s$ is numerically less or greater than 1.

Analyse the first hourly means (the shorter series) for A_1, B_1.

Find ζ' from $\tan\zeta' = \dfrac{B_1}{A_1}$.

Find R′ from $R' = A_1 \sec\zeta'$, or $B_1 \operatorname{cosec}\zeta'$, according as $\tan\zeta'$ is numerically less or greater than 1.

Harmonic analysis of O.

Analyse the hourly means for A_1, B_1.

Find ζ_o from $\tan \zeta_o = \frac{B_1}{A_1}$.

Find R_o from $R_o = A_1 \sec \zeta_o \times 1{\cdot}0029$ or $B_1 \operatorname{cosec} \zeta_o \times 1{\cdot}0029$, according as $\tan \zeta_o$ is numerically less or greater than 1.

N.B. Log 1·0029 = ·0013.

Angles and Factors for Reduction.

N.A. stands for *Nautical Almanac.*

Call local mean noon of day 0 of the series of observations to be reduced, or of the tide table to be computed *the Epoch.* N.A., p. 1: Find ☊ the mean longitude of the ascending node of ☽'s orbit at epoch. Find sin ☊, cos ☊, sin 2 ☊, cos 2 ☊, and compute the following small angles (+ when ☊ lies between 0° and 180°, – when ☊ lies between 180° and 360°), and numerical factors

Angles to be determined as + or –

$$\begin{cases} \nu = 12^\circ{\cdot}9 \sin ☊ - 1^\circ{\cdot}3 \sin 2☊ \\ \xi = 11^\circ{\cdot}8 \sin ☊ - 1^\circ{\cdot}3 \sin 2☊ \\ \nu' = 8^\circ{\cdot}9 \sin ☊ - 0^\circ{\cdot}7 \sin 2☊ \\ 2\nu'' = 17^\circ{\cdot}7 \sin ☊ - 0^\circ{\cdot}7 \sin 2☊ \end{cases}$$

Factors

$$\begin{cases} f = 1 \quad - {\cdot}037 \cos ☊ \\ f' = 1{\cdot}006 + {\cdot}115 \cos ☊ - {\cdot}009 \cos 2☊ \\ f'' = 1{\cdot}024 + {\cdot}286 \cos ☊ + {\cdot}008 \cos 2☊ \\ f_o = 1{\cdot}009 + {\cdot}187 \cos ☊ - {\cdot}015 \cos 2☊ \end{cases}$$

In the N.A. find the ☉'s parx. at the middle of the fortnight or month of observation, subtract from it the ☉'s mean parx. (*see* Preface to N.A.). Multiply the result by 19⅓, and considering the product as degrees, look out the sine of the angle and add 1, the result is $p_,$; *e.g.*, if ☉'s parx. be 8″·85, and if the mean parx. be 8″·95*, we get

$$\text{diff.} - 0''{\cdot}10, \text{ and } -19\tfrac{1}{3} \times {\cdot}10 = -1^\circ{\cdot}93 = -1^\circ\ 56'$$

$$\sin(-1^\circ\ 56') = -{\cdot}034, \quad p_, = 1 - {\cdot}034 = {\cdot}966$$

This is a short way of finding the cube of the ratio of ☉'s parx. to ☉'s mean parx.

* [The sun's mean parallax is now taken as 8″·80, but the rule remains sufficiently exact.]

Arguments at Epoch.

From N.A., find ☽, the moon's mean longitude at epoch. From N.A. find ☉, the sun's mean longitude at epoch, by converting sidereal time to angle at 15° per hour. The diurnal increase of ☽ = 13° 11′. The corrections for longitude of port are subtracted for E. long. and added for W. long. The correction to ☽ is 0°·549 for each hour of longitude, and 0°·041 to ☉ for each hour of longitude.

Find $-2(☽-\xi)$, $☉-\nu$, $2(☉-\nu)$, $☉-\nu'$, and compute the following "arguments at epoch,"

$$V=2(☉-\nu)-2(☽-\xi);\quad V'=☉-\nu'+270°;\quad V_o=☉-\nu-2(☽-\xi)+90°$$

Find a mean value for ☉ for the period under reduction, by adding to ☉ seven days' motion for a fortnight's observation, or 15 days' motion for a month's observation. The motion for a day may be taken as 1°, and thus we add 7° or 15°; with this mean ☉ compute, $2☉-\nu'$; $V''=2☉-2\nu''$.

Final Reduction.

Principal Lunar Tide called M_2.

Let mean semi-range $= H_m$, and constant angle of retardation or lag $= \kappa_m$.

The angle of retardation is hereafter called *the lag.*

Then from M sheet take R_m, ζ_m, and compute

$$H_m=\frac{R_m}{f},\quad \kappa_m=\zeta_m+V$$

Principal Solar Tide called S_2, *and Lunisolar Semi-diurnal Tide called* K_2*.

For S_2, let mean semi-range $= H_s$, lag $= \kappa_s$.

For K_2, let mean semi-range $= H''$, lag $= \kappa''$.

Find ψ, as a positive or negative angle, for a fortnight, from

$$\tan\psi=\frac{f''\sin V''}{3{\cdot}71p_{,}+f''\cos V''}$$

Then from S sheet take R_s, ζ_s, and compute

$$H_s=\frac{3{\cdot}71\cos\psi}{3{\cdot}71p_{,}+f''\cos V''}R_s,\quad H''=\frac{1}{3{\cdot}67}H_s,\quad \kappa_s=\kappa''=\zeta_s+\psi$$

For a month replace the 3·71 in these formulæ by 3·84.

* [An error in the *Manual* has been corrected.]

Lunisolar Diurnal Tide called K_1, *and Solar Diurnal Tide called* P*.

For K_1, let mean semi-range $= H'$, lag $= \kappa'$.

For P, let mean semi-range $= H_p$, lag $= \kappa_p$.

Find ϕ, as a positive or negative angle, from

$$\tan\phi = \frac{\sin(2\odot - \nu')}{3f' - \cos(2\odot - \nu')}$$

Then from S sheet take R′, ζ', and for a fortnight compute

$$H' = \frac{3{\cdot}007\cos\phi}{3f' - \cos(2\odot - \nu')} R', \quad H_p = \frac{1}{3} H', \quad \kappa' = \kappa_p = \zeta' + V' + \phi + 6^\circ{\cdot}9$$

For a month replace 3·007 by 3·027 and 6°·9 by 13°·3.

Lunar Diurnal Tide called O.

For O, let mean semi-range $= H_o$, lag $= \kappa_o$.

Then from O sheet take R_o, ζ_o, and compute

$$H_o = \frac{R_o}{f_o}, \quad \kappa_o = \zeta_o + V_o$$

For rough results all the diurnal tides may be omitted, unless the diurnal inequality is known to be large.

Collect Results.

These constants express six of the most important tides, and A_0 gives the height of mean water mark from the zero of the tide gauge.

* [An error in the *Manual* has been corrected.]

EXAMPLE OF REDUCTION OF A FORTNIGHT'S OBSERVATION AT PORT BLAIR, ANDAMAN ISLANDS.

[N.B. The horizontal lines by which the computer will divide the paper into squares have been omitted.]

(Portion of) S Sheet.

Date, 1880	Day	0h.	1h.	2h.	3h.	4h.	5h.	6h.	7h.	8h.	9h.	10h.	11h.	12h.	13h.	14h.	15h.	16h.	17h.	18h.	19h.	20h.	21h.	22h.	23h.
April 19th	0d.	4·4	4·7	5·1	5·4	5·7	5·8	5·7	5·4	4·8	4·2	3·7	3·4	3·3	3·5	4·0	4·5	5·0	5·4	5·6	5·5	5·2	4·7	4·2	3·9
„ 20th	1d.	3·7	3·9	4·3	4·9	5·5	6·0	6·2	6·1	5·7	4·9	4·1	3·4	2·9	2·9	3·2	3·9	4·7	5·4	6·1	6·3	6·1	5·6	4·9	4·1
„ 21st	2d.	3·5	3·3	3·5	4·1	4·8	5·7	6·3	6·6	6·4	5·8	4·9	3·8	2·9	2·4	2·5	3·1	4·0	5·1	6·1	6·8	6·9	6·5	5·6	4·6
„ 22nd	3d.	3·7	3·0	2·8	&c.	&c.																			
							*			*		*		*			*		*						
	&c.																					&c.	3·7	4·2	4·9
May 1st .	12d.	5·6	6·4	6·7	6·7	6·3	5·7	4·9	4·1	3·5	3·2	3·2	3·6	4·2	4·8	5·3	5·6	5·6	5·4	5·0	4·6	4·2	4·0	3·9	4·1
„ 2nd .	13d.	4·6	5·2	5·8	6·2	6·4	6·2	5·7	5·1	4·5	3·8	3·4	3·3	3·5	3·9	4·6	5·2	5·6	5·8	5·7	5·4	5·0	4·5	4·1	4·0
Sum .		78·7	74·0	67·0	61·0	57·1	57·4	59·6	64·2	68·5	71·9	72·5	70·3	65·8	60·4	55·0	52·3	50·5	53·4	59·4	67·3	74·9	80·6	83·2	82·4
	Divisor 14.																								
Mean for diurnal	}	5·62	5·29	4·79	4·36	4·08	4·10	4·26	4·59	4·89	5·14	5·18	5·02	4·70	4·31	3·93	3·74	3·61	3·81	4·24	4·81	5·35	5·76	5·94	5·89
May 3rd .	14d.	4·1	4·5	5·1	5·7	6·2	6·4	6·3	5·9	5·2	4·4	3·7	3·3	3·1	3·4	4·0	4·8	5·5	6·1	6·4	6·3	5·9	5·3	4·6	4·0
Sum .		82·8	78·5	72·1	66·7	63·3	63·8	65·9	70·1	73·7	76·3	76·2	73·6	68·9	63·8	59·0	57·1	56·0	59·5	65·8	73·6	80·8	85·9	87·8	86·4
	Divisor 15.																								
Mean for semid.		5·52	5·23	4·81	4·45	4·22	4·25	4·39	4·67	4·91	5·09	5·08	4·91	4·59	4·25	3·93	3·81	3·73	3·97	4·39	4·91	5·39	5·73	5·85	5·76

Harmonic Analysis for Diurnal Component of Means from 14 Days on S Sheet.

I.		II.		III.	IV.	V.	VI.	VII.	VIII.
Hours	Hourly Values, 0 to 11	Hours	Hourly Values, 12 to 23	I. − II.	Middle Four of III. with opposite Sign	Last Four of III.	III. + IV. + V.	See III., IV., and V.	See III., IV., and V.
0	5·62	12	4·70	(a) +·92	−·47	−·46	(c) −·01	M +3·22	M +3·22
1	5·29	13	4·31	+·98	−·29	−·62	(d) +·07	−N +2·93	N −2·93
2	4·79	14	3·93	+·86	−·02	−·76	(e) +·08	(a) + ·92	(b) + ·02
3	4·36	15	3·74	Sum +·62	+·22	−·87	(f) −·03	$\frac{1}{3}\alpha$ + ·03	$-\frac{1}{3}\beta$ − ·04
4	4·08	16	3·61	+·47				Sum +7·10	Sum + ·27
5	4·10	17	3·81	+·29	$\alpha=(c)+(d)-(f)=+\cdot09$			×·0658	×·0658
				M = +3·22	$\beta=(d)+(e)+(f)=+\cdot12$			$A_1=+\cdot4672$	$B_1=+\cdot0178$
6	4·26	18	4·24	(b) +·02			A_1+, B_1+, ζ' lies in 1st quadrant.		
7	4·59	19	4·81	−·22					
8	4·89	20	5·35	−·46	$\log B_1=8\cdot2504$			$\log\sec\zeta'=\cdot0003$	
9	5·14	21	5·76	Sum −·62	$\text{colog } A_1=\cdot3305$			$\log A_1=9\cdot6695$	
10	5·18	22	5·94	−·76	$\log\tan\zeta'=8\cdot5809$			$\log R'=9\cdot6698$	
11	5·02	23	5·89	−·87	$\zeta'=2°\ 11'=2°\cdot2$			$R'=\cdot468$	
				N = −2·93					

Harmonic Analysis for Semi-Diurnal Component of Means from 15 Days on S Sheet.

I.		II.		IX.	X.	XI.	XII.	XIII.	XIV.	XV.
Hours	Hourly Values, 0 to 11	Hours	Hourly Values, 12 to 23	I. + II.	Second Half of IX.	IX. − X.	Second Half of XI.	XI. − XII. omitting top entry of XII.	XI. + XII. omitting top entry of XI.	IX. + X.
0	5·52	12	4·59	10·11	8·78	+1·33	−2·56	+1·33	−2·56	18·89
1	5·23	13	4·25	9·48	9·58	− ·10	−2·98	+2·88	−3·08	19·06
2	4·81	14	3·93	8·74	10·30	−1·56	−2·45	+ ·89	−4·01	19·04
3	4·45	15	3·81	8·26	10·82	−2·56		Sum = +5·10	Sum = −9·65	19·08
4	4·22	16	3·73	7·95	10·93	−2·98		× ·067	× ·067	18·88
5	4·25	17	3·97	8·22	10·67	−2·45		A_2 = +·3417	B_2 = −·6466	18·89
										Sum 113·84
6	4·39	18	4·39	8·78						÷24
7	4·67	19	4·91	9·58						A_0 = 4·743
8	4·91	20	5·39	10·30						
9	5·09	21	5·73	10·82						
10	5·08	22	5·85	10·93						
11	4·91	23	5·76	10·67						

A_2+, B_2-, ζ_s lies in 4th quadrant.

$\log B_2 = 9·8106$
$\text{colog } A_2 = ·4664$
$\log \tan \zeta_s = ·2770$
$\zeta_s = 360° - 62° 9'$
$= 297° 51'$
$= 297°·9$

$\log \text{cosec } \zeta_s = ·0535$
$\log B_2 = 9·8106$
$\log R_s = 9·8641$
$R_s = ·731$

(Portion of) M Sheet.

(Decimal points omitted for brevity; the dotted squares are those holding two entries.)

[N.B. The horizontal lines by which the computer will divide the paper into squares have been omitted.]

	0h.	1h.	2h.	3h.	4h.	5h.	6h.	7h.	8h.	9h.	10h.	11h.	12h.	13h.	14h.	15h.	16h.	17h.	18h.	19h.	20h.	21h.	22h.	23h.
0d.	44	47	51	54	57	58	57	54	48	42	37	34	33	35	40 45	50	54	56	55	52	47	42	39	37
1d.	39	43	49	55	60	62	61	57	49	41	34	29	29	32	39	47	54	61	63 61	56	49	41	35	33
2d.	35	41	48	57	63	66	64	58	49	38	29	24	25	31	40	51	61	68	69	65	56	46	37	30 28
3d.	31	39	49	60	67	&c.																		
*			*			*			*			*			*			*			*			
																					&c.	45	38	34
13d.	33	35	39	46	52	56	58	57	54	50	45	41	40	41	45	51	57 62	64	63	59	52	44	37	33
Sums	407	505	691	845	886	912	872	817	667	469	376	393	438	548	748	862	1030	1091	1026	851	695	559	406	388
Divisors	14	14	15	15	14	14	14	15	15	14	14	15	15	14	15	14	15	15	15	14	14	15	14	15
Means	291	361	461	563	633	651	623	545	445	335	269	262	292	391	499	616	687	727	684	608	496	373	290	259

The harmonic analysis with columns I., II., IX., X., XI., XII., XIII., XIV. gives

$A_2 = -1{\cdot}8251$, $B_2 = +1{\cdot}1437$; ζ_m lies in 2nd quadrant.

$\log B_2 = {\cdot}0583$

$\text{colog } A_2 = 9{\cdot}7387$

$\log \tan \zeta_m = 9{\cdot}7970$

$\zeta_m = 180° - 32° \, 4' = 147° \, 56' = 147°{\cdot}9$

$\log \sec \zeta_m = {\cdot}0719$

$\log A_2 = {\cdot}2613$

$\log 1{\cdot}0115 = {\cdot}0050$

$\log R_m = {\cdot}3382$

$R_m = 2{\cdot}178$

(Portion of) O Sheet.

(Decimal points omitted for brevity.)

[N.B. The horizontal lines by which the computer will divide the paper into squares have been omitted.]

	0h.	1h.	2h.	3h.	4h.	5h.	6h.	7h.	8h.	9h.	10h.	11h.	12h.	13h.	14h.	15h.	16h.	17h.	18h.	19h.	20h.	21h.	22h.	23h.
0d.	44	47	51	54	57	58	57 54	48	42	37	34	33	35	40	45	50	54	56	55	52 47	42	39	37	39
1d.	43	49	55	60	62	61	57	49	41 34	29	29	32	39	47	54	61	63	61	56	49	41	35	33 35	41
2d.	48	57	63	66	64	58	49	38	29	24	25	31 40	51	61	68	69	65	56	46	37	30	28	31	39
3d.	49 60	67	69	&c.																				
*		*			*			*			*			*			*			*				
																					&c.	46	42	40
12d.	39	41	46	52	58	62	64	62	57 51	45	38	34	33	35	39	46	52	56	58	57	54	50	45 41	40
Sums	701	648	676	583	635	599	721	657	738	659	752	706	698	702	629	588	605	602	578	685	662	631	720	699
Divisors	14	14	14	13	14	14	15	14	15	13	14	14	14	14	14	13	14	14	14	15	14	13	15	14
Means	501	463	483	448	454	428	481	469	492	507	537	504	499	501	449	452	432	430	413	457	473	485	480	499

The harmonic analysis with columns I., II., III., IV., V., VI., VII., VIII. gives

$A_1 = -{\cdot}0658$, $B_1 = +{\cdot}1303$; ζ_o lies in 2nd quadrant.

$\log B_1 = 9{\cdot}1149$
$\text{colog } A_1 = 1{\cdot}1818$
$\log \tan \zeta_o = {\cdot}2967$
$\zeta_o = 180° - 63°\ 12' = 116°{\cdot}8$

$\log \text{cosec } \zeta_o = {\cdot}0494$
$\log B_1 = 9{\cdot}1149$
$\log 1{\cdot}0029 = {\cdot}0013$
$\log R_o = 9{\cdot}1656$
$R_o = {\cdot}146$

PORT BLAIR, ANDAMAN ISLANDS; lat. 11° 41′ N., long. 92° 45′ E.

Observation commences at epoch, 0^h mean time, April 19, 1880. Long. 92° 45′ E. = $6^h{\cdot}183$ E.

$$\text{N.A. p. 1} \quad ☊ = 280°; \quad \sin ☊ = -\cdot985; \quad \sin 2☊ = -\cdot34$$
$$\cos ☊ = +\cdot174; \quad \cos 2☊ = -\cdot94$$

Angles.

$$\nu = 12°{\cdot}9 \sin ☊ - 1°{\cdot}3 \sin 2☊ = -12°{\cdot}26$$
$$\xi = 11°{\cdot}8 \sin ☊ - 1°{\cdot}3 \sin 2☊ = -11°{\cdot}18$$
$$\nu' = 8°{\cdot}9 \sin ☊ - 0°{\cdot}7 \sin 2☊ = -8°{\cdot}46$$
$$2\nu'' = 17°{\cdot}7 \sin ☊ - 0°{\cdot}7 \sin 2☊ = -17°{\cdot}36$$

Factors.

$$f = 1{\cdot}000 - \cdot037 \cos ☊ = \cdot994$$
$$f' = 1{\cdot}006 + \cdot115 \cos ☊ - \cdot009 \cos 2☊ = 1{\cdot}035$$
$$f'' = 1{\cdot}024 + \cdot286 \cos ☊ + \cdot008 \cos 2☊ = 1{\cdot}066$$
$$f_o = 1{\cdot}009 + \cdot187 \cos ☊ - \cdot015 \cos 2☊ = 1{\cdot}056$$

The mean value of ☉'s parx. for the fortnight commencing April 19 is ☉'s parx. on April 25 = 8″·89 (N.A. 1880); mean parx. from Preface to N.A. = 8″·95; difference = – 0″·06; multiply by $19\frac{1}{3}$ = – 1″·16; change ″ to ° = – 1° 10′; sin (– 1° 10′) = – ·020; $p_{,} = 1 - \cdot020 = \cdot980$.

Arguments at Epoch.

☽'s daily motion 13° 11′; hourly 0°·55; ☽'s motion in $6^h{\cdot}18$ = 3°·40

☉'s hourly motion 0°·041; ☉'s motion in $6^h{\cdot}18$ = 0°·25

N.A. (Moon's Libration)—

0^h G.M.T. Ap. 20 ☽ =	159° 27′
– one day's motion = –	13° 11′
0^h G.M.T. Ap. 19 ☽ =	146° 16′
=	146°·27
– $6^h{\cdot}18$ E. long. = –	3 ·40
☽ =	142 ·87
– ξ = +	11 ·18
☽ – ξ =	154°·05
2 (☽ – ξ) =	308°·10
– 2 (☽ – ξ) =	51°·90

Sid^l time 0^h G.M.T. Ap. 19 = $1^h\,51^m\,51^s$
= $1^h\,51^m{\cdot}85$
= $1^h{\cdot}864$
$\frac{1}{2}$ sid^l time = 932

$\odot$ 0^h G.M.T. Ap. 19 = $27^\circ{\cdot}96$
$-\,6^h{\cdot}18$ E. long. = $-\,{\cdot}25$

$\odot = 27^\circ{\cdot}71$
$-\nu = +\,12\,{\cdot}26$

$\odot - \nu = 39\,{\cdot}97$

$2(\odot - \nu) = 79\,{\cdot}94$
$-\,2(☽ - \xi) = 51\,{\cdot}90$

Sum $V = 131^\circ{\cdot}84$

$\odot = 27^\circ{\cdot}7$
$-\nu' = 8\,{\cdot}5$
$270^\circ = 270$

Sum $V' = 306^\circ{\cdot}2$

$\odot - \nu = 39^\circ{\cdot}97$
$-\,2(☽ - \xi) = 51\,{\cdot}90$
$+\,90^\circ = 90$

Sum $V_o = 181^\circ{\cdot}87$

Compute $2\odot - \nu'$ and V'' with mean values of $\odot$. The reduction is to cover a fortnight, and therefore a week later than April 19 is the middle of the period, and $\odot$ has increased by seven days' motion or 7°.

April 19, $\odot = 28^\circ$
a week's increase = 7°
mean $\odot = 35^\circ$

$2\odot = 70^\circ$
$-\nu' = +\,8^\circ$
$2\odot - \nu' = 78^\circ$

$2\odot = 70^\circ$
$-\,2\nu'' = 17^\circ$
Sum $V'' = 87^\circ$

Reductions.

M_2, *Principal Lunar Semi-diurnal.*

From harmon. anal. $\zeta_m = 147^\circ{\cdot}93$
$+\,V = 131\,{\cdot}82$
Sum $\kappa_m = 280^\circ$

$$H_m = \frac{R_m}{f} = \frac{2{\cdot}178}{{\cdot}994} = 2^{ft}{\cdot}191$$

K_2, *Lunisolar Semi-diurnal, and* S_2 *Principal Solar Semi-diurnal.*

$f'' = 1{\cdot}066$; $p_{,} = {\cdot}980$; $R_s = {\cdot}731$; $\zeta_s = 297^\circ{\cdot}9$; $V'' = 87^\circ$

$$H_s = \frac{3{\cdot}71 \cos \psi}{3{\cdot}71 p_{,} + f'' \cos V''} R_s$$

Compute—

$\log f'' = {\cdot}0278$
$\log \cos V'' = 8{\cdot}7188$
$\log f'' \cos V'' = 8{\cdot}7466$

$f'' \cos V'' = +\,0{\cdot}056$
$3{\cdot}71 p_{,} = +\,3{\cdot}636$
$3{\cdot}71 p_{,} + f'' \cos V'' = 3{\cdot}692$
$\log 3{\cdot}692 = {\cdot}5673$

$\log f'' = \cdot0278$
$\log\sin V'' = 9\cdot9994$
$\text{colog } 3\cdot692 = 9\cdot4327$

$\log\tan\psi = 9\cdot4599$
$\sin V''$ is +, therefore ψ is +, and
$\psi = +16^\circ 5' = +16^\circ\cdot1$
$\zeta_s = 297\ \cdot9$
Sum $\kappa_s = 314^\circ = \kappa''$

$\log\cos\psi = 9\cdot9827$
$\log 3\cdot71 = \cdot5694$
$\log R_s = 9\cdot8639$
$\text{colog } 3\cdot692 = 9\cdot4327$

$\log H_s = 9\cdot8487$
$H_s = \cdot706$
$$H'' = \frac{H_s}{3\cdot67} = \cdot192$$

K₁, *Lunisolar Diurnal, and* P *Solar Diurnal.*

$2\odot - \nu' = 78^\circ$; $\cos(2\odot - \nu') = +\cdot208$; $\sin(2\odot - \nu') = +\cdot978$
$f' = 1\cdot035$; $R' = \cdot468$; $\zeta' = 2^\circ\cdot2$; $V' = 306^\circ\cdot2$

$$H' = \frac{3\cdot007\cos\phi}{3f' - \cos(2\odot - \nu')}R'$$

Compute—

$3f' = 3\cdot105$
$-\cos(2\odot - \nu') = -\ \cdot208$
$3f' - \cos(2\odot - \nu') = 2\cdot897$

$$\tan\phi = \frac{\sin(2\odot - \nu')}{3f' - \cos(2\odot - \nu')} = +\frac{\cdot978}{2\cdot897} = +\cdot338$$

ϕ is +, and $\phi = +18^\circ 40' = 18^\circ\cdot7$
$V' = 306\ \cdot2$
$\zeta' = 2\ \cdot2$
$6^\circ\cdot9 = 6\ \cdot9$
Sum $\kappa' = 334^\circ = \kappa_p$

$\log\cos\phi = 9\cdot9765$
$\log 3\cdot007 = \cdot4781$
$\text{colog } 2\cdot897 = 9\cdot5381$
$\log R' = 9\cdot6702$

$\log H' = 9\cdot6629$
$H' = \cdot460$; $H_p = \frac{1}{3}H' = \cdot153$

O, *Lunar Diurnal.*

$V_o = 181^\circ\cdot9$; $f_o = 1\cdot056$; $R_o = \cdot146$; $\zeta_o = 116^\circ\cdot8$

Compute—

$V_o = 181^\circ\cdot9$
$\zeta_o = 116\ \cdot8$

Sum $\kappa_o = 299^\circ$

$$H_o = \frac{R_o}{f_o} = \frac{\cdot146}{1\cdot056} = \cdot138$$

RESULTS OF HARMONIC ANALYSIS of 15 days' hourly observations at Port Blair, commencing 0^h, April 19, 1880.

			Mean of Three Years' Hourly Observation.
	A_0	$= 4{\cdot}74$ ft.	$4{\cdot}740$ ft.
M_2	H_m	$= 2{\cdot}19$ ft.	$2{\cdot}022$ ft.
	κ_m	$= 280°$	$278°$
S_2	H_s	$= 0{\cdot}71$ ft.	$0{\cdot}968$ ft.
	κ_s	$= 314°$	$315°$
K_2	H''	$= 0{\cdot}19$ ft.	$0{\cdot}282$ ft.
	κ''	$= 314°$	$311°$
K_1	H'	$= 0{\cdot}46$ ft.	$0{\cdot}397$ ft.
	κ'	$= 334°$	$327°$
P	H_p	$= 0{\cdot}15$ ft.	$0{\cdot}134$ ft.
	κ_p	$= 334°$	$326°$
O	H_o	$= 0{\cdot}14$ ft.	$0{\cdot}160$ ft.
	κ_o	$= 299°$	$302°$

The second column is inserted for the sake of comparison, and gives the results of three years of continuous hourly observation by the Tidal Department of the Survey of India. The concordance between the two affords evidence of the utility of even so short a series of observations as a fortnight.

IV. THE CONSTANTS TO BE USED IN COMPUTING A TIDE TABLE.

The possibility of computing a tide table depends on the knowledge of certain tidal constants appropriate to the port. In the preceding example we have shown how these constants are derivable from a short series of observations. The constants are there presented in what is called the harmonic method, and an example is worked out below for Port Blair, with such constants as have been derived above from a fortnight of observation. The values used, however, are taken from the extended series of observations made by the Indian Survey*.

The harmonic notation is, however, rather recent, and is not adopted in the tide tables of the Admiralty. We must, therefore, show how the principal constants of the harmonic method are derivable from the other

* The incompleteness of the data, with which we are supposed to be working, necessitates the use of certain approximations which would not have been used if "the elliptic tides" had been evaluated.

notation, and thus the present method of computation will be made available, wherever anything is known of the tides.

In the Admiralty tide tables the tides are specified by giving the time of high water at full and change of moon, and the rise at spring and neap. The semidiurnal constants of the harmonic method are derivable from these very easily. Spring rise is the average height between low and high water marks at spring tide; neap rise the average height between high water-mark at neap tide and low water-mark at spring tide; neap range is the average height between high and low water marks at neap tide. The average should be taken from a great many springs and neaps.

Then
$$\mathrm{H}_m + \mathrm{H}_s = \tfrac{1}{2} \text{ spring rise}$$
$$\mathrm{H}_m - \mathrm{H}_s = \tfrac{1}{2} \text{ neap range}$$
$$\mathrm{H}_m \quad = \tfrac{1}{2} \text{ neap rise}$$

If a the age of the tide be known, it may be expressed in hours. Then reading the hours as degrees a may be treated as an angle; and if D be the ratio of the neap rise to the excess of spring rise above neap rise, we have

$$D = \frac{\mathrm{H}_m}{\mathrm{H}_s}$$

If T be the time of H.W. at full and change expressed in hours,

$$\kappa_m \text{ (in degrees)} = 29°T - \tan^{-1}\frac{\sin a}{D + \cos a}, \text{ and } \kappa_s = \kappa_m + a$$

If the age be unknown, we may take a as 36°, and

$$\tan^{-1}\frac{\sin a}{D + \cos a} = \tan^{-1}\frac{3}{5D + 4}$$

For example, at Dungeness, Straits of Magellan (Adm. Tide Table) H.W. at full and change is 8h. 30m. $= 8^{\text{h}}{\cdot}5$; spring rise is 36 ft. to 44 ft., or say, 40 ft., neap rise is 30 ft

Hence $\mathrm{H}_m + \mathrm{H}_s = 20$; $\mathrm{H}_m = 15$; therefore $\mathrm{H}_s = 5$, and $D = \dfrac{\mathrm{H}_m}{\mathrm{H}_s} = 3$.

The age of the tide being unknown, we assume 36 h. as a likely value, so that $a = 36°$, and

$$\tan^{-1}\frac{3}{5D+4} = \tan^{-1}\frac{3}{19} = \tan^{-1}\frac{1}{6{\cdot}33} = 8°$$

Again multiplying the time of H.W. at full and change by 29°, we have $8{\cdot}5 \times 29° = 247°$, so that $\kappa_m = 247° - 8° = 239°$, and $\kappa_s = 239° + 36° = 275°$.

The diurnal inequality is complex, and it seems unnecessary to enter into details excepting in the harmonic notation.

Where it is stated that the tides are "affected by diurnal inequality," it is not possible to predict the tides from the information contained in the so-called tide table.

A tide table is first computed with reference to mean water-mark, but it is usual in navigational works to refer to "the mean level of low water of ordinary spring tides." The datum level may be taken as $H_m + H_s + H' + H_o$ below mean water-mark, and hence to refer to the datum level we must add $H_m + H_s + H' + H_o$ to both H.W. and L.W. heights.

This datum has been defined for the first time in the prefaces to the Indian Tide Tables for 1887, and is called "Indian spring low-water mark." It has been chosen so as to agree as a general rule with "Low water of ordinary spring tides." Accurate agreement was out of the question, since the Admiralty datum does not appear susceptible of an exact scientific definition.

In many estuaries and rivers the water rises much more rapidly than it falls, and we sometimes find a double H.W. To take account of these phenomena we should have to include, according to the schedule for Harmonic Analysis, the terms A_4, B_4, both from the M sheet and S sheet. It is not possible, without devoting too much space to the subject, to show how these "over-tides" are to be included in the computation of the table. It is proper to remark that in such an estuary either the H.W. or L.W., as computed by the method below, may be found considerably in error.

V. The Computation of a Tide Table.

The method of computation will be explained most easily by an actual numerical example*. The computation is divided into a number of sections and schedules, each line of each schedule is independent of all the others, and thus a single tide may be computed as easily as a complete table. The numerical value of any quantity required in the computation of any column of a schedule is written at the top of the column, but outside the boundary line of the schedule. In several cases explanatory headings are also put outside the boundary line, but the process of derivation of each column from what goes before is accurately stated inside the boundary line. It will be stated below in § VI. how the computations may be abridged where accuracy is not desired.

Tide Table for Port Blair E. long. $6^h{\cdot}183$, commencing Feb. 1, 1885.

Tidal constants serving as basis of table—

$$\left.\begin{matrix} H_m = 2{\cdot}022 \\ \kappa_m = 278^\circ \end{matrix}\right\} \quad \left.\begin{matrix} H_s = 0{\cdot}968 \\ \kappa_s = 315^\circ \end{matrix}\right\} \quad \left.\begin{matrix} H'' = 0{\cdot}282 \\ \kappa'' = 310^\circ \end{matrix}\right\}$$

$$\left.\begin{matrix} H' = 0{\cdot}397 \\ \kappa' = 327^\circ \end{matrix}\right\} \quad \left.\begin{matrix} H_p = 0{\cdot}134 \\ \kappa_p = 326^\circ \end{matrix}\right\} \quad \left.\begin{matrix} H_o = 0{\cdot}160 \\ \kappa_o = 302^\circ \end{matrix}\right\}$$

* The reasoning on which the following processes are based is given in a report to the British Association, 1886. [The preceding paper in this volume.]

A. *Computation of Constants for a Fortnight, commencing Feb.* 1, 1885.

N.A. $☊ = 187°$; $\sin ☊ = -\cdot 122$; $\sin 2☊ = +\cdot 24$

$\cos ☊ = -\cdot 993$; $\cos 2☊ = +\cdot 97$

Compute from the formula

$$\Delta = 16°\cdot 51 + 3°\cdot 44 \cos ☊ - 0°\cdot 19 \cos 2☊$$

Therefore

$$\Delta = 16°\cdot 51 - 3°\cdot 416 - 0°\cdot 184 = 12°\cdot 91;\ 2\Delta = 25°\cdot 82 = 25°\ 49'$$

$$\log\cos 2\Delta = 9\cdot 9543$$

By the formulæ in reduction § III. with above value of ☊, we find

$$\nu = -1°\cdot 89;\ \xi = -1°\cdot 75;\ f = 1\cdot 037;\ f' = \cdot 882;\ f_o = \cdot 807$$

$$\nu' = -1°\cdot 22$$

Find mean value of ☉'s parx. and decl. for a fortnight, beginning Feb. 1.

Parx. on Feb. 10 is (N.A.) $8''\cdot 96$; mean parx. (Pref. N.A.) $8''\cdot 85$; diff. $+0''\cdot 11$; multiply by $19\frac{1}{3}$, and read as degrees $= +2°\cdot 13 = 2°\ 8'$; $\sin 2°\ 8' = \cdot 037$; $p_{,} = 1\cdot 037$.

Decl. on Feb. 8 :—

$$\delta_{,} = 14°\ 50';\ 2\delta_{,} = 29°\ 40';\ \cos^2\delta_{,} = \tfrac{1}{2}(1 + \cos 2\delta_{,}) = \tfrac{1}{2} \times 1\cdot 8689 = \cdot 935$$

$$1\cdot 086 p_{,} \cos^2\delta_{,} = \cdot 970;\ H_s = 0\cdot 968;\ 1\cdot 086 p_{,} \cos^2\delta_{,} H_s = 1^{ft}\cdot 020 = S$$

Note that the 1·086 which occurs here is an absolute constant for all times and places.

Compute "age of declinational inequality" as below :—

$$\text{'Age'} = 52^h\cdot 2 \tan(\kappa'' - \kappa_m),\ \text{and}\ \kappa'' - \kappa_m = 310° - 278° = 32°$$

Therefore $\text{'Age'} = 52^h\cdot 2 \tan 32° = 32^h\cdot 6$

The $52^h\cdot 2$ which occurs here is an absolute constant for all ports.

With constants, absolute for all ports, C_1, C_2 (whose logarithms are given below) compute $\alpha = C_1 H'' \cos(\kappa'' - \kappa_m)$; $A = \alpha \cos 2\Delta$; $\beta = C_2 \frac{H''}{H_m} \sin(\kappa'' - \kappa_m)$; $B = \beta \cos 2\Delta$, as follows :—

$\log C_1 = \cdot 6344$	$\log C_2 = 2\cdot 3925$
$\log\cos(\kappa'' - \kappa_m) = 9\cdot 9284$	$\log\sin(\kappa'' - \kappa_m) = 9\cdot 7242$
$\log H'' = 9\cdot 4502$	$\log H'' = 9\cdot 4502$
	$\text{colog}\, H_m = 9\cdot 6942$
$\log \alpha = \cdot 0130$	
$\log\cos 2\Delta = 9\cdot 9543$	$\log \beta = 1\cdot 2611$
	$\log\cos 2\Delta = 9\cdot 9543$
$\log A = 9\cdot 9673$	
$A = 0^{ft}\cdot 927$	$\log B = 1\cdot 2154$
	$B = 16°\cdot 42$

Computation of Angles determining the Position of ☉ and ☽ at 0^h Feb. 1, 1885, Port Blair M.T.

N.A. (Moon's Libration)*—

☽ Jan. 31 =	138°·6	E. long.	6^h·18
1 day's motion =	+ 13·2	Hourly increase of ☽	0°·55
☽ Feb. 1 =	151·8		
Corr^n for E. long. =	− 3·4	Corr^n for E. long.	− 3°·399
☽ =	148·4		
− ξ =	+ 1·8		
☽ − ξ =	150°·2		
2 (☽ − ξ) =	300°·4		
− 2 (☽ − ξ) =	59°·6		

N.A. Sid^l time at 0^h G.M.T. Feb. 1 =	20^h·8	At epoch ☉ =	312°
½ sid^l time =	10·4	$-\nu'$ =	+ 1
		+ 270° =	− 90
Sum, ☉ 0^h Feb. 1 =	312°	V' =	223°
$-\nu$ =	+ 2		
☉ $-\nu$ =	314		
− 2 (☽ − ξ) =	60		
+ 90° =	90		
V_o =	104°		

2☉ $-\nu'$ is to be computed with the mean value of ☉ for a fortnight; add therefore 7° to ☉ at epoch:—

☉ at epoch =	312°	mean 2☉ =	278°
motion for 1 week =	+ 7	$-\nu'$ =	+ 1
mean ☉ =	319°	2☉ $-\nu'$ =	279°

* [The moon's mean longitude is no longer to be found under the heading of 'Moon's Libration' but is on p. 1 of the *Nautical Almanac*.]

Compute as follows:—

$\cos(2\odot - \nu') = +\cdot156$; $3f' = 2\cdot646$; $3f' - \cos(2\odot - \nu') = 2\cdot490$

$\sin(2\odot - \nu') = -\cdot988$

$$\tan\phi = \frac{\sin(2\odot - \nu')}{3f' - \cos(2\odot - \nu')} = -\frac{\cdot988}{2\cdot490} = -\cdot397$$

$$\phi \text{ is } -, \text{ and } \phi = -21^\circ 40' = -21^\circ\cdot7$$

$$R' = \frac{3f' - \cos(2\odot - \nu')}{3\cos\phi} H'; \quad H' = 0\cdot397$$

Compute—

$\kappa' =$	327°	$\log\sec\phi =$	$\cdot0318$
$-\phi =$	$+22$	$\text{colog } 3 =$	$9\cdot5229$
		$\log 2\cdot490 =$	$\cdot3962$
$\kappa' - \phi =$	349	$\log H' =$	$9\cdot5988$
$-V' =$	-223		
		$\log R' =$	$9\cdot5497$
$\kappa' - \phi - V' = \zeta' =$	126°	$R' =$	$0\cdot355$

$$R_o = f_o H_o = \cdot807 \times \cdot160 = \cdot129$$

$\kappa_o =$	302°	$\log f =$	$\cdot0158$
$-V_o =$	-104	$\log H_m =$	$\cdot3058$
Sum $\zeta_o =$	198°	$\log R_m = \log f H_m =$	$\cdot3216$
		$\log 3 =$	$\cdot4771$
		$\log 3 R_m =$	$\cdot7987$
		$R_m =$	$2\cdot097$

Collecting constants.

$R_m = 2^{ft}\cdot097$; $S = 1^{ft}\cdot020$

$\log 3 R_m = \cdot7987$; 'Age' $= 32^h\cdot6$

$A = 0^{ft}\cdot927$; $\log\alpha = \cdot0130$; $R' = 0^{ft}\cdot355$; $R_o = 0^{ft}\cdot129$

$B = 16^\circ\cdot42$; $\log\beta = 1\cdot2611$; $\zeta' = 126^\circ$; $\zeta_o = 198^\circ$

Compute also—

the mean interval $i = \frac{\kappa_m}{29} = 9^h\cdot59$; and $\gamma = \kappa_s - \frac{30}{29}\kappa_m = 27^\circ\cdot4$

S, R′, ζ' must be recomputed for each month, the remaining constants would serve for six months, or perhaps a year of continuous tidal computation. The value of R_o would serve for six months, but care must be taken in computing each month that ζ_o be computed by reference to the first noon of the month as a new epoch.

B. *Parallactic Correction of Lunar Semi-diurnal Tide.*

The semi-range of the lunar semi-diurnal tide is $R_m (= 2^{ft}{\cdot}097)$ and it has to be corrected for the ☽'s parx. The parallactic correction is found by multiplying R_m by the factor p, where

$$p = (☽\text{'s parx.})^3 \div (☽\text{'s mean parx.})^3$$

The ☽'s mean parx. is 57′ 2″, but the ☽'s parx. in question must be taken at a time anterior to H.W. by the "age" ($-32^{h}{\cdot}6$).

To find p (approximately) subtract 57′ 2″ from the ☽'s parx., substitute ° ′ for ′ ″, look out the sine of the angle, multiply it by 3, and add 1 to the result; then p is less than 1 if the ☽'s parx. is below its mean value, because the sine of a − angle is −, and *vice versâ*.

We begin by making a table of $\delta_1 R_m$ (the parallactic corrections to R_m) for each 0′·5 of parx. above or below the mean, to be applied + when the parx. is greater than 57′, and − when it is below.

Tables B and C serve for all time, so long as the same tidal constants are used.

B. Auxiliary Table for Parallactic Corrections, denoted by $\delta_1 R_m$.

log 3 R_m = ·7987

I.	II.	III.	IV.
Minutes of Parx. in excess or defect above or below 57′ 2″	Read Degrees for Minutes in I., and enter log sin (I.)	II. + log 3 R_m	Natural Number of III. $\pm\ \delta_1 R_m$
′ 0·5	7·9408	8·7395	·05
1	8·2419	9·0406	·11
1·5	8·4179	9·2166	·16
2	8·5428	9·3415	·22
2·5	8·6397	9·4384	·27
3	8·7188	9·5175	·33
3·5	8·7857	9·5844	·38
4	8·8436	9·6423	·44
4·5	8·8946	9·6933	·49
5	8·9403	9·7390	·55

C. *Declinational Correction to Lunar Semi-diurnal Tide.*

A declinational correction has also to be applied to R_m, say $\delta_2 R_m$; to κ_m, say $\delta_2\kappa_m$; to i, say $\delta_2 i$; to γ, say $\delta_2\gamma$.

If δ be the ☽'s decl. at a time anterior to H.W. by the 'age';

$$\delta_2 R_m = \alpha \cos 2\delta - A\,;\quad \delta_2\kappa_m = \beta \cos 2\delta - B\,;\quad \delta_2 i = i + \tfrac{1}{29}\,\delta_2\kappa_m\,;\quad \delta_2\gamma = \gamma - \delta_2\kappa_m$$

We begin by forming a table of declinational corrections for each degree of decl. either N. or S.

C. Auxiliary Table for Declinational Corrections, viz., $\delta_2 R_m$, $\delta_2\kappa_m$, and corrected i and γ.

$\log \alpha = \cdot 0130$ $A = \cdot 93$ $\log \beta = 1\cdot 2611$ $B = 16^\circ\cdot 42$ $i = 9^h\cdot 59$ $\gamma = 27^\circ\cdot 4$

I.	II.	III.	IV.	V.	VI.	VII.	VIII.	IX.	X.	XI.
Degrees of N. or S. decl. of ☽	log cos (2 × I.)	log α + II.	Nat. No. of III.	Corrⁿ to R_m IV. − A ($\delta_2 R_m$)	log β + II.	Nat. No. of VI.	Corrⁿ to κ_m VII. − B ($\delta_2\kappa_m$)	VIII. ÷ 29 $= \frac{1}{30}$ VIII. + $\frac{1}{30^2}$ VIII.	Corrᵈ i i + IX.	Corrᵈ γ γ − VIII.
0°	0·0000	·0130	1·03	+·10	1·2611	18°·24	+1°·82	+·061 + ·002 = +·06	9ʰ·65	25°·6
1	9·9997	·0127	1·03	·10	1·2608	18·23	1·81	+·060 + ·002 = ·06	9·65	25·6
2	9·9989	·0119	1·03	·10	1·2600	18·20	1·78	+·059 + ·002 = ·06	9·65	25·6
3	9·9976	·0106	1·02	·09	1·2587	18·14	1·72	+·057 + ·002 = ·06	9·65	25·7
4	9·9958	·0088	1·02	·09	1·2569	18·07	1·65	+·055 + ·002 = ·06	9·65	25·7
5	9·9934	·0064	1·01	·08	1·2545	17·97	1·55	+·052 + ·002 = ·05	9·64	25·8
6	9·9904	·0034	1·01	·08	1·2515	17·84	1·42	+·047 + ·002 = ·05	9·64	26·0
7	9·9869	9·9999	1·00	·07	1·2480	17·70	1·28	+·043 + ·001 = ·04	9·63	26·1
8	9·9828	9·9958	·99	·06	1·2439	17·54	1·12	+·037 + ·001 = ·04	9·63	26·3
9	9·9782	9·9912	·98	·05	1·2393	17·35	0·93	+·031 + ·001 = ·03	9·62	26·5
10	9·9730	9·9860	·97	·04	1·2341	17·14	0·72	+·024 + ·001 = ·03	9·62	26·7
11	9·9672	9·9802	·96	·03	1·2283	16·92	0·50	+·017 + ·001 = ·02	9·61	26·9
12	9·9607	9·9737	·94	+·01	1·2218	16·67	+0·25	+·008 + ·000 = +·01	9·60	27·1
13	9·9537	9·9667	·93	·00	1·2148	16·40	−0·02	−·001 − ·000 = −·00	9·59	27·4
14	9·9459	9·9589	·91	−·02	1·2070	16·11	0·31	−·010 − ·000 = ·01	9·58	27·7
15	9·9375	9·9505	·89	·04	1·1986	15·80	0·62	−·021 − ·001 = ·02	9·57	28·0
16	9·9284	9·9414	·87	·06	1·1895	15·47	0·95	−·032 − ·001 = ·03	9·56	28·4
17	9·9186	9·9316	·85	·08	1·1797	15·13	1·29	−·043 − ·001 = ·04	9·55	28·7
18	9·9080	9·9210	·83	·10	1·1691	14·76	1·66	−·055 − ·002 = ·06	9·53	29·1
19	9·8965	9·9095	·81	−·12	1·1576	14·38	−2·04	−·068 − ·002 = −·07	9·52	29·4

D. *Parallactic and Declinational Corrections to Lunar Semi-diurnal Tide.*

Each H.W. follows a ☽'s transit at the port, approximately, by the interval i ($9^{h}\cdot6$), and we require the ☽'s parx. and decl. at a moment anterior to H.W. by *age of tide.* Times are to be reduced to G.M.T. and round numbers used. To reduce to G.M.T. subtract E. long., $= 6^{h}\cdot2$ for Port Blair. (Add for W. long.) The local time of ☽'s transit is G.M.T. of transit less 2^{m} for each hour of E. long.; for Port Blair less $12^{m} = 0^{h}\cdot2$ (for W. long. add this corrn.)

G.M.T. at which we want ☽'s parx. and decl. $\left\{ \begin{array}{l} = \text{G.M.T. of ☽'s transit} - \text{long. corrn. for transit } (0^{h}\cdot2) - \text{E. long. in time } (6^{h}\cdot2) + \text{mean interval } i\ (9^{h}\cdot6) \dot{-} \text{age of tide } (32^{h}\cdot6) \end{array} \right.$

= G.M.T. of ☽'s transit $- 29^{h}\cdot4$

In the following table we determine roughly this moment of time, in correspondence with each transit of ☽, look out parx. and decl. and find corrected R_m from auxiliary tables B and C. These corrected values we shall call m_0, M_0; m_1, M_1, &c., large letters being associated with upper, and small with lower transits and the subscript numbers being the numbers of the days of the tide table, the first day being numbered zero.

The corrected intervals, also from tables B and C, we call i_0, I_0; i_1, I_1; &c., and the corrected values of $\kappa_s - \frac{30}{29}\kappa_m$ (or γ) we call γ_0, Γ_0; γ_1, Γ_1; &c.

D.

Determine the times at which to find ☽'s parx. and decl.			Find excess of parx. above mean		Parx. corrⁿ to R_m, $R_m=2·10$		Decl. corrⁿ to R_m		Corr^d heights	Corr^d intervals	Corr^d γ's
I.	II.	III.	IV.	V.	VI.	VII.	VIII.	IX.	X.	XI.	XII.
Date of ☽'s upper or lower transit	G.M.T. to nearest hour, from N.A.	Date of II. – 29^h	☽'s parx. at the Greenwich noon or midn. nearest to III. from N.A.	IV. – 57′	$\delta_1 R_m$ interpolated from IV. of B	R_m + VI.	☽'s decl. at III. from N.A.	$\delta_2 R_m$ interpolated from V. of C	VII. + IX.	Corr^d interval i interpolated from X. of C	Corr^d γ interpolated from XI. of C
	h.	h.	′	′	ft.	ft.	°	ft.	ft.	h.	°
Feb. 1 L	2	Jan. 30, 21	60·0	+3·0	+·33	2·43	N 11·0	+·03	$m_0=2·46$	$i_0=9·61$	$\gamma_0=26·9$
U	14	31, 9	59·6	2·6	·28	2·38	9·1	·05	$M_0=2·43$	$I_0=9·62$	$\Gamma_0=26·5$
2 L	3	22	59·2	2·2	·24	2·34	6·9	·07	$m_1=2·41$	$i_1=9·63$	$\gamma_1=26·1$
U	15	Feb. 1, 10	58·8	1·8	·20	2·30	4·7	·08	$M_1=2·38$	$I_1=9·64$	$\Gamma_1=25·8$
3 L	4	23	58·4	1·4	·15	2·25	2·4	·10	$m_2=2·35$	$i_2=9·65$	$\gamma_2=25·6$
U	16	2, 11	57·9	0·9	·10	2·20	N 0·3	·10	$M_2=2·30$	$I_2=9·65$	$\Gamma_2=25·6$
4 L	4	23	57·5	+0·5	+·05	2·15	S 1·9	·10	$m_3=2·25$	$i_3=9·65$	$\gamma_3=25·6$
U	17	3, 12	57·0	0·0	·00	2·10	4·1	·09	$M_3=2·19$	$I_3=9·65$	$\Gamma_3=25·7$
5 L	5	4, 0	56·6	–0·4	–·04	2·06	6·1	·08	$m_4=2·14$	$i_4=9·64$	$\gamma_4=26·0$
U	17	12	56·2	0·8	·09	2·01	7·9	·06	$M_4=2·07$	$I_4=9·63$	$\Gamma_4=26·3$
6 L	6	5, 1	55·8	1·2	·13	1·97	9·8	·04	$m_5=2·01$	$i_5=9·62$	$\gamma_5=26·7$
U	18	13	55·4	1·6	·17	1·93	11·4	+·02	$M_5=1·95$	$I_5=9·61$	$\Gamma_5=27·0$
7 L	7	6, 2	55·1	1·9	·21	1·89	13·0	·00	$m_6=1·89$	$i_6=9·59$	$\gamma_6=27·4$
U	19	14	54·8	2·2	·24	1·86	14·3	–·03	$M_6=1·83$	$I_6=9·58$	$\Gamma_6=27·8$
8 L	7	7, 2	54·6	2·4	·26	1·84	15·4	·05	$m_7=1·79$	$i_7=9·57$	$\gamma_7=28·2$
U	20	15	54·4	–2·6	–·28	1·82	S 16·5	–·07	$M_7=1·75$	$I_7=9·55$	$\Gamma_7=28·6$

E. *Determination of Local Mean and apparent Times of Moon's Transit, and of Angles for Computing the Fortnightly Inequality.*

It is convenient to treat the upper and lower transits separately in schedules of similar forms. The angle x_0, X_0, x_1, X_1, &c., on which the fortnightly inequality of time and height depends, is twice the apparent time of transit converted to angle at 15° per hour, and with the corresponding angles γ_0, Γ_0, γ_1, Γ_1, &c. subtracted.

The corrections for long. of port are − for E., + for W. long.

E. Lower Transits.

Nautical Almanac		Port Blair M.T. of transit	Port Blair appt. time of transit Equation of time = +14m		Twice appt. time converted to angle		Required angles
I.	II.	III.	IV.	V.	VI.	VII.	VIII.
Date	G.M.T. of ☽'s transit	II. − 2m per hour of E. long.	III. − equation of time	IV. in decimals	V. × 30	Enter γ from XII. of D	VI. − VII.
1885	h. m.	h. m.	h. m.	h.			
Feb. 1	1 50	1 38	1 24	1·40	42°·0	26°·9	x_0= 15°·1
" 2	2 41	2 29	2 15	2·25	67·5	26·1	x_1= 41·4
" 3	3 30	3 18	3 4	3·07	92·1	25·6	x_2= 66·5
" 4	4 17	4 5	3 51	3·85	115·5	25·6	x_3= 89·9
" 5	5 4	4 52	4 38	4·63	138·9	26·0	x_4=112·9
" 6	5 50	5 38	5 24	5·40	162·0	26·7	x_5=135·3
" 7	6 36	6 24	6 10	6·17	185·1	27·4	x_6=157·7
" 8	7 23	7 11	6 57	6·95	208·5	28·2	x_7=180·3

E. Upper Transits.

Nautical Almanac		Port Blair M.T. of transit	Port Blair appt. time of transit Equation of time = +14m		Twice appt. time converted to angle		Required angles
I.	II.	III.	IV.	V.	VI.	VII.	VIII.
Date	G.M.T. of ☽'s transit	II. − 2m per hour of E. long.	III. − equation of time	IV. in decimals	V. × 30	Enter Γ from XII. of D	VI. − VII. − 360°
1885	h. m.	h. m.	h. m.	h.			
Feb. 1	14 15	14 3	13 49	13·82	414°·6	26°·5	X_0= 28°·1
" 2	15 5	14 53	14 39	14·65	439·5	25·8	X_1= 53·7
" 3	15 53	15 41	15 27	15·45	463·5	25·6	X_2= 77·9
" 4	16 40	16 28	16 14	16·23	486·9	25·7	X_3=101·2
" 5	17 27	17 15	17 1	17·02	510·6	26·3	X_4=124·3
" 6	18 13	18 1	17 47	17·78	533·4	27·0	X_5=146·4
" 7	18 59	18 47	18 33	18·55	556·5	27·8	X_6=168·7
" 8	19 46	19 34	19 20	19·33	579·9	28·6	X_7=191·3

F. *Figure for Times and Heights of Semi-diurnal Tide.*

The next step is to find the resultant of the lunar and solar tides. Draw the straight line OA. Produce AO to S, and take OS = S (for Port Blair S = 1ft·020) on any convenient scale. With OA as initial line set off the angles x_0, X_0, x_1, X_1, found in VIII. of E, for a fortnight; a new figure is desirable for the next fortnight. On Om_0, OM_0, Om_1, OM_1, &c. set off to adopted scale the heights m_0, M_0, m_1, M_1, &c. found in X. of D.

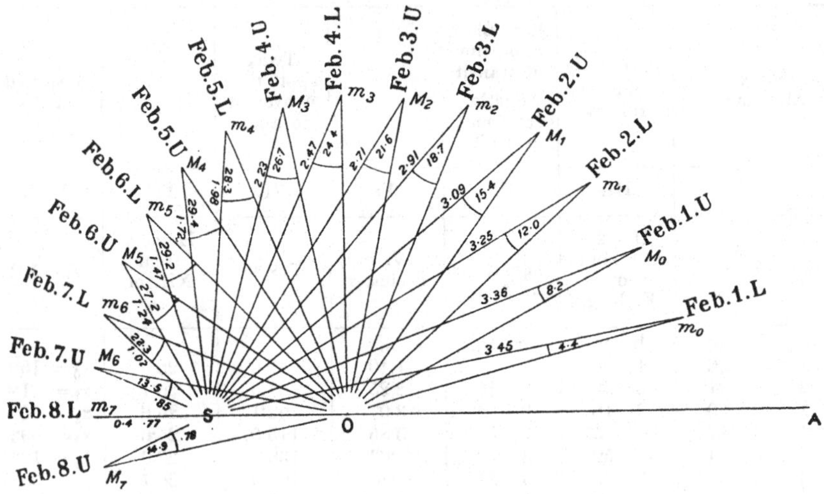

Join m_0S, M_0S, m_1S, M_1S, &c., and measure all the lengths m_0S, M_0S, m_1S, M_1S, &c. on adopted scale. These are the successive heights of H.W.

Measure all the angles Om_0S, OM_0S, &c. and count them as + in the upper half of the figure, and − in the lower half. Each height and each angle is associated with one upper or lower transit of ☽.

G. *Formation of Tide Table for Semi-diurnal Tides.*

The successive heights Om_0, OM_0, Om_1, OM_1, &c., of H.W. have been found graphically in F. The angles Om_0S, OM_0S, Om_1S, OM_1S, &c. found in the figure F must be reduced to time at the rate of 29° per hour, and subtracted from the mean time of ☽'s transit found in III. of E, to these have to be added the corrected intervals i_0, I_0, i_1, I_1, &c. from XI. of D. Next a time half-way between consecutive H.W.'s is taken as the time of L.W.; and the height of L.W. is taken as the mean of the H.W. before and after with a − prefixed.

These processes are carried out in the following schedule. If the time comes out greater than 24^h, the tide in question really belongs to the next day.

G.

						H.W. times	L.W. times	H.W. heights	L.W. heights
I.	II.	III.	IV.	V.	VI.	VII.	VIII.	IX.	X.
Day, and upper or lower transit	Angles OMS from figure	II.÷29	Mean time of transit from III. of E. in decimals	IV. − III.	Corrected *i* from XI. of D	V. + VI.	Interpolate in VII.	Hts. from figure	Interpolate in IX.
1885		h.	h.	h.	h.	h.	h.	ft.	ft.
Feb. 1. L	+ 4·4	+0·15	1·63	1·48	9·61	11·09		3·45	
							17·24		−3·41
U	8·2	0·28	14·05	13·77	9·62	23·39		3·36	
							5·55		−3·31
„ 2. L	12·0	0·41	2·48	2·07	9·63	11·70		3·25	
							17·85		−3·17
U	15·4	0·53	14·88	14·35	9·64	24·00		3·09	
							6·15		−3·00
„ 3. L	18·7	0·64	3·30	2·66	9·65	12·31		2·91	
							18·46		−2·81
U	21·6	0·74	15·69	14·95	9·65	24·60		2·71	
							6·75		−2·59
„ 4. L	24·4	0·84	4·08	3·24	9·65	12·89		2·47	
							19·05		−2·35
U	26·7	0·92	16·47	15·55	9·65	25·20		2·23	
							7·36		−2·10
„ 5. L	28·3	0·98	4·86	3·88	9·64	13·52		1·98	
							19·70		−1·85
U	29·4	1·01	17·25	16·24	9·63	25·87		1·72	
							8·06		−1·59
„ 6. L	29·2	1·01	5·63	4·62	9·62	14·24		1·47	
							20·47		−1·35
U	27·2	0·94	18·02	17·08	9·61	26·69		1·24	
							8·96		−1·13
„ 7. L	22·3	0·77	6·40	5·63	9·59	15·22		1·02	
							21·56		− ·94
U	+13·5	+0·47	8·78	18·31	9·58	27·89		·85	
							10·33		− ·81
„ 8. L	− 0·4	−0·01	7·19	7·20	9·57	16·77		·77	
							23·20		− ·77
U	−14·9	−0·51	19·57	20·08	9·55	29·63		·78	

H. *Correction for Diurnal Inequality.*

We first enter the tide table as found in the final columns of G, bringing, however, the times greater than 24^h into the succeeding day.

The successive processes are then adequately explained in the following schedule.

Columns VII., VIII., and XIII., XIV. are conveniently taken from a traverse table, for disregarding the decimal points, R' or R_o may be considered as distances run, the angle in VI. or XI., as the "course," and R' cos VI. or R_o cos XI. is the "Diff. lat.," whilst R' sin VI. or R_o sin XI. is the "Dep."

Tide Table without diurnal corrections				$\zeta' = 126°$	Angles on which first diurnal tide depends		First correction to heights R′ = 0·36 ft.		
I.	II.	III.	IV.	V.	VI.		VII.	VIII.	IX.
No. of Day, Date, and H.W. or L.W.	Heights from IX. and X. of G	Times from VII. and VIII. of G	First entry of III. for each day multiplied by 15	IV. − ζ′	1. V. + No. of Day	2. Fill blanks in 1, by adding once, twice, thrice, 93°	R′ cos VI.	R′ sin VI.	Subtract 1° from IV. for each hour of III.
	ft.	h.					ft.		
0. Feb. 1. H.W.	+3·45	11·09	166°	40°	40°		+·27	+·23	155°
L.W.	−3·41	17·24				133°	−·25	+·27	
H.W.	+3·36	23·39				226°	−·25	−·26	
1. Feb. 2. L.W.	−3·31	5·55	83	317	318		+·27	−·24	77
H.W.	+3·25	11·70				51	+·23	+·28	
L.W.	−3·17	17·85				144	−·29	+·21	
2. Feb. 3. H.W.	+3·09	0·00	0	234	236		−·20	−·30	0
L.W.	−3·00	6·15				329	+·31	−·19	
H.W.	+2·91	12·31				62	+·17	+·32	
L.W.	−2·81	18·46				155	−·33	+·15	
3. Feb. 4. H.W.	+2·71	·60	9	243	246		−·15	−·33	8
L.W.	−2·59	6·75				339	+·34	−·12	
H.W.	+2·47	12·89				72	+·11	+·34	
L.W.	−2·35	19·05				165	−·35	+·09	
4. Feb. 5. H.W.	+2·23	1·20	18	252	256		−·09	−·35	17
L.W.	−2·10	7·36				349	+·35	−·07	
H.W.	+1·98	13·52				82	+·05	+·36	
L.W.	−1·85	19·70				175	−·36	+·03	
5. Feb. 6. H.W.	+1·72	1·87	20	262	267		−·02	−·36	26
L.W.	−1·59	8·06				0	+·36	+·00	
H.W.	+1·47	14·24				93	−·02	+·36	
L.W.	−1·35	20·47				186	−·30	−·04	
6. Feb. 7. H.W.	+1·24	2·69	40	274	280		+·06	−·35	37
L.W.	−1·13	8·96				13	+·35	+·08	
H.W.	+1·02	15·22				106	−·10	+·35	
L.W.	− ·94	21·56				199	−·34	−·12	
7. Feb. 8. H.W.	+ ·85	3·89	58	294	299		+·17	−·32	54
L.W.	− ·81	10·33				32	+·31	+·19	
H.W.	+ ·77	16·77				125	−·21	+·29	
L.W.	− ·77	23·20				218	−·28	−·22	
8. Feb. 9. H.W.	+ ·78	5·63	84	318	326		+·30	−·20	78

	Angles on which second diurnal tide depends $\zeta_o = 198°$		Second correction to heights $R_o =$ 0·13 ft.			Total correction to heights	Corrected heights	Reduction to "Indian Spring Low Water." $H_m + H_s + H' + H_o = 3·55$ ft.	Total correction to time	Corrected times
X.	XI.		XII.	XIII.	XIV.	XV.	XVI.	XVII.	XVIII.	XIX.
Subtract from IX. $25\frac{1}{3}$ multiplied by No. of day	1. $X. - \zeta_o$	2. Fill blanks in 1, by adding once, twice, thrice, 87°	R_o cos XI.	R_o sin XI.	VIII. + XIII.	VII. + XII.	XV. + II.	XVI. + H_m + H_s + H′ + H_o	XIV. ÷ II.	III. − XVIII.
			ft.		h.	ft.	ft.	ft.	h.	h.
155°	317°		+·10	−·09	+·14	+·37	+3·82	7·37	+·04	11·05
		44°	+·09	+·09	+·36	−·16	−3·57	−·02	−·11	17·35
		131	−·09	+·10	−·16	−·34	+3·02	6·57	−·05	23·44
52	214		−·11	−·07	−·31	+·16	−3·15	·40	+·09	5·46
		301	+·07	−·11	+·17	+·30	+3·55	7·10	+·05	11·65
		28	+·12	+·06	+·27	−·17	−3·34	·21	−·09	17·94
309	111		−·05	+·12	−·18	−·25	+2·84	6·39	−·06	·05
		198	−·12	−·04	−·23	+·19	−2·81	·74	+·08	6·07
		285	+·03	−·13	+·19	+·20	+3·11	6·66	+·07	12·24
		12	+·13	+·03	+·18	−·20	−3·01	·54	−·06	18·52
292	94		−·01	+·13	−·20	−·16	+2·55	6·10	−·07	·67
		181	−·13	−·00	−·12	+·21	−2·38	1·17	+·05	6·70
		268	−·00	−·13	+·21	+·11	+2·58	6·13	+·09	12·80
		355	+·13	−·01	+·08	−·22	−2·57	·98	−·03	19·08
276	78		+·03	+·13	−·22	−·06	+2·17	5·72	−·10	1·30
		165	−·13	+·03	−·04	+·22	−1·88	1·67	+·02	7·34
		252	−·04	−·12	+·24	+·01	+1·99	5·54	+·12	13·40
		339	+·12	−·02	+·01	−·24	−2·09	1·46	+·01	19·69
259	61		+·06	+·11	−·25	+·04	+1·76	5·31	−·14	2·01
		148	−·11	+·07	+·07	+·25	−1·34	2·21	−·04	8·10
		235	−·08	−·11	+·25	−·10	+1·37	4·92	+·17	14·07
		322	+·10	−·08	−·12	−·26	−1·61	1·94	+·08	20·39
245	47		+·09	+·10	−·25	+·15	+1·39	4·94	−·20	2·89
		134	−·09	+·09	+·17	+·26	− ·87	2·68	−·15	9·11
		221	−·10	−·09	+·26	−·20	+ ·82	4·37	+·25	14·97
		308	+·10	−·09	−·21	−·24	−1·18	2·37	+·22	21·34
237	39		+·10	+·08	−·24	+·27	+1·12	4·67	−·28	4·17
		126	−·08	+·11	+·30	+·23	− ·58	2·97	−·37	10·70
		213	−·11	−·07	+·22	−·32	+ ·45	4·00	+·29	16·48
		300	+·07	−·11	−·33	−·21	− ·98	2·57	+·43	22·77
235	37		+·10	+·08	−·12	+·40	+1·18	4·73	−·15	5·78

K. *Final Tide Table.*

It remains to reduce the decimals of an hour to minutes, and to change from the astronomical to the civil date at the port. It will be found more convenient to keep the heights in decimals of a foot, and not reduce to inches. The times and heights are given in XIX. and XVII. of table H.

K. TIDE TABLE for Port Blair, the Heights being referred to "Indian Spring Low Water Mark."

Civil Date		Times of H.W. and L.W.		Heights of H.W. and L.W.		Civil Date		Times of H.W. and L.W.		Heights of H.W. and L.W.	
1885		h.	m.		ft.	1885		h.	m.		ft.
Feb. 1.	p.m.	11	3	H.W.	7·4	Feb. 6.	a.m.	1	24	H.W.	5·5
" 2.	a.m.	5	21	L.W.	·0		a.m.	7	41	L.W.	1·5
	a.m.	11	26	H.W.	6·6		p.m.	2	1	H.W.	5·3
	p.m.	5	28	L.W.	·4		p.m.	8	6	L.W.	2·2
	p.m.	11	39	H.W.	7·1	" 7.	a.m.	2	4	H.W.	4·9
" 3.	a.m.	5	56	L.W.	·2		a.m.	8	23	L.W.	1·9
	p.m.	0	3	H.W.	6·4		p.m.	2	53	H.W.	4·9
	p.m.	6	4	L.W.	·7		p.m.	9	7	L.W.	2·7
" 4.	a.m.	0	14	H.W.	6·7	" 8.	a.m.	2	58	H.W.	4·4
	a.m.	6	31	L.W.	·5		a.m.	9	20	L.W.	2·4
	p.m.	0	40	H.W.	6·1		p.m.	4	10	H.W.	4·7
	p.m.	6	42	L.W.	1·2		p.m.	10	42	L.W.	3·0
" 5.	a.m.	0	48	H.W.	6·1	" 9.	a.m.	4	29	H.W.	4·0
	a.m.	7	5	L.W.	1·0		a.m.	10	46	L.W.	2·6
	p.m.	1	18	H.W.	5·7		p.m.	5	47	H.W.	4·7
	p.m.	7	20	L.W.	1·7						

In the official Indian Tide Tables the tides of Port Blair are referred to a datum 3·13 ft. below mean water, that is to say 0·42 ft. higher than the datum here used. To effect a comparison then subtract 0·42 ft. from all these heights, and the concordance will be found fairly satisfactory.

The Indian Tide Tables are formed by the tide-predicting instrument, by which the approximations here used are avoided, and are based on much wider data than those supposed to be here available.

VI. On Abridgements which may be adopted in Computing a Tide Table.

For navigational purposes a very rough tide table will often suffice. Such a table may be computed as follows:

H_m, H_s, κ_m, κ_s, and mean establishment may be derived from spring and neap rise, age, and establishment as shown in § IV. If the "age" be unknown it may be assumed as 36 h., and $\kappa_s - \kappa_m$ may be taken as 36°.

Then let A be the apparent time of any ☽'s transit reduced to angle at 30° per hour, and we have for the height of H.W. from spring L.W. mark

$$2H_m + H_s[1 + \cos(A - \kappa_s + \kappa_m)]$$

and for the height of L.W. from same level

$$H_s[1 - \cos(A - \kappa_s + \kappa_m)]$$

The time of H.W. is

$$\text{M.T. of ☽'s tr.} + \text{mean estab.} - 2^h \frac{H_s}{H_m} \sin(A - \kappa_s + \kappa_m)$$

And the time of L.W. is 6 h. 12 m. later, or half-way between two consecutive H.W.'s computed by above rule.

For example:

At Port Blair $H_m = 2{\cdot}0$ ft., $H_s = 1{\cdot}0$ ft., $\kappa_s - \kappa_m = 37°$, mean establishment $= 9^h{\cdot}6$; and we found M.T. of ☽'s lower transit on Feb. 5, 1885 = 4 h. 52 m. $= 4^h{\cdot}9$, and appt. time of transit reduced to angle at 30° per hour is 139°, so that A = 139°.

Then $A - \kappa_s + \kappa_m = 102°$; $\cos 102° = -0{\cdot}2$; $\sin 102° = +1{\cdot}0$

$$H_s[1 + \cos 102°] = 0{\cdot}8; \quad 2H_m = 4{\cdot}0$$

$$H_s[1 - \cos 102°] = 1{\cdot}2; \quad 2^h \frac{H_s}{H_m} \sin 102° = 2 \times \tfrac{1}{2} \times 1 = 1^h{\cdot}0$$

$$\text{Time of H.W.} = 4^h{\cdot}9 + 9^h{\cdot}6 - 1^h{\cdot}0 = 13^h{\cdot}5$$

$$\text{Time of L.W.} = 13^h{\cdot}5 + 6^h{\cdot}2 = 19^h{\cdot}7$$

Hence H.W., Feb. 6, at 1 h. 30 m. a.m., height 4·8 ft.

L.W., Feb. 6, at 7 h. 42 m. a.m., height 1·2 ft.

It must be noticed that we are here supposed to know nothing of the diurnal tides, and the datum level being $H_m + H_s$ or 3·0 ft. below mean water is considerably higher than that used above.

The results are more nearly in accordance with the complete value as found in the preceding section than would usually be the case.

A graphical method of using the same data would be more accurate. The figure would be the same as that of the last section, but the m's and M's would be determined by sweeping a circle with radius H_m about O as centre, and OS would be taken as equal to H_s

The further step in accuracy would be to proceed as in computation of § V., but to compute auxiliary tables B. and C. for each minute of parx. and each 2° of decl. only. Table D. may be abridged by computing corrections for parx. and decl. for upper transits only, and columns XI., XII. may be omitted entirely. Table E. for lower transit may be omitted. Figure F. may be drawn for upper transits only, and the entries in G. for lower transits may be filled in by interpolation. In Table H. for diurnal tides only the first entry for each day in VI. (2) and XI. (2) need be made, and only the first two entries for each day of VII., VIII., XII., XIII., XIV., XVIII. computed. The third and fourth entries for each day of XV. and XIX. may be taken as respectively numerically equal to the first and second ones, but with the opposite signs. These abridgements would reduce the computation by nearly a half. Other abridgements will doubtless occur to the computer, but they will all involve loss of accuracy.

VII. Works of Reference.

A general account of the theory of tides will be found in most Popular Astronomies, but we are not aware of any book which gives a complete exposition of tidal theory and practice. Airy's well known article on "Tides and Waves" in the *Encyclopædia Metropolitana* may be referred to, but as great advances have been made since the time of its publication, it would seem preferable to refer to the article by the present writer, which is about to be contributed to the *Encyclopædia Britannica**.

A complete list of all papers on the tides published since the time of Newton will be found in the *Bibliographie Astronomique,* Houzeau and Lancaster, Brussels, 1882. For an account of the harmonic method and its connexion with the method of hour angles, &c., see the Reports to the British Association for 1883 and 1885, and 1886, for an explanation of the methods here used [Papers 1, 2, 3 above].

Tables of the harmonic tidal constants at a considerable number of ports are given in a paper by A. W. Baird and G. H. Darwin in the *Proceedings of the Royal Society,* 1885.

Computation forms for the reduction of a long series of tidal observations, and copies of the *British Association Report,* 1883, may be purchased of the Cambridge Scientific Instrument Company.

A manual of practical tidal observation by Major A. W. Baird, R.E., will shortly be on sale by Messrs Taylor and Francis, Red Lion Court, Fleet Street.

* [Since this time my own *Tides and Kindred Phenomena in the Solar System* (Murray) has been published.]

5.

ON THE HARMONIC ANALYSIS OF TIDAL OBSERVATIONS OF HIGH AND LOW WATER.

[*Proceedings of the Royal Society*, XLVIII. (1890), pp. 278—340.]

§ 1. *Introduction.*

EXTENSIVE use of the tide-gauge has only been made in recent years, and by far the largest number of tidal records consist only of observations of high and low water (H. and L.W.). Such observations have usually been reduced by determining the law governing the relationship between the times and heights of H. and L.W. and the positions of the moon and sun. This method is satisfactory so long as the diurnal inequalities are small, but it becomes both complex and unsatisfactory when the diurnal inequality is large. In such cases the harmonic notation for the tide is advantageous, and as, except in the North Atlantic Ocean, the diurnal inequality is generally considerable, a proper method of evaluating the harmonic constants from H. and L.W. observations is desirable.

The essential difference between the method here proposed and that followed by Laplace and his successors is that they introduced astronomical considerations from the first and applied them to each H. and L.W., whereas the positions of the sun and moon will only be required here at a single instant of time. In their method, the time of moon's transit, and hence the interval, was found for each tide; the age of the moon, and the moon's and sun's parallaxes and declinations were also required. An extensive table from the astronomical ephemeris was thus necessary, and there still remained the classification of heights and intervals according to the age of moon, and two parallaxes, and two declinations. The classification could hardly be less laborious, and was probably less mechanical, than the sorting processes employed below. There is probably, therefore, a considerable saving of labour in the present method, and, besides, I conceive that the results are more satisfactory when expressed in the harmonic notation.

My object has been to make the whole process a purely mechanical one, and, although nothing can render the reduction of tidal observations a light piece of work, I believe that it is here presented in a form which is nearly as short as possible.

The analytical difficulties to be encountered in such a task are small, but the arrangement of a heavy mass of arithmetic, so as to involve a minimum of labour and therefore of expense, is by no means easy. How far I have succeeded must be left to the decision of those who will, I hope, use the methods here devised.

When a question of this kind is attacked, the solution cannot be deemed complete unless the investigation is left in such a state that an ordinary trained computer is able to use it as a code of instructions by which to reduce a series of observations, without any knowledge of tidal theory.

An actual numerical example is thus essential, both to test the method and to serve as instructions to a computer. The Appendix contains so much of the reduction of three months of observation at Bombay as will serve as such a code. If the series be longer than three months, or in such cases as the proper treatment of gaps in the series, it is necessary to refer back to the body of the paper for instructions.

I now pass to the theoretical reasons for the rules for reduction.

§ 2. *Notation.*

The notation of the Report to the British Association for 1883 [Paper 1], and in use in the Indian tidal work and elsewhere, is here followed.

The earth's angular velocity is denoted by γ; the hourly mean motions of the moon, sun, and lunar perigee by σ, η, ϖ (γῆ, σελήνη, ἥλιος); the mean longitudes of moon, sun, and lunar perigee by s, h, p, and the mean solar hour angle by t. The R.A. and longitude in the lunar orbit of the intersection of the equator with the lunar orbit are ν, ξ; and N is the longitude of the moon's node.

The several harmonic tides are denoted by arbitrarily chosen initial letters. Those with which we shall principally have to deal are—

Semi-diurnal.

Name	Initial	Speed	Equilibrium argument
Principal lunar ...	M_2	$2(\gamma-\sigma)$	$2t+2(h-\nu)-2(s-\xi)$
,, solar ...	S_2	$2(\gamma-\eta)$	$2t$
Luni-solar	K_2	2γ	$2t+2(h-\nu'')$
Larger elliptic ...	N	$2\gamma-3\sigma+\varpi$	$2t+2(h-\nu)-2(s-\xi)-(s-p)$
Smaller ,, ...	L	$2\gamma-\sigma-\varpi$	$2t+2(h-\nu)-2(s-\xi)+(s-p)+\pi$

Diurnal.

Name	Initial	Speed	Equilibrium argument
Luni-solar	K_1	γ	$t+(h-\nu')-\frac{1}{2}\pi$
Lunar..............	O	$\gamma-2\sigma$	$t+(h-\nu)-2(s-\xi)+\frac{1}{2}\pi$
Solar	P	$\gamma-2\eta$	$t-h+\frac{1}{2}\pi$

The symbol H denotes the mean semi-range of any one of the tides, and κ its retardation of phase behind what it would be according to the equilibrium theory; f denotes a certain factor of augmentation of the lunar and luni-solar tides depending on the value of N.

The particular tide to which H, κ, f refer will in general be indicated by a subscript small letter, the same as the letter constituting the initial of the tide. Thus, for example, the M_2 tide is expressed by

$$f_m H_m \cos\{2t+2(h-\nu)-2(s-\xi)-\kappa_m\}$$

I have allowed a departure from this notation in the case of the tides K_2 and K_1, where I write H″, κ'', f″ for the first, and H′, κ', f′ for the second. The angles $2\nu''$ and ν' (which, like ν and ξ, are functions of N) are also involved in the arguments* (or angle under the cosine in the expression for the height of the particular tide) of these two tides.

It is obviously necessary to suppose the reader to have some acquaintance with the harmonic notation, or it would be necessary to repeat the Report on Tides above referred to.

§ 3. *The General Method of Treating H. and L.W. Observations.*

Noon of the day on which the observations begin is to be taken as the epoch, and the mean solar time elapsed since epoch is denoted by t. V with the proper subscript letter denotes the increase of argument since epoch; for example, $V_m = 2(\gamma-\sigma)t$.

Then the height of the water h, estimated from mean sea-level, is expressed by a number of terms of the form $A\cos V + B\sin V$, or, in an alternative form, $R\cos(V-\zeta)$.

In order to explain the principle of the method proposed, let us take two typical terms involving V_p and V_q, and let the rates of increase of V_p be p, and of V_q be q.

Then we have

$$h = A_p\cos V_p + B_p\sin V_p + A_q\cos V_q + B_q\sin V_q \;\ldots\ldots\ldots\ldots(1)$$

Since at H. or L.W. h is a maximum or a minimum, we must have

$$0 = A_p\sin V_p - B_p\cos V_p + \frac{q}{p}A_q\sin V_q - \frac{q}{p}B_q\cos V_q \;\ldots\ldots\ldots(2)$$

* It is well to explain that I have sometimes elsewhere used argument to denote the argument according to the equilibrium theory, that is to say, with κ equal to zero. In this paper I call the latter the equilibrium argument.

Let us write $$\frac{q}{p} = k_q \quad \text{.................................(3)}$$

Then multiply (1) by $\cos V_p$ and (2) by $\sin V_p$, and add; and again multiply (1) by $\sin V_p$ and (2) by $\cos V_p$, and subtract, and we have

$$\left.\begin{aligned} h\cos V_p &= A_p + A_q(\cos V_p \cos V_q + k_q \sin V_p \sin V_q) \\ &\quad + B_q(\cos V_p \sin V_q - k_q \sin V_p \cos V_q) \\ h\sin V_p &= B_p + A_q(\sin V_p \cos V_q - k_q \cos V_p \sin V_q) \\ &\quad + B_q(\sin V_p \sin V_q + k_q \cos V_p \cos V_q) \end{aligned}\right\} \text{.........(4)}$$

Let
$$\left.\begin{aligned} \Sigma &= \tfrac{1}{2}\cos(V_p - V_q) + \tfrac{1}{2}\cos(V_p + V_q) = \cos V_p \cos V_q \\ \Delta &= \tfrac{1}{2}\cos(V_p - V_q) - \tfrac{1}{2}\cos(V_p + V_q) = \sin V_p \sin V_q \\ \sigma &= \tfrac{1}{2}\sin(V_p - V_q) + \tfrac{1}{2}\sin(V_p + V_q) = \sin V_p \cos V_q \\ \delta &= \tfrac{1}{2}\sin(V_p - V_q) - \tfrac{1}{2}\sin(V_p + V_q) = -\cos V_p \sin V_q \end{aligned}\right\} \text{......(5)}$$

Also let
$$\left.\begin{aligned} \mathrm{F} &= \Sigma + k_q\Delta, & \mathrm{f} &= \sigma + k_q\delta \\ \mathrm{G} &= -\delta - k_q\sigma, & \mathrm{g} &= \Delta + k_q\Sigma \end{aligned}\right\} \text{..................(6)}$$

Then our equations are

$$\left.\begin{aligned} h\cos V_p &= A_p + \mathrm{F}A_q + \mathrm{G}B_q \\ h\sin V_p &= B_p + \mathrm{f}A_q + \mathrm{g}B_q \end{aligned}\right\} \text{.....................(7)}$$

A similar pair of equations will result from each H. and L.W. When a series of tides is considered, we may take the mean of the equations and substitute a mean F, G, f, g.

The general principle here adopted is to take the means over such periods that the mean F, G, f, g become very small. In fact, we shall, in several cases, be able to reduce them so far that these terms are negligible, and get simply $\frac{1}{n+1}\Sigma h \frac{\cos}{\sin} V_p = \frac{A_p}{B_p}$; but in other cases, where what is typified as the p tide is a small one, whilst one or more of the tides typified as q is large, it will be necessary to find F, G, f, g. The finding of these coefficients is clearly reducible to the finding of the mean values of $\frac{\cos}{\sin}(V_p \pm V_q)$.

Another useful principle may be illustrated thus: if the q tide does not differ much in speed from the p tide, we may put $V_q = V_p + \nu t$, where ν is a small speed. Then we write

$$\begin{aligned} h &= R_p \cos(V_p - \zeta_p) + R_q \cos(V_p + \nu t - \zeta_q) \\ &= \cos V_p \{R_p \cos \zeta_p + R_q \cos(\nu t - \zeta_q)\} \\ &\quad + \sin V_p \{R_p \sin \zeta_p - R_q \sin(\nu t - \zeta_q)\} \end{aligned}$$

If we neglect ν/p, the condition for maximum and minimum in conjunction with this gives

$$\begin{aligned} h\cos V_p &= R_p \cos \zeta_p + R_q \cos(\nu t - \zeta_q) \\ h\sin V_p &= R_p \sin \zeta_p - R_q \sin(\nu t - \zeta_q) \end{aligned}$$

Then taking the mean of these equations over a period beginning with $t=0$ and ending when $t=\pi/\nu$, we have (writing $A_p = \mathrm{R}_p \cos \zeta_p$, $B_p = \mathrm{R}_p \sin \zeta_p$)

$$\frac{1}{n+1} \Sigma h \cos V_p = A_p + \lambda R_q \cos(\alpha - \zeta_q)$$

$$\frac{1}{n+1} \Sigma h \sin V_p = B_p - \lambda R_q \sin(\alpha - \zeta_q)$$

where λ and α are certain constants, depending on the sum of a trigonometrical series.

Again, if we take means from $t=\pi/\nu$ to $t=2\pi/\nu$, the second terms have their signs changed.

Hence the difference between these two successive sums will give $\lambda R_q \cos(\alpha - \zeta_q)$ and $\lambda R_q \sin(\alpha - \zeta_q)$. There will be usually two terms such as those typified by q, and we shall then have to take two other means, viz., one beginning at $\pi/2\nu$ and ending at $3\pi/2\nu$, and the other beginning at $3\pi/2\nu$ and ending at $5\pi/2\nu$. From the difference of these sums we get $-\lambda R_q \sin(\alpha - \zeta_q)$ and $\lambda R_q \cos(\alpha - \zeta_q)$. From these four equations the two R_q's and the two ζ_q's are found. The solution is a little complicated in reality by the fact that it is not possible to take $t=0$ exactly at the beginning of the series, because the first tide does not occur exactly at noon, but this is a detail which will become clear below.

When all the A's and B's or R's and ζ's have been found, the position of the sun and moon at the epoch, found from the *Nautical Almanac*, and certain constants found from the Auxiliary Tables in Baird's *Manual of Tidal Observations**, are required to complete the evaluation of the H's and κ's.

The details of the processes will become clear when we consider the various tides.

It may be worth mentioning that I have almost completely evaluated the F's and G's, which give the perturbation of one tide on another, in the case considered in the Appendix. Without giving any of the details of the laborious arithmetic involved, it may suffice to say that the conclusion fully justifies the omission of all those terms, which are neglected in the computation as presented below.

§ 4. *The tides* N *and* L.

These are the two lunar elliptic tides.

For the sake of brevity all the tides excepting M_2, N, L are omitted from the analytical expressions.

Since $\quad V_n = V_m - (\sigma - \varpi)t, \qquad V_l = V_m + (\sigma - \varpi)t$

* Taylor and Francis, Fleet Street, 1886.

the expression becomes

$$\begin{aligned} h &= A_m \cos V_m + B_m \sin V_m + R_n \cos [V_m - (\sigma - \varpi) t - \zeta_n] \\ &\qquad\qquad + R_l \cos [V_m + (\sigma - \varpi) t - \zeta_l] \\ &= \cos V_m \{A_m + R_n \cos [(\sigma - \varpi) t + \zeta_n] + R_l \cos [(\sigma - \varpi) t - \zeta_l]\} \\ &\quad + \sin V_m \{B_m + R_n \sin [(\sigma - \varpi) t + \zeta_n] - R_l \sin [(\sigma - \varpi) t - \zeta_l]\} \end{aligned}$$

Hence, taking into account the equation which expresses that h is a maximum or minimum, and neglecting the variation of $s - p$ compared with that of V_m, we have

$$\left.\begin{aligned} h \cos V_m &= A_m + R_n \cos [(\sigma - \varpi) t + \zeta_n] + R_l \cos [(\sigma - \varpi) t - \zeta_l] \\ h \sin V_m &= B_m + R_n \sin [(\sigma - \varpi) t + \zeta_n] - R_l \sin [(\sigma - \varpi) t - \zeta_l] \end{aligned}\right\} \quad \ldots(8)$$

The mean interval between each tide and the next is 6·210 hours. Then if e be the increment of $s - p$ in that period (so that with $\sigma - \varpi$ equal to $0^\circ{\cdot}54437$ per hour, e is equal to $3^\circ{\cdot}3807$), and if a, b be the values of $(\sigma - \varpi) t + \zeta_n$ and $(\sigma - \varpi) t - \zeta_l$ at the time of the first tide under consideration, the equations corresponding to the $(r+1)^{\text{th}}$ tide are approximately

$$\left.\begin{aligned} h \cos V_m &= A_m + R_n \cos (a + re) + R_l \cos (b + re) \\ h \sin V_m &= B_m + R_n \sin (a + re) - R_l \sin (b + re) \end{aligned}\right\} \quad \ldots\ldots\ldots(9)$$

If we take the mean of $n + 1$ successive tides, the two latter terms on the right of (9) will be multiplied by $\dfrac{\sin \frac{1}{2} (n+1) e}{(n+1) \sin \frac{1}{2} e}$, and the r in the arguments $a + re$, $b + re$, will be equal to $\frac{1}{2} n$. If the $(n+2)^{\text{th}}$ tide falls exactly a semi-lunar-anomalistic period later than the first, $(n+1) e = \pi$. On account of the incommensurability of the angular velocity $\sigma - \varpi$ this condition cannot be rigorously satisfied, but if the whole series of observations be broken up into such semi-periods, then on the average of many such summations it may be taken as true.

Then, since $\frac{1}{2} e$ is a small angle,

$$(n+1) \sin \tfrac{1}{2} e = \tfrac{1}{2} \pi, \quad \text{and} \quad \sin \tfrac{1}{2} (n+1) e = 1\,;$$

hence the factor is $2/\pi$.

Again $\frac{1}{2} ne = \frac{1}{2} \pi - \frac{1}{2} e$; thus, if $n + 1$ is the mean number of tides in a semi-anomalistic period, our mean equations are

$$\left.\begin{aligned} \frac{\pi}{2(n+1)} \{\Sigma h \cos V_m - A_m\} &= - R_n \sin (a - \tfrac{1}{2} e) - R_l \sin (b - \tfrac{1}{2} e) \\ \frac{\pi}{2(n+1)} \{\Sigma h \sin V_m - B_m\} &= R_n \cos (a - \tfrac{1}{2} e) - R_l \cos (b - \tfrac{1}{2} e) \end{aligned}\right\} \quad \ldots(10)$$

where the summations Σ are carried out over the first semi-lunar-anomalistic period, which may be designated as 1.

In applying these equations to the next semi-period 2, the result is got by writing $a+(n+1)e$ or $a+\pi$ for a, and $b+\pi$ for b.

Thus the equations are simply the same as (10), with the signs on the *left* changed.

The equations for semi-periods 3, 4, &c., will be all identical on the right, with alternately + and − signs on the left.

Let the observations run over m semi-lunar-anomalistic periods; then double the equations appertaining to periods 2, 3, ... $(m-1)$, and add all the m equations together, and divide by $2(m-1)$, and we have

$$\left.\begin{aligned}\frac{\pi}{4(n+1)(m-1)}\Sigma h\cos V_m &= -R_n\sin(a-\tfrac{1}{2}e)-R_l\sin(b-\tfrac{1}{2}e)\\ \frac{\pi}{4(n+1)(m-1)}\Sigma h\sin V_m &= \quad R_n\cos(a-\tfrac{1}{2}e)-R_l\cos(b-\tfrac{1}{2}e)\end{aligned}\right\}\ldots(11)$$

where Σ now denotes summation of the following kind:

$$\{\Sigma(1)-\Sigma(2)\}+\{\Sigma(3)-\Sigma(2)\}+\{\Sigma(3)-\Sigma(4)\}+\{\Sigma(5)-\Sigma(4)\}+\&c.$$

the numbers (1), (2), &c., indicating the number of the semi-lunar-anomalistic periods over which the partial sums are taken.

Suppose the whole series of observations to be reduced covers $2m+1$ *quarter*-lunar-anomalistic periods, which we denote by i, ii, iii, &c.

First suppose that the semi-period denoted previously by 1 consists of i + ii, that 2 consists of iii + iv, and so on.

Let t_o be the time of the first tide of the series, and since we take noon of the first day as epoch, t_o cannot be more than a few hours.

Let $\qquad j=\tfrac{1}{2}e-(\sigma-\varpi)t_o=1^\circ\cdot 6903-(\sigma-\varpi)t_o$, a small angle

Then

$$\left.\begin{aligned}a-\tfrac{1}{2}e&=(\sigma-\varpi)t_o+\zeta_n-\tfrac{1}{2}e=\quad\zeta_n-j\\ b-\tfrac{1}{2}e&=(\sigma-\varpi)t_o-\zeta_l-\tfrac{1}{2}e=-(\zeta_l+j)\end{aligned}\right\}\ldots\ldots\ldots\ldots(12)$$

Then denoting the operation $\dfrac{\pi}{4(n+1)(m-1)}\Sigma$ by S° (the mark ° indicating that the first tide included is nearly at epoch, when $(\sigma-\varpi)t=0$), we have from (11) and (12)

$$\left.\begin{aligned}S^\circ h\cos V_m &= -R_n\sin(\zeta_n-j)+R_l\sin(\zeta_l+j)\\ S^\circ h\sin V_m &= \quad R_n\cos(\zeta_n-j)-R_l\cos(\zeta_l+j)\end{aligned}\right\}\ldots\ldots\ldots(13)$$

Secondly, suppose the semi-lunar-anomalistic period indicated by 1 consists of ii + iii, that 2 consists of iv + v, and so on.

Obviously the result is got by writing $t_o+\tfrac{1}{2}\pi/(\sigma-\varpi)$ for t_o, or, what amounts to the same thing, by putting $j-\tfrac{1}{2}\pi$ in place of j; but we must also write $S^{\frac{1}{2}\pi}$ for S°, so as to show that the summation begins when $(\sigma-\varpi)t$ is nearly equal to $\frac{1}{2}\pi$. Then

$$\left.\begin{aligned} S^{\frac{1}{2}\pi}h\cos V_m &= -R_n\cos(\zeta_n-j)-R_l\cos(\zeta_l+j) \\ S^{\frac{1}{2}\pi}h\sin V_m &= -R_n\sin(\zeta_n-j)-R_l\sin(\zeta_l+j) \end{aligned}\right\} \quad \text{.........(14)}$$

Hence
$$\left.\begin{aligned} R_n\sin(\zeta_n-j) &= -S^{\circ}h\cos V_m - S^{\frac{1}{2}\pi}h\sin V_m \\ R_n\cos(\zeta_n-j) &= S^{\circ}h\sin V_m - S^{\frac{1}{2}\pi}h\cos V_m \\ R_l\sin(\zeta_l+j) &= S^{\circ}h\cos V_m - S^{\frac{1}{2}\pi}h\sin V_m \\ R_l\cos(\zeta_l+j) &= -S^{\circ}h\sin V_m - S^{\frac{1}{2}\pi}h\cos V_m \end{aligned}\right\} \quad \text{.........(15)}$$

These four equations give the four unknowns R_n, ζ_n, R_l, ζ_l, and j is equal to $1^{\circ}\cdot 69-(\sigma-\varpi)t_o$.

Then if u_n, u_l denote the equilibrium arguments of the tides N and L at epoch, we have

$$u_n = 2(h_o-\nu)-2(s_o-\xi)-(s_o-p_o)$$
$$u_l = 2(h_o-\nu)-2(s_o-\xi)+(s_o-p_o)+\pi$$

where h_o, s_o, p_o are the mean longitudes of moon, sun, and lunar perigee at epoch, and ν and ξ are small angles, functions of the longitude of the moon's node (tabulated in Baird's *Manual*)*.

Then if $\mathfrak{f}_m$ is the factor of reduction (also tabulated by Baird) for the tides M_2, N, L,

$$\kappa_n = \zeta_n + u_n, \qquad \kappa_l = \zeta_l + u_l$$
$$H_n = \frac{R_n}{\mathfrak{f}_m}, \qquad H_l = \frac{R_l}{\mathfrak{f}_m}$$

In this investigation the interferences of the solar and diurnal tides are neglected, on the assumption that they are completely eliminated.

The difference between a lunar period and an anomalistic period is so small that the elimination of the diurnal tides will be satisfactory, but the effect of the solar tide will probably be sensible, unless we have under reduction 13 quarter-lunar-anomalistic periods, which only exceed 6 semi-lunations by about 25 hours.

The evaluation of the elliptic tides N and L from a series of observations shorter than a quarter year would be very unsatisfactory, and it is not likely that such an evaluation will be attempted. But if such a case is undertaken, the solar disturbance may be found by a plan strictly analogous to that pursued below in the case of the tides K_1, O, P. The reader may be left to deduce the requisite formulæ from the theory in § 3.

In the case of a long series of observations, each quarter year should be reduced independently, and the mean values of $H_n\cos\kappa_n$ and $H_n\sin\kappa_n$ should

* [Given also in algebraic form at the end of the first paper in this volume.]

be adopted as the values of the functions; whence H_n and κ_n are easily found. The L tide is, of course, to be treated similarly.

§ 5. *The Tide* M_2.

This is the principal lunar tide.

If we take the mean of $n+1$ successive tides, the equations (9) give us approximately

$$\frac{1}{n+1}\Sigma h \cos V_m = A_m, \qquad \frac{1}{n+1}\Sigma h \sin V_m = B_m \quad \text{......(16)}$$

We here assume that in taking this mean over an exact number of semi-lunations, the lunar elliptic tides, the solar tides, and the diurnal tides are eliminated.

With respect to the elliptic tides, this condition can only be approximately satisfied, because no small number of semi-lunations is equal to a number of anomalistic periods, and the like is true of the diurnal tides. In the example given below the diurnal tides are much larger than the elliptic tides, and I have found by actual computation (the details of which are not, however, given) that the disturbance in the value of the M_2 tide arising from the diurnal tides is quite insensible, and it may be safely accepted that the same is true of the disturbance from the elliptic tides.

With respect to the disturbance arising from the principal solar tide S_2, I find that it is adequately, although not completely, eliminated by making the number $n+1$ of tides under summation Σ cover an exact number of semi-lunations.

If the whole series of observations be short, it would be pedantic to attempt a close accuracy in results, and we may accept these formulæ; if the series be long, the residual errors will be gradually completely eliminated.

We have then

$$R_m \cos \zeta_m = A_m, \qquad R_m \sin \zeta_m = B_m$$

If u_m be the equilibrium argument at epoch, we have

$$u_m = 2(h_o - \nu) - 2(s_o - \xi)$$

Whence

$$\kappa_m = \zeta_m + u_m, \quad \text{and} \quad H_m = \frac{R_m}{f_m}$$

The meanings of h_o, s_o, ν, ξ, f_m, have been explained in the last section.

§ 6. *The Tides* S_2 *and* K_2.

These are the principal solar and luni-solar semi-diurnal tides.

If the tide S_2 is in the same phase as K_2 at any time, three months later they are in opposite phases. Hence, for a short series of observations, the two tides cannot be separated, and both must be considered together. It is proposed to treat a long series of observations as made up of a succession of short series; hence I begin with a short series.

For the sake of brevity all the tides excepting S_2 and K_2 are omitted from the analytical expressions.

Since $$V'' = V_s + 2\eta t$$

$$\begin{aligned} h &= R_s \cos (V_s - \zeta_s) + R'' \cos (V_s + 2\eta t - \zeta'') \\ &= \cos V_s \{R_s \cos \zeta_s + R'' \cos (2\eta t - \zeta'')\} \\ &\quad + \sin V_s \{R_s \sin \zeta_s - R'' \sin (2\eta t - \zeta'')\} \end{aligned}$$

Hence, taking into account the equation which expresses that h is a maximum or minimum, and neglecting the variation of $2h$ or $2\eta t$ compared with that of V_s, we have

$$\begin{aligned} h \cos V_s &= R_s \cos \zeta_s + R'' \cos (2\eta t - \zeta'') \\ h \sin V_s &= R_s \sin \zeta_s - R'' \sin (2\eta t - \zeta'') \end{aligned}$$

The mean interval between each tide and the next is $6^h{\cdot}210$. Then if g be the increment of $2h$ in that period (so that with 2η equal to $0^\circ{\cdot}082$ per hour, g is equal to $0^\circ{\cdot}510$), the equations corresponding to the $(r+1)^{\text{th}}$ tide are approximately

$$\left.\begin{aligned} h \cos V_s &= R_s \cos \zeta_s + R'' \cos (rg - \zeta'') \\ h \sin V_s &= R_s \sin \zeta_s - R'' \sin (rg - \zeta'') \end{aligned}\right\} \quad \ldots\ldots\ldots\ldots\ldots(17)$$

Now, if P be the cube of the ratio of the sun's parallax to its mean parallax, the expression for S_2, together with its parallactic inequality (the tides T, R of harmonic notation), is $PH_s \cos (2t - \kappa_s)$.

Since t is the mean solar hour angle, 2t is the same thing as V_s.

Hence $$R_s = PH_s, \quad \zeta_s = \kappa_s$$

Also if P_0 be the value of P at epoch, then for a period of two or three months we may take $P = P_0 (1 + pt)$, where $P_0 p$ is equal to dP/dt.

Again, if we put $\gamma = \dfrac{H''}{H_s}$, we have

$$R'' = f''H'' = f''\gamma H_s$$

Also since the argument of the K_2 tide is $2t + 2h - 2\nu'' - \kappa''$, where $2\nu''$ is a certain function of the longitude of the moon's node (tabulated by Baird), and since $t = 0$, $h = h_0$ at epoch, it follows that

$$-\zeta'' = 2h_0 - 2\nu'' - \kappa''$$

Now, when the means of the equations (17) are taken for $n+1$ successive tides, the latter terms become $\frac{\lambda_n}{\gamma} R'' \frac{\cos}{\sin} (\frac{1}{2}ng - \zeta'')$, where

$$\lambda_n = \gamma \cdot \frac{\sin \frac{1}{2}(n+1)g}{(n+1)\sin \frac{1}{2}g} \quad \text{..........................(18)}$$

Also, if we write

$$\left.\begin{aligned} \omega &= 2h_0 - 2\nu'' + \tfrac{1}{2}ng \\ \Pi &= P_0(1 + \tfrac{1}{2}np \times 6^{\text{h}}\cdot 21) \\ A_s &= \frac{1}{n+1}\Sigma h \cos V_s \\ B_s &= \frac{1}{n+1}\Sigma h \sin V_s \end{aligned}\right\} \quad \text{.....................(19)}$$

our equations become

$$\left.\begin{aligned} A_s &= \Pi \mathrm{H}_s \cos \kappa_s + \mathrm{f}''\lambda_n \mathrm{H}_s \cos(\omega - \kappa'') \\ B_s &= \Pi \mathrm{H}_s \sin \kappa_s - \mathrm{f}''\lambda_n \mathrm{H}_s \sin(\omega - \kappa'') \end{aligned}\right\} \quad \text{..............(20)}$$

It may be observed that Π is the mean value of P during the interval embraced by the $n+1$ tides.

In reducing a short series of observations we have to assume what is usually nearly true, viz., that $\kappa'' = \kappa_s$ and $\gamma = 0{\cdot}272$, as would be the case in the equilibrium theory of tides.

With this hypothesis, put

$$U \cos \phi = \Pi + \lambda_n \mathrm{f}'' \cos \omega$$

$$U \sin \phi = \qquad \lambda_n \mathrm{f}'' \sin \omega$$

from which to find U and ϕ. Then

$$A_s = \mathrm{H}_s U \cos(\kappa_s - \phi)$$

$$B_s = \mathrm{H}_s U \sin(\kappa_s - \phi)$$

from which to find H_s and κ_s.

Lastly, $\kappa'' = \kappa_s, \quad \mathrm{H}'' = \gamma \mathrm{H}_s = 0{\cdot}272\, \mathrm{H}_s$

In order to minimise the disturbance due to the lunar tide M_2, we have to make the $n+1$ tides cover an exact number of semi-lunations, namely, the same period as that involved in the evaluation of M_2. The elimination of the M_2 tide is adequate, although not so complete as the elimination of the effect of the S_2 tide on M_2, because M_2 is nearly three times as large as S_2.

A Long Series of Observations. Suppose that there is a half year of observation, or two periods of six semi-lunations, each of which periods contains exactly the same number of tides.

Then each of these periods is to be reduced independently with the assumption that $\gamma = 0{\cdot}272$ and $\kappa_s = \kappa''$. If this assumption is found subsequently to be very incorrect, it might be necessary to amend these reductions by multiplying λ_n by $\mathrm{H}'' \div 0{\cdot}272\mathrm{H}_s$, and by adding $\kappa_s - \kappa''$ to ω; but such repetition will not usually be necessary. From these reductions we get independent values of $\mathrm{H}_s \cos\kappa_s$, $\mathrm{H}_s \sin\kappa_s$ from each quarter year, and the mean of these is to be adopted, from which to compute H_s and κ_s. It remains to evaluate H'' and κ''.

The factor f'' and the angle $2\nu''$ vary so slowly that the change may be neglected from one quarter to the next, although each quarter is supposed to have been reduced with its proper values.

Let h_0 and h_0' be the sun's mean longitude at the two epochs; they will clearly differ by nearly 90°, and we put $2h_0' = 2h_0 + \pi + 2\delta h$. Hence it is clear that the value of ω in the second quarter is $\omega + 2\delta h + \pi$.

Thus the four equations, such as (20), appertaining to the two quarters, may be written

$$\left.\begin{aligned} A_s &= \Pi \mathrm{H}_s \cos\kappa_s + \frac{\lambda_n}{\gamma}\,.\,\mathrm{f}''\mathrm{H}''\cos(\omega - \kappa'') \\ B_s &= \Pi \mathrm{H}_s \sin\kappa_s - \frac{\lambda_n}{\gamma}\,.\,\mathrm{f}''\mathrm{H}''\sin(\omega - \kappa'') \\ A_s' &= \Pi' \mathrm{H}_s \cos\kappa_s - \frac{\lambda_n}{\gamma}\,.\,\mathrm{f}''\mathrm{H}''\cos(\omega + 2\delta h - \kappa'') \\ B_s' &= \Pi' \mathrm{H}_s \sin\kappa_s + \frac{\lambda_n}{\gamma}\,.\,\mathrm{f}''\mathrm{H}''\sin(\omega + 2\delta h - \kappa'') \end{aligned}\right\} \quad \ldots\ldots\ldots\ldots(21)$$

where the accented symbols apply to the second quarter, and where

$$\frac{\lambda_n}{\gamma} = \frac{\sin\frac{1}{2}(n+1)g}{(n+1)\sin\frac{1}{2}g} = 0{\cdot}656, \text{ a constant.}$$

From (21),

$$A_s - A_s' - (\Pi - \Pi')\,\mathrm{H}_s \cos\kappa_s = 2\frac{\lambda_n}{\gamma}\,.\,\mathrm{f}''\mathrm{H}''\cos\delta h\cos(\omega + \delta h - \kappa'')$$

$$-B_s + B_s' + (\Pi - \Pi')\,\mathrm{H}_s \sin\kappa_s = 2\frac{\lambda_n}{\gamma}\,.\,\mathrm{f}''\mathrm{H}''\cos\delta h\sin(\omega + \delta h - \kappa'')$$

From these two equations, H'' and κ'' may be computed, and since $\Pi - \Pi'$ is very small, approximate values of $\mathrm{H}_s \cos\kappa_s$, $\mathrm{H}_s \sin\kappa_s$ suffice.

§ 7. *The Diurnal Tides* K_1, O, P.

Amongst the diurnal tides I shall only consider K_1 the luni-solar diurnal, O the principal lunar diurnal, and P the principal solar diurnal tides.

There is the same difficulty in separating P from K_1 as in the case of K_2 and S_2, and therefore in a short series of observations P and K_1 have to be treated together. It is proposed to treat a long series of observations as made up of a succession of short series; hence I begin with a short series.

For the sake of brevity all the tides excepting K_1, O, P are omitted from the analytical expressions.

If $\frac{1}{2}V_m$ denotes $(\gamma - \sigma)t$, we have

$$V' = \tfrac{1}{2}V_m + \sigma t, \quad V_o = \tfrac{1}{2}V_m - \sigma t, \quad V_p = \tfrac{1}{2}V_m + (\sigma - 2\eta)t, \text{ and}$$

$$\begin{aligned} h &= R' \cos(\tfrac{1}{2}V_m + \sigma t - \zeta') + R_o \cos(\tfrac{1}{2}V_m - \sigma t - \zeta_o) \\ &\qquad + R_p \cos\{\tfrac{1}{2}V_m + (\sigma - 2\eta)t - \zeta_p\} \\ &= \cos\tfrac{1}{2}V_m\{R'\cos(\sigma t - \zeta') + R_o\cos(\sigma t + \zeta_o) + R_p\cos[(\sigma - 2\eta)t - \zeta_p]\} \\ &\quad + \sin\tfrac{1}{2}V_m\{-R'\sin(\sigma t - \zeta') + R_o\sin(\sigma t + \zeta_o) - R_p\sin[(\sigma - 2\eta)t - \zeta_p]\} \end{aligned}$$

Hence, taking account of the equation which expresses that h is a maximum or minimum, and neglecting the variation of σt compared with that of $\frac{1}{2}V_m$*, we have

$$\begin{aligned} h\cos\tfrac{1}{2}V_m &= R'\cos(\sigma t - \zeta') + R_o\cos(\sigma t + \zeta_o) + R_p\cos\{(\sigma - 2\eta)t - \zeta_p\} \\ h\sin\tfrac{1}{2}V_m &= -R'\sin(\sigma t - \zeta') + R_o\sin(\sigma t + \zeta_o) - R_p\sin\{(\sigma - 2\eta)t - \zeta_p\} \end{aligned}$$

The mean interval between each tide and the next is $6^{h}\cdot210$.

Then if e be the increment of s, and z the increment of $s - 2h$ in that period (so that with σ equal to $0^\circ\cdot5490$ per hour and $\sigma - 2\eta$ equal to $0^\circ\cdot4669$ per hour, e is equal to $3^\circ\cdot4095$ and z equal to $2^\circ\cdot8994$); and if a, b, c denote the values of $\sigma t - \zeta'$, $\sigma t + \zeta_o$, $(\sigma - 2\eta)t - \zeta_p$ at the time of the first tide under consideration, the equations corresponding to the $(r+1)$th tide are approximately

$$\left.\begin{aligned} h\cos\tfrac{1}{2}V_m &= R'\cos(a + re) + R_o\cos(b + re) + R_p\cos(c + rz) \\ h\sin\tfrac{1}{2}V_m &= -R'\sin(a + re) + R_o\sin(b + re) - R_p\sin(c + rz) \end{aligned}\right\} \quad \ldots(22)$$

If we take the mean of $n + 1$ successive tides, the first pair of terms will be multiplied by $\dfrac{\sin\frac{1}{2}(n+1)e}{(n+1)\sin\frac{1}{2}e}$ and the last term by the similar function with z in place of e; also the r in the arguments must be put equal to $\frac{1}{2}n$.

If the $(n+2)$th tide falls exactly a semi-lunar period later than the first, $(n+1)e = \pi$. On account of the incommensurability of the angular velocity σ, this condition cannot be rigorously satisfied, but if the whole series of observations be broken up into such semi-periods, then, on the average of many such summations, it may be taken as true.

Since $\frac{1}{2}e$ is a small angle, $(n+1)\sin\frac{1}{2}e = \frac{1}{2}\pi$, and $\sin\frac{1}{2}(n+1)e = 1$; hence the first factor is equal to $2/\pi$.

Again,

$$\tfrac{1}{2}(n+1)z = \tfrac{1}{2}(n+1)e \cdot \frac{z}{e} = \tfrac{1}{2}\pi \cdot \frac{\sigma - 2\eta}{\sigma} = 76^\circ\ 32' \text{ in degrees}$$

and

$$(n+1)\sin\tfrac{1}{2}z = \tfrac{1}{2}\pi \cdot \frac{\sigma - 2\eta}{\sigma}$$

* I have satisfied myself by analysis, which I do not reproduce, that on taking means this error becomes very small.

Therefore

$$\frac{\sin \frac{1}{2}(n+1)z}{(n+1)\sin \frac{1}{2}z} = \frac{2}{\pi} \cdot \frac{\sigma}{\sigma - 2\eta} \sin 76^\circ\ 32' = \frac{2}{\pi} \times 1{\cdot}1436 = \frac{2}{\pi} \times \lambda, \text{ suppose}$$

Again $\frac{1}{2}ne = \frac{1}{2}\pi - \frac{1}{2}e = \frac{1}{2}\pi - 1^\circ{\cdot}7048$

$$\frac{1}{2}nz = \frac{1}{2}\pi - 13^\circ{\cdot}4647 - 1^\circ{\cdot}4497 = \frac{1}{2}\pi - 14^\circ{\cdot}9144$$

Now let

$$\left.\begin{aligned} \alpha &= a - 1^\circ{\cdot}7048 \\ \beta &= b - 1^\circ{\cdot}7048 \\ \gamma &= c - 14^\circ{\cdot}9144 \end{aligned}\right\} \quad \text{.........................(23)}$$

and we have

$$\left.\begin{aligned} a + \tfrac{1}{2}ne &= \tfrac{1}{2}\pi + \alpha \\ b + \tfrac{1}{2}ne &= \tfrac{1}{2}\pi + \beta \\ c + \tfrac{1}{2}nz &= \tfrac{1}{2}\pi + \gamma \end{aligned}\right\} \quad \text{.........................(24)}$$

Thus, if $n+1$ is the mean number of tides in a semi-lunar period, the means of equations (22) become

$$\left.\begin{aligned} \frac{\pi}{2(n+1)} \Sigma h \cos \tfrac{1}{2} V_m &= -R' \sin \alpha - R_o \sin \beta - \lambda R_p \sin \gamma \\ \frac{\pi}{2(n+1)} \Sigma h \sin \tfrac{1}{2} V_m &= -R' \cos \alpha + R_o \cos \beta - \lambda R_p \cos \gamma \end{aligned}\right\} \quad \text{...(25)}$$

where the summations are carried out over the first semi-lunar period, which may be designated as 1.

In applying these equations to the next semi-period 2, the result is obtained by writing $a+(n+1)e$ for a, $b+(n+1)e$ for b, and $c+(n+1)z$ for c; that is to say, $a+\pi$ for a, $b+\pi$ for b, and $c+153^\circ{\cdot}0706$ or $c+\pi-26^\circ{\cdot}9294$ for c.

If, therefore, we put $\epsilon = 26^\circ{\cdot}9294$, we obtain the result from (25) by changing the signs on the left and writing $\gamma - \epsilon$ for γ.

The equations for semi-periods 3, 4, 5, &c., will be alternately + and − on the left, and identical as regards the terms in α and β, but with $\gamma - 2\epsilon$, $\gamma - 3\epsilon$, $\gamma - 4\epsilon$, &c., successively in place of γ.

Let the observations run over m semi-lunar periods; then double the equations appertaining to periods 2, 3 ... $(m-1)$, add all the m equations together, and divide by $2(m-1)$.

The terms in R_p will involve the series

$$\frac{\sin}{\cos}\gamma + 2\frac{\sin}{\cos}(\gamma - \epsilon) + 2\frac{\sin}{\cos}(\gamma - 2\epsilon) + \ldots + \frac{\sin}{\cos}\{\gamma - (m-1)\epsilon\}$$

This is equal to $2\dfrac{\sin \frac{1}{2}(m-1)\epsilon}{\tan \frac{1}{2}\epsilon} \dfrac{\sin}{\cos}\{\gamma - \tfrac{1}{2}(m-1)\epsilon\}$

Then if we put

$$\mu = \frac{\lambda \sin \frac{1}{2}(m-1)\epsilon}{(m-1)\tan \frac{1}{2}\epsilon}, \text{ where } \lambda = 1{\cdot}1436$$

our equations (25) become

$$\left.\begin{aligned}&\frac{\pi}{4(n+1)(m-1)}\Sigma h\cos\tfrac{1}{2}V_m\\&\qquad=-R'\sin\alpha-R_o\sin\beta-\mu R_p\sin\{\gamma-\tfrac{1}{2}(m-1)\epsilon\}\\&\frac{\pi}{4(n+1)(m-1)}\Sigma h\sin\tfrac{1}{2}V_m\\&\qquad=-R'\cos\alpha+R_o\cos\beta-\mu R_p\cos\{\gamma-\tfrac{1}{2}(m-1)\epsilon\}\end{aligned}\right\}\quad\text{...(26)}$$

where Σ now denotes summation of the following kind:

$$\{\Sigma(1)-\Sigma(2)\}+\{\Sigma(3)-\Sigma(2)\}+\{\Sigma(3)-\Sigma(4)\}+\dots$$

Suppose the whole series of observations to be reduced covers exactly $2m+1$ *quarter*-lunar periods, which we denote by I, II, III, &c.

First suppose that the semi-period denoted previously by 1 consists of I + II, that 2 consists of III + IV, and so on.

Let t_o denote the time of the first tide of the series, and since noon of the first day is epoch, t_o cannot be more than a few hours.

$$\left.\begin{aligned}\text{Let}\qquad &i=\tfrac{1}{2}e-\sigma t_o=1^\circ{\cdot}7048-\sigma t_o\\ \text{and}\qquad &k=\tfrac{1}{2}z-(\sigma-2\eta)t_o=1^\circ{\cdot}4497-(\sigma-2\eta)t_o\end{aligned}\right\}\quad\text{...............(27)}$$

i and k are clearly small angles.

Since $\epsilon=26^\circ{\cdot}9294$, $\tfrac{1}{2}\epsilon+1^\circ{\cdot}4497=14^\circ{\cdot}9144$; and from (23)

$$\left.\begin{aligned}\alpha&=\sigma t_o-\zeta'-1^\circ{\cdot}7048=-(\zeta'+i)\\ \beta&=\sigma t_o+\zeta_o-1^\circ{\cdot}7048=\;\;(\zeta_o-i)\\ \gamma&=(\sigma-2\eta)t_o-\zeta_p-14^\circ{\cdot}9144=-\zeta_p-k-\tfrac{1}{2}\epsilon\end{aligned}\right\}\quad\text{............(28)}$$

If the same notation be adopted as that explained in § 4 (the only difference being that we now deal with quarter-lunar instead of quarter-anomalistic periods), we have

$$\left.\begin{aligned}S^\circ h\cos\tfrac{1}{2}V_m&=\;\;R'\sin(\zeta'+i)-R_o\sin(\zeta_o-i)+\mu R_p\sin(\zeta_p+k+\tfrac{1}{2}m\epsilon)\\ S^\circ h\sin\tfrac{1}{2}V_m&=-R'\cos(\zeta'+i)+R_o\cos(\zeta_o-i)-\mu R_p\cos(\zeta_p+k+\tfrac{1}{2}m\epsilon)\end{aligned}\right\}\quad\text{...(29)}$$

Secondly, suppose the semi-lunar period indicated by 1 consists of II + III, that 2 consists of IV + V, and so on. Then, obviously, the result is got by writing $t_o+\tfrac{1}{2}\pi/\sigma$ for t_o; that is to say, write $i-\tfrac{1}{2}\pi$ for i and $k-\tfrac{1}{2}(\sigma-2\eta)\pi/\sigma$, or $k-\tfrac{1}{2}\pi+\eta\pi/\sigma$ for k. But $\eta\pi/\sigma$ is equal to $\tfrac{1}{2}\epsilon$, and we write $k-\tfrac{1}{2}\pi+\tfrac{1}{2}\epsilon$ for k. Therefore, following the notation used in § 4 for N and L,

$$\left.\begin{aligned}S^{\frac{1}{2}\pi}h\cos\tfrac{1}{2}V_m&=-R'\cos(\zeta'+i)-R_o\cos(\zeta_o-i)\\&\qquad-\mu R_p\cos\{\zeta_p+k+\tfrac{1}{2}(m+1)\epsilon\}\\ S^{\frac{1}{2}\pi}h\sin\tfrac{1}{2}V_m&=-R'\sin(\zeta'+i)-R_o\sin(\zeta_o-i)\\&\qquad-\mu R_p\sin\{\zeta_p+k+\tfrac{1}{2}(m+1)\epsilon\}\end{aligned}\right\}\quad\text{......(30)}$$

These four S's require correction for the disturbance due to the semidiurnal terms M_2 and S_2, and I shall return to this point later. In the meantime write

$$\left.\begin{matrix}W\\X\end{matrix}\right\} = S^{\circ}h \begin{matrix}\cos\\ \sin\end{matrix} \tfrac{1}{2}V_m + \text{corr.}, \qquad \left.\begin{matrix}Y\\Z\end{matrix}\right\} = S^{\frac{1}{2}\pi}h \begin{matrix}\cos\\ \sin\end{matrix} \tfrac{1}{2}V_m + \text{corr.} \quad\text{......(31)}$$

and we have

$$\left.\begin{aligned}\tfrac{1}{2}(W+Z) &= -R_o\sin(\zeta_o - i) - \mu R_p \sin\tfrac{1}{4}\epsilon\cos\{\zeta_p + k + \tfrac{1}{4}(2m+1)\epsilon\}\\ \tfrac{1}{2}(X-Y) &= R_o\cos(\zeta_o - i) - \mu R_p \sin\tfrac{1}{4}\epsilon\sin\{\zeta_p + k + \tfrac{1}{4}(2m+1)\epsilon\}\end{aligned}\right\}\text{...(32)}$$

$$\left.\begin{aligned}\tfrac{1}{2}(W-Z) &= R'\sin(\zeta' + i) + \mu R_p \cos\tfrac{1}{4}\epsilon\sin\{\zeta_p + k + \tfrac{1}{4}(2m+1)\epsilon\}\\ \tfrac{1}{2}(X+Y) &= -R'\cos(\zeta' + i) - \mu R_p \cos\tfrac{1}{4}\epsilon\cos\{\zeta_p + k + \tfrac{1}{4}(2m+1)\epsilon\}\end{aligned}\right\}\text{...(33)}$$

If we put

$$\left.\begin{aligned}L &= \{\tfrac{1}{2}(X+Y) + R'\cos(\zeta' + i)\}\tan\tfrac{1}{4}\epsilon\\ M &= \{\tfrac{1}{2}(W-Z) - R'\sin(\zeta' + i)\}\tan\tfrac{1}{4}\epsilon\end{aligned}\right\}\text{...............(34)}$$

the equations (33) may be written

$$\left.\begin{aligned}\tfrac{1}{2}(W+Z) - L &= -R_o\sin(\zeta_o - i)\\ \tfrac{1}{2}(X-Y) + M &= R_o\cos(\zeta_o - i)\end{aligned}\right\}\text{..................(35)}$$

The four equations (32), (33) involve six unknown quantities, R', ζ', R_o, ζ_o, R_p, ζ_p, and are insufficient for their determination.

In reducing a short series of observations it is necessary to assume what is usually nearly true, viz., that $\kappa_p = \kappa'$, and $H_p/H' = 0{\cdot}3309$, as would be the case in the equilibrium theory of tides.

Then, writing q for 0·3309, we have approximately $R_p = H_p = qH'$. The argument of the K_1 tide is $t + (h - \nu') - \frac{1}{2}\pi - \kappa'$, where ν' is a certain function of the longitude of the moon's node (tabulated by Baird); and the argument of the P tide is $t - h + \frac{1}{2}\pi - \kappa_p$.

At the noon which is taken as epoch $t = 0$, $h = h_o$, and the two arguments are equal to $-\zeta'$ and $-\zeta_p$.

Hence

$$\begin{aligned}-\zeta' &= h_o - \nu' - \tfrac{1}{2}\pi - \kappa'\\ -\zeta_p &= -h_o + \tfrac{1}{2}\pi - \kappa_p\end{aligned}$$

Therefore

$$\zeta_p = \zeta' + 2h_o - \nu' - \pi + (\kappa_p - \kappa')$$

Putting $\kappa_p = \kappa'$ as explained above,

$$\zeta_p + k = \zeta' + i + 2h_o - \nu' - \pi + l$$

where

$$\begin{aligned}l = k - i &= [1^{\circ}{\cdot}450 - (\sigma - 2\eta)\,t_o] - [1^{\circ}{\cdot}705 - \sigma t_o]\\ &= -0^{\circ}{\cdot}255 + 2\eta t_o, \text{ a small angle} \quad\text{..........................(36)}\end{aligned}$$

Then, if

$$\left.\begin{aligned}\theta &= 2h_o - \nu' + l + \tfrac{1}{4}(2m+1)\,\epsilon\\ \rho_m &= q\mu\cos\tfrac{1}{4}\epsilon\end{aligned}\right\}\text{.....................(37)}$$

we have

$$\left.\begin{aligned}\tfrac{1}{2}(W-Z) &= f'H'\sin(\zeta' + i) - \rho_m H'\sin(\zeta' + i + \theta)\\ \tfrac{1}{2}(X+Y) &= -f'H'\cos(\zeta' + i) + \rho_m H'\cos(\zeta' + i + \theta)\end{aligned}\right\}\text{......(38)}$$

Let

$$\left.\begin{aligned} T\cos\psi &= f' - \rho_m\cos\theta \\ T\sin\psi &= \qquad \rho_m\sin\theta \end{aligned}\right\} \quad \ldots\ldots\ldots\ldots\ldots\ldots\ldots\ldots(39)$$

whence T and ψ may be computed; and

$$\left.\begin{aligned} \tfrac{1}{2}(W - Z) &= \quad H'T\sin(\zeta' + i - \psi) \\ \tfrac{1}{2}(X + Y) &= -H'T\cos(\zeta' + i - \psi) \end{aligned}\right\} \quad \ldots\ldots\ldots\ldots\ldots(40)$$

From these we compute H′ and ζ' and $\zeta' + i$.

Then if $u' = h_o - \nu' - \frac{1}{2}\pi$, the equilibrium argument at epoch of K_1,

$$\kappa' = \zeta' + u'$$

We have also $H_p = qH' = 0{\cdot}3309H'$, $\kappa_p = \kappa'$.

Returning to equations (34) and (35), we compute $R' = f'H'$, and hence $R' \begin{smallmatrix}\cos\\ \sin\end{smallmatrix} (\zeta' + i)$, and then L and M.

Having these, we compute R_o and ζ_o from (35).

Then, if $u_o = h_o - \nu - 2(s_o - \xi) + \frac{1}{2}\pi$, the equilibrium argument at epoch of O,

$$\kappa_o = \zeta_o + u_o, \quad \text{and} \quad R_o = \frac{H_o}{f_o}$$

where f_o is a certain function of the longitude of the moon's node, tabulated in Baird's *Manual.*

A Long Series of Observations. Suppose that there is a half year of observation, or two periods of thirteen quarter-lunar periods, each of which contains exactly the same number of tides.

Then each of these periods is to be reduced independently with the assumption that $q = 0{\cdot}3309$ and $\kappa_p = \kappa'$. If this assumption be found subsequently to be very incorrect, it might be necessary to amend these reductions by adding $\kappa_p - \kappa'$ to the value of θ, and by multiplying ρ_m by $H_p \div 0{\cdot}3309H'$, but such repetition will not usually be necessary.

From these reductions we get independent values of $H'\cos\kappa'$, $H'\sin\kappa'$, $H_o\cos\kappa_o$, $H_o\sin\kappa_o$ from each quarter year, and the means of these are to be adopted from which to compute H', κ', H_o, κ_o.

It remains to evaluate H_p and κ_p.

The factor f′ and the angle ν' vary so slowly that the change from one quarter to the next may be neglected, although each quarter is supposed to have been reduced with its proper values.

Let h_o, h_o' be the values of the sun's mean longitude at the two epochs; then since the second epoch is nearly a quarter year later than the first, h_o' will exceed h_o by about 90°.

Let $h_o' = h_o + \frac{1}{2}\pi + \delta h$, so that δh is small.

If $\zeta' + \delta\zeta'$, $\zeta_p + \delta\zeta_p$ be the values of ζ', ζ_p at the second epoch, we have $\zeta' + \delta\zeta' = -h_o' + \nu' + \frac{1}{2}\pi + \kappa'$, $\zeta' = -h_o + \nu' + \frac{1}{2}\pi + \kappa'$, and therefore $\delta\zeta' = -\frac{1}{2}\pi - \delta h$.

Again, $\zeta_p + \delta\zeta_p = h_o' - \frac{1}{2}\pi + \kappa_p$, $\zeta_p = h_o - \frac{1}{2}\pi + \kappa_p$

and therefore $\delta\zeta_p = \frac{1}{2}\pi + \delta h$

Let $i + \delta i$, $k + \delta k$ be the values of i and k corresponding to the second epoch, and let W′, X′, Y′, Z′ be the values of those quantities in the second quarter. Then, replacing $\frac{1}{4}(2m+1)\epsilon$ by 87°·5, since that is its value when $2m+1$ is 13, we have from (33)

$$\frac{1}{2}(\mathrm{W}' - \mathrm{Z}') = -R'\cos(\zeta' + i + \delta i - \delta h) + \mu R_p \cos\tfrac{1}{4}\epsilon\cos(\zeta_p + k + \delta k + \delta h + 87^\circ{\cdot}5)$$

$$\frac{1}{2}(\mathrm{X}' + \mathrm{Y}') = -R'\sin(\zeta' + i + \delta i - \delta h) + \mu R_p \cos\tfrac{1}{4}\epsilon\sin(\zeta_p + k + \delta k + \delta h + 87^\circ{\cdot}5)$$

$$\frac{1}{2}(\mathrm{W} - \mathrm{Z}) = R'\sin(\zeta' + i) + \mu R_p \cos\tfrac{1}{4}\epsilon\sin(\zeta_p + k + 87^\circ{\cdot}5)$$

$$\frac{1}{2}(\mathrm{X} + \mathrm{Y}) = -R'\cos(\zeta' + i) - \mu R_p \cos\tfrac{1}{4}\epsilon\cos(\zeta_p + k + 87^\circ{\cdot}5)$$

Hence

$$\left.\begin{aligned}\tfrac{1}{4}(\mathrm{W}'-\mathrm{Z}') - \tfrac{1}{4}(\mathrm{X}+\mathrm{Y}) &= R'\sin\tfrac{1}{2}(\delta i - \delta h)\sin\{\zeta' + i + \tfrac{1}{2}(\delta i - \delta h)\}\\ &\quad + \mu R_p\cos\tfrac{1}{4}\epsilon\cos\tfrac{1}{2}(\delta k + \delta h)\cos\{\zeta_p + k + \tfrac{1}{2}(\delta k + \delta h) + 87^\circ{\cdot}5\}\\ \tfrac{1}{4}(\mathrm{W}-\mathrm{Z}) + \tfrac{1}{4}(\mathrm{X}'+\mathrm{Y}') &= -R'\sin\tfrac{1}{2}(\delta i - \delta h)\cos\{\zeta' + i + \tfrac{1}{2}(\delta i - \delta h)\}\\ &\quad + \mu R_p\cos\tfrac{1}{4}\epsilon\cos\tfrac{1}{2}(\delta k + \delta h)\sin\{\zeta_p + k + \tfrac{1}{2}(\delta k + \delta h) + 87^\circ{\cdot}5\}\end{aligned}\right\}\ldots(41)$$

In these equations R' is equal to f′H′ and R_p is equal to H_p.

The terms involving R' are clearly small, and approximate values of R' and ζ', as derived from the first quarter, will be sufficient to compute them. Afterwards we can compute R_p or H_p and ζ_p; then if u_p denotes $-h_o + \frac{1}{2}\pi$, the equilibrium argument of P at the first epoch, $\kappa_p = \zeta_p + u_p$.

The values of H_p, κ_p thus deduced ought not to differ very largely from those assumed in the two independent reductions.

The same investigation serves for the evaluation of the P tide from any two sets of observations, each consisting of thirteen quarter-lunar periods, and with a small change in the analysis we need not suppose each to consist of thirteen such periods. But the two epochs must be such that $\sin\delta h$ is small and $\cos\delta h$ is large, or the formulæ, although analytically correct, will fail in their object.

§ 8. *The Disturbance of* K_1, O, P *due to* M_2 *and* S_2.

It has been remarked in § 7 that the diurnal tides are perturbed by the semi-diurnal. The general method has been given in § 3, by which to calculate the effect on any one tide, whose increment of argument since epoch is V_p and speed is p, due to a tide whose increment is V_q and speed q.

Since in the present instance all the diurnal tides have been consolidated into one of speed $\gamma-\sigma$, we have to calculate the effect of the tides whose speeds are $2(\gamma-\sigma)$ and $2(\gamma-\eta)$ on the tide whose speed is $\gamma-\sigma$. It follows, therefore, that the factor q/p or k_q of (3) is in the first case equal to $2(\gamma-\sigma)/(\gamma-\sigma)$ or 2, and in the second case is $2(\gamma-\eta)/(\gamma-\sigma)$ or 2·070; or $k_m=2$, $k_s=2{\cdot}070$.

The coefficients F, G, f, g, as due to the tide M_2 of speed $2(\gamma-\sigma)$, will be written with suffix m, and as due to the tide S_2 of speed $2(\gamma-\eta)$, with suffix s. The sums and means have also to be taken in the two ways denoted by S° and $S^{\frac{1}{2}\pi}$. Hence we have altogether to compute sixteen coefficients, which by an easily intelligible notation may be written

$$F_m{}^{(0)},\ G_m{}^{(0)},\ \dots\ f_s{}^{(\frac{1}{2}\pi)},\ g_s{}^{(\frac{1}{2}\pi)}$$

In order to compute the sixteen coefficients, it is necessary to find the mean cosines and sines of the four following angles, viz.:

$\frac{1}{2}V_m \pm V_m$, $\frac{1}{2}V_m \pm V_s$, and the means have to be taken in the two ways denoted S° and $S^{\frac{1}{2}\pi}$.

These means are exactly the same in form as what the means of h cos and h sin (which had to be evaluated in S° and $S^{\frac{1}{2}\pi}$) would be if all the heights were regarded as positive unity, irrespective of whether they are H.W. or L.W. Hence the same plan of computation serves here as elsewhere; the plan is explained in the following section.

By comparison of equation (7) and the definitions (31) of W, X, Y, Z in the last section, we have

$$\left.\begin{aligned}
W &= S^\circ h\cos\tfrac{1}{2}V_m - \{A_m F_m{}^{(0)} + B_m G_m{}^{(0)} + A_s F_s{}^{(0)} + B_s G_s{}^{(0)}\}\\
X &= S^\circ h\sin\tfrac{1}{2}V_m - \{A_m f_m{}^{(0)} + B_m g_m{}^{(0)} + A_s f_s{}^{(0)} + B_s g_s{}^{(0)}\}\\
Y &= S^{\frac{1}{2}\pi} h\cos\tfrac{1}{2}V_m - \{A_m F_m{}^{(\frac{1}{2}\pi)} + B_m G_m{}^{(\frac{1}{2}\pi)} + A_s F_s{}^{(\frac{1}{2}\pi)} + B_s G_s{}^{(\frac{1}{2}\pi)}\}\\
Z &= S^{\frac{1}{2}\pi} h\sin\tfrac{1}{2}V_m - \{A_m f_m{}^{(\frac{1}{2}\pi)} + B_m g_m{}^{(\frac{1}{2}\pi)} + A_s f_s{}^{(\frac{1}{2}\pi)} + B_s g_s{}^{(\frac{1}{2}\pi)}\}
\end{aligned}\right\}\quad\dots(42)$$

The four quantities A_m, B_m, A_s, B_s are known from the evaluations of the tides M_2 and S_2; whence the corrections referred to in § 7 are calculable.

§ 9. *On the Summations.*

It will be seen from the preceding sections that sums have to be found of the following functions:

$$h \overset{\cos}{\underset{\sin}{.}} V_m,\ h \overset{\cos}{\underset{\sin}{.}} V_s,\ h \overset{\cos}{\underset{\sin}{.}} \tfrac{1}{2} V_m$$

and also of

$$\overset{\cos}{\underset{\sin}{.}} \tfrac{1}{2} V_m,\ \overset{\cos}{\underset{\sin}{.}} \tfrac{3}{2} V_m,\ \overset{\cos}{\underset{\sin}{.}} (\tfrac{1}{2} V_m \pm V_s)$$

It is necessary to calculate the five angles $\frac{1}{2}V_m$, V_s, $\frac{1}{2}V_m \pm V_s$, and V_m, for each tide, and the reader will easily see, by the example in the Appendix, how they may be computed with considerable rapidity, by aid of an auxiliary table A.

The computation of sines and cosines and multiplication by heights, may, with sufficient accuracy, be abridged, by regarding the cosine or sine of any angle lying within a given 5° of the circumference as equal to the cosine or sine of the middle of that 5°.

The process then consists in the grouping of the heights according to the values of their V's (V_m, V_s, $\frac{1}{2}V_m$, as the case may be). The heights in each group are then summed. Since the L.W. heights are all negative, they are treated in a separate table, and are considered as positive until their combination with the H.W. at a later stage. We shall, for the present, only speak of one of these groupings, taking it as a type of both.

Since $\overset{\cos}{\underset{\sin}{.}} (\alpha + 180°) = - \overset{\cos}{\underset{\sin}{.}} \alpha$, the eighteen groups forming the 3rd quadrant may be thrown in with the 1st quadrant by a mere change of sign; and the like is true of the 4th and 2nd quadrants.

Since $\cos(180° - \alpha) = -\cos\alpha$ and $\sin(180° - \alpha) = \sin\alpha$, it follows that we have to go through the 2nd quadrant in reversed order, in order to fall in with the succession which holds in the 1st quadrant, and, moreover, the cosine changes its sign, whilst the sine does not do so. Hence the following schemes will give us the eighteen groups which all have the same cosines and sines:

for cosines (1st − 3rd) − (2nd − 4th) reversed

for sines (1st − 3rd) + (2nd − 4th) reversed

Thus, one grouping of the heights serves for both cosines and sines, and, save for the last step, the additions are the same.

The combination of the H.W. and L.W. results is best made at the stage where 1st − 3rd and 2nd − 4th have been formed.

The negative signs for the L.W. results are introduced before addition to the H.W. results, and total $1^{st} - 3^{rd}$ and $2^{nd} - 4^{th}$ are thus formed.

After the eighteen cosine and sine total numbers are thus formed, they are to be multiplied by the cosines or sines of 2° 30′, 7° 30′, 12° 30′, ... 87° 30′. The products are then summed so as to give $\Sigma h \begin{smallmatrix}\cos\\ \cdot \\ \sin\end{smallmatrix}$.

It was noted at the beginning of this section that we also have sums of the form $\Sigma \begin{smallmatrix}\cos\\ \cdot \\ \sin\end{smallmatrix}$. These sums are obviously made by entering unity in place of each height, and, of course, not treating the L.W. as negative. Thus, where the H.W. and L.W. are combined it is not necessary to change the sign of the L.W., as was done in the combination of H.W. and L.W. for $\Sigma h \begin{smallmatrix}\cos\\ \cdot \\ \sin\end{smallmatrix}$. These summations are considerably less laborious than the others.

In the case of the tides M_2 and S_2, the division of the sums $\Sigma h \begin{smallmatrix}\cos\\ \cdot \\ \sin\end{smallmatrix}$ by the total number of entries gives the required results. But for N, L, and similarly for the diurnal tides K_1, O, P, the grouping and summations have to be broken into a number of subordinate periods, which are to be operated on to form S° and $S^{\frac{1}{2}\pi}$. The multiplication by the eighteen mean cosines and sines is best deferred to a late stage in the computation.

Thus, for example for N and L, the quarter-lunar-anomalistic periods, i, ii, iii, &c., are treated independently, and we find $(1^{st} - 3^{rd}) \pm (2^{nd} - 4^{th}$ reversed) for each. There are thus eighteen cosine numbers and eighteen sine numbers for each of i, ii, iii, &c.

We next form the sums two and two, i + ii, iii + iv, &c.; next find the differences (i + ii) − (iii + iv), (v + vi) − (iii + iv), &c.; add the differences together; then multiply by the eighteen cosines or sines of $2\frac{1}{2}°$, $7\frac{1}{2}°$, &c., and finally multiply by $\frac{\pi}{4(n+1)(m-1)}$, and so find $S^{\circ} h \begin{smallmatrix}\cos\\ \cdot \\ \sin\end{smallmatrix}$.

We next go through exactly the same process, but beginning with ii instead of i, and so find $S^{\frac{1}{2}\pi} h \begin{smallmatrix}\cos\\ \cdot \\ \sin\end{smallmatrix}$.

The same process applies, *mutatis mutandis*, for finding S° and $S^{\frac{1}{2}\pi} \begin{smallmatrix}\cos\\ \cdot \\ \sin\end{smallmatrix}$.

There are two cases which merit attention in particular. The sorting of heights in quarter-lunar-anomalistic periods, according to values of V_m, serves, in the first instance, for the evaluation of N and L, but it serves, secondly, to evaluate M_2, for we then simply neglect the subdivision into quarter periods and treat the whole as one series, but stop at the end of a semi-lunation.

The sorting of heights in quarter-lunar periods, according to the values of $\frac{1}{2}V_m$, also serves several purposes.

We first find from it S° and $S^{\frac{1}{2}\pi} h \overset{\cos}{\underset{\sin}{.}} \frac{1}{2}V_m$, and secondly, by merely counting the entries in each group for each quarter period, instead of adding up the heights, we arrive at S° and $S^{\frac{1}{2}\pi} \cos \frac{1}{2}V_m$. (It may be noted in passing that what is wanted, according to preceding analysis, is the sum of $\overset{\cos}{\underset{\sin}{.}} (\frac{1}{2}V_m - V_m)$, so that there will be a change of sign in the sine sum to get the desired result.)

But, besides these, S° and $S^{\frac{1}{2}\pi} \overset{\cos}{\underset{\sin}{.}} (\frac{1}{2}V_m + V_m)$ can be obtained with sufficient accuracy from the same sorting.

The angles $\frac{1}{2}V_m$ were sorted in four times eighteen groups, for each quarter-lunar period. If each angle were multiplied by three, the eighteen entries of the 1st quadrant would be converted into three groups of six, lying in three quadrants, viz., Ist, IInd, IIIrd; the 2nd quadrant is changed to IVth, Ist, IInd; the 3rd to IIIrd, IVth, Ist; and the 4th to IInd, IIIrd, IVth. Hence eighteen entries of 1st – 3rd are converted into three sixes, Ist – IIIrd, IInd – IVth, – {Ist – IIIrd}; and eighteen entries of 2nd – 4th are converted into three sixes, – {IInd – IVth}, Ist – IIIrd, IInd – IVth.

Hence a new Ist – IIIrd of six entries is made up thus:

first six of former 1st – 3rd
\+ second six of former 2nd – 4th
– third six of former 1st – 3rd

And a new IInd – IVth of six entries is made up of

– first six of former 2nd – 4th
\+ second six of former 1st – 3rd
\+ third six of former 2nd – 4th

These Ist – IIIrd and IInd – IVth may now be treated just like the other ones. Thus, without calculating $\frac{3}{2}V_m$, we have from the former 1st – 3rd and 2nd – 4th the results of a fresh grouping according to values of $\frac{3}{2}V_m$.

It is true that there is a considerable loss of accuracy, because all angles within 15° are now treated as having the same sine and cosine.

§ 10. *Rules for the Partition of the Observations into Groups.*

It appears from the preceding investigations that it is required to divide up the observations into groups. This may be done, with all necessary accuracy, and with great convenience, by dividing the tides just as they would be divided if every H.W. followed L.W., and *vice versâ*, at the mean interval of $6^h{\cdot}2103$.

Now a quarter-lunar-anomalistic period is $165^h{\cdot}3272$, a quarter-lunar period is $163^h{\cdot}9295$, and semi-lunation is $354^h{\cdot}3670$. Hence, dividing these numbers by $6^h{\cdot}2103$, we find that there are 26·62145 tides in a quarter-anomalistic period, 26·3964 in a quarter period, and 57·0612 in a semi-lunation.

It may be remarked in passing that these results show that the $n+1$ of (10), § 4, is 53·243, and the $n+1$ of (25), § 7, is 52·793.

It is, of course, impossible to have a fractional number of tides, and, therefore, we make a small multiplication table of these numbers, and take the nearest integer in each case. For example, in the case of the semi-lunations, we have

1.	57·0612	57	4.	228·2448	228
2.	114·1224	114	5.	285·3060	285
3.	171·1836	171	6.	342·3672	342

These have to be divided between H. and L.W. For the sake of convenience, I suppose that we always begin the series with a H.W., then when the integer is odd we put in one more H.W. than L.W., and thus have the following rule:

No. of semi-lunation ...	1	2	3	4	5	6
No. of last H.W. in the semi-lunation.........	29	57	86	114	143	171
No. of last L.W. in the semi-lunation.........	28	57	85	114	142	171

The H.W. and L.W. are here supposed to be numbered consecutively from 1 onwards in separate tables.

The other rules of partition given in Appendix E are found in the same way.

§ 11. *On the Over-Tides.*

Observations of H. and L.W. are very inappropriate for the determination of these tides (of which the most important are M_4, M_6, S_4, S_6), because they express the departure of the wave from the simple harmonic shape, and we are supposed to have no information as to what occurs between two tides.

These tides make the interval from H. to L.W. longer than from L. to H.W., and there is no doubt that, assuming the existence in the expression for h of a term of the form $A_{2m} \cos 2V_m + B_{2m} \sin 2V_m$, we shall get an approximation to A_{2m} and B_{2m} by finding the mean of $h \cos 2V_m$ and $h \sin 2V_m$. But the computation of the F, G, f, g, coefficients for the perturbation of M_4 by M_2 would be essential, and thus the amount of additional computation would be very great, whereas in the analysis of continuous observation the over-tides are found almost without any additional work. I am inclined to think that it would be best to obtain hourly observations for several days at several parts of a lunation, and by some methods of interpolation to construct a typical semi-diurnal tide-wave, from which, by the ordinary methods of harmonic analysis, we could find the ratio of the heights of the over-tides to the fundamental, and the relationship of their phases.

I make no attempt at such an investigation in this place.

§ 12. *On the Annual and Semi-annual Tides.*

These tides are frequently of much importance, so that they ought not to be neglected from a navigational point of view. It is obviously impossible to obtain any results from a series of observations of less than a year's duration.

Rules for the partition of tides into months or 12th parts of a year are given in the Appendix E. The mean of all the H. and L.W. observations for each month may be taken as the height of mean water at the middle of the month, and the 12 values for the year may be submitted to the ordinary processes of harmonic analysis for the evaluation of these two tides.

We have supposed in the previous investigation that the tide heights are measured from mean sea-level, and although it is not necessary that this condition should be rigorously satisfied, it might be well, where there is a large annual tide, to refer the heights to different datum levels in the different quarters of the year.

§ 13. *On Gaps in the Series of Observations.*

It often happens in actual observations that a few tides are missing through some accident, or are obviously vitiated by heavy weather. Now the present method depends for its applicability on the evanescence of terms in the averages. It is true that it is rigorously applicable even for scattered observations, but if applied to such a case all the F, G, f, g coefficients have to be calculated, and, as every tide reacts on every other, the computation would be so extensive as to make the method almost impracticable. Thus, where there is a gap, observations must be fabricated (of course noting that they *are* fabrications) by some sort of interpolation, and even values which are very incorrect are better than none*. If the interpolation is extensive, it might be well to test its correctness in a few places when the reduction is done. If a whole week or fortnight be missing, and if the computer cannot find a plausible method of interpolation, I can only suggest a preliminary reduction from the continuous parts, and the computation of a tide table for the hiatus. Each such case must be treated on its merits, and it is hardly possible to formulate general rules.

APPENDIX.

Tables and Rules of General Applicability.

A. *To find* $\frac{1}{2}V_m$.

The following table is for finding what would be the mean moon's hour-angle, if the moon had been on the meridian at the epoch. This angle is denoted by $\frac{1}{2}V_m$ or $(\gamma - \sigma)t$, and is equal to the angle through which the earth has turned relatively to the mean moon (at 14°·4920521 per mean solar hour) since epoch.

It would be advantageous to extend the table up to 90 days, but it can be used as it is for periods greater than 30 days by the division of the time into sets of 30 days. In the second period of 30 days 5°·7 must be *subtracted* from the tabular entry, for the third period 11°·4, and so on†.

For example: Find $\frac{1}{2}V_m$ for $78\frac{1}{2}^d\ 11^h\ 23^m$. The day is $18\frac{1}{2}$ of the third 30, and the tabular entry for $18\frac{1}{2}^d\ 11^h$ is 113°·9, and subtracting 11°·4 we have 102°·5; 23^m gives 5°·6, so that $\frac{1}{2}V_m = 108°·1$. The correct result is 107°·99, and it is obvious that an error of 0°·1 may easily be incurred by the use of the table.

The row for day $-\frac{1}{2}$ is given because it may be necessary to use one tide before epoch; this row is used in the example below.

* Fabricated times and heights would very likely be no worse than real observations during a few days of rough weather. A perfect tide table only claims to predict the tide apart from the influence of wind and atmospheric pressure; and, conversely, tidal observations must be sufficiently numerous to eliminate these influences by averages.

† Observe that the decimals run thus, ·0, ·5, ·0, &c., then ·4, ·9, ·4, &c., then ·8, ·3, ·8, &c., and so on. The first entry in which the sequence alters I call a "change." The incidence of "changes" may be found thus: $\gamma - \sigma$ is $14\frac{1}{2} - ·00795$; take Crelle's multiplication table for 795, and note where the last digit but three, having been 4, becomes 5; I say that this is a "change." For example, $559 \times 795 = 444405$, and $560 \times 795 = 445200$; then a change occurs at the 560th hour, or at the $(12 \times 46 + 8)^{th}$ hour, or at $23^d\ 8^h$. If the table be continued to 60 and 90 days, &c., by subtracting 5°·7, 11°·4, &c., the changes will fall a little wrong, but they may easily be corrected by means of Crelle's table, as here shown.

Table of $(\gamma-\sigma)t$ or $\frac{1}{2}V_m$.

Days	0h	1h	2h	3h	4h	5h	Days	6h	7h	8h	9h	10h	11h
−½	186°·1	200°·6	215°·1	229°·6	244°·1	258°·6	−½	273°·0	287°·5	302°·0	316°·5	331°·0	345°·5
0	0	14·5	29·0	43·5	58·0	72·5	0	87·0	101·4	115·9	130·4	144·9	159·4
½	173·9	188·4	202·9	217·4	231·9	246·4	½	260·9	275·3	289·8	304·3	318·8	333·3
1	347·8	2·3	16·8	31·3	45·8	60·3	1	74·8	89·3	103·7	118·2	132·7	147·2
½	161·7	176·2	190·7	205·2	219·7	234·2	½	248·7	263·2	277·7	292·1	306·6	321·1
2	335·6	350·1	4·6	19·1	33·6	48·1	2	62·6	77·1	91·6	106·0	120·5	135·0
½	149·5	164·0	178·5	193·0	207·5	222·0	½	236·5	251·0	265·5	280·0	294·4	308·9
3	323·4	337·9	352·4	6·9	21·4	35·9	3	50·4	64·9	79·4	93·9	108·3	122·8
½	137·3	151·8	166·3	180·8	195·3	209·8	½	224·3	238·8	253·3	267·8	282·3	296·7
4	311·2	325·7	340·2	354·7	9·2	23·7	4	38·2	52·7	67·2	81·7	96·2	110·6
½	125·1	139·6	154·1	168·6	183·1	197·6	½	212·1	226·6	241·1	255·6	270·1	284·6
5	299·0	313·5	328·0	342·5	357·0	11·5	5	26·0	40·5	55·0	69·5	84·0	98·5
½	113·0	127·4	141·9	156·4	170·9	185·4	½	199·9	214·4	228·9	243·4	257·9	272·4
6	286·9	301·3	315·8	330·3	344·8	359·3	6	13·8	28·3	42·8	57·3	71·8	86·3
½	100·8	115·3	129·7	144·2	158·7	173·2	½	187·7	202·2	216·7	231·2	245·7	260·2
7	274·7	289·2	303·6	318·1	332·6	347·1	7	1·6	16·1	30·6	45·1	59·6	74·1
½	88·6	103·1	117·6	132·0	146·5	161·0	½	175·5	190·0	204·5	219·0	233·5	248·0
8	262·5	277·0	291·5	305·9	320·4	334·9	8	349·4	3·9	18·4	32·9	47·4	61·9
½	76·4	90·9	105·4	119·9	134·3	148·8	½	163·3	177·8	192·3	206·8	221·3	235·8
9	250·3	264·8	279·3	293·8	308·3	322·7	9	337·2	351·7	6·2	20·7	35·2	49·7
½	64·2	78·7	93·2	107·7	122·2	136·6	½	151·1	165·6	180·1	194·6	209·1	223·6
10	238·1	252·6	267·1	281·6	296·1	310·6	10	325·0	339·5	354·0	8·5	23·0	37·5
½	52·0	66·5	81·0	95·5	110·0	124·5	½	138·9	153·4	167·9	182·4	196·9	211·4
11	225·9	240·4	254·9	269·4	283·9	298·4	11	312·9	327·3	341·8	356·3	10·8	25·3
½	39·8	54·3	68·8	83·3	97·8	112·3	½	126·8	141·3	155·7	170·2	184·7	199·2
12	213·7	228·2	242·7	257·2	271·7	286·2	12	300·7	315·2	329·6	344·1	358·6	13·1
½	27·6	42·1	56·6	71·1	85·6	100·1	½	114·6	129·1	143·6	158·0	172·5	187·0
13	201·5	216·0	230·5	245·0	259·5	274·0	13	288·5	303·0	317·5	331·9	346·4	0·9
½	15·4	29·9	44·4	58·9	73·4	87·9	½	102·4	116·9	131·4	145·9	160·3	174·8
14	189·3	203·8	218·3	232·8	247·3	261·8	14	276·3	290·8	305·3	319·8	334·2	348·7
½	3·2	17·7	32·2	46·7	61·2	75·7	½	90·2	104·7	119·2	133·7	148·2	162·6

mins.	
1	0°·2
2	0·5
3	0·7
4	1·0
5	1·2
6	1·4
7	1·7
8	1·9
9	2·2
10	2·4
11	2·7
12	2·9
13	3·1
14	3·4
15	3·6
16	3·9
17	4·1
18	4·3
19	4·6
20	4·8
21	5·1
22	5·3
23	5·6
24	5·8
25	6·0
26	6·3
27	6·5
28	6·8
29	7·0

15	177·1	191·6	206·1	220·6	235·1	249·6
½	351·0	5·5	20·0	34·5	49·0	63·5
16	164·9	179·4	193·9	208·4	222·9	237·4
½	338·9	353·3	7·8	22·3	36·8	51·3
17	152·8	167·2	181·7	196·2	210·7	225·2
½	326·7	341·2	355·6	10·1	24·6	39·1
18	140·6	155·1	169·5	184·0	198·5	213·0
½	314·5	329·0	343·5	357·9	12·4	26·9
19	128·4	142·9	157·4	171·9	186·3	200·8
½	302·3	316·8	331·3	345·8	0·2	14·7
20	116·2	130·7	145·2	159·7	174·2	188·6
½	290·1	304·6	319·1	333·6	348·1	2·5
21	104·0	118·5	133·0	147·5	162·0	176·5
½	277·9	292·4	306·9	321·4	335·9	350·4
22	91·8	106·3	120·8	135·3	149·8	164·3
½	265·7	280·2	294·7	309·2	323·7	338·2
23	79·6	94·1	108·6	123·1	137·6	152·1
½	253·5	268·0	282·5	297·0	311·5	326·0
24	67·4	81·9	96·4	110·9	125·4	139·9
½	241·3	255·8	270·3	284·8	299·3	313·8
25	55·2	69·7	84·2	98·7	113·2	127·7
½	229·1	243·6	258·1	272·6	287·1	301·6
26	43·0	57·5	72·0	86·5	101·0	115·5
½	216·9	231·4	245·9	260·4	274·9	289·4
27	30·8	45·3	59·8	74·3	88·8	103·3
½	204·8	219·2	233·7	248·2	262·7	277·2
28	18·7	33·1	47·6	62·1	76·6	91·1
½	192·6	207·1	221·5	236·0	250·5	265·0
29	6·5	21·0	35·5	49·9	64·4	78·9
½	180·4	194·9	209·4	223·8	238·3	252·8
30	− 5·7					
60	− 11·4					
90	− 17·2					
120	− 22·9					
150	− 28·6					

15	264·1	278·6	293·1	307·6	322·1	336·6
½	78·0	92·5	107·0	121·5	136·0	150·5
16	251·9	266·4	280·9	295·4	309·9	324·4
½	65·8	80·3	94·8	109·3	123·8	138·3
17	239·7	254·2	268·7	283·2	297·7	312·2
½	53·6	68·1	82·6	97·1	111·6	126·1
18	227·5	242·0	256·5	271·0	285·5	300·0
½	41·4	55·9	70·4	84·9	99·4	113·9
19	215·3	229·8	244·3	258·8	273·3	287·8
½	29·2	43·7	58·2	72·7	87·2	101·7
20	203·1	217·6	232·1	246·6	261·1	275·6
½	17·0	31·5	46·0	60·5	75·0	89·5
21	190·9	205·4	219·9	234·4	248·9	263·4
½	4·9	19·3	33·8	48·3	62·8	77·3
22	178·8	193·2	207·7	222·2	236·7	251·2
½	352·7	7·2	21·6	36·1	50·6	65·1
23	166·6	181·1	195·5	210·0	224·5	239·0
½	340·5	355·0	9·5	23·9	38·4	52·9
24	154·4	168·9	183·4	197·8	212·3	226·8
½	328·3	342·8	357·3	11·8	26·2	40·7
25	142·2	156·7	171·2	185·7	200·2	214·6
½	316·1	330·6	345·1	359·6	14·1	28·5
26	130·0	144·5	159·0	173·5	188·0	202·5
½	303·9	318·4	332·9	347·4	1·9	16·4
27	117·8	132·3	146·8	161·3	175·8	190·3
½	291·7	306·2	320·7	335·2	349·7	4·2
28	105·6	120·1	134·6	149·1	163·6	178·1
½	279·5	294·0	308·5	323·0	337·5	352·0
29	93·4	107·9	122·4	136·9	151·4	165·9
½	267·3	281·8	296·3	310·8	325·3	339·8

30	7·2
31	7·5
32	7·7
33	8·0
34	8·2
35	8·5
36	8·7
37	8·9
38	9·2
39	9·4
40	9·7
41	9·9
42	10·1
43	10·4
44	10·6
45	10·9
46	11·1
47	11·4
48	11·6
49	11·8
50	12·1
51	12·3
52	12·6
53	12·8
54	13·0
55	13·3
56	13·5
57	13·8
58	14·0
59	14·3

To find V_s.

No table is necessary for the conversion of time into angle at 30° per hour to find V_s, or $2(\gamma-\eta)t$, since we multiply the hours by 30, and add half the number of minutes. This rule is the same for every day.

B. *The tides* S_2 *and* K_2.

It is required to compute U and ϕ from

$$U\cos\phi = \Pi + \lambda_n f'' \cos\omega$$

$$U\sin\phi = \lambda_n f'' \sin\omega$$

where

$$\omega = 2h_0 - 2\nu'' + \alpha_n$$

and

$$\Pi = 1 + 3\left(\frac{\text{sun's parx.} - \text{mean parx.}}{\text{mean parx.}}\right)$$

the sun's parallax referred to being its value at the middle of the period under reduction.

If, for example, February 14 is the middle of the period, Π is found thus:

Sun's parx. Feb. 14 = 8″·95, mean parx. = 8″·85, diff. = + 0″·10

Then

$$\Pi = 1 + \frac{3 \times 0{\cdot}10}{8{\cdot}85} = 1{\cdot}034$$

The period under reduction consists in this case of an exact number of semi-lunations. The following table gives λ_n and α_n, according to the number of semi-lunations:

No. of semi-lunations ...	1.	2.	3.	4.	5.	6.
$\log \lambda_n$	9·4300	9·4159	9·3920	9·3575	9·3113	9·2517
α_n	14°·28	28°·82	43°·36	57°·90	72°·43	86°·97

h_0 is the sun's mean longitude at epoch, found from *Nautical Almanac*; and $2\nu''$, f″ are found from Baird's *Manual* in the tables applicable to the tide K_2 [or from the formulæ at the end of Paper 1].

C. *The Tides* N *and* L.

Summations are carried out over quarter-lunar-anomalistic periods, numbered i, ii, iii, &c. Grand totals are then made in two different ways, viz.

$$[\Sigma(\text{i}+\text{ii}) - \Sigma(\text{iii}+\text{iv})] + [\Sigma(\text{v}+\text{vi}) - \Sigma(\text{iii}+\text{iv})]$$
$$+ [\Sigma(\text{v}+\text{vi}) - \Sigma(\text{vii}+\text{viii})] + \text{\&c., to find } S^{\circ}$$

and

$$[\Sigma(\text{ii}+\text{iii}) - \Sigma(\text{iv}+\text{v})] + [\Sigma(\text{vi}+\text{vii}) - \Sigma(\text{iv}+\text{v})]$$
$$+ [\Sigma(\text{vi}+\text{vii}) - \Sigma(\text{viii}+\text{ix})] + \text{\&c., to find } S^{\frac{1}{2}\pi}$$

where, for example, Σ (i + ii) denotes summation carried over the half period made up of i and ii. These totals are multiplied by certain mean cosines and sines (whose values are given in F), and are summed. The next process is multiplication by a factor Φ ($\pi/4(n+1)(m-1)$ of § 4), of which the value depends on the number of quarter-lunar-anomalistic periods under treatment. The following table gives the value of this factor:

No. of ¼-lunar-anom. periods ...	iii.	v.	vii.	ix.	xi.	xiii.
Φ	0·02950	0·01475	0·00738	0·00492	0·00369	0·00295

The angle j is also required; it depends on the time of the first tide under reduction. If t_o be the time in hours since epoch to the first tide,

$$j = 1^\circ{\cdot}690 - 0^\circ{\cdot}5444\, t_o$$

For instance, in the example below the first tide is at $3^h\ 14^m$ of day $-\frac{1}{2}$; this is $8^h\ 46^m$, or $8^h{\cdot}77$, *before* epoch, so that $t_o = -8^h{\cdot}77$; then

$$j = 1^\circ{\cdot}690 + 0^\circ{\cdot}5444 \times 8{\cdot}77 = +6^\circ{\cdot}46$$

D. *The Tides* K_1, O, P.

Summations are carried out over quarter-lunar periods numbered I, II, III, &c., and totals are formed like those mentioned in C, and a factor Ψ (which differs slightly from Φ) is required in the formation of S° and $S^{\frac{1}{2}\pi}$. This factor depends on the number of quarter-lunar periods under treatment, and the following table gives its value:

No. of ¼-lunar periods ...	III.	V.	VII.	IX.	XI.	XIII.
Ψ	0·02976	0·01488	0·00744	0·00496	0·00372	0·00298

The angles i and l are required; they depend on the time of the first tide under reduction. If t_o be the time in hours of the first tide since epoch,

$$i = \quad 1^\circ{\cdot}705 - 0^\circ{\cdot}549\, t_o$$

$$l = -0^\circ{\cdot}255 + 0^\circ{\cdot}082\, t_o$$

For instance, in the example below we have, as shown in C, $t_o = -8^h{\cdot}77$, and

$$i = +6^\circ{\cdot}52$$

$$l = -0^\circ{\cdot}97$$

It is required to compute T and ψ from

$$T\cos\psi = f' - \rho_n \cos\theta$$

$$T\sin\psi = \qquad \rho_n \sin\theta$$

where $$\theta = 2h_o - \nu' + l + \beta_n$$

The period under reduction consists in this case of an exact number of quarter-lunar periods, and the following table gives the values of ρ_n and β_n, according to the number of quarter-lunar periods:

No. of $\frac{1}{4}$-lunar periods ...	III.	V.	VII.	IX.	XI.	XIII.
$\log \rho_n$	9·5749	9·5628	9·5508	9·5303	9·5009	9·4618
β_n	20°·20	33°·66	47°·13	60°·59	74°·06	87°·52

h_0 is the sun's mean longitude at epoch, the formula for l is given above, and ν', f$'$ are found from Baird's *Manual* in the tables applicable to the tide K_1 [or from the formulæ at the end of Paper 1].

E. *Rules for the Partition of the Observations into Groups.*

If the first event after epoch is a L.W., either omit it from the reductions, or let the first tide be the H.W. which precedes epoch. Thus we are to begin with a H.W.*

The H.W. and L.W. are treated apart in separate tables.

Each tide (H.W. or L.W., as the case may be) is numbered consecutively, from 1 onwards.

The following are rules for partitions:

* This is not necessary, but it makes the statements of the subsequent rules simpler, as they have not to be given in an alternative form.

For Tides N and L.

Quarter-lunar-anomalistic periods, numbered i, ii, iii, &c.

No. of $\frac{1}{4}$-lunar-anomalistic periods	i.	ii.	iii.	iv.	v.	vi.	vii.	viii.	ix.	x.	xi.	xii.	xiii.
No. of last H.W. in the $\frac{1}{4}$ period	14	27	40	53	67	80	93	107	120	133	147	160	173
No. of last L.W. in the $\frac{1}{4}$ period	13	26	40	53	66	80	93	106	120	133	146	159	173
Total No. of tides up to end of each $\frac{1}{4}$ period	27	53	80	106	133	160	186	213	240	266	293	319	346

For Tides K_1, O, P.

Quarter-lunar periods, numbered I, II, III, &c.

No. of $\frac{1}{4}$-lunar periods	I.	II.	III.	IV.	V.	VI.	VII.	VIII.	IX.	X.	XI.	XII.	XIII.
No. of last H.W. in the $\frac{1}{4}$ period	13	27	40	53	66	79	93	106	119	132	145	159	172
No. of last L.W. in the $\frac{1}{4}$ period	13	26	39	53	66	79	92	105	119	132	145	158	171
Total No. of tides up to end of each $\frac{1}{4}$ period	26	53	79	106	132	158	185	211	238	264	290	317	343

For M_2 and S_2.

Semi-lunations numbered 1, 2, 3, &c.

No. of semi-lunation	1	2	3	4	5	6
No. of last H.W. in the semi-lunation	29	57	86	114	143	171
No. of last L.W. in the semi-lunation	28	57	85	114	142	171
Total No. of tides up to end of each semi-lunation	57	114	171	228	285	342

For Annual and Semi-annual Tides.

Months, or $\frac{1}{12}$th parts of a year, numbered 1, 2, 3.

No. of month	1	2	3
No. of last H.W. in month	59	118	176
No. of last L.W. in month	59	117	176
Total No. of tides up to the end of each month	118	235	352
No. of tides in each month	118	117	117

The epoch for the second quarter year should be 91 days after first epoch, that for the third 92 days after the second, for the fourth 91 days after the third, except in leap year, when the last should also be 92 days.

There are six tides (or about thirty-seven hours) more in a quarter year than in xiii quarter-lunar-anomalistic periods; the times of these six tides (or ten tides in one of the quarters) are to be omitted from the reduction, and their heights are only required when the annual or semi-annual tides are to be found.

F. *Cosine and Sine Factors for all the Tides.*

These are the cosines and sines of 2° 30′, 7° 30′, 12° 30′, &c. They are as follows:

Cosine and Sine Factors.

Read downwards for cosines, upwards for sines.

1. 0·999	10. 0·676
2. 0·991*	11. 0·609*
3. 0·976	12. 0·537
4. 0·954	13. 0·462
5. 0·924*	14. 0·383*
6. 0·887	15. 0·301
7. 0·843	16. 0·216
8. 0·793*	17. 0·130*
9. 0·737	18. 0·044

In the evaluation of $S^{\circ} \frac{\cos}{\sin} \frac{3}{2} V_m$ and $S^{\frac{1}{4}\pi} \frac{\cos}{\sin} \frac{3}{2} V_m$, only the factors marked * are required.

G. *Increments of Arguments in Various Times.*

The following table gives the increments of arguments of the several tides in various periods, multiples of 360° being subtracted. This table facilitates verification of the calculation of the harmonic constants.

	M_2	S_2	K_2	N
1 hour	28°·984104	30°·00000	30°·08214	28°·43973
1 day	− 24 ·3815	0	+ 1 ·9713	− 37 ·4465
10 days	+116 ·185	0	+ 19 ·713	− 14 ·465
100 days	+ 81 ·85	0	−162 ·87	−144 ·65
	L	K_1	O	P
1 hour	29°·52848	15°·04107	13°·94304	14°·95893
1 day	− 11 ·3165	+ 0 ·9856	− 25 ·3671	− 0 ·9856
10 days	−113 ·165	+ 9 ·856	+106 ·329	− 9 ·856
100 days	− 51 ·65	+98 ·56	− 16 ·71	− 98 ·56

Example.

(a.) *Place, Time, Datum Level, and Unit of Length.*

The case chosen is three months of observation (in reality the tidal predictions of the Indian Government) at Bombay, and the epoch is 0^h, January 1, 1887.

A datum at or very near mean water-mark is taken, so that all the H.W. are positive and the L.W. negative. This datum is found by taking the mean of all the H.W. and L.W. of the original observations. In this case 99 inches was subtracted from all the tide heights. I might more advantageously have subtracted 102 or 103 inches, but 99 inches was chosen from considerations applicable to my earlier attempts, but which do not apply to the computation in its present form.

At places where there is a large annual inequality in the height of water, it would be advisable to use a different datum for each quarter of a year. It is not, however, important that the datum should conform rigorously to mean water-mark, for even the discrepancy of 3½ inches, which occurs in my example, does not materially affect the result.

In recording the heights, a convenient unit of length is to be used, and it is advantageous that the H.W. and the L.W. should be expressible by two figures, so that the larger H.W. and L.W. shall fall into the eighties and nineties. The unit of length is here the inch.

(b.) *Times and Angles.*

The times of H.W., numbered consecutively, are entered in a table, as shown on p. 191. Since 0^h astronomical time is the epoch, the P.M. tides will come in the half days which are numbered with integrals, and the A.M. tides in the half days which fall between the integral numbers.

From time to time there will be a half day with no H.W.; this row in the table should be left blank, but there happens to be no such row in the sample shown. A computation form for times and angles might be printed, for, although the exigencies of the printer have not allowed the entries to be equally spaced in the sample below, yet the computation form might be printed with equal spaces, and the dividing lines are to be filled in by hand.

The L.W. table is similar.

Both H. and L.W. are to be divided into quarter-lunar-anomalistic and quarter-lunar periods, and semi-lunations, according to the rules given in E. These partitions and the numbering of the entries could not be printed, because of the occasional blank rows.

The formation of $\frac{1}{2}V_m$ and of V_s, by means of Table A and the rule following it, is obvious. In the subtractions and additions under the headings $\frac{1}{2}V_m - V_s$ and $\frac{1}{2}V_m + V_s$, 360° is added or subtracted where necessary. V_m is found by doubling $\frac{1}{2}V_m$.

(c.) *The Heights.*

The H.W. heights are written in columns, as shown in the column of figures on the margin of the table on the next page (with the same blanks as in the table of times and angles), and are so arranged, either on strips of paper, or by folding the paper, that the heights may be pinned to the times, bringing each height opposite to an angle on the same row with the time corresponding to that height. The heights will on one occasion have to be pinned opposite the V_m column, on a second occasion opposite the V_s column, and on a third occasion opposite the $\frac{1}{2}V_m$ column.

The L.W. heights are written in similar columns, but the minus signs should be omitted.

It is well to divide the columns, or to put fiducial marks in the table for easy verification of the proper allocation of the heights with the times. Any marks suffice, but the division into quarter-anomalistic periods, as shown in the column of heights printed at the margin of the table on the next page, seems to be as good as any other.

If it is proposed to evaluate the annual and semi-annual tides, it is necessary to carry on the heights beyond the times by 3 (or 5) H.W. and

(*Continued on* p. 195)

H.W

No. of Entry	Day	Time	min.	hr.	$\frac{1}{2}V_m$	min.	hr.	V_s	$\frac{1}{2}V_m - V_s$	$\frac{1}{2}V_m + V_s$	V_m	$\frac{1}{4}$-Anom. periods	$\frac{1}{4}$-Lunar periods	Semi-lunations
		h. m.	°	°	°	°	°	°	°	°	°			
1	$-\frac{1}{2}$	3 14	3·4	229·6	233·0	7·0	90	97·0	136·0	330·0	106·0	i begins	I begins	1 begins
2	0	3 38	9·2	43·5	52·7	19·0	90	109·0	303·7	161·7	105·4			
3	$\frac{1}{2}$	3 49	11·8	217·4	229·2	24·5	90	114·5	114·7	343·7	98·4			
4	1	4 31	7·5	45·8	53·3	15·5	120	135·5	277·8	188·8	106·6			
5	$\frac{1}{2}$	4 30	7·2	219·7	226·9	15·0	120	135·0	91·9	1·9	93·8			
6	2	5 46	11·1	48·1	59·2	23·0	150	173·0	246·2	232·2	118·4			
7	$\frac{1}{2}$	5 20	4·8	222·0	226·8	10·0	150	160·0	66·8	26·8	93·6			
8	3	7 15	3·6	64·9	68·5	7·5	210	217·5	211·0	286·0	137·0			
9	$\frac{1}{2}$	6 29	7·0	224·3	231·3	14·5	180	194·5	36·8	65·8	102·6			
10	4	8 28	6·8	67·2	74·0	14·0	240	254·0	180·0	328·0	148·0			
11	$\frac{1}{2}$	7 40	9·7	226·6	236·3	20·0	210	230·0	6·3	106·3	112·6			
12	5	9 22	5·3	69·5	74·8	11·0	270	281·0	153·8	355·8	149·6			
13	$\frac{1}{2}$	8 47	11·4	228·9	240·3	23·5	240	263·5	36·8	143·8	120·6		I ends	
14	6	10 9	2·2	71·8	74·0	4·5	300	304·5	129·5	18·5	148·0	i ends	II begins	
15	$\frac{1}{2}$	9 43	10·4	231·2	241·6	21·5	270	291·5	310·1	173·1	123·2	ii begins		
	&c.			&c.			&c.		&c.	&c.	&c.		&c.	
169	$\frac{1}{2}$	1 22	5·3	220·0	225·3	11·0	30	41·0	184·3	266·3	90·6			
170	87	2 7	1·7	48·4	50·1	3·5	60	63·5	346·6	113·6	100·2			
171	$\frac{1}{2}$	1 53	12·8	207·9	220·7	26·5	30	56·5	164·2	277·2	81·4			6 ends
172	88	2 49	11·8	36·2	48·0	24·5	60	84·5	323·5	132·5	96·0		XIII ends	
173	$\frac{1}{2}$	2 30	7·2	210·1	217·3						74·6	xiii ends		

Heights
H.W.
55 i
23
47
19
41
19
37
26
35
37
37
52
42
66 i

50 ii
80
57
91
63
None
99
67
101
66
99
65
92
58 ii

81 iii
&c., &c.

H.W. V_m.

1st quad.		Angle	0°	5°	10°	15°	20°	25°	30°	35°	40°	45°	50°	55°	60°	65°	70°	75°	80°	85°	No. of entries.
	v		.	.	.	.	.	.	.	.	.	.	.	.	.	.	.	39	45	53	4
			.	.	.	.	.	.	.	.	.	.	.	.	.	.	.	.	.	34	
	ix		.	.	.	.	.	.	.	.	.	.	.	.	.	.	38	45	53	.	3
	x		.	.	.	.	.	.	.	.	.	.	.	.	.	.	.	31	.	.	1
	xiii		.	.	.	.	.	.	.	.	.	.	.	.	.	.	.	.	50	.	1
		......6 semi-lunations end......																			
			.	.	.	.	.	.	.	.	.	.	.	.	.	.	42	.	.	.	1
2nd quad.		Angle	90°	95°	100°	105°	110°	115°	120°	125°	130°	135°	140°	145°	150°	155°	160°	165°	170°	175°	
	i		41	47	35	55	37	19	42	.	.	26	.	37	.	.	.	.	.	.	14
			37	.	.	23	.	.	.	.	.	.	.	52	.	.	.	.	.	.	
			.	.	.	19	.	.	.	.	.	.	.	66	.	.	.	.	.	.	
	ii		.	.	.	.	.	66	50	99	99	91	80	.	.	.	.	.	.	.	13
			.	.	.	.	.	65	57	.	101	.	.	.	.	.	.	.	.	.	
			.	.	.	.	.	58	63	.	.	.	.	.	.	.	.	.	.	.	
			.	.	.	.	.	.	67	.	.	.	.	.	.	.	.	.	.	.	
			.	.	.	.	.	.	92	.	.	.	.	.	.	.	.	.	.	.	
	iii &c.			&c.					&c.				&c.				&c.				&c.

	xi	.	.	.	45	61	76	96	85	99	.	.	.	.	39	.	.	.	.	14
		.	.	.	.	30	.	88	79	87		.	.	.	.	.	.	.	.	
		.	.	.	.	.	.	.	69	45	.	.	.	.	.	.	.	.	.	
		.	.	.	.	.	.	.	57	.	.	.	.	.	.	.	.	.	.	
	xii	.	.	.	.	.	.	.	.	20	.	.	56	66	51	20	44	27	42	11
		.	.	.	.	.	.	.	.	.	.	.	.	.	.	64	.	36	.	
		.	.	.	.	.	.	.	.	.	.	.	.	.	.	.	.	61	.	
	xiii	57	.	61	59	.	65	.	67	.	60	67	.	.	.	.	.	.	.	11
		.	.	55	.	.	62	.	62	.	.	.	.	.	.	.	.	.	.	
		....									6 semi-lunations end									
		.	49	.	.	.	.	.	.	.	.	.	.	.	.	.	.	.	.	
3rd quad.	Angle	180°	185°	190°	195°	200°	205°	210°	215°	220°	225°	230°	235°	240°	245°	250°	255°	260°	265°	
	viii	56	.	.	.	.	.	.	.	.	.	.	.	.	.	.	.	.	.	1
	xii	49	.	.	.	.	.	.	.	.	.	.	.	.	.	.	.	.	.	2
		55	.	.	.	.	.	.	.	.	.	.	.	.	.	.	.	.	.	
		....								6 semi-lunations end										
4th quad.		Nil																		Total 173

L.W. V_m.

																				No. of entries.
1st quad.		Angle	0°	5°	10°	15°	20°	25°	30°	35°	40°	45°	50°	55°	60°	65°	70°	75°	80°	85°	
	viii		30	.	.	.	.	.	.	.	.	.	.	.	.	.	.	.	.	.	1
	xii		36	19	28	.	.	.	.	.	.	.	.	.	.	.	.	.	.	.	3
		......6 semi-lunations end......																			
2nd quad.			Nil																		
3rd quad.		Angle	180°	185°	190°	195°	200°	205°	210°	215°	220°	225°	230°	235°	240°	245°	250°	255°	260°	265°	
	i		.	.	.	.	.	.	.	.	.	.	.	.	.	.	.	26	16	.	2
	v		.	.	.	.	.	.	.	.	.	.	.	.	.	22	32	.	41	.	4
			.	.	.	.	.	.	.	.	.	.	.	.	.	.	12	.	.	.	
	ix		.	.	.	.	.	.	.	.		.	.		.	26	35	48	51	44	7
			.	.	.	.	.	.	.	.	.	.	.	.	.	.	15	.	.	45	
	xiii		.	.	.	.	.	.	.	.	.	.	.	.	.	.	.	36	.	.	4
			.	.	.	.	.	.	.	.	.	.	.	.	.	.	.	62	.	.	
		......6 semi-lunations end......																			
			.	.	.	.	.	.	.	.	.	.	.	.	.	.	.	27	.	.	
			.	.	.	.	.	.	.	.	.	.	.	.	.	.	.	56	.	.	
4th quad.		Angle	270°	275°	280°	285°	290°	295°	300°	305°	310°	315°	320°	325°	330°	335°	340°	345°	350°	355°	
	&c.			&c.					&c.				&c.				&c.				&c.
	xii		.	.	.	.	.	.	.	.	.	.	.	55	.	42	50	13	.	44	10
			.	.	.	.	.	.	.	.	.	.	.	55	.	53	.	44	.	.	
			.	.	.	.	.	.	.	.	.	.	.	.	.	.	.	47	.	.	
			.	.	.	.	.	.	.	.	.	.	.	.	.	.	.	50	.	.	
	xiii		67	50	.	66	54	.	55	.	60	56	.	.	.	.	.	.	.	.	8
			.	.	.	.	.	.	63	.	.	.	.	.	.	.	.	.	.	.	
		......6 semi-lunations end......																			Total 173

3 (or 5) L.W., and to partition them into months. The mean for each month is evaluated, and if the successive quarters of the year are referred to different data, the mean monthly heights must all be referred to a common datum. This process is not carried out in my example, because it is useless to attempt the evaluation of these tides of long period from three months of observation.

The sum of all the H.W. entries to the end of xiii is 9791. The sum of all the L.W. entries to the end of xiii is 8577. There are 173 H.W. and 173 L.W. Hence mean sea-level is $\frac{1}{346}(9791-8577)$, which is equal to $+3{\cdot}51$. It would have been better, therefore, to have subtracted 103 inches instead of 99 inches from all the heights. This would have given mean sea-level at $-0{\cdot}49$ from the datum adopted.

(d.) *Sorting the Heights according to the Values of the Angles.*

In the tables on pp. 192, 193, 194 the column 0° belongs to all angles between 0° and 5°; 5° belongs to 5° to 10°; and so on.

Where an angle falls *exactly* on a multiple of 5°, an arbitrary rule of classification is required, and it is easiest to deem it to belong to the next succeeding 5°, rather than to the preceding 5°.

(e.) *Sorting according to Values of V_m.* (See pp. 192, 193, 194.)

The H. and L.W. are treated in separate tables, similar in form save that the − signs of the L.W. heights are omitted.

The sheets of heights (c) are pinned opposite to the V_m's on the tables of angles (b), and the heights are then entered successively into the columns corresponding to their V_m's in a table like that on pp. 192, 193, 194.

The division into quarter-lunar-anomalistic periods is maintained, but as this sorting is to serve a double purpose, it is necessary to mark the end of the last semi-lunation. In these tables there are two H.W. and two L.W., which fall after the end of 6 semi-lunations, and before the end of xiii quarter-lunar-anomalistic periods.

Nearly all the entries fall into one quadrant for H.W., and into another for L.W. Thus there are no H.W. entries in the 4th quadrant, and no L.W. entries in the 2nd quadrant; there are altogether only 10 H.W. entries in 1st quadrant, and 3 in 3rd quadrant; and there are only 4 L.W. entries in the 1st quadrant, and 17 in 3rd quadrant. Something like this would hold true at all ports.

(f.) *Table of Sums for* N *and* L. (See p. 197.)

Maintaining the divisions i, ii, iii, &c., it is now necessary to sum each of the four times 18 vertical columns in each of the xiii divisions, to subtract the 18 columns of the 3rd quadrant from the 18 columns of the 1st, and to subtract the 18 columns of the 4th quadrant from the 18 of the 2nd. The 2nd − 4th columns have then to be reversed.

Since in this case nearly all the H.W. fall in the 2nd quadrant, and nearly all the L.W. fall in the 3rd quadrant, it is easy to write down at once 2nd − 4th, and 1st − 3rd, as shown on the next page.

In this the − signs of the L.W. entries have to be reintroduced, but as the L.W. lie mostly in 3rd quadrant, which enters with negative sign, they become positive again. It thus happens that nearly all the columns come out +; there are, however, a few − in xii.

(g.) *Table of Sums for* M_2. (See p. 198.)

We now disregard the sub-divisions i, ii, iii, &c., and sum the 4 times 18 columns into grand totals, stopping the summations, however, at the end of 6 semi-lunations (*i.e.* at 171 H.W. and 171 L.W.).

It would hardly be wise to attempt in this case the subtractions 1st − 3rd, 2nd − 4th, without the intermediate steps.

The following table (p. 198) gives the results.

(h.) *General Rule for Cosine and Sine Summations.*

For 'cosines' the 18 numbers required are derived from (1st − 3rd) − (2nd − 4th, reversed).

For 'sines' the 18 numbers required are derived from (1st − 3rd) + (2nd − 4th, reversed).

(i.) *Evaluations of* N *and* L (continued). (See p. 199.)

Cosines. From Table (f) of Sums enter the 18 'cosine' numbers, in accordance with (h) in xiii vertical columns, and perform the operations indicated in the example on page 199.

The column of 'cosine factors' are those given in F, and Φ is given in C for xiii $\frac{1}{2}$-lunar-anomalistic periods.

(f.) *Sums for* N *and* L.

i	(2nd – 4th) H.W.	78	47	35	97	37	19	42	.	.	26	.	155	.	.	.	.	.	.
	(2nd – 4th) L.W.	9	.	.	.	62	40	216	11	18	.	.	.	.	.	.	.	.	.
	(2nd – 4th) Total	87	47	35	97	99	59	258	11	18	26	.	155	.	.	.	.	.	.
	(1st – 3rd) H.W.	.	.	.	.	.	.	.	.	.	.	.	.	.	.	.	.	.	.
	(1st – 3rd) L.W.	.	.	.	.	.	.	.	.	.	.	.	.	.	.	.	26	16	.
	(1st – 3rd) Total	.	.	.	.	.	.	.	.	.	.	.	.	.	.	.	26	16	.
	(2nd – 4th) Reversed	.	.	.	.	.	.	155	.	26	18	11	258	59	99	97	35	47	87
	&c.	&c.					&c.				&c.					&c.			
xii	2nd – 4th H.W.	.	.	.	.	.	.	.	.	20	.	.	56	66	51	84	44	124	42
	2nd – 4th L.W.	.	.	.	.	.	.	.	.		.	.	110	.	95	50	154	.	44
	(2nd – 4th) Total	.	.	.	.	.	.	.	.	20	.	.	166	66	146	134	198	124	86
	1st – 3rd H.W.	−104	.	.	.	.	.	.	.	.	.	.	.	.	.	.	.	.	.
	1st – 3rd L.W.	−36	−19	−28	.	.	.	.	.	.	.	.	.	.	.	.	.	.	.
	(1st – 3rd) Total	−140	−19	−28	.	.	.	.	.	.	.	.	.	.	.	.	.	.	.
	(2nd – 4th) Reversed	86	124	198	134	146	66	166	.	.	20	.	.	.	.	.	.	.	.
xiii	2nd – 4th H.W.	57	49	116	59	.	127	.	129	.	60	67	.	.	.	.	.	.	.
	2nd – 4th L.W.	67	50	.	66	54	.	118	.	60	56	.	.	.	.	.	.	.	.
	(2nd – 4th) Total	124	99	116	125	54	127	118	129	60	116	67	.	.	.	.	.	.	.
	1st – 3rd H.W.	.	.	.	.	.	.	.	.	.	.	.	.	.	.	42	.	50	.
	1st – 3rd L.W.	.	.	.	.	.	.	.	.	.	.	.	.	.	.	.	181	.	.
	(1st – 3rd) Total	..	..	..	..	..	..	..	..	..	..	..	..	..	..	42	181	50	..
	(2nd – 4th) Reversed	..	..	..	..	..	..	..	67	116	60	129	118	127	54	125	116	99	124

(g.) *Sums for* M_2.

1st H.W.	.	.	.	.	.	.	.	.	.	.	.	.	.	.	38	115	148	87
1st L.W.	−66	−19	−28	.	.	.	.	.	.	.	.	.	.	.	.	.	.	.
1st Total	−66	−19	−28	.	.	.	.	.	.	.	.	.	.	.	38	115	148	87
2nd H.W.	262	247	210	452	292	976	1056	1120	941	678	598	696	439	135	395	122	317	216
2nd L.W.	.	.	.	.	.	.	.	.	.	.	.	.	.	.	.	.	.	.
2nd Total	262	247	210	452	292	976	1056	1120	941	678	598	696	439	135	395	122	317	216
3rd H.W.	160	.	.	.	.	.	.	.	.	.	.	.	.	.	.	.	.	.
3rd L.W.	.	.	.	.	.	.	.	.	.	.	.	.	.	−48	−94	−172	−108	−199
3rd Total	160	.	.	.	.	.	.	.	.	.	.	.	.	−48	−94	−172	−108	−199
4th H.W.	.	.	.	.	.	.	.	.	.	.	.	.	.	.	.	.	.	.
4th L.W.	−264	−188	−54	−481	−548	−810	−1453	−1095	−532	−518	−436	−240	−353	−220	−209	−241	−52	−66
4th Total	−264	−188	−54	−481	−548	−810	−1453	−1095	−532	−518	−436	−240	−353	−220	−209	−241	−52	−66
(2nd − 4th) Total......	526	435	264	933	840	1786	2509	2215	1473	1196	1034	936	792	355	604	363	369	282
(1st − 3rd) Total	−226	−19	−28	.	.	.	.	.	.	.	.	.	.	48	132	287	256	286
(2nd − 4th) Reversed.	282	369	363	604	355	792	936	1034	1196	1473	2215	2509	1786	840	933	264	435	526

Cosines				Sums in pairs			Differences					Sum of 5 preceding cols.	× Cosine factors		
i	ii &c.	xii	xiii	i+ii	iii+iv &c.	xi+xii	i+ii −(iii+iv)	v+vi −(iii+iv)	v+vi −(vii+viii)	ix+x −(vii+viii)	ix+x −(xi+xii)			+	−
.	.	−226	.	.	−62	−226	+62	+62	+220	+220	+226	+790 ×	·999 =	790	.
.	.	−143	.	.	−123	−143	+123	+123	+122	+122	+143	+633	·991	628	.
.	.	−226	.	.	−165 &c.	−226	+165	+165	.	.	+226	+556	·976	542	.
.	. &c.	−134	.	.	−182	−134	+182	+182	+288	+288	+134	+1074	·954	1025	.
.	.	−146	.	.	.	−185	.	.	+170	+170	+185	+525	·924	485	.
.	.	−66	.	.	−410	−66	+410	+410	+316	+316	+66	+1518	·887	1346	.
−155	.	−166	.	−155	−182 &c.	−166	+27	−11	−139	−132	−20	−275	·843	.	232
.	−80 &c.	.	−67	−80	−334	−15	+254	+207	+114	+71	−155	+491	·793	389	.
−26	−91	.	−116	−117	−178	−46	+61	−39	−106	−300	−365	−749	·737	.	551
−18	−311	−20	−60	−329	−235 &c.	−336	−94	+55	+11	+49	+194	+215	·676	145	.
−11	−211	.	−129	−222	−166	−378	−56	−567	−654	−429	−130	−1836	·609	.	1118
−258	−529 &c.	.	−118	−787	−225	−479	−562	−27	+310	+476	+393	+590	·537	317	.
−59	−371	.	−127	−430	−394	−350	−36	+179	−68	+24	+227	+326	·462	151	.
−99	−93	.	−54	−192	.	−91	−192	−69	+233	+218	+7	+197	·383	75	.
−97	−163	.	−83	−260	−163 &c.	−45	−97	+58	−105	−103	−58	−305	·301	.	92
−9	. &c.	.	+65	−9	.	.	−9	−20	−20	+70	+70	+91	·216	20	.
−31	.	.	−49	−31	.	.	−31	−94	−94	−54	−54	−327	·130	.	43
−87	.	.	−124	−87	.	.	−87	−70	−70	−69	−69	−365	·044	.	16
														+5913	−2052
														−2052	
												Total.........		+3861	
												factor Φ......		× ·00295	
												$S°h \cos V_m$		= +11·38	

Cosines (continued). Form columns ii + iii, iv + v, vi + vii, viii + ix, x + xi, xii + xiii. Form difference columns (ii + iii) − (iv + v), (vi + vii) − (iv + v), (vi + vii) − (viii + ix), (x + xi) − (viii + ix), (x + xi) − (xii + xiii); add the 5 difference columns together; multiply by cosine factors; sum and multiply by Φ or 0·00295.

The result is $S^{\frac{1}{2}\pi} h \cos V_m = - 8{\cdot}38$

Sines. From Table (f) of Sums, enter 18 "sine" numbers, in accordance with (h) in xiii vertical columns.

Perform all the same operations as those on "cosine" numbers, save that we use sine factors, which are the same as cosine factors in inverse order, viz., beginning with 0·044 and ending with 0·999.

The two results are

$S^{\circ} h \sin V_m = + 5{\cdot}94,$ $S^{\frac{1}{2}\pi} h \sin V_m = + 10{\cdot}06$

Collecting results, proceed thus:

$S^{\circ} h \cos V_m$	$= + 11{\cdot}38.$	$S^{\frac{1}{2}\pi} h \cos V_m$	$= - 8{\cdot}38$
$S^{\frac{1}{2}\pi} h \sin V_m$	$= + 10{\cdot}06.$	$S^{\circ} h \sin V_m$	$= + 5{\cdot}94$
Sum	$= + 21{\cdot}44.$	Sum	$= - 2{\cdot}44$
Diff.	$= + 1{\cdot}32.$	Diff.	$= - 14{\cdot}32$
$P = \frac{1}{2}$ sum	$= + 10{\cdot}72.$	$R = \frac{1}{2}$ sum	$= - 1{\cdot}22$
$Q = \frac{1}{2}$ diff.	$= + 0{\cdot}66.$	$S = \frac{1}{2}$ diff.	$= - 7{\cdot}16$

(j.) *Evaluation of* M_2 (continued).

From the Table (g) of Sums for M_2 enter in one vertical column 18 cosine numbers, in accordance with (h); multiply them by cosine factors; add up and divide by the total number of entries for 6 semi-lunations, viz., 342.

The result is $A_m = \frac{1}{342} \Sigma h \cos V_m = - 30{\cdot}58$

Then enter in vertical column 18 sine numbers, in accordance with (h); multiply them by sine factors, add up, and divide by 342.

The result is $B_m = \frac{1}{342} \Sigma h \sin V_m = + 38{\cdot}47$

(k.) *Sorting according to Values of* V_s, *and Evaluation of* S_2, K_2.

The H. and L.W. are treated in separate tables, similar in form save that the − signs of the L.W. heights are omitted.

The sheets of heights (c) are pinned opposite to the V_s's on the Tables of Angles (b), and the heights are entered successively into the columns corresponding to their V_s's in a table like (e), which was used for sorting according to values of V_m. The sorting is carried as far as the end of an exact multiple of a semi-lunation,—in this case to the end of 6 semi-lunations. No sub-division is *necessary*, but for the purpose of verification it is useful to break the entries into groups of about 40. This is conveniently done by a division after each third $\frac{1}{4}$-lunar-anomalistic period, so that i, ii, iii would be the first group; iv, v, vi the second; vii, viii, ix the third; and x, xi, xii, and all but the end of xiii, the last.

In this case the entries fall into all the four quadrants with about equal frequency.

We next sum the four times 18 columns, just as with M_2 in (g), and form $1^{st} - 3^{rd}$ and $2^{nd} - 4^{th}$, reversed, in the same way.

Next we write the 18 cosine numbers, $(1^{st} - 3^{rd}) - (2^{nd} - 4^{th}$, reversed) in vertical column, multiply by cosine factors, add, and divide by the total number of entries, which is 342. Afterwards write the sine-numbers $(1^{st} - 3^{rd}) + (2^{nd} - 4^{th}$, reversed), multiply by sine factors, add, and divide by 342.

The results are

$$A_s = \tfrac{1}{342}\Sigma h \cos V_s = +21{\cdot}08. \qquad B_s = \tfrac{1}{342}\Sigma h \sin V_s = +3{\cdot}62$$

(l.) *Sorting according to Values of $\frac{1}{2}V_m$.*

The whole process is precisely parallel to the sorting according to values of V_m in (e); the thirteen divisions are, however, given by the quarter-lunar-periods I, II, ... XIII. The only difference lies in the substitution of the factor Ψ (for XIII equal to 0·00298) for Φ. It is unnecessary to give an example.

The results are

$$S^{\circ}h \cos \tfrac{1}{2}V_m = -10{\cdot}50, \qquad S^{\circ}h \sin \tfrac{1}{2}V_m = +8{\cdot}04$$

$$S^{\frac{1}{2}\pi}h \cos \tfrac{1}{2}V_m = +\ 0{\cdot}40, \qquad S^{\frac{1}{2}\pi}h \sin \tfrac{1}{2}V_m = +3{\cdot}74$$

(m.) *Sorting of $\frac{1}{2}V_m$.*

It is required to find what the sums in (l) would be if every H.W. height had been unity, and every L.W. the same both in magnitude *and sign*; in fact to find $S^{\circ} \cos \frac{1}{2}V_m$, $S^{\frac{1}{2}\pi} \cos \frac{1}{2}V_m$, &c.

This is done by counting the entries in the preceding sorting in (l) without regard to magnitude, taking the L.W. entries as actually positive, instead of being (as they are) negative quantities with the negative sign suppressed.

Since in this case we have simply to count entries which are all treated as positive, the table of sums of H. and L.W. may be written together. The following example gives part of the work:

H. and L.W. $\frac{1}{2}V_m$.

I.																		
2nd	.	.	.	.	.	.	.	1	1	1	.	1	1	1	.	.	.	.
4th................	.	.	.	.	.	.	.	.	.	.	.	3	4	.	.	.	.	.
2nd – 4th	.	.	.	.	.	.	.	+1	+1	+1	.	−2	−3	+1	.	.	.	.
1st................	.	.	.	.	.	.	.	.	.	.	2	1	.	1	2	.	.	.
3rd................	.	.	.	.	.	.	.	.	.	3	2	1	1	.	.	.	.	.
1st – 3rd..............	.	.	.	.	.	.	.	.	.	−3	0	0	−1	+1	+2	.	.	.
2nd – 4th, rev.	.	.	.	.	+1	−3	−2	.	+1	+1	+1	.	.	.	.	.	.	.
II.																		
2nd	.	.	.	.	.	.	.	.	.	.	.	.	4	3	.	.	.	.
4th................	.	.	.	.	.	.	.	.	.	.	2	3	1	.	.	.	.	.
2nd – 4th	.	.	.	.	.	.	.	.	.	.	−2	−3	+3	+3	.	.	.	.
1st................	.	.	.	.	.	.	.	.	.	.	.	.	2	3	2	.	.	.
3rd................	.	.	.	.	.	.	.	.	.	.	.	3	4	.	.	.	.	.
1st – 3rd..............	.	.	.	.	.	.	.	.	.	.	.	−3	−2	+3	+2		.	.
2nd – 4th, rev.	.	.	.	.	+3	+3	−3	−2	.	.	.	.	.	.	.	.	.	.
III.	&c.						&c.						&c.					

We next proceed to form XIII columns of cosine numbers, and generally to operate exactly as though these numbers were heights; and then proceed with XIII columns of sine numbers in the same way.

The results are

$$S^{\circ} \cos \tfrac{1}{2} V_m = -0{\cdot}0522, \qquad S^{\circ} \sin \tfrac{1}{2} V_m = +0{\cdot}0117$$

$$S^{\frac{1}{2}\pi} \cos \tfrac{1}{2} V_m = -0{\cdot}0168, \qquad S^{\frac{1}{2}\pi} \sin \tfrac{1}{2} V_m = -0{\cdot}0129$$

(n.) *Formation of the Mean Sums of* $\cos \frac{3}{2} V_m$ *and* $\sin \frac{3}{2} V_m$.

These may be found with sufficient accuracy from the last Table (m) of Sums, part of which is given. In that table lines are drawn dividing the columns into three divisions of six each. These are treated in the way shown in the following example:

H. and L.W. $\frac{3}{2} V_m$.

							Refer to preceding sorting (m)
I.	.	.	.	.	.	.	− 1st six of 2nd − 4th
	.	.	.	−3	0	0	+2nd six of 1st − 3rd
	−3	+1	.	.	.	.	+3rd six of 2nd − 4th
2nd − 4th	−3	+1	.	−3	0	0	
	.	.	.	.	.	.	+1st six of 1st − 3rd
	.	+1	+1	+1	.	−2	+2nd six of 2nd − 4th
	+1	−1	−2	.	.	.	− 3rd six of 1st − 3rd
1st − 3rd...........	+1	0	−1	+1	.	−2	
2nd − 4th, rev. ...	0	0	−3	.	+1	−3	
II.	.	.	.	.	.	.	− 1st six of 2nd − 4th
	.	.	.	.	.	−3	+2nd six of 1st − 3rd
	+3	+3	.	.	.	.	+3rd six of 2nd − 4th
2nd − 4th	+3	+3	.	.	.	−3	
	.	.	.	.	.	.	+1st six of 1st − 3rd
	.	.	.	.	−2	−3	+2nd six of 2nd − 4th
	+2	−3	−2	.	.	.	− 3rd six of 1st − 3rd
1st − 3rd...........	+2	−3	−2	.	−2	−3	
2nd − 4th, rev. ...	−3	.	.	.	+3	+3	
III.			&c.				&c.

We have now only 6 instead of 18 sub-divisions of the quadrant, but the cosine and sine numbers are found in exactly the same way as before.

The following example shows part of the treatment, and the cosine factors are those marked * in F.

Cosines				Sums in pairs				Differences			Sum of 5 columns		Cosine factors	+	−
I.	II.		XIII.	I.+II.	III.+IV.		XI.+XII.	(I.+II.) −(III.+IV.)		(IX.+X.) −(XI.+XII.)					
+1	+5	&c.	0	+6	−2	&c.	+3	+8	&c.	0	+23	×	0·991	22·973	.
0	−3		−1	−3	+3		−1	−6		+1	− 8		0·924	.	7·392
+2	−2		0	0	0		−2	0		−4	−16		0·793	.	12·688
+1	.		+1	+1	−5	&c.	+1	+6	&c.	−1	+20		0·609	12·180	.
−1	−5	&c.	−1	−6	+2		−3	−8		+1	−17		0·383	.	6·511
+1	−6		+1	−5	−1		−3	−4		+7	+20		0·130	3·120	.
														38·273	26·591
														26·591	
														+11·682	
													Factor Ψ...	× 0·00298	
													$S^\circ \cos \frac{3}{2} V_m =$	+ 0·0347	

The remaining process is exactly like that pursued before, and the four results are

$$S^{\circ} \cos \tfrac{3}{2} V_m = + 0{\cdot}0347, \qquad S^{\circ} \sin \tfrac{3}{2} V_m = - 0{\cdot}1830$$

$$S^{\frac{1}{2}\pi} \cos \tfrac{3}{2} V_m = - 0{\cdot}0479, \qquad S^{\frac{1}{2}\pi} \sin \tfrac{3}{2} V_m = - 0{\cdot}0173$$

(o.) *The Sorting of* $\frac{1}{2}V_m + V_s$ *and of* $\frac{1}{2}V_m - V_s$.

These angles have to be sorted without reference to the heights, or just as though all the heights were unity. Every entry is to be regarded as unity. The following example shows part of the sorting of $\frac{1}{2}V_m + V_s$, and 1 denotes a H.W., † a L.W.; by this device H. and L.W. may be sorted on the same paper.

We may also, if it is found convenient, put on it the sorting of $\frac{1}{2}V_m - V_s$ by adopting, say 0, to denote a H.W. and * a L.W., each one of these four signs denoting simply unity.

H. and L.W. $\frac{1}{2}V_m + V_s$. 1 for H.W., † for L.W.

1st quad.	Angles	0°	5°	10°	15°	20°	25°	30°	35°	40°	45°	50°	55°	60°	65°	70°	75°	80°	85°
I.		1	.	†	.	.	1	.	†	†	.	.	.	1	1†	.	.	.	.
II.		†	.	.	1	†	.	.	1	.	†	.	1	.	.	1	.	.	†
				&c.						&c.					&c.				

2nd quad.	Angles	90°	95°	100°	105°	110°	115°	120°	125°	130°	135°	140°	145°	150°	155°	160°	165°	170°	175°
I.		.	.	.	1†	.	.	.	.	.	.	1	.	.	.	1	†	.	.
II.		1	.	.	†	1	.	.	1†	.	.	.	†	.	.	.	†	1	.
				&c.						&c.					&c.				

3rd quad.	Angles 180°, &c.
	&c.

4th quad.	Angles 270°, &c.
	&c.

We then proceed to count these 1's and †'s just as was done with the number of entries in the sorting of $\frac{1}{2}V_m$, and to operate on them in the same way.

The results are

$$S^{\circ}\cos(\tfrac{1}{2}V_m - V_s) = -\cdot 0078, \qquad S^{\circ}\sin(\tfrac{1}{2}V_m - V_s) = -\cdot 0060$$
$$S^{\frac{1}{2}\pi}\cos(\tfrac{1}{2}V_m - V_s) = +\cdot 0280, \qquad S^{\frac{1}{2}\pi}\sin(\tfrac{1}{2}V_m - V_s) = +\cdot 0078$$
$$S^{\circ}\cos(\tfrac{1}{2}V_m + V_s) = +\cdot 1244, \qquad S^{\circ}\sin(\tfrac{1}{2}V_m + V_s) = +\cdot 0094$$
$$S^{\frac{1}{2}\pi}\cos(\tfrac{1}{2}V_m + V_s) = +\cdot 0147, \qquad S^{\frac{1}{2}\pi}\sin(\tfrac{1}{2}V_m + V_s) = +\cdot 0834$$

(p.) *Evaluation of* $F_m^{(o)}$, $G_m^{(o)}$, $f_m^{(o)}$, $g_m^{(o)}$, $F_m^{(\frac{1}{2}\pi)}$, $G_m^{(\frac{1}{2}\pi)}$, $f_m^{(\frac{1}{2}\pi)}$, $g_m^{(\frac{1}{2}\pi)}$, $F_s^{(o)}$, $G_s^{(o)}$, $f_s^{(o)}$, $g_s^{(o)}$, $F_s^{(\frac{1}{2}\pi)}$, $G_s^{(\frac{1}{2}\pi)}$, $f_s^{(\frac{1}{2}\pi)}$, $g_s^{(\frac{1}{2}\pi)}$.

These 16 coefficients are required to correct the four sums $S^{\circ}h \,{}^{\cos}_{\sin}\, \frac{1}{2}V_m$, $S^{\frac{1}{2}\pi}h \,{}^{\cos}_{\sin}\, \frac{1}{2}V_m$, for the influence of the tides M_2 and S_2.

I call $S^{\circ}h\cos\frac{1}{2}V_m$ + corrn., W, $S^{\circ}h\sin\frac{1}{2}V_m$ + corrn., X, and the other two Y and Z.

The correction to be applied to $S^{\circ}h\cos\frac{1}{2}V_m$ to get W is

$$-[F_m^{(o)}A_m + G_m^{(o)}B_m + F_s^{(o)}A_s + G_s^{(o)}B_s]$$

and the correction to be applied to $S^{\circ}h\sin\frac{1}{2}V_m$ to get X is

$$-[f_m^{(o)}A_m + g_m^{(o)}B_m + f_s^{(o)}A_s + g_s^{(o)}B_s]$$

and the two other corrections are given by symmetrical formulæ with $(\frac{1}{2}\pi)$ in place of (o).

These coefficients are computed from S° and $S^{\frac{1}{2}\pi}$ of ${}^{\cos}_{\sin}(\frac{1}{2}V_m \pm V_m)$ and of ${}^{\cos}_{\sin}(\frac{1}{2}V_m \pm V_s)$, as given in (m) (n) (o). It must be especially noticed that we have above in (m) computed S° and $S^{\frac{1}{2}\pi}$ of $\sin\frac{1}{2}V_m$; but $\frac{1}{2}V_m - V_m = -\frac{1}{2}V_m$, so that the signs of our previous results must be changed in these two cases.

If we remark that k_m and k_s are constants found by theoretical considerations, that A_m, B_m, A_s, B_s, are already found, and that in the first column we are compelled to omit the affixes to the letters S, k, and the F's and G's, because they indicate various sorts of S's and k's and F's and G's in the different columns, the computations in the following table are easily followed:

	S°	S°	$S^{\frac{1}{2}\pi}$	$S^{\frac{1}{2}\pi}$
	$k_m=2$	$k_s=2{\cdot}07$	$k_m=2$	$k_s=2{\cdot}07$
	$V_p=V_m$	$V_p=V_s$	$V_p=V_m$	$V_p=V_s$
$\frac{1}{2}S\cos(\frac{1}{2}V_m-V_p)$	−·0261	−·0039	−·0084	+·0140
$\frac{1}{2}S\cos(\frac{1}{2}V_m+V_p)$	+·0174	+·0622	+·0240	+·0074
Sum Σ	−·0087	+·0583	−·0324	+·0214
Diff. Δ	−·0435	−·0661	+·0156	+·0066
$k\Sigma$	−·0174	+·1207	−·0648	+·0443
$k\Delta$	−·0870	−·1368	+·0312	+·0137
$\Sigma+k\Delta=F$	−·0957	−·0785	−·0012	+·0351
$\Delta+k\Sigma=g$	−·0609	+·0546	−·0492	+·0509
$\frac{1}{2}S\sin(\frac{1}{2}V_m-V_p)$	−·0059	−·0030	+·0065	+·0039
$\frac{1}{2}S\sin(\frac{1}{2}V_m+V_p)$	−·0915	+·0047	−·0087	+·0417
Sum σ	−·0974	+·0017	−·0022	+·0456
Diff. δ	+·0856	−·0077	+·0152	−·0378
$k\sigma$	−·1948	+·0035	−·0044	+·0944
$k\delta$	+·1712	−·0159	+·0304	−·0782
$\sigma+k\delta=f$	+·0738	−·0142	+·0282	−·0326
$-(\delta+k\sigma)=G$	+·1092	+·0042	−·0108	−·0566

Coefficients......		$-A_m$	$-B_m$	$-A_s$	$-B_s$
$(S^{\circ}h\cos\frac{1}{2}V_m)$	−10·50	(F_m°) −·0957	(G_m°) +·1092	(F_s°) −·0785	(G_s°) +·0042 = W
$(S^{\circ}h\sin\frac{1}{2}V_m)$	+ 8·04	(f_m°) +·0738	(g_m°) −·0609	(f_s°) −·0142	(g_s°) +·0546 = X
$(S^{\frac{1}{2}\pi}h\cos\frac{1}{2}V_m)$	+ 0·40	$(F_m^{\frac{1}{2}\pi})$ −·0012	$(G_m^{\frac{1}{2}\pi})$ −·0108	$(F_s^{\frac{1}{2}\pi})$ +·0351	$(G_s^{\frac{1}{2}\pi})$ −·0326 = Y
$(S^{\frac{1}{2}\pi}h\sin\frac{1}{2}V_m)$	+ 3·74	$(f_m^{\frac{1}{2}\pi})$ +·0282	$(g_m^{\frac{1}{2}\pi})$ −·0492	$(f_s^{\frac{1}{2}\pi})$ −·0566	$(g_s^{\frac{1}{2}\pi})$ +·0509 = Z

Multiply by $-A_m=+30{\cdot}58$, $-B_m=-38{\cdot}47$, $-A_s=-21{\cdot}08$, $-B_s=-3{\cdot}53$

−10·50	−2·91	−4·23	+1·66	−0·01	= W
+ 8·04	+2·24	+2·36	+0·30	−0·19	= X
+ 0·40	−0·04	+0·42	−0·74	+0·12	= Y
+ 3·74	+0·86	+1·91	+1·20	−0·18	= Z

W		X		Y		Z	
+	−	+	−	+	−	+	−
	10·50	8·04		0·40		3·74	
	2·91	2·24			0·04	0·86	
	4·23	2·36		0·42		1·91	
1·66		0·30			0·74	1·20	
	0·01		0·19	0·12			0·18
	17·65	12·94		0·94	0·78	7·71	
	1·66	0·19		0·78		0·18	
	−15·99	+12·75		+0·16		+7·53	

$$W = -15{\cdot}99 \qquad X = +12{\cdot}75$$
$$Z = +\ 7{\cdot}53 \qquad Y = +\ 0{\cdot}16$$
$$W + Z = -\ 8{\cdot}46 \qquad X + Y = +12{\cdot}91$$
$$W - Z = -23{\cdot}52 \qquad X - Y = +12{\cdot}59$$
$$\tfrac{1}{2}(W + Z) = -\ 4{\cdot}23 \qquad \tfrac{1}{2}(X + Y) = +\ 6{\cdot}46$$
$$\tfrac{1}{2}(W - Z) = -11{\cdot}76 \qquad \tfrac{1}{2}(X - Y) = +\ 6{\cdot}30$$

(q.) *Computation of Astronomical and other Constants.*

Find s_o, the moon's mean longitude (see *Nautical Almanac*), and h_o the sun's mean longitude (sidereal time reduced to angle) from the *Nautical Almanac*, and p_o the longitude of moon's perigee, from Baird's *Manual**, Appendix Table XII (there called π), at the epoch 0^h, January 1, 1887, Bombay mean time, in E. Longitude $4^h{\cdot}855$.

From Baird, Tables XIV, XV, XVIII, find N the longitude of Moon's node, and I, ν, ξ at mid-period, February 14, 1887† [or see the *Nautical Almanac* and the formulæ at end of Paper 1].

With the value of I find f_m from XIX (1) for the tides M_2, N, L; from XIX (3) find f_o for the tide O; from XIX (8) find f' for the tide K_1; from XIX (9) find f'' for the tide K_2; from XX find ν' for the tide K_1; and from XXI find $2\nu''$ for the tide K_2; [or use the formulæ at the end of Paper 1].

The results are

$$s_o = 359^\circ{\cdot}43, \qquad h_o = 280^\circ{\cdot}63, \qquad p_o = 165^\circ{\cdot}36$$
$$\nu = 9^\circ{\cdot}60, \qquad \xi = 9^\circ{\cdot}00$$
$$1/f_m = 0{\cdot}9709, \qquad 1/f_o = 1{\cdot}161, \qquad f' = 0{\cdot}915, \qquad f'' = 0{\cdot}802$$
$$\nu' = 6^\circ{\cdot}30, \qquad 2\nu'' = 11^\circ{\cdot}75$$

Then compute initial equilibrium arguments, in the symbol for which the subscript letters indicate the tides referred to,

$$u_m = 2(h_o - \nu) - 2(s_o - \xi), \qquad u_o = (h_o - \nu) - 2(s_o - \xi) + \tfrac{1}{2}\pi, \qquad u_s = 0^\circ$$
$$= 201^\circ{\cdot}20, \qquad = 20^\circ{\cdot}17$$

$$\text{for } K_1,\ u' = h_o - \nu' - \tfrac{1}{2}\pi, \qquad \text{for } K_2,\ u'' = 2h_o - 2\nu''$$
$$= 184^\circ{\cdot}33, \qquad = 189^\circ{\cdot}51$$

$$u_n = u_m - (s_o - p_o), \qquad u_l = u_m + (s_o - p_o) + \pi$$
$$= 7^\circ{\cdot}13, \qquad = 215^\circ{\cdot}27$$

$$u_p = -h_o + \tfrac{1}{2}\pi = 169^\circ{\cdot}37$$

We have already shown in B the way of computing Π, and $\Pi = 1{\cdot}034$‡.

* *Manual for Tidal Observations*, by Major Baird. Taylor and Francis, Fleet Street, 1886.

† In making these reductions I have really used the value of N for July 1, 1887, because I am operating on tidal *predictions* made for the whole year 1887, which were doubtless made with mean N for that year. The difference is almost insensible.

‡ As the Indian tide-predicting instrument takes no account of solar parallax, I should in reality have done better to take Π as unity. But of course this consideration does not apply to real observations.

In C and D we have shown how to compute j, i, l, and $j = +6^\circ{\cdot}46$, $i = +6^\circ{\cdot}52$, $l = -0^\circ{\cdot}97$.

By the formula in B, with $\alpha_n = 86^\circ{\cdot}97$ for 6 semi-lunations,

$$\begin{aligned}\omega &= 2h_o - 2\nu'' + \alpha_n = u'' + \alpha_n \\ &= 276^\circ{\cdot}48 \\ &= -83^\circ{\cdot}52\end{aligned}$$

By the formula in D, with $\beta_n = 87^\circ{\cdot}52$ for XIII quarter-lunar periods,

$$\begin{aligned}\theta &= 2h_o - \nu' + l + \beta_n \\ &= 281^\circ{\cdot}51 = -78^\circ{\cdot}49\end{aligned}$$

By the formula in B, viz.:

$$\begin{aligned}U\cos\phi &= \Pi + \lambda_n f''\cos\omega \\ U\sin\phi &= \qquad \lambda_n f''\sin\omega\end{aligned}$$

With $\log\lambda_n = 9{\cdot}2517$ for 6 semi-lunations, and with the above values of Π, f'', ω:

$$\phi = -7^\circ{\cdot}72, \qquad (+)\log U = 0{\cdot}0251$$

By the formula in D, viz.:

$$\begin{aligned}T\cos\psi &= f' - \rho_n\cos\theta \\ T\sin\psi &= \qquad \rho_n\sin\theta\end{aligned}$$

with $\log\rho_n = 9{\cdot}4618$ for XIII quarter-lunar periods, and with the above values of f' and θ:

$$\psi = -18^\circ{\cdot}32, \qquad (+)\log T = 9{\cdot}9557$$

(r.) *Final Evaluation of* M_2.

From (j) $B_m = +38{\cdot}47$, $\quad A_m = -30{\cdot}58$, $\quad \tan\zeta_m = \dfrac{B_m}{A_m}$

B_m is + and A_m is −, so that ζ_m lies in second quadrant; whence

$$\zeta_m = \pi - 51^\circ{\cdot}51 = 128^\circ{\cdot}49$$

Then

$$H_m = \frac{1}{f_m} \cdot B_m \operatorname{cosec}\zeta_m$$

whence, on reducing from inches to feet,

$$H_m = 3{\cdot}98 \text{ ft.}$$

Also

$$\kappa_m = \zeta_m + u_m = 128^\circ{\cdot}49 + 201^\circ{\cdot}20 = 329^\circ{\cdot}69$$

where the value of u_m is taken from (q).

(s.) *Final Evaluation of* N *and* L.

Taking the values of P, Q, R, S from (i),

$$f_m H_n \sin(\zeta_n - j) = -P = -10{\cdot}72, \qquad f_m H_l \sin(\zeta_l + j) = +Q = +0{\cdot}66$$

$$f_m H_n \cos(\zeta_n - j) = -S = +\ 7{\cdot}16, \qquad f_m H_l \cos(\zeta_l + j) = -R = +1{\cdot}22$$

$\zeta_n - j$ lies in 4th quad., $\zeta_l + j$ lies in 1st quad.

whence $$\zeta_n - j = -56^\circ{\cdot}27$$

Then $$H_n = \frac{1}{f_m} \operatorname{cosec}(\zeta_n - j) \times (-P)$$

whence, on reducing from inches to feet,

$$H_n = 1{\cdot}04 \text{ ft.}$$

Again, since from (q) $j = +6^\circ{\cdot}54$, we have $\zeta_n = -49^\circ{\cdot}73 = 310^\circ{\cdot}27$, and $\kappa_n = \zeta_n + u_n = 310^\circ{\cdot}27 + 7^\circ{\cdot}13 = 317^\circ{\cdot}40$, where the value of u_n is taken from (q).

Turning to the second pair of equations,

$$\zeta_l + j = 28^\circ{\cdot}4$$

Then $$H_l = \frac{1}{f_m} \sec(\zeta_l + j) \times (-R)$$

whence, on reducing from inches to feet,

$$H_l = 0{\cdot}11 \text{ ft.}$$

Again, since $j = +6^\circ{\cdot}5$, we have $\zeta_l = 21^\circ{\cdot}9$, and

$$\kappa_l = \zeta_l + u_l = 21^\circ{\cdot}9 + 215^\circ{\cdot}3 = 237^\circ{\cdot}2$$

where the value of u_l is taken from (q).

(t.) *Final Evaluation of* S_2 *and* K_2.

From (k) $B_s = +3{\cdot}62$, $A_s = +21{\cdot}08$; $\tan \zeta_s = \dfrac{B_s}{A_s}$

B_s and A_s are +, so that ζ_s lies in 1st quadrant;

whence $$\zeta_s = 9^\circ{\cdot}71$$

Then $$H_s = \frac{A_s \sec \zeta_s}{U}$$

whence, with $\log U$ already found in (q) as 0·0251, and, reducing inches to feet,

$$H_s = 1{\cdot}68 \text{ ft.}$$

Again $$\kappa_s = \zeta_s + \phi = 9^\circ{\cdot}71 - 7^\circ{\cdot}72 = 1^\circ{\cdot}99$$

where the value of ϕ is taken from (q).

Lastly, $H'' = 0{\cdot}272 H_s = 0{\cdot}46$ ft., and $\kappa'' = \kappa_s = 2^\circ$

The factor 0·272 is an absolute constant.

(u.) *Final Evaluation of* K_1, O, P.

Taking the values of $\frac{1}{2}(W-Z)$, $\frac{1}{2}(X+Y)$ from (p),

$$TH' \sin(\zeta'+i-\psi) = \tfrac{1}{2}(W-Z) = -11{\cdot}76$$

$$TH' \cos(\zeta'+i-\psi) = -\tfrac{1}{2}(X+Y) = -6{\cdot}47$$

$\zeta'+i-\psi$ lies in third quadrant, and

$$\zeta'+i-\psi = \pi + 61^\circ{\cdot}2 = 241^\circ{\cdot}2$$

Then since, from (q), $\psi = -18^\circ{\cdot}32$, we have $\zeta'+i = 222^\circ{\cdot}9$; and since from (q) $i = 6^\circ{\cdot}52$, therefore $\zeta' = 216^\circ{\cdot}4$; whence

$$\kappa' = \zeta' + u' = 216^\circ{\cdot}4 + 184^\circ{\cdot}3 = 40^\circ{\cdot}7$$

where the value of u' is taken from (q).

Then $$H' = \frac{\frac{1}{2}(W-Z)}{T} \operatorname{cosec}(\zeta'+i-\psi)$$

whence, with log T already found in (q) as 9·9557, and reducing from inches to feet,

$$H' = 1{\cdot}24 \text{ ft.}$$

Also $$H_p = 0{\cdot}331 H' = 0{\cdot}41, \quad \text{and} \quad \kappa_p = \kappa' = 41^\circ$$

The factor 0·331 is an absolute constant.

We now have to compute

$$L = \tfrac{1}{2}(X+Y)\tan\tfrac{1}{4}\epsilon + f'H'\cos(\zeta'+i)\tan\tfrac{1}{4}\epsilon$$

$$M = \tfrac{1}{2}(W-Z)\tan\tfrac{1}{4}\epsilon - f'H'\sin(\zeta'+i)\tan\tfrac{1}{4}\epsilon$$

where $\log\tan\frac{1}{4}\epsilon = 9{\cdot}0677$, an absolute constant for all times and places. With the values of f′ and $\frac{1}{2}(X+Y)$ and $\frac{1}{2}(W-Z)$ given above in (q) and (p), and with the values of H′ and $\zeta'+i$ just found, there results

$$L = -0{\cdot}410 \qquad M = -{\cdot}281$$

Now $$f_o H_o \sin(\zeta_o - i) = -\tfrac{1}{2}(W+Z) + L$$

$$f_o H_o \cos(\zeta_o - i) = \tfrac{1}{2}(X-Y) + M$$

We have found in (p)

$$\tfrac{1}{2}(W+Z) = -4{\cdot}23 \qquad \tfrac{1}{2}(X-Y) = +6{\cdot}30$$

so that $f_o H_o \sin(\zeta_o - i) = +3{\cdot}82$ $\qquad f_o H_o \cos(\zeta_o - i) = +6{\cdot}02$

Whence $\zeta_o - i$ lies in the first quadrant, and

$$\zeta_o - i = 32^\circ{\cdot}40$$

Then $$H_o = \frac{1}{f_o}[\tfrac{1}{2}(X-Y) + M]\sec(\zeta_o - i)$$

whence, reducing from inches to feet,

$$H_o = 0{\cdot}69 \text{ ft.}$$

Again,

$$\kappa_o = \zeta_o + u_o = (\zeta_o - i) + i + u_o = 32^\circ\cdot40 + 6^\circ\cdot52 + 20^\circ\cdot17 = 59^\circ\cdot09$$

where the value of u_o is taken from (q).

(v.) *Final Reduction of Mean Water Mark.*

We subtracted 99 inches from all the heights before using them, and the mean of the heights was then +3·51 inches. Hence mean water is 102·51 inches, or 8·54 feet above the datum of the original tidal observations.

(w.) *Results of Reduction.*

		Mean of 9 yrs. obs.	Error of present calc. in inches and minutes
Mean water, 8·54 ft.		8·223	4 in.
M_2	H = 3·98 ft.	4·043	$\frac{3}{4}$ in. too small
	κ = 330°	330°	nil
S_2	H = 1·68 ft.	1·625	$\frac{2}{3}$ in. too large
	κ = 2°	3°	2^m too slow
K_2	H = 0·46 ft.	0·405	$\frac{2}{3}$ in. too large
	κ = 2°	352°	20^m too fast
N	H = 1·04 ft.	0·997	$\frac{1}{2}$ in. too large
	κ = 317°	313°	8^m too fast
L	H = 0·11 ft.	0·088	$\frac{1}{4}$ in. too large
	κ = 237°	308°	$2^h\ 21^m$ too slow
K_1	H = 1·24 ft.	1·396	$1\frac{3}{4}$ in. too small
	κ = 41°	45°	16^m too slow
O	H = 0·69 ft.	0·658	$\frac{1}{3}$ in. too large
	κ = 59°	48°	44^m too fast
P	H = 0·41 ft.	0·404	$\frac{1}{10}$ in. too large
	κ = 41°	43°	8^m too slow

The second column is given because, if the calculation had been conducted by rigorous methods instead of approximately, my results should have agreed very nearly* with these. The causes of several of the discrepancies are explicable. The error of mean water mark is due to the necessity for neglecting the annual and semi-annual tides in a short series of observations. The error in phase in K_2 is a necessary incident of the shortness of the series of observations. The tide L is only about an inch in height, and accuracy of result could not be expected.

* I do not know the exact values of the constants used in the Bombay Tide Table, which has been used as representing observation.

The magnitude of the error in time in the diurnal tides is rather disappointing, but it is clear that the length of observation has not been sufficient to disentangle the O tide from the K_1 tide. It may be remarked also that an error of 1° in phase makes twice as much difference in time with the diurnal tides as with the semi-diurnal.

Lastly it is probable that all these errors would have been sensibly diminished if I had subtracted 103 inches from the heights all through instead of 99, and I know that this is to some extent the case.

(x.) *Verification.*

In a calculation of this kind some gross error of principle may have been committed, such, for example, as imputing to some of the κ's a wrong sign; and this is the kind of mistake which is easily overlooked in a mere verification of arithmetical processes. It is well, therefore, to test whether the tide heights and times are actually given by the computed constants. This is conveniently done by selecting some three or four tides from amongst those from which the reductions have been made, and it makes the calculation much shorter if we pick out cases in which it is H. or L.W. within a few minutes of noon.

For example, in the present case it was L.W. on February 16 (day 46) at $0^h\ 7^m$ P.M., and the height was 4 ft. 0 in.

Now, if U denotes the value of any equilibrium argument whose value at the epoch, 0^h, January 1, was denoted in (q) by u, and if A_o denotes the height of mean sea-level above datum, the expression for the height of water is

$$h = A_o + f_m H_m \cos(U_m - \kappa_m) + H_s \cos(U_s - \kappa_s) + f'' H'' \cos(U'' - \kappa'')$$
$$+ f_m H_n \cos(U_n - \kappa_n) + f_m H_l \cos(U_l - \kappa_l) + f' H' \cos(U' - \kappa')$$
$$+ f_o H_o \cos(U_o - \kappa_o) + H_p \cos(U_p - \kappa_p)$$

The time of H.W. depends on a formula involving the sines of the same angles in place of cosines.

Since we have chosen cases where it is H. or L.W. at noon, the U's exceed the u's by an exact number of days' motion.

The evaluation of the separate terms may be conveniently made by means of an ordinary nautical traverse table, where (neglecting the decimal point) fH is represented by the "Distance," and $fH\cos(U-\kappa)$ is given by "Latitude," and $fH\sin(U-\kappa)$ by "Departure."

If we know the time of H. or L.W. within 20^m or so, the following calculation will give the true time and height. In this case we know that there should be a L.W. at about 0^h of day 46. The increments of argument are computed from the Table G, and the κ's are subtracted either by actual subtraction or by addition of $2\pi-\kappa$.

	M_2	S_2	K_2	N	L	K_1	O	P
Increment in 40^d	$464^\circ\cdot7$		$78^\circ\cdot9$	$-57^\circ\cdot9$	$-452\cdot7$	$39\cdot4$	$425\cdot3$	
Ditto 6^d	$-146\cdot3$		$11\cdot8$	$-224\cdot7$	$-67\cdot9$	$5\cdot9$	$-152\cdot2$	(see K_1)
Ditto 46^d	$318\cdot4$		$90\cdot7$	$-282\cdot6$	$-520\cdot6$	$45\cdot3$	$273\cdot1$	$-45\cdot3$
$u=$	$201\cdot2$		$189\cdot5$	$7\cdot1$	$215\cdot3$	$196\cdot9$	$3\cdot4$	$169\cdot4$
$U=$	$519\cdot6$		$280\cdot2$	$-275\cdot5$	$-305\cdot3$	$242\cdot2$	$276\cdot5$	$124\cdot1$
$-\kappa=$	$-329\cdot7$	$-2\cdot0$	$-2\cdot0$	$+42\cdot6$	$+122\cdot8$	$-40\cdot7$	$-59\cdot1$	$-40\cdot7$
$U-\kappa=$	$189\cdot9$	$-2\cdot0$	$278\cdot2$	$-232\cdot9$	$-182\cdot5$	$201\cdot5$	$217\cdot4$	$83\cdot4$
$U-\kappa=$	$\pi+10$	-2	-82	$\pi-53$	$\pi-2$	$\pi+22$	$\pi+37$	83

		f	H	fH	$U-\kappa$	fH cos $(U-\kappa)$ +	fH cos $(U-\kappa)$ −	fH sin $(U-\kappa)$ +	fH sin $(U-\kappa)$ −
Semidiurnals	M_2	1·03	3·98	4·10	$\pi+10^\circ$		4·04		0·71
	S_2	1·0	1·68	1·68	-2	1·68			0·06
	K_2	0·80	0·44	0·37	-82	0·05			0·37
	N	1·03	1·04	1·07	$\pi-53$		0·64	0·86	
	L	1·03	0·11	0·11	$\pi-2$		0·11	0·00	
						+1·73	−4·79	+0·86	−1·14
							+1·73		+0·86
							$A_2=-3\cdot06$		$B_2=-0\cdot28$
Diurnals	K_1	0·915	1·24	1·14	$\pi+22$		1·06		0·43
	O	0·86	0·69	0·59	$\pi+37$		0·47		0·36
	P	1·00	0·41	0·41	83	0·05		0·41	
						+0·05	−1·53	+0·41	−0·79
							+0·05		+0·41
							$A_1=-1\cdot48$		$B_1=-0\cdot38$
							$\frac{1}{4}A_1=-0\cdot37$		$\frac{1}{2}B_1=-0\cdot19$
							$A_2+\frac{1}{4}A_1=-3\cdot43$		$B_2+\frac{1}{2}B_1=-0\cdot47$

$$\text{Time}=-120^m\left(\frac{B_2+\frac{1}{2}B_1}{A_2+\frac{1}{4}A_1}\right)=-120^m\times\frac{47}{343}=-16^m=11^h\ 44^m\ \text{A.M.}$$

Tabular time $=0^h\ 7^m$ P.M.
Error $=-23^m$

$A_2+A_1=-4\cdot54$
Mean water $+8\cdot54$

Height $=4\cdot00$

Height L.W.	4 ft.	0 in.
Tabular height......	4	0
Error		0

This result is as good as might be expected, and, considered as a prediction, would be amply sufficient for navigational purposes.

6.

ON AN APPARATUS FOR FACILITATING THE REDUCTION OF TIDAL OBSERVATIONS.

[*Proceedings of the Royal Society*, LII. (1892), pp. 345—389.]

§ 1. *Introduction.*

THE tidal oscillation of the ocean may be represented as the sum of a number of simple harmonic waves which go through their periods approximately once, twice, thrice, four times in a mean solar day. But these simple harmonic waves may be regarded as being rigorously diurnal, semi-diurnal, ter-diurnal, and so forth, if the length of the day referred to be adapted to suit the particular wave under consideration. The idea of a series of special scales of time is thus introduced, each time-scale being appropriate to a special tide. For example, the mean interval between successive culminations of the moon is $24^h\ 50^m$, and this interval may be described as the mean lunar day. Now there is a series of tides, bearing the initials M_1, M_2, M_3, M_4, &c., which go through their periods rigorously once, twice, thrice, four times, &c., in a mean lunar day. The solar tides, S, proceed according to mean solar time; but, besides mean lunar and mean solar times, there are special time scales appropriate to the larger (N) and smaller (L) lunar elliptic tides, to the evectional (ν), to the diurnal (K_1) and semi-diurnal (K_2) luni-solar tides, to the lunar diurnal (O), &c.

The process of reduction consists of the determination of the mean height of the water at each of 24 special hours, and subsequent harmonic analysis. The means are taken over such periods of time that the influence of all the tides governed by other special times is eliminated.

The process by which the special hourly heights have hitherto been obtained is the entry of the heights observed at the mean solar hours in a schedule so arranged that each entry falls into a column appropriate to the nearest special hour. Schedules of this kind were prepared by Mr Roberts for the Indian Government*. The successive rearrangements for each sort of special time were made by recopying the whole of the observations time after time into a series of appropriate schedules. The mere clerical labour of this work is enormous, and great care is required to avoid mistakes.

All this copying might be avoided if the observed heights were written on movable pieces. But a year of observation gives 8,760 hourly heights, and the orderly sorting and re-sorting of nearly 9,000 pieces of paper or tablets might prove more laborious and more treacherous than recopying the figures.

It occurred to me, however, that the marshalling of movable pieces might be reduced to manageable limits if all the 24 observations pertaining to a single mean solar day were moved together, for the movable pieces would be at once reduced to 365, and each piece might be of a size convenient to handle.

The realisation of this plan affords the subject of this paper, and it will appear that not only is all desirable accuracy attainable, but that the other requisite of such a scheme is satisfied, namely, that the whole computing apparatus shall serve any number of times and for any number of places.

The first idea which naturally occurred was to have narrow sliding tablets which should be thrown into their places by a number of templates. It is unnecessary to recount all my trials and failures, but it will suffice to say that the slides and templates require the precision of a mathematical instrument if they are to work satisfactorily, and that the manufacture would be so expensive as to make the price of the instrument prohibitive.

The idea of making the tablets or strips to slide into their places was then abandoned, and the strips are now made with short pins on their under sides, so that they can be stuck on to a drawing board in any desired position. The templates, which were also troublesome to make, are replaced by large sheets of paper with numbered marks on them to show how the strips are to be set. The guide sheet is laid on a drawing board, and the pins on the strips pierce the paper and fix them in their proper positions.

The shifting of the strips from one arrangement to the next is certainly slower than when they slid into their places automatically, but I find that even without practice it only takes about 7 or 8 minutes to shift 74 of them from any one arrangement to a new one.

* An edition of these computation forms was reprinted by aid of a grant from the Royal Society, and is sold by the Cambridge Scientific Instrument Company, but only about a dozen copies now remain. In the course of the preparation of the "guide sheets" of the method proposed in this paper, I found that there are many small mistakes in these Indian forms, but they are fortunately not of such a kind as to produce a sensible vitiation of results. I learn that the mistakes arose from a misunderstanding on the part of a computer employed to draw up the forms. [The whole of this edition of computation forms has, now in 1907, been sold.]

The accuracy of my guide sheets was controlled by aid of Mr Roberts's forms, and it was the occasional discrepancy between my results and the forms which led to the detection of the errors referred to.

The strip belonging to each mean solar day is divided by black lines into 24 equal spaces, intended for the entry of the hourly heights of water. The strip is 9 in. long by $\frac{1}{5}$ in. wide and the divisions ($\frac{3}{8}$ by $\frac{1}{5}$) are of convenient size for the entries. There was much difficulty in discovering a good material, but after various trials artificial ivory, or xylonite, was found to serve the purpose. Xylonite is white, will take writing with Indian ink or pencil, and can easily be cleaned with a damp cloth. It is just as easy to write with liquid Indian ink as with ordinary ink, which must not be used, because it stains the surface.

The strips have a great tendency to warp, and I have two methods of overcoming this. A veneer of xylonite on hard wood serves well, or solid xylonite may be stiffened by sheet brass let into a slot on the under side. In the first plan the pins are fixed in the wood, and in the second the brass is filed to a spike at each end*. Whichever plan is adopted, the strips are expensive, costing about £7 for a set, and I do not at present see any way of making them cheaper.

The observations are to be treated in groups of two and a half lunations or 74 days. A set of strips, therefore, consists of 74, numbered from 0 to 73 in small figures on their flat ends.

If a set be pinned horizontally on a drawing board in vertical column, we have a form consisting of rows for each mean solar day and columns for each hour. The observed heights of the water are then written on the strips.

When the 24 columns are summed and divided by the number of entries we obtain the mean solar hourly mean heights. The harmonic analysis of these means gives the mean solar tides. But for evaluating the other tides the strips must be rearranged, and to this point we turn our attention.

Let us consider a special case, that of mean lunar time. A mean lunar hour is about $1^h\ 2^m$ m.s. time; hence the 12^h of each m.s. day must lie within 31^m m.s. time of a mean lunar hour. The following sample gives the incidence to the nearest lunar hour of the first few days in a year:

Mean solar time			Mean lunar time	
0^d	12^h	=	0^d	12^h
1	12	=	1	11
2	12	=	2	10
3	12	=	3	9
4	12	=	4	8
5	12	=	5	8
6	12	=	6	7
7	12	=	7	6
&c.			&c.	

* [The first of these two plans was found to be the more convenient one.]

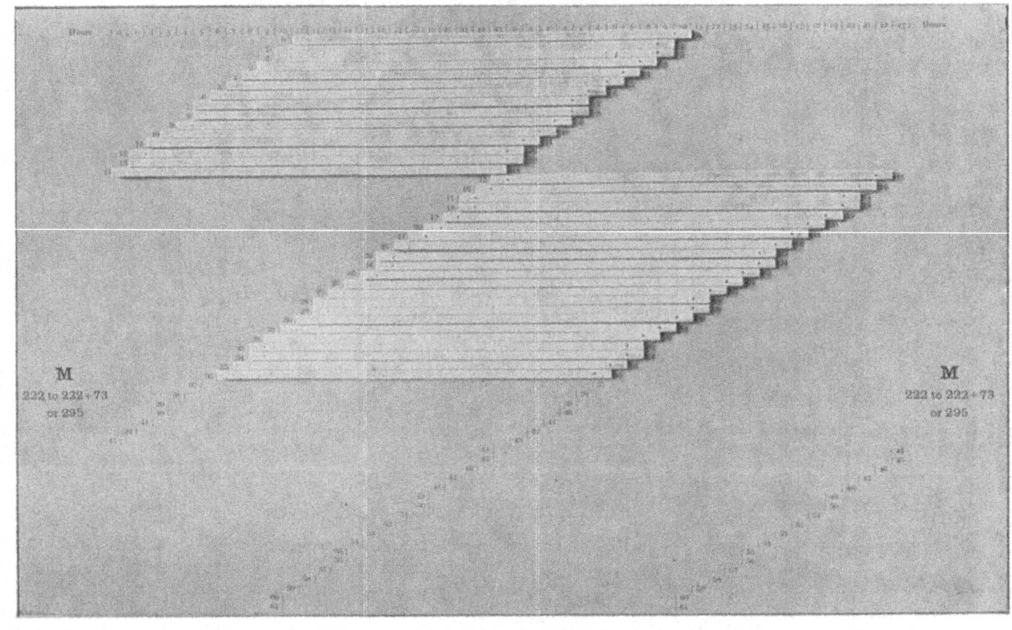

Abacus for the harmonic analysis of tidal observations.

The successive 12^h of mean solar time will march retrogressively through all the 24 hours of mean lunar time.

Now, if starting from strip 0, we push strip 1 one division to the left, strip 2 two divisions to the left, and so on, the entries on the strips will be arranged in columns of approximately lunar time.

The rule for this arrangement is given by marks on a sheet of paper 18 in. broad; these marks consist of parallel numbered steps or zigzags showing where the ends of each strip are to be placed so as to bring the hourly values into their proper places.

At the end of a lunation mean solar time has gained a whole day over mean lunar time and the 12^h solar again agrees with the 12^h lunar. On the guide sheet we see that the zigzag which takes its origin at the left end of strip 0 has descended diagonally from right to left until it has reached the left margin of the paper, and a new zigzag then begins on the right margin.

When the strips are pinned out following the zigzags on the sheet marked M, the entries are arranged in 48 columns, but the number of entries in each column is different. The 48 columns are to be regarded as appertaining to $0^h, 1^h, \ldots, 22^h, 23^h, 0^h, 1^h, \ldots, 22^h, 23^h$. Thus, the number of entries in the left-hand column of any hour added to the number of entries in the right-hand column of the same hours is, in each case, 74. The 48 incomplete columns may, in fact, be regarded as 24 complete ones.

The 24 complete columns are then summed; the 24 sums would, if divided by the total number of strips, give the 24 mean lunar hourly heights. The harmonic analysis of these sums gives certain constants, which, when divided by the number of strips, are the required tidal constants. It must be remarked, however, that, as the incidence of the entries is not exact in lunar time, investigation must be made of the corrections arising out of this inexactness.

The explanation of the guide sheet for lunar time will serve, *mutatis mutandis*, for all the others.

The zigzags have to be placed so as to bring the columns into exact alignment, and printers' types provide all the accuracy requisite. Accordingly, the computing strips are made to suit a chosen type. The standard length for one of the 24 divisions on the strips was chosen as that of a "2-em English quadrate"; 24 of these come to 9 inches, which is the length of a strip. I found the English quadrate a little too narrow, and accordingly between each line of quadrates there is a "blind rule," of 42 to the inch. The depth of the guide sheet is that of 74 quadrates and 74 rules, making $15\frac{1}{2}$ in. The computing strips are $\frac{1}{5}$ in. broad, and 74 of them occupy $14\frac{4}{5}$ in. The excess of $15\frac{1}{2}$ above $14\frac{4}{5}$, or $\frac{7}{10}$ in., is necessary to permit the easy arrangement of the strips.

[The plate opposite p. 219 * shows portion of one of the guide-sheets (that applicable to 222^d to 295^d) for M or mean-lunar time. Thirty-seven of the strips are shown in position, and the lower part of the sheet is still bare.]

To guard against the risk of the computer accidentally using the wrong sheet, the sheets are printed on coloured paper, the sequence of colours being that of the rainbow. The sheets for days 0 to 73 are all red; those for days 74 to 74 + 73, or 147, are all yellow; those for days 148 to 148 + 73, or 221, are green; those for days 222 to 222 + 73, or 295, are blue; and those for days 296 to 296 + 73, or 369, are violet.

Thus, when the observations for the first 74 days of the year are written on the strips all the sheets will be red; the strips will then be cleaned, and the observations for the second 74 days written in, when all the guide sheets will be yellow, and so on.

I must now refer to another considerable abridgment of the process of harmonic analysis. It is independent of the method of arrangement just sketched.

In the Indian computation forms the mean solar hourly heights have been found for the whole year, and the observations have been rearranged for the evaluation of certain other tides governed by a time scale which differs but little from the mean solar scale. I now propose to break the mean solar heights into sets of 30 days, and to analyse them, and next to harmonically analyse the 12 sets of harmonic constituents for annual and semi-annual inequalities. By this plan the harmonic constants for 11 different tides are obtained by one set of additions. In fact, we now get the annual, semi-annual, and solar elliptic tides, which formerly demanded much troublesome extra computation. A great saving is secured by this alone, and the results are in close agreement with those derived from the old method.

The guide sheets marked S and the computation forms are arranged so that the observations are broken up into the proper groups of 30 days, and they show the computer how to make the subsequent calculations.

I have also devised an abridged method of evaluating the tides of long period MSf, Mf, Mm. The method is less accurate than that followed hitherto, but it appears to give fairly good results, and reduces the work to very small dimensions.

Before entering on the details of my plan it is proper to mention that Dr Börgen has devised and used a method for attaining the same end. He has prepared sheets of tracing paper with diagonal lines on them, so arranged that when any sheet is laid on the copy of the observations written in daily rows and hourly columns, the numbers to be summed are found written between a pair of lines. This plan is excellent, but I fear that the difficulty of adding correctly in diagonal lines is considerable, and the comparative faintness of figures seen through tracing paper may be fatiguing to the eyes. Dr Börgen's

* [From my *Tides and Kindred Phenomena in the Solar System.*]

plan* is simple and inexpensive, and had I not thought that the plan now proposed has considerable advantages I should not have brought it forward.

In the investigations which follow the notation of the Report of 1883 to the British Association on Harmonic Analysis [Paper 1, above] is used without further explanation.

§ 2. *Evaluation of* A_0, Sa, Ssa, S_1, S_2, S_4, S_6, T, R, K_2, K_1, P.

The 24 mean solar hourly heights of water are entered in a schedule of 24 columns, with one row for each day, extending to n days; the 24 columns are summed, and the sums divided by n; the 24 means are harmonically analysed; it is required to find from the results the values of the harmonic constituents.

The speed of any one of the tides differs from a multiple of 15° per hour by a small angle; thus, any one of the tides is expressible in the form $H\cos[(15^\circ q-\beta)t-\zeta]$, where q is 0, 1, 2, 3, &c., and β is small.

When t lies between 0^h and 24^h this formula expresses the oscillation of level due to this tide on the day 0 of the series of days.

If multiples of 24^h from 1 to $n-1$ be added to t, the expression gives the height at the same hour, t, of mean solar time on each of the succession of days.

Then if $\mathfrak{h}$ denotes the mean height of water, as due to this tide alone, at the hour t, we have

$$\mathfrak{h}=\frac{1}{n}\sum_{1}^{n} H\cos[(15^\circ q-\beta)t-\zeta+(15^\circ q-\beta)\,24\,(n-1)]$$

$$=\frac{1}{n}H\frac{\sin 12n\beta}{\sin 12\beta}\cos[(15^\circ q-\beta)t-\zeta-12\beta(n-1)] \quad \text{.........(1)}$$

When t is put successively equal to 0^h, 1^h, ..., 23^h we get the 24 values of $\mathfrak{h}$ which are to be submitted to harmonic analysis.

The mean value of $\mathfrak{h}$, say A_0 (not to be confused with A_0 as written at the head of this section, where is denoted the mean sea level above datum) is found by taking the mean of the 24 values of $\mathfrak{h}$.

By the formula for the summation of a series of cosines it is easy to prove that

$$A_0=\frac{1}{24n}H\frac{\sin 12n\beta}{\sin(\frac{1}{2}\beta-\frac{15}{2}q)}\cos[\zeta+(12n-\tfrac{1}{2})\beta+\tfrac{15}{2}q] \text{.........(2)}$$

We will now find the pth harmonic constituents A_p, B_p. By the ordinary rules

$$\left.\begin{matrix}A_p\\B_p\end{matrix}\right\}=\tfrac{1}{12}\sum_{0}^{23}\mathfrak{h}\begin{matrix}\cos\\\sin\end{matrix}15^\circ pt \quad \text{...........................(3)}$$

* *Annalen der Hydrographie und Maritimen Meteorologie.* June, July, Aug., 1894.

Now $$2 \begin{matrix}\cos\\ \cdot\\ \sin\end{matrix} 15^\circ pt \cos[(15q-\beta)t-\zeta-12\beta(n-1)]$$

$$= \begin{matrix}\cos\\ \sin\end{matrix}[\{15^\circ(q+p)-\beta\}t-\zeta-12\beta(n-1)]$$

$$\begin{matrix}+\cos\\ -\sin\end{matrix}[\{15^\circ(q-p)-\beta\}t-\zeta-12\beta(n-1)]$$

and $\frac{1}{12}$ of the sum of the 24 values corresponding to $t=0, 1, \ldots, 23$ is

$$\tfrac{1}{24}\frac{\sin 12[15^\circ(q+p)-\beta]}{\sin\frac{1}{2}[15^\circ(q+p)-\beta]}\begin{matrix}\cos\\ \sin\end{matrix}[\tfrac{23}{2}\{15^\circ(q+p)-\beta\}-\zeta-12\beta(n-1)]$$

$\pm$ the same with sign of p changed.

This expression admits of simplification, because $12\times 15^\circ = 180^\circ$; making this simplification, and introducing the result into (3), we obtain

$$\left.\begin{matrix}A_p\\ B_p\end{matrix}\right\} = \frac{1}{24n}H\sin 12n\beta\left\{\frac{\begin{matrix}-\cos\\ +\sin\end{matrix}[\zeta+(12n-\frac{1}{2})\beta+\frac{15^\circ}{2}(q+p)]}{\sin[\frac{15^\circ}{2}(q+p)-\frac{1}{2}\beta]} - \frac{\begin{matrix}-\cos\\ -\sin\end{matrix}[\zeta+(12n-\frac{1}{2})\beta+\frac{15^\circ}{2}(q-p)]}{\sin[\frac{15^\circ}{2}(q-p)-\frac{1}{2}\beta]}\right\} \quad \ldots\ldots(4)$$

In the particular case where $p=q$, we have

$$\left.\begin{matrix}A_q\\ B_q\end{matrix}\right\} = \frac{1}{24n}H\sin 12n\beta\left\{\frac{\begin{matrix}-\cos\\ +\sin\end{matrix}[\zeta+(12n-\frac{1}{2})\beta+15^\circ q]}{\sin(15^\circ q-\frac{1}{2}\beta)} + \frac{\begin{matrix}+\cos\\ +\sin\end{matrix}[\zeta+(12n-\frac{1}{2})\beta]}{\sin\frac{1}{2}\beta}\right\} \quad \ldots\ldots\ldots\ldots(5)$$

If the number of days n be large, A_p, B_p will be small unless the denominator of one of the two terms in (4) be very small. This last case can only occur when $p=q$ and when β is small. Hence, in the analysis of a term of the form under consideration, we may neglect all the harmonics except the qth one. Accordingly (2) and (5) are the only formulæ required.

A case, however, which there will be occasion to use hereafter is when $n=30$, $q=2$, when (4) becomes

$$\left.\begin{matrix}A_2\\ B_2\end{matrix}\right\} = \tfrac{1}{720}H\sin 360\beta\left\{\frac{\begin{matrix}\cos\\ \sin\end{matrix}(\zeta+359\frac{1}{2}\beta)}{\sin\frac{1}{2}\beta}\;\frac{\begin{matrix}-\cos\\ +\sin\end{matrix}(\zeta+30^\circ+359\frac{1}{2}\beta)}{\sin(30^\circ-\frac{1}{2}\beta)}\right\} \quad \ldots(6)$$

For the present we have to apply (5) in the two cases $q=1$, $\beta=0^\circ{\cdot}0410686$ and $q=2$, $\beta=0^\circ{\cdot}0821372$; now the ratios of cosec $\frac{1}{2}\beta$ to cosec $(15^\circ q-\frac{1}{2}\beta)$ in these two cases are 722 to 1 and 697 to 1. In both cases the first term of (5) is negligible compared with the second.

Now write $$\mathfrak{F} = \frac{24n \sin \frac{1}{2}\beta}{\sin 12n\beta} \quad \text{.............................(7)}$$

and (5) becomes, with sufficient exactness,

$$\left.\begin{matrix} A_q \\ B_q \end{matrix}\right\} = \frac{H}{\mathfrak{F}} \begin{matrix} \cos \\ \sin \end{matrix} [\zeta + (12n - \tfrac{1}{2})\beta] \quad \text{....................(8)}$$

If this be compared with (2), we see that when $q = 0$ this formula also comprises (2).

In the applications to be made β is very small, so that $\mathfrak{F}$ is approximately a function of the form $\theta \operatorname{cosec} \theta$. This function increases very rapidly when θ passes 90°, but for considerable values less than 90° it only slightly exceeds unity; for example, when $\theta = 60°$, $\mathfrak{F} = 1{\cdot}2$, but when $\theta = 180°$, $\mathfrak{F} = \text{infinity}$.

It follows, therefore, that if the number n of days in the series is such that $12n\beta$ is less than say 60°, the magnitudes of A_q, B_q are but little diminished by division by $\mathfrak{F}$; but if $12n\beta$ is nearly 180°, A_q, B_q become vanishingly small.

If the typical tide here considered be the principal lunar tide M_2, and if the number of days be as nearly as possible an exact multiple of a semi-lunation, $12n\beta$ is nearly 180°, and the corresponding A_2, B_2 become very small. No number of whole days can be an exact multiple of a semi-lunation, so that A_2, B_2 corresponding to M_2 cannot be made to vanish completely. For the present they may be treated as negligible, and we return to this point in the next section.

The above investigation shows that in the expression for the whole oscillation of sea level upon which the proposed analysis is performed all those tides may be omitted from which β is not very small, and also all those whose frequencies are such that the period under consideration $12n\beta$ is nearly 180°.

Since the period under consideration will be a lunation, it follows that, as far as is now material, the general expression for sea level may be written as follows, t denoting mean solar hour angle equal to 15° t:

m. w., annual..............	$A_o + H_{sa} \cos(h - \kappa_{sa})$
semi-annual	$+ H_{ssa} \cos(2h - \kappa_{ssa})$
Solar tides, S_1, S_2.........	$+ H_{\frac{1}{2}s} \cos(\mathrm{t} - \kappa_{\frac{1}{2}s}) + H_s \cos(2\mathrm{t} - \kappa_s)$
S_3, S_4.........	$+ H_{2s} \cos(4\mathrm{t} - \kappa_{2s}) + H_{3s} \cos(6\mathrm{t} - \kappa_{3s})$
Solar elliptic, T...........	$+ H_t \cos(2\mathrm{t} - h + p_{,} - \kappa_t)$
R...........	$+ H_r \cos(2\mathrm{t} + h - p_{,} + \pi - \kappa_r)$
Luni-solar, K_2	$+ \mathrm{f}''H'' \cos(2\mathrm{t} + 2h - 2\nu'' - \kappa'')$
K_1	$+ \mathrm{f}'H' \cos(\mathrm{t} + h - \nu' - \frac{1}{2}\pi - \kappa')$
Solar diurnal, P	$+ H_p \cos(\mathrm{t} - h + \frac{1}{2}\pi - \kappa_p)$ ....(9)

This includes all the tides whose initials are written at the head of this section.

It is now necessary to break up the year into 12 equidistant lunations of 30 days. This can be done by the omission of 5 days in ordinary years, and of 6 days in leap years.

If the days of the year are numbered 0 to 364 (365 in leap year), the twelve months are as follows:

0, 0^d to 29^d; 1, 30^d to 59^d; *omit* 60^d; 2, 61^d to 90^d; 3, 91^d to 120^d; *omit* 121^d; 4, 122^d to 151^d; 5, 152^d to 181^d; *omit* 182^d; 6, 183^d to 212^d; 7, 213^d to 242^d; 8, 243^d to 272^d; *omit* 273^d; 9, 274^d to 303^d; 10, 304^d to 333^d; *omit* 334^d; 11, 335^d to 364^d; *in leap year omit* 365^d.

The increments of sun's mean longitude from $0^d\ 0^h$ of month 0 up to 0^h of the day numbered 0 of each group of days or month are as follows:

0, 0°; 1, 30° − 0°·431; 2, 60°·124; 3, 90° − 0°·306; 4, 120°·249; 5, 150° − 0°·182; 6, 180°·373; 7, 210° − 0°·057; 8, 240° − 0°·488; 9, 270°·068; 10, 300° − 0°·364; 11, 330°·191.

Thus if h_o be the sun's mean longitude at $0^d\ 0^h$ of month 0, the sun's mean longitude at $0^d\ 0^h$ of month τ is $h_o + 30°\tau$, with sufficient approximation.

Now let V with appropriate suffix denote the initial "equilibrium argument" at $0^d\ 0^h$ of month 0, so that

$$V_{sa} = h_o, \quad V_{ssa} = 2h_o, \quad V_t = -h_o + p_{,}, \quad V_r = h_o - p_{,} + \pi, \quad V'' = 2h_o - 2\nu''$$
$$V' = h_o - \nu' - \tfrac{1}{2}\pi, \quad V_p = -h_o + \tfrac{1}{2}\pi$$

then the general expression (9) for the tide in the month τ becomes

$$\begin{aligned}
&A_o + H_{sa}\cos(\eta t + V_{sa} + 30°\tau - \kappa_{sa}) + H_{ssa}\cos(2\eta t + V_{ssa} + 60°\tau - \kappa_{ssa})\\
&\quad + H_{\frac{1}{2}s}\cos(15°t - \kappa_{\frac{1}{2}s}) + H_s\cos(30°t - \kappa_s) + H_{2s}\cos(60°t - \kappa_{2s})\\
&\qquad\qquad + H_{3s}\cos(90°t - \kappa_{3s})\\
&\quad + H_t\cos[(30° - \eta)t + V_t - 30°\tau - \kappa_t]\\
&\qquad\qquad + H_r\cos[(30° + \eta)t + V_r + 30°\tau - \kappa_r]\\
&\quad + f''H''\cos[(30° + 2\eta)t + V'' + 60°\tau - \kappa'']\\
&\qquad\qquad + f'H'\cos[(15° + \eta)t + V' + 30°\tau - \kappa']\\
&\quad + H_p\cos[(15° - \eta)t + V_p - 30°\tau - \kappa_p] \quad \text{.........(10)}
\end{aligned}$$

Each of these terms falls into the type $\cos[(15°q - \beta)t \sim \zeta]$, and β is in every case either $\pm\eta$, -2η, or 0.

Now, when harmonic analysis of the mean of 30 days is carried out, coefficients $\mathfrak{F}$ are introduced.

Write therefore

$$\mathfrak{F}_1 = \frac{24 \times 30 \sin \frac{1}{2}\eta}{\sin 360\eta}, \qquad \mathfrak{F}_2 = \frac{24 \times 30 \sin \eta}{\sin 720\eta}$$

With the known value of η,

$$\log \mathfrak{F}_1 = 0{\cdot}00483, \qquad \log \mathfrak{F}_2 = 0{\cdot}01945$$

In applying the method investigated above, it will be observed that a term of any frequency $15^\circ q - \beta$ only contributes to the harmonic constituent of order q.

Then applying our general rule (8) term by term*, and observing that $359\frac{1}{2}\eta = 14^\circ{\cdot}76$, and $719\eta = 29^\circ{\cdot}53$, the result may be written as follows:

$$\mathfrak{A}_0^{(\tau)} = A_0 + \frac{H_{sa}}{\mathfrak{F}_1} \cos(\kappa_{sa} - \mathrm{V}_{sa} - 30^\circ\tau - 14^\circ{\cdot}76) + \frac{H_{ssa}}{\mathfrak{F}_2} \cos(\kappa_{ssa} - \mathrm{V}_{ssa} - 60^\circ\tau - 29^\circ{\cdot}53)$$

$$\left.\begin{matrix}\mathfrak{A}_1^{(\tau)} \\ \mathfrak{B}_1^{(\tau)}\end{matrix}\right\} = H_{\frac{1}{2}s} \begin{matrix}\cos \\ \sin\end{matrix} \kappa_{\frac{1}{2}s} + \frac{\mathrm{f}'H'}{\mathfrak{F}_1} \begin{matrix}\cos \\ \sin\end{matrix} (\kappa' - \mathrm{V}' - 30^\circ\tau - 14^\circ{\cdot}76) + \frac{H_p}{\mathfrak{F}_1} \begin{matrix}\cos \\ \sin\end{matrix} (\kappa_p - \mathrm{V}_p + 30^\circ\tau + 14^\circ{\cdot}76)$$

$$\left.\begin{matrix}\mathfrak{A}_2^{(\tau)} \\ \mathfrak{B}_2^{(\tau)}\end{matrix}\right\} = H_s \begin{matrix}\cos \\ \sin\end{matrix} \kappa_s + \frac{H_t}{\mathfrak{F}_1} \begin{matrix}\cos \\ \sin\end{matrix} (\kappa_t - \mathrm{V}_t + 30^\circ\tau + 14^\circ{\cdot}76) + \frac{H_r}{\mathfrak{F}_1} \begin{matrix}\cos \\ \sin\end{matrix} (\kappa_r - \mathrm{V}_r - 30^\circ\tau - 14^\circ{\cdot}76) + \frac{\mathrm{f}''H''}{\mathfrak{F}_2} \begin{matrix}\cos \\ \sin\end{matrix} (\kappa'' - \mathrm{V}'' - 60^\circ\tau - 29^\circ{\cdot}53)$$

$$\left.\begin{matrix}\mathfrak{A}_4^{(\tau)} \\ \mathfrak{B}_4^{(\tau)}\end{matrix}\right\} = H_{2s} \begin{matrix}\cos \\ \sin\end{matrix} \kappa_{2s}; \qquad \left.\begin{matrix}\mathfrak{A}_6^{(\tau)} \\ \mathfrak{B}_6^{(\tau)}\end{matrix}\right\} = H_{3s} \begin{matrix}\cos \\ \sin\end{matrix} \kappa_{3s} \quad \text{.....................(11)}$$

With the meaning of the term month in the present context, the sun has a mean motion of 30° per month, and each of the first five $\mathfrak{A}$'s and $\mathfrak{B}$'s is a function with a constant part and with annual and semi-annual inequalities.

When τ has successively the 12 values $0, 1, \ldots, 11$, we have 12 equidistant values of the $\mathfrak{A}$'s and $\mathfrak{B}$'s. These may be harmonically analysed for annual and semi-annual inequalities.

* [The formulæ in the text are devised for the case when a whole year's observations are under reduction. When a short period is being treated, such as a fortnight or a month (as in Papers 3 and 4 above), it is advisable to treat 15 or 30 days for semi-diurnal tides, but only 14 or 27 days for diurnal tides. I therefore give here, in the form of a schedule, certain numerical values which have been used in the previous papers:

n	$(12n-\frac{1}{2})\,2\eta$	$\log \mathfrak{F}_2$	n	$(12n-\frac{1}{2})\,\eta$	$\log \mathfrak{F}_1$
15	14°·76	·00483	14	6°·83	·00105
30	29°·53	·01945	27	13°·29	·00391

.]

Suppose that the several coefficients to be determined by harmonic analysis are defined by the following equations:

$$\mathfrak{A}_0^{(\tau)} = A_0 + A_1 \cos 30^\circ\tau + B_1 \sin 30^\circ\tau + A_2 \cos 60^\circ\tau + B_2 \sin 60^\circ\tau$$

$$\left.\begin{matrix}\mathfrak{A}_1^{(\tau)}\\ \mathfrak{B}_1^{(\tau)}\\ \mathfrak{A}_2^{(\tau)}\\ \mathfrak{B}_2^{(\tau)}\end{matrix}\right\} = \left.\begin{matrix}\mathrm{C}_0\\ \mathrm{c}_0\\ \mathrm{E}_0\\ \mathrm{e}_0\end{matrix}\right\} + \left.\begin{matrix}\mathrm{C}_1\\ \mathrm{c}_1\\ \mathrm{E}_1\\ \mathrm{e}_1\end{matrix}\right\} \cos 30^\circ\tau + \left.\begin{matrix}\mathrm{D}_1\\ \mathrm{d}_1\\ \mathrm{F}_1\\ \mathrm{f}_1\end{matrix}\right\} \sin 30^\circ\tau + \left.\begin{matrix}\\ \\ \mathrm{E}_2\\ \mathrm{e}_2\end{matrix}\right\} \cos 60^\circ\tau + \left.\begin{matrix}\\ \\ \mathrm{F}_2\\ \mathrm{f}_2\end{matrix}\right\} \sin 60^\circ\tau$$

$$\tfrac{1}{12}\Sigma\mathfrak{A}_4^{(\tau)} = A_4,\quad \tfrac{1}{12}\Sigma\mathfrak{B}_4^{(\tau)} = B_4,\quad \tfrac{1}{12}\Sigma\mathfrak{A}_6^{(\tau)} = A_6,\quad \tfrac{1}{12}\Sigma\mathfrak{B}_6^{(\tau)} = B_6 \quad \ldots(12)$$

Then on comparing (12) with (11) we see that:

$$\mathrm{A}_0 = A_0$$

$$\left.\begin{matrix}A_1\\ B_1\end{matrix}\right\} = \frac{H_{sa}}{\mathfrak{F}_1} \begin{matrix}\cos\\ \sin\end{matrix} (\kappa_{sa} - \mathrm{V}_{sa} - 14^\circ{\cdot}76), \qquad \left.\begin{matrix}A_2\\ B_2\end{matrix}\right\} = \frac{H_{ssa}}{\mathfrak{F}_2} \begin{matrix}\cos\\ \sin\end{matrix} (\kappa_{ssa} - \mathrm{V}_{ssa} - 29^\circ{\cdot}53)$$

$$\left.\begin{matrix}\mathrm{C}_0\\ \mathrm{c}_0\end{matrix}\right\} = H_{\frac{1}{2}s} \begin{matrix}\cos\\ \sin\end{matrix} \kappa_{\frac{1}{2}s}$$

$$\left.\begin{matrix}\mathrm{C}_1\\ \mathrm{D}_1\end{matrix}\right\} = \frac{\begin{matrix}+\\ +\end{matrix}\,\mathrm{f}'H'}{\mathfrak{F}_1} \begin{matrix}\cos\\ \sin\end{matrix} (\kappa' - \mathrm{V}' - 14^\circ{\cdot}76) \frac{\begin{matrix}+\\ -\end{matrix}\,H_p}{\mathfrak{F}_1} \begin{matrix}\cos\\ \sin\end{matrix} (\kappa_p - \mathrm{V}_p + 14^\circ{\cdot}76)$$

$$\left.\begin{matrix}\mathrm{c}_1\\ \mathrm{d}_1\end{matrix}\right\} = \frac{\begin{matrix}+\\ -\end{matrix}\,\mathrm{f}'H'}{\mathfrak{F}_1} \begin{matrix}\sin\\ \cos\end{matrix} (\kappa' - \mathrm{V}' - 14^\circ{\cdot}76) \frac{\begin{matrix}+\\ +\end{matrix}\,H_p}{\mathfrak{F}_1} \begin{matrix}\sin\\ \cos\end{matrix} (\kappa_p - \mathrm{V}_p + 14^\circ{\cdot}76)$$

$$\left.\begin{matrix}\mathrm{E}_0\\ \mathrm{e}_0\end{matrix}\right\} = H_s \begin{matrix}\cos\\ \sin\end{matrix} \kappa_s$$

$$\left.\begin{matrix}\mathrm{E}_1\\ \mathrm{F}_1\end{matrix}\right\} = \frac{\begin{matrix}+\\ -\end{matrix}\,H_t}{\mathfrak{F}_1} \begin{matrix}\cos\\ \sin\end{matrix} (\kappa_t - \mathrm{V}_t + 14^\circ{\cdot}76) \frac{\begin{matrix}+\\ +\end{matrix}\,H_r}{\mathfrak{F}_1} \begin{matrix}\cos\\ \sin\end{matrix} (\kappa_r - \mathrm{V}_r - 14^\circ{\cdot}76)$$

$$\left.\begin{matrix}\mathrm{e}_1\\ \mathrm{f}_1\end{matrix}\right\} = \frac{\begin{matrix}+\\ +\end{matrix}\,H_t}{\mathfrak{F}_1} \begin{matrix}\sin\\ \cos\end{matrix} (\kappa_t - \mathrm{V}_t + 14^\circ{\cdot}76) \frac{\begin{matrix}+\\ -\end{matrix}\,H_r}{\mathfrak{F}_1} \begin{matrix}\sin\\ \cos\end{matrix} (\kappa_r - \mathrm{V}_r - 14^\circ{\cdot}76)$$

$$\left.\begin{matrix}\mathrm{E}_2\\ \mathrm{F}_2\end{matrix}\right\} = \frac{\begin{matrix}+\\ +\end{matrix}\,\mathrm{f}''H''}{\mathfrak{F}_2} \begin{matrix}\cos\\ \sin\end{matrix} (\kappa'' - \mathrm{V}'' - 29^\circ{\cdot}53)$$

$$\left.\begin{matrix}\mathrm{e}_2\\ \mathrm{f}_2\end{matrix}\right\} = \frac{\begin{matrix}+\\ -\end{matrix}\,\mathrm{f}''H''}{\mathfrak{F}_2} \begin{matrix}\sin\\ \cos\end{matrix} (\kappa'' - \mathrm{V}'' - 29^\circ{\cdot}53) \quad \ldots\ldots(13)$$

From these equations we get

$$\left.\begin{matrix}\frac{1}{2}(\mathrm{C}_1 - \mathrm{d}_1)\\ \frac{1}{2}(\mathrm{c}_1 + \mathrm{D}_1)\end{matrix}\right\} = \frac{\mathrm{f}'H'}{\mathfrak{F}_1} \begin{matrix}\cos\\ \sin\end{matrix} (\kappa' - \mathrm{V}' - 14^\circ{\cdot}76)$$

$$\left.\begin{matrix}\frac{1}{2}(\mathrm{C}_1 + \mathrm{d}_1)\\ \frac{1}{2}(\mathrm{c}_1 - \mathrm{D}_1)\end{matrix}\right\} = \frac{H_p}{\mathfrak{F}_1} \begin{matrix}\cos\\ \sin\end{matrix} (\kappa_p - \mathrm{V}_p + 14^\circ{\cdot}76)$$

$$\left.\begin{matrix}\frac{1}{2}(\mathrm{E}_1 - \mathrm{f}_1)\\ \frac{1}{2}(\mathrm{e}_1 + \mathrm{F}_1)\end{matrix}\right\} = \frac{H_r}{\mathfrak{F}_1} \begin{matrix}\cos\\ \sin\end{matrix} (\kappa_r - \mathrm{V}_r - 14^\circ{\cdot}76)$$

$$\left.\begin{matrix}\frac{1}{2}(\mathrm{E}_1 + \mathrm{f}_1)\\ \frac{1}{2}(\mathrm{e}_1 - \mathrm{F}_1)\end{matrix}\right\} = \frac{H_t}{\mathfrak{F}_1} \begin{matrix}\cos\\ \sin\end{matrix} (\kappa_t - \mathrm{V}_t + 14^\circ{\cdot}76)$$

$$\left.\begin{matrix}\frac{1}{2}(\mathrm{E}_2 - \mathrm{f}_2)\\ \frac{1}{2}(\mathrm{e}_2 + \mathrm{F}_2)\end{matrix}\right\} = \frac{\mathrm{f}''H''}{\mathfrak{F}_2} \begin{matrix}\cos\\ \sin\end{matrix} (\kappa'' - \mathrm{V}'' - 29^\circ{\cdot}53) \quad \ldots\ldots(14)$$

The tidal constants of the several tides enumerated at the head of this section are determinable from these equations.

Our rule is accordingly to analyse in twelve groups of 30 days, and then to analyse the resulting harmonic constituents for annual and semi-annual inequalities, combining the final results according to the formula just found.

The edition of "guide sheets" and computation forms which I have drawn up are so arranged as to facilitate the whole process and to render it quite straightforward. By this single set of additions it is thus possible to evaluate eleven tides and mean water.

§ 3. *Clearance of* T, R *from perturbation by* M_2.

The method of the last section was designed to render all the tides insensible excepting those enumerated. But M_2 is so much larger than any other tide that there is a small residual disturbance which ought to be corrected.

I have made computations, which I do not give, but which show that the disturbance of all the harmonic constituents except $\mathfrak{A}_2$, $\mathfrak{B}_2$ is insensible. It is required then to determine the correction to be applied to $\mathfrak{A}_2$, $\mathfrak{B}_2$, and thence those for E, F, e, f.

Suppose that, when time t is counted from $0^d\ 0^h$ of month τ, the M_2 tide is expressed by $M\cos[(30° - \beta)t - \zeta]$, where $\beta = 1°{\cdot}0158958$.

When means taken over 30 days are harmonically analysed the formula (6) gives the contributions to $\mathfrak{A}_2$, $\mathfrak{B}_2$. As it is now required to obliterate these contributions, the signs must be changed, and the corrections are

$$\left.\begin{matrix}\delta\mathfrak{A}_2^{(\tau)}\\ \delta\mathfrak{B}_2^{(\tau)}\end{matrix}\right\} = -\tfrac{1}{720}M\sin 360\beta\left\{\frac{\begin{matrix}\cos\\ \sin\end{matrix}(\zeta + 359\tfrac{1}{2}\beta)}{\sin\tfrac{1}{2}\beta} \quad \frac{\begin{matrix}-\cos\\ +\sin\end{matrix}(\zeta + 30° + 359\tfrac{1}{2}\beta)}{\sin(30° - \tfrac{1}{2}\beta)}\right\} \quad \text{...(15)}$$

For reasons which will appear below I now write

$$\zeta = \zeta_m^{(\tau)} - 0°{\cdot}5258$$

Then introducing the value of β into (15), I find

$$\left.\begin{matrix}\delta\mathfrak{A}_2^{(\tau)}\\ \delta\mathfrak{B}_2^{(\tau)}\end{matrix}\right\} = -\tfrac{1}{720}M\sin 5°\ 43'{\cdot}35\left\{\frac{\begin{matrix}\cos\\ \sin\end{matrix}(\zeta_m^{(\tau)} + 4°{\cdot}689)}{\sin 0°\ 30'\ 28''{\cdot}6} \quad \frac{\begin{matrix}-\cos\\ +\sin\end{matrix}(\zeta_m^{(\tau)} + 34°{\cdot}689)}{\sin 29°\ 29'{\cdot}52}\right\} \quad \text{...(16)}$$

Let ζ_m denote the value of $\zeta_m^{(\tau)}$ at $0^d\ 0^h$ of month 0, and let $\zeta_m^{(\tau)} = \zeta_m + \theta^{(\tau)}$, and let $\mathfrak{F}_2$ denote a certain factor whose logarithm is 0·00849, and let $M = \mathrm{f}H_m$.

In the harmonic analysis for the M_2 tide, considered below in § 6, we shall have

$$A_2 = \frac{fH_m}{\mathfrak{F}_2}\cos\zeta_m, \qquad B_2 = \frac{fH_m}{\mathfrak{F}_2}\sin\zeta_m$$

Accordingly $$M\begin{smallmatrix}\cos\\ \sin\end{smallmatrix}\zeta_m^{(\tau)} = \mathfrak{F}_2\left[A_2\begin{smallmatrix}\cos\\ \sin\end{smallmatrix}\theta^{(\tau)}\begin{smallmatrix}-\\ +\end{smallmatrix}B_2\begin{smallmatrix}\sin\\ \cos\end{smallmatrix}\theta^{(\tau)}\right]$$

These values of $M\begin{smallmatrix}\cos\\ \sin\end{smallmatrix}\zeta_m^{(\tau)}$ must now be introduced into (16), but the algebraic process need not be given in detail. If we write

$$\left.\begin{matrix}\frac{1}{2}(S+P)\\ \frac{1}{2}(R+Q)\end{matrix}\right\} = \tfrac{1}{720}\mathfrak{F}_2\frac{\sin 5^\circ\, 43'\!\cdot\!35 \begin{smallmatrix}\cos\\ \sin\end{smallmatrix}(4^\circ\, 41'\!\cdot\!32)}{\sin 0^\circ\, 30'\, 28''\!\cdot\!6}$$

$$\left.\begin{matrix}\frac{1}{2}(S-P)\\ \frac{1}{2}(R-Q)\end{matrix}\right\} = \tfrac{1}{720}\mathfrak{F}_2\frac{\sin 5^\circ\, 43'\!\cdot\!35 \begin{smallmatrix}\cos\\ \sin\end{smallmatrix}(34^\circ\, 41'\!\cdot\!32)}{\sin 29^\circ\, 29'\, 31''\!\cdot\!4}$$

it follows that

$$P = 0\!\cdot\!01564, \quad Q = 0\!\cdot\!00114, \quad R = 0\!\cdot\!00147, \quad S = 0\!\cdot\!01611$$

Then, when the substitution of the values of $M\begin{smallmatrix}\sin\\ \cos\end{smallmatrix}\zeta_m^{(\tau)}$ is carried out, we find

$$\left.\begin{matrix}\delta\mathfrak{A}_2^{(\tau)}\\ \delta\mathfrak{B}_2^{(\tau)}\end{matrix}\right\} = \cos\theta^{(\tau)}\left[\begin{matrix}-P\\ -R\end{matrix}A_2\begin{matrix}+Q\\ -S\end{matrix}B_2\right] + \sin\theta^{(\tau)}\left[\begin{matrix}+Q\\ -S\end{matrix}A_2\begin{matrix}+P\\ +R\end{matrix}B_2\right]$$

By the definition of $\theta^{(\tau)}$ it appears that $-\theta^{(\tau)}$ is the increment of twice the mean moon's hour angle during the time from $0^d\ 0^h$ of month 0 up to $0^d\ 0^h$ of month τ, that is to say $\theta^{(\tau)} = -2(\gamma-\sigma)t$ for the time specified. The following table gives the values of $\theta^{(\tau)}$ and of its cosine and sine for each month:

Month (τ)	No. of days from epoch 0 to epoch τ	$\theta^{(\tau)}$	$\cos\theta^{(\tau)}$	$\sin\theta^{(\tau)}$
0	0	0° 0′	1·000	0·000
1	30	11 27	0·980	0·199
2	61	47 16	0·678	0·735
3	91	58 43	0·519	0·855
4	122	π − 85 28	− 0·079	0·997
5	152	π − 74 1	− 0·275	0·961
6	183	π − 38 11	− 0·786	0·618
7	213	π − 26 44	− 0·893	0·450
8	243	π − 15 18	− 0·965	0·264
9	274	π + 20 32	− 0·937	− 0·351
10	304	π + 31 59	− 0·848	− 0·530
11	335	π + 67 49	− 0·378	− 0·926

If $\cos\theta^{(\tau)}$, $\sin\theta^{(\tau)}$ are regarded as quantities having annual and semi-annual inequalities, we may write

$$\cos\theta^{(\tau)} = \alpha_0 + \alpha_1 \cos 30°\tau + \beta_1 \sin 30°\tau + \alpha_2 \cos 60°\tau + \beta_2 \sin 60°\tau + \dots$$

$$\sin\theta^{(\tau)} = \gamma_0 + \gamma_1 \cos 30°\tau + \delta_1 \sin 30°\tau + \gamma_2 \cos 60°\tau + \delta_2 \sin 60°\tau + \dots$$

On analysing the numerical values of $\cos\theta^{(\tau)}$, $\sin\theta^{(\tau)}$ by the ordinary processes, I find

$$\alpha_0 = -0{\cdot}165, \qquad \gamma_0 = +0{\cdot}273$$
$$\alpha_1 = +0{\cdot}626, \qquad \gamma_1 = -0{\cdot}500$$
$$\beta_1 = +0{\cdot}756, \qquad \delta_1 = +0{\cdot}642$$
$$\alpha_2 = +0{\cdot}159, \qquad \gamma_2 = -0{\cdot}046$$
$$\beta_2 = +0{\cdot}199, \qquad \delta_2 = +0{\cdot}166$$

But in § 2 the harmonic constituents of $\mathfrak{A}_2$ when analysed for annual and semi-annual inequality were denoted by E_0, E_1, F_1, E_2, F_2, and the constituents of $\mathfrak{B}_2$ were denoted by e_0, e_1, f_1, e_2, f_2. Hence the ten corrections to the E's and F's are (with an easily intelligible alternative notation)

$$\delta E_{0,1,2} = (-P\alpha_{0,1,2} + Q\gamma_{0,1,2}) A_2 + (Q\alpha_{0,1,2} + P\gamma_{0,1,2}) B_2$$

$$\delta F_{1,2} = (-P\beta_{1,2} + Q\delta_{1,2}) A_2 + (Q\beta_{1,2} + P\delta_{1,2}) B_2$$

$$\delta e_{0,1,2} = (-R\alpha_{0,1,2} - S\gamma_{0,1,2}) A_2 + (-S\alpha_{0,1,2} + R\gamma_{0,1,2}) B_2$$

$$\delta f_{1,2} = (-R\beta_{1,2} - S\delta_{1,2}) A_2 + (-S\beta_{1,2} + R\delta_{1,2}) B_2$$

On substituting the numerical values of α, β, γ, δ, P, Q, R, S, I find

	Coeffit. of A_2	Coeffit. of B_2
$\delta E_0 =$	$+0{\cdot}0029$	$+0{\cdot}0041$
$\delta e_0 =$	$-0{\cdot}0042$	$+0{\cdot}0031$
$\delta\frac{1}{2}(E_1 + f_1) =$	$-0{\cdot}0109$	$-0{\cdot}0091$
$\delta\frac{1}{2}(E_1 - f_1) =$	$+0{\cdot}0006$	$+0{\cdot}0020$
$\delta\frac{1}{2}(e_1 + F_1) =$	$-0{\cdot}0020$	$+0{\cdot}0000$
$\delta\frac{1}{2}(e_1 - F_1) =$	$+0{\cdot}0091$	$-0{\cdot}0108$
$\delta\frac{1}{2}(E_2 + f_2) =$	$-0{\cdot}0028$	$-0{\cdot}0018$
$\delta\frac{1}{2}(E_2 - f_2) =$	$+0{\cdot}0002$	$+0{\cdot}0012$
$\delta\frac{1}{2}(e_2 + F_2) =$	$-0{\cdot}0012$	$+0{\cdot}0001$
$\delta\frac{1}{2}(e_2 - F_2) =$	$+0{\cdot}0017$	$-0{\cdot}0027$

Most of these corrections are negligible, but the four which affect the solar elliptic tides T, R must be included, because those tides are so small that a small error affects them sensibly. Hence we may take, with sufficient accuracy,

$$\delta\tfrac{1}{2}(e_1 - F_1) = +0{\cdot}009A_2 - 0{\cdot}011B_2, \quad \delta\tfrac{1}{2}(e_1 + F_1) = -0{\cdot}002A_2$$

$$\delta\tfrac{1}{2}(E_1 + f_1) = -0{\cdot}011A_2 - 0{\cdot}009B_2, \quad \delta\tfrac{1}{2}(E_1 - f_1) = +0{\cdot}0006A_2 + 0{\cdot}002B_2 \qquad \dots\dots(16^*)$$

where A_2, B_2 are the components of the M_2 derived from the reduction of that tide by the process of § 6.

Provision for these corrections is made in the computation forms.

§ 4. *Evaluation of* A_0, Sa, S_1, S_2, S_4, S_6, K_2, K_1, P, *when a complete year of observation is not available.*

It is now proposed to consider the case where the period of observation is as much as six complete months and less than a complete year.

The method of the last section apparently depends on the completeness of the year, yet, with certain modifications, it may be rendered available for shorter periods.

We suppose that so much of the year as is available is broken into sets of 30 days by the rules of the last section, and that the means are harmonically analysed. The results of such harmonic analysis for month (τ) are given in (11) of § 3, but for the purpose in hand they now admit of some simplification. It is clear that it is not worth while to evaluate the very small solar elliptic tides T and R from a short period of observation. If then, we denote by $P^{(\tau)}$ the ratio of the cube of the sun's parallax to its mean parallax at the middle of the month (τ), the first three terms of the third of (11) may be included in the expression $P^{(\tau)} H_s \begin{matrix}\cos\\ \sin\end{matrix} \kappa_s$. The last term of this equation really does involve the solar parallax to some extent and we may, with sufficient approximation, write the third pair of equations

$$\left.\begin{matrix}\mathfrak{A}_2^{(\tau)} \div P^{(\tau)} \\ \mathfrak{B}_2^{(\tau)} \div P^{(\tau)}\end{matrix}\right\} = H_s \begin{matrix}\cos\\ \sin\end{matrix} \kappa_s + \frac{f'' H''}{\mathfrak{F}_2} \begin{matrix}\cos\\ \sin\end{matrix} (\kappa'' - V'' - 60^\circ\tau - 29^\circ{\cdot}53)$$

Let us now consider the value of $P^{(\tau)}$. The longitude of the solar perigee is 281° or −79°, and the ratio of the sun's parallax to its mean parallax is approximately $1+e_{,}\cos(h+79^\circ)$, and the cube of that ratio is $1+3e_{,}\cos(h+79^\circ)$ or $1+0{\cdot}0504\cos(h+79^\circ)$. Now h, the sun's longitude at the middle of month (τ), is $h_0 + 15^\circ + 30^\circ\tau$; hence

$$P^{(\tau)} = 1 + 0{\cdot}0504 \cos(h_0 + 30^\circ\tau + 94^\circ)$$

and

$$\frac{1}{P^{(\tau)}} = 1 - 0{\cdot}0504 \cos(h_0 + 30^\circ\tau + 94^\circ)$$

Thus it is easy to compute the values of $1/P^{(\tau)}$ for the successive months, when we know h_0 the sun's mean longitude at $0^d\ 0^h$ of the month 0.

The semi-annual tide, being usually small, may be neglected in these incomplete observations, and the equations (11) now become

$$\mathfrak{A}_0^{(\tau)} = A_0 + \frac{H_{sa}}{\mathfrak{F}_1} \cos(\kappa_{sa} - V_{sa} - 30^\circ\tau - 14^\circ{\cdot}76)$$

$$\left.\begin{matrix}\mathfrak{A}_1{}^{(\tau)}\\ \mathfrak{B}_1{}^{(\tau)}\end{matrix}\right\} = H_{\frac{1}{2}8}\begin{matrix}\cos\\ \sin\end{matrix}\kappa_{\frac{1}{2}8} + \frac{f'H'}{\mathfrak{F}_1}\begin{matrix}\cos\\ \sin\end{matrix}(\kappa' - V' - 30^\circ\tau - 14^\circ{\cdot}76)$$

$$+\frac{H_p}{\mathfrak{F}_1}\begin{matrix}\cos\\ \sin\end{matrix}(\kappa_p - V_p + 30^\circ\tau + 14^\circ{\cdot}76)$$

$$\left.\begin{matrix}\mathfrak{A}_2{}^{(\tau)} \div P^{(\tau)}\\ \mathfrak{B}_2{}^{(\tau)} \div P^{(\tau)}\end{matrix}\right\} = H_8\begin{matrix}\cos\\ \sin\end{matrix}\kappa_8 + \frac{f''H''}{\mathfrak{F}_2}\begin{matrix}\cos\\ \sin\end{matrix}(\kappa'' - V'' - 60^\circ\tau - 29^\circ{\cdot}53)$$

$$\left.\begin{matrix}\mathfrak{A}_4{}^{(\tau)}\\ \mathfrak{B}_4{}^{(\tau)}\end{matrix}\right\} = H_{28}\begin{matrix}\cos\\ \sin\end{matrix}\kappa_{28}$$

$$\left.\begin{matrix}\mathfrak{A}_6{}^{(\tau)}\\ \mathfrak{B}_6{}^{(\tau)}\end{matrix}\right\} = H_{38}\begin{matrix}\cos\\ \sin\end{matrix}\kappa_{38}, \qquad \frac{1}{P^{(\tau)}} = 1 - 0{\cdot}0504\cos(h_0 + 30^\circ\tau + 94^\circ) \quad \ldots\ldots(17)$$

When the series of successive values of the $\mathfrak{A}$'s and $\mathfrak{B}$'s are harmonically analysed (by processes which we shall consider shortly) the several coefficients resulting from such analysis will be defined by

$$\mathfrak{A}_0{}^{(\tau)} = A_0 + A_1\cos 30^\circ\tau + B_1\sin 30^\circ\tau$$

$$\left.\begin{matrix}\mathfrak{A}_1{}^{(\tau)}\\ \mathfrak{B}_1{}^{(\tau)}\end{matrix}\right\} = \left.\begin{matrix}\mathrm{C}_0\\ \mathrm{c}_0\end{matrix}\right\} + \left.\begin{matrix}\mathrm{C}_1\\ \mathrm{c}_1\end{matrix}\right\}\cos 30^\circ\tau + \left.\begin{matrix}\mathrm{D}_1\\ \mathrm{d}_1\end{matrix}\right\}\sin 30^\circ\tau$$

$$\left.\begin{matrix}\mathfrak{A}_2{}^{(\tau)} \div P^{(\tau)}\\ \mathfrak{B}_2{}^{(\tau)} \div P^{(\tau)}\end{matrix}\right\} = \left.\begin{matrix}\mathrm{E}_0\\ \mathrm{e}_0\end{matrix}\right\} + \left.\begin{matrix}\mathrm{E}_2\\ \mathrm{e}_2\end{matrix}\right\}\cos 60^\circ\tau + \left.\begin{matrix}\mathrm{F}_2\\ \mathrm{f}_2\end{matrix}\right\}\sin 60^\circ\tau$$

$$\text{Mean } \mathfrak{A}_4{}^{(\tau)} = A_4, \qquad \text{Mean } \mathfrak{B}_4{}^{(\tau)} = B_4$$

$$\text{Mean } \mathfrak{A}_6{}^{(\tau)} = A_6, \qquad \text{Mean } \mathfrak{B}_6{}^{(\tau)} = B_6 \quad \ldots\ldots\ldots\ldots\ldots(18)$$

Then the subsequent procedure as given in (13) and (14) holds good, the only difference being that we do not obtain the semi-annual and solar elliptic tides.

We shall now consider the harmonic analysis of an imperfect series of values.

It must be premised that each monthly value of $\mathfrak{A}_2{}^{(\tau)}$, $\mathfrak{B}_2{}^{(\tau)}$ is to be divided by its corresponding $P^{(\tau)}$ before the analysis is made.

Suppose that $C^{(\tau)}$ denotes a function which is subject to semi-annual inequality, and that

$$C^{(\tau)} = \mathrm{A}_0 + \mathrm{A}_2\cos 60^\circ\tau + \mathrm{B}_2\sin 60^\circ\tau$$

Then it is clear that

$$C^{(0)} = \mathrm{A}_0 + \mathrm{A}_2$$

$$C^{(1)} = \mathrm{A}_0 + \tfrac{1}{2}\mathrm{A}_2 + \tfrac{1}{2}\sqrt{3}\,\mathrm{B}_2$$

$$C^{(2)} = \mathrm{A}_0 - \tfrac{1}{2}\mathrm{A}_2 + \tfrac{1}{2}\sqrt{3}\,\mathrm{B}_2$$

&c. &c.

I now define D_0, D_1, D_2 thus:

$$D_0 = C^{(0)} + C^{(1)} + C^{(2)} + \ldots$$

$$D_1 = C^{(0)} + \tfrac{1}{2}C^{(1)} - \tfrac{1}{2}C^{(2)} \ldots$$

$$D_2 = 0 \, . \, C^{(0)} + \tfrac{1}{2}\sqrt{3}\, C^{(1)} + \tfrac{1}{2}\sqrt{3}\, C^{(2)} \ldots$$

If there be n equations and if they be treated by the method of least squares, we get

$$D_0 = n\mathrm{A}_0 + \mathrm{A}_2 (1 + \tfrac{1}{2} - \tfrac{1}{2} \ldots) + \mathrm{B}_2 (0 + \tfrac{1}{2}\sqrt{3} + \tfrac{1}{2}\sqrt{3} \ldots)$$

$$D_1 = \mathrm{A}_0 (1 + \tfrac{1}{2} - \tfrac{1}{2} \ldots) + \mathrm{A}_2 (1 + \tfrac{1}{4} + \tfrac{1}{4} \ldots) + \mathrm{B}_2 (0 + \tfrac{1}{4}\sqrt{3} - \tfrac{1}{4}\sqrt{3} \ldots)$$

$$D_2 = \mathrm{A}_0 (0 + \tfrac{1}{2}\sqrt{3} + \tfrac{1}{2}\sqrt{3} \ldots) + \mathrm{A}_2 (0 + \tfrac{1}{4}\sqrt{3} - \tfrac{1}{4}\sqrt{3} \ldots) + \mathrm{B}_2 (0 + \tfrac{3}{4} + \tfrac{3}{4} \ldots)$$

These are the three equations from which A_0, A_2, B_2 are to be found.

A schedule is given below for the formation of D_0, D_1, D_2, and a table of the solutions of these equations according to the number of months available.

Next, suppose $C^{(\tau)}$ denotes a function which is subject to annual inequality, and that

$$C^{(\tau)} = \mathrm{A}_0 + \mathrm{A}_1 \cos 30^\circ\tau + \mathrm{B}_1 \sin 30^\circ\tau$$

Then

$$C^{(0)} = \mathrm{A}_0 + \mathrm{A}_1$$

$$C^{(1)} = \mathrm{A}_0 + \tfrac{1}{2}\sqrt{3}\, \mathrm{A}_1 + \tfrac{1}{2}\mathrm{B}_1$$

$$C^{(2)} = \mathrm{A}_0 + \tfrac{1}{2}\mathrm{A}_1 + \tfrac{1}{2}\sqrt{3}\, \mathrm{B}_1$$

&c., &c.

In this case the method of least squares gives

$$D_0 = C^{(0)} + C^{(1)} + C^{(2)} \ldots$$
$$= n\mathrm{A}_0 + \mathrm{A}_1 (1 + \tfrac{1}{2}\sqrt{3} + \tfrac{1}{2} \ldots) + \mathrm{B}_1 (0 + \tfrac{1}{2} + \tfrac{1}{2}\sqrt{3} \ldots)$$

$$D_1 = C^{(0)} + \tfrac{1}{2}\sqrt{3}\, C^{(1)} + \tfrac{1}{2}C^{(2)} \ldots$$
$$= \mathrm{A}_0 (1 + \tfrac{1}{2}\sqrt{3} + \tfrac{1}{2} \ldots) + \mathrm{A}_1 (1 + \tfrac{3}{4} + \tfrac{1}{4} \ldots) + \mathrm{B}_1 (0 + \tfrac{1}{4}\sqrt{3} + \tfrac{1}{4}\sqrt{3} \ldots)$$

$$D_2 = 0 \, . \, C^{(0)} + \tfrac{1}{2}C^{(1)} + \tfrac{1}{2}\sqrt{3}\, C^{(2)} \ldots$$
$$= \mathrm{A}_0 (0 + \tfrac{1}{2} + \tfrac{1}{2}\sqrt{3} \ldots) + \mathrm{A}_1 (0 + \tfrac{1}{4}\sqrt{3} + \tfrac{1}{4}\sqrt{3} \ldots) + \mathrm{B}_1 (0 + \tfrac{1}{4} + \tfrac{3}{4} \ldots)$$

Tables are given below for the formation of D_0, D_1, D_2, and of the solutions of the equations according to the number of months available.

Reduction of incomplete series of results.

Form for finding D_0, D_1, D_2 where the monthly values are subject to semi-annual inequality.

	i.		ii.	iii.	iv.	v.		vi.		
	Monthly values			i.+ii.	Last 2 of iii. revd.	iii.+iv.	Factor M	$M\times$v.	iii.−iv.	
0	a	6	a'	$a+a'$	......	$a+a'$	1	$a+a'$	....	
1	b	7	b'	$b+b'$	$f+f'$	$b+\ldots+f'$	$\frac{1}{2}$	$\frac{1}{2}(b+\ldots+f')$	$b+\ldots-f'$	$S_4=0{\cdot}866$
2	c	8	c'	$c+c'$	$e+e'$	$c+\ldots+e'$	$-\frac{1}{2}$	$-\frac{1}{2}(c+\ldots+e')$	$c+\ldots-e'$	
3	d	9	d'	$d+d'$	......	$d+d'$	-1	$-(d+d')$	....	
4	e	10	e'	$e+e'$						
5	f	11	f'	$f+f'$		Sum D_0		Sum D_1	Sum $\times S_4$	
									D_2	

Form for finding D_0, D_1, D_2 where the monthly values are subject to annual inequality.

	i.		ii.	iii.	iv.	v.	vi.		vii.	viii.			
	Monthly values			i.+ii.	i.−ii.	Last 2 of iv. revd.	iv.−v.	Factor M	$M\times$vi.	iv.+v.	Factor M	$M\times$viii.	
0	a	6	a'	$a+a'$	$a-a'$	......	$a-a'$	1	$a-a'$	$a-a'$	0	0	
1	b	7	b'	$b+b'$	$b-b'$	$f-f'$	$b-\ldots+f'$	S_4	$S_4(b-\ldots+f')$	$b-\ldots-f'$	$\frac{1}{2}$	$\frac{1}{2}(b-\ldots-f')$	
2	c	8	c'	$c+c'$	$c-c'$	$e-e'$	$c-\ldots+e'$	$\frac{1}{2}$	$\frac{1}{2}(c-\ldots+e')$	$c-\ldots-e'$	S_4	$S_4(c-\ldots-e')$	$S_4=0{\cdot}866$
3	d	9	d'	$d+d'$	$d-d'$	......	$d-d'$	0	0	$d-d'$	1	$d-d'$	
4	e	10	e'	$e+e'$	$e-e'$				Sum D_1			Sum D_2	
5	f	11	f'	$f+f'$	$f-f'$								
				Sum D_0									

In both these forms those monthly values which are unknown are to be treated as being zero. For example, if 9 months of observation are available, d', e', f' will be zero.

Rule for finding semi-annual inequality from an incomplete series.

Number of months available		Coefft. of D_0	Coefft. of D_1	Coefft. of D_2
6	$A_0=$	+0·167		
	$A_2=$		+0·333	
	$B_2=$			+0·333
7	$A_0=$	+0·148	−0·037	
	$A_2=$	−0·037	+0·259	
	$B_2=$			+0·333
8	$A_0=$	+0·136	−0·045	−0·026
	$A_2=$	−0·045	+0·253	−0·019
	$B_2=$	−0·026	−0·019	+0·275
9	$A_0=$	+0·123	−0·027	−0·047
	$A_2=$	−0·027	+0·228	+0·011
	$B_2=$	−0·047	+0·011	+0·241
10	$A_0=$	+0·107		−0·041
	$A_2=$		+0·182	
	$B_2=$	−0·041		+0·238

Rule for finding annual inequality from an incomplete series.

Number of months available		Coefft. of D_0	Coefft. of D_1	Coefft. of D_2
6	$A_0=$	+0·977	−0·326	−1·215
	$A_1=$	−0·326	+0·442	+0·405
	$B_1=$	−1·215	+0·405	+1·845
7	$A_0=$	+0·424		−0·528
	$A_1=$	+0·250		
	$B_1=$	−0·528		+0·990
8	$A_0=$	+0·226	+0·062	−0·233
	$A_1=$	+0·062	+0·230	−0·093
	$B_1=$	−0·233	−0·093	+0·552
9	$A_0=$	+0·146	+0·057	−0·098
	$A_1=$	+0·057	+0·230	−0·083
	$B_1=$	−0·098	−0·083	+0·326
10	$A_0=$	+0·110	+0·036	−0·036
	$A_1=$	+0·036	+0·218	−0·048
	$B_1=$	−0·036	−0·048	+0·218

We thus get the following rule for the evaluation of A_0, Ssa, S_1, S_2, S_4, S_6, K_2, K_1, P from 6, 7, 8, 9, or 10 months of observation:

Proceed as though the year were complete and find the $\mathfrak{A}$*'s and* $\mathfrak{B}$*'s for as many months as are available. Reduce the* $\mathfrak{A}_2$, $\mathfrak{B}_2$ *by multiplication by* $1/P^{(\tau)}$ *or* $1 - 0{\cdot}0504 \cos (h_0 + 30^\circ \tau + 94^\circ)$.

Analyse $\mathfrak{A}_0^{(\tau)}$, $\mathfrak{A}_1^{(\tau)}$, $\mathfrak{B}_1^{(\tau)}$, *for annual inequality, and* $\mathfrak{A}_2^{(\tau)}/P^{(\tau)}$, $\mathfrak{B}_2^{(\tau)}/P^{(\tau)}$ *for semi-annual inequality according to the rules for reduction of incomplete series just given.*

Complete the reduction as in § 3.

These rules for reduction do not include the case of 11 months, nor the case where any month in the series is incomplete (*e.g.*, if a fortnight's observation were wanting in one of the months), because these cases may be treated thus:—the $\mathfrak{A}$'s and $\mathfrak{B}$'s return to the same value at the end of a year, and therefore the case of eleven months is the same as that of a missing month at any other part of the year. In both these cases we may interpolate the missing $\mathfrak{A}$'s and $\mathfrak{B}$'s and treat the year as complete.

If three or more weeks of observation were missing they might fall so as to spoil two months, and in this case we should have an incomplete series. It is then to be recommended that the equations of least squares be formed and the equations solved. So many similar cases may arise that it does not seem worth while to solve the equations until the case arises.

§ 5. *Evaluation of* A_0, S_2, S_4, K_2, K_1, P *from a short period of observation**.

If the available tidal observations only extend over a few months, it is useless to attempt the independent evaluation of those tides which we have hitherto found by means of annual and semi-annual inequalities in the monthly harmonic constants. We will suppose that 30 days of observations are available. Then when we neglect the annual tide, and the solar (meteorological) tide S_1, we have from (11) or (17), which give the analysis of 30 days,

$$\mathfrak{A}_0 = A_0$$

$$\left.\begin{matrix}\mathfrak{A}_1\\ \mathfrak{B}_1\end{matrix}\right\} = \frac{f'H'}{\mathfrak{F}_1} \begin{matrix}\cos\\ \sin\end{matrix} (\kappa' - V' - 14^\circ{\cdot}76) + \frac{H_p}{\mathfrak{F}_1} \begin{matrix}\cos\\ \sin\end{matrix} (\kappa_p - V_p + 14^\circ{\cdot}76)$$

$$\left.\begin{matrix}\mathfrak{A}_2\\ \mathfrak{B}_2\end{matrix}\right\} = PH_s \begin{matrix}\cos\\ \sin\end{matrix} \kappa_s + \frac{f''H''}{\mathfrak{F}_2} \begin{matrix}\cos\\ \sin\end{matrix} (\kappa'' - V'' - 29^\circ{\cdot}53)$$

$$\left.\begin{matrix}\mathfrak{A}_4\\ \mathfrak{B}_4\end{matrix}\right\} = H_{2s} \begin{matrix}\cos\\ \sin\end{matrix} \kappa_{2s}, \qquad P = 1 + 0{\cdot}0504 \cos (h_0 + 15^\circ)$$

It is now necessary to assume that the P tide has the same amount of retardation as the K_1, and that the ratio of their amplitudes is the same as

* [See footnote to § 2 above for changes to be made in this section when only 15 days are available, and when the diurnal tides are analysed over either 27 or 14 days.]

in the equilibrium theory. We also make the like assumption with respect to the K_2 and S_2 tides.

Accordingly we put

$$H_p = \tfrac{1}{3}H', \qquad \kappa_p = \kappa'; \qquad H'' = \tfrac{3}{11}H_s, \qquad \kappa'' = \kappa_s$$

Now since

$$V' = h_0 - \tfrac{1}{2}\pi - \nu', \qquad V_p = -h_0 + \tfrac{1}{2}\pi, \qquad V'' = 2h_0 - 2\nu''$$

we have

$$\kappa_p - V_p + 14^\circ{\cdot}76 = \kappa' - V' - 14^\circ{\cdot}76 + (2h_0 - \nu' + 29^\circ{\cdot}53) + \pi$$

$$\kappa'' - V'' - 29^\circ{\cdot}53 = \kappa_s - (2h_0 - 2\nu'' + 29^\circ{\cdot}53)$$

Therefore

$$\left.\begin{matrix}\mathfrak{A}_1\\ \mathfrak{B}_1\end{matrix}\right\} = \frac{f'H'}{\mathfrak{F}_1}\begin{matrix}\cos\\ \sin\end{matrix}(\kappa' - V' - 14^\circ{\cdot}76)$$

$$- \frac{\tfrac{1}{3}H'}{\mathfrak{F}_1}\begin{matrix}\cos\\ \sin\end{matrix}(\kappa' - V' - 14^\circ{\cdot}76 + 2h_0 - \nu' + 29^\circ{\cdot}53)$$

$$\left.\begin{matrix}\mathfrak{A}_2\\ \mathfrak{B}_2\end{matrix}\right\} = PH_s\begin{matrix}\cos\\ \sin\end{matrix}\kappa_s + \frac{\tfrac{3}{11}f''H_s}{\mathfrak{F}_2}\begin{matrix}\cos\\ \sin\end{matrix}(\kappa_s - 2h_0 + 2\nu'' - 29^\circ{\cdot}53)$$

Let us put $\tan\phi = \dfrac{\sin(2h_0 - \nu' + 29^\circ{\cdot}53)}{3f' - \cos(2h_0 - \nu' + 29^\circ{\cdot}53)}$

$$\tan\psi = \frac{f''\sin(2h_0 - 2\nu'' + 29^\circ{\cdot}53)}{\tfrac{11}{3}P\mathfrak{F}_2 + f''\cos(2h_0 - 2\nu'' + 29^\circ{\cdot}53)} \quad \text{...........(19)}$$

Then

$$\left.\begin{matrix}\mathfrak{A}_1\\ \mathfrak{B}_1\end{matrix}\right\} = \frac{H'}{\mathfrak{F}_1}\,\frac{3f' - \cos(2h_0 - \nu' + 29^\circ{\cdot}53)}{3\cos\phi}\begin{matrix}\cos\\ \sin\end{matrix}(\kappa' - V' - 14^\circ{\cdot}76 - \phi)$$

$$\left.\begin{matrix}\mathfrak{A}_2\\ \mathfrak{B}_2\end{matrix}\right\} = H_s\,\frac{\tfrac{11}{3}P\mathfrak{F}_2 + f''\cos(2h_0 - 2\nu'' + 29^\circ{\cdot}53)}{\tfrac{11}{3}\mathfrak{F}_2\cos\psi}\begin{matrix}\cos\\ \sin\end{matrix}(\kappa_s - \psi) \quad \text{...........(20)}$$

If therefore

$$\left.\begin{matrix}\mathfrak{A}_1\\ \mathfrak{B}_1\end{matrix}\right\} = R_1\begin{matrix}\cos\\ \sin\end{matrix}\zeta_1, \qquad \left.\begin{matrix}\mathfrak{A}_2\\ \mathfrak{B}_2\end{matrix}\right\} = R_2\begin{matrix}\cos\\ \sin\end{matrix}\zeta_2$$

we have $\kappa' = \zeta_1 + V' + 14^\circ{\cdot}76 + \phi = \kappa_p$

$$H' = \frac{3\mathfrak{F}_1 R_1\cos\phi}{3f' - \cos(2h_0 - \nu' + 29^\circ{\cdot}53)}, \qquad H_p = \tfrac{1}{3}H$$

$$\kappa_s = \zeta_2 + \psi = \kappa''$$

$$H_s = \frac{\tfrac{11}{3}\mathfrak{F}_2 R_2\cos\psi}{\tfrac{11}{3}P\mathfrak{F}_2 + f''\cos(2h_0 - 2\nu'' + 29^\circ{\cdot}53)}, \qquad H'' = \tfrac{3}{11}H_s \quad \text{...(21)}$$

If there be several months available it is recommended that each 30 days be treated quite independently, so that from each group of days we shall get H', κ' and H_s, κ_s. Then the mean value of $H'\cos\kappa'$ is to be taken as the final value of that function, and $H'\sin\kappa'$ is to be treated similarly; finally H', κ' are to be found. The several values of H_s, κ_s may be treated in the same way. Of course we assume throughout that $\kappa_p = \kappa'$, $H_p = \frac{1}{3}H'$, $\kappa'' = \kappa_s$, $H'' = \frac{3}{11}H_s$, assumptions which are usually nearly correct.

The mean value of $\mathfrak{A}_0$ must be taken as giving A_0, but at places with a considerable annual tide it is impossible to obtain a good value of mean water mark from a short series of observations.

§ 6. *On the evaluation of the several tides by grouping of mean solar days.*

Let $n(\gamma - \chi)$ denote the speed in degrees per m.s. hour of any one tide, n being equal to 1, 2, 3, 4, 5, or 6. Then $15^\circ/(\gamma - \chi)$ may be called one "special hour." Since $15^\circ/(\gamma - \eta)$ is one m.s. hour, the ratio of the m.s. to the special hour is $(\gamma - \chi)/(\gamma - \eta)$.

Let one m.s. hour be equal to $1 - \beta$ special hour, then

$$\beta = 1 - \frac{\gamma - \chi}{\gamma - \eta}, \text{ special hours}$$

Let it be required to express the 12^h of any m.s. day of a series of days by reference to special time. It is clear that 12^h m.s. time will be specified by one of the 24 special hours, with something less than half a special hour added or subtracted.

Having fixed the 12^h of m.s. time of a particular m.s. day in the special time scale, let us treat that m.s. day as a whole, and consider the incidence of the other 23 m.s. hours in special time. It is clear that in m.s. time we work backwards and forwards from 12^h by subtracting or adding unity, and that in special time we subtract or add $1 - \beta$.

If 12^h m.s. time be $x^h + \alpha$, where α lies between $\pm\frac{1}{2}$ special time, the following is a schedule of equivalence:

Mean solar time		Special time
0^h	=	$(x^h - 12^h) + (\alpha + 12\beta)$
1^h	=	$(x^h - 11^h) + (\alpha + 11\beta)$
2^h	=	$(x^h - 10^h) + (\alpha + 10\beta)$
......		......
11^h	=	$(x^h - 1^h) + (\alpha + \beta)$
12^h	=	$x^h + \alpha$
13^h	=	$(x^h + 1^h) + (\alpha - \beta)$
......		......
22^h	=	$(x^h + 10^h) + (\alpha - 10\beta)$
23^h	=	$(x^h + 11^h) + (\alpha - 11\beta)$

In the column of special time it is supposed that 24^h is added or subtracted, so that the result is less than 24^h. For example, if x is 10, the hour column of special time will run 22^h, 23^h, 0^h, ..., 9^h, 10^h, 11^h, ..., 20^h, 21^h.

If the series of days be long x will have all integral values between 0 and 23 with equal frequency, and since α has all values between $+\frac{1}{2}$ and $-\frac{1}{2}$ with equal frequency, the excess of the solar hour above the nearest exact special hour (which may be called the error) will have all its possible values with equal frequency. If the mean solar hours be arranged in a schedule of columns headed 0^h, 1^h, ..., 23^h of special time, each column will be subject to errors which follow the same law of frequency.

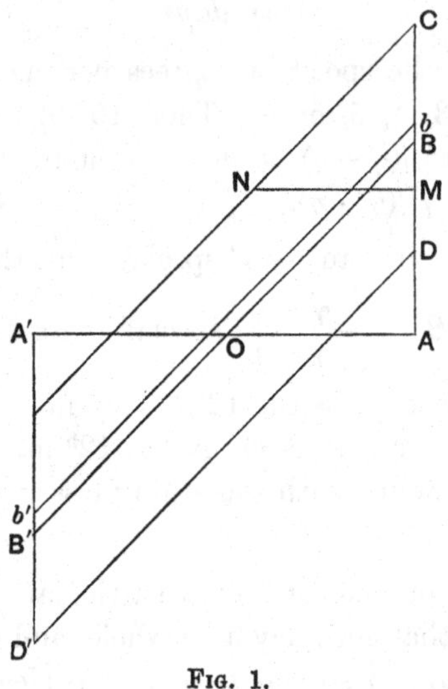

FIG. 1.

Let abscissæ (fig. 1) measured from O along A′OA represent magnitude of α.

Since α lies between $\pm\frac{1}{2}$, the limit of the figure is given by $OA = OA' = \frac{1}{2}$.

If magnitude of error (*i.e.* m.s. − special hour), measured in special time, be represented by ordinates, a line BOB′ at 45° to AOA′ represents all the errors which can arise in the incidence of the m.s. 12^h in the schedule of special time.

If a line bb' be drawn parallel to and above BB′ by a distance β, we have a representation of all the errors of incidence of the m.s. 11^h. If a series of equidistant parallel lines be drawn above and below BB′ until there are 12 above and 11 below, then the errors of all the m.s. hours are represented, the top one showing the errors of the m.s. 0^h and the bottom one the errors of the m.s. 23^h.

Any special hour corresponds with equal frequency with each solar hour, and hence each mode of error occurs with equal frequency.

It is now necessary to consider in how many ways an error of given magnitude can occur. If in the figure AM represents an error of given magnitude, then wherever MN cuts a diagonal line, it shows that an error may arise in one way.

It is thus clear that there are no + errors greater than $\frac{1}{2} + 12\beta$, and no − errors greater than $\frac{1}{2} + 11\beta$, and

Errors of magnitude		
$\frac{1}{2} + 12\beta$ to $\frac{1}{2} + 11\beta$	may arise in	1 way
$\frac{1}{2} + 11\beta$ to $\frac{1}{2} + 10\beta$	"	2 ways
$\frac{1}{2} + 10\beta$ to $\frac{1}{2} + 9\beta$	"	3 ways
......		
$\frac{1}{2} - 10\beta$ to $\frac{1}{2} - 11\beta$	"	23 ways
$\frac{1}{2} - 11\beta$ to $-(\frac{1}{2} - 12\beta)$	"	24 ways
$-(\frac{1}{2} - 12\beta)$ to $-(\frac{1}{2} - 11\beta)$	"	23 ways
......		
$-(\frac{1}{2} + 9\beta)$ to $-(\frac{1}{2} + 10\beta)$	"	2 ways
$-(\frac{1}{2} + 10\beta)$ to $-(\frac{1}{2} + 11\beta)$	"	1 way

The frequency of error is represented graphically in fig. 2. The slope of the two staircases is drawn at 45°, but any other slope would have done equally well.

A frequency curve of this form is not very convenient, and, as there are many steps in the ascending and descending slopes, I substitute the frequency curve shown in fig. 3. This is clearly equivalent to the former one. In fig. 3 all the times shown in fig. 2 are converted to angle at 15° to the hour; ϵ accordingly denotes $15°\beta$.

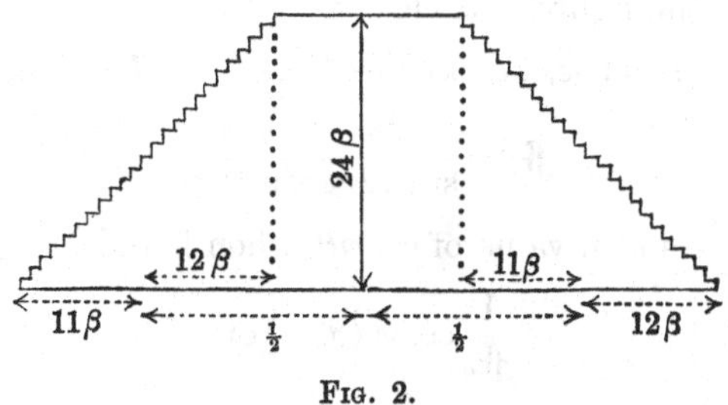

Fig. 2.

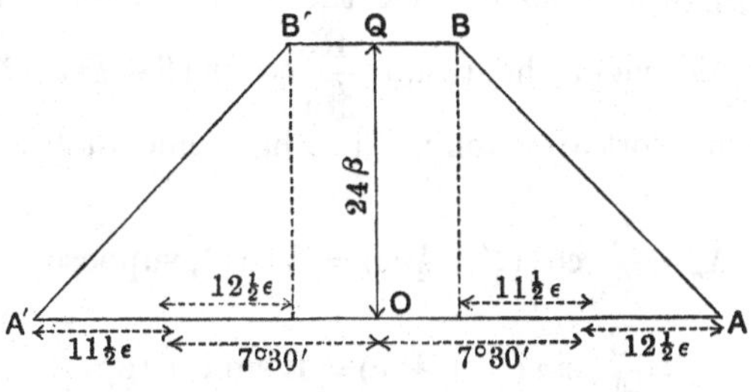

Fig. 3.

Now let $\cos n(\theta - x)$ be the observed value of a function whose true value is $\cos n\theta$, and suppose that x, the error of θ, has a frequency $f(x)$; then the mean value of the function deduced from many observations will be

$$\int_{-\infty}^{+\infty} f(x) \cos n(\theta - x)\, dx \div \int_{-\infty}^{+\infty} f(x)\, dx$$

In our case $f(x)$ is the ordinate of the frequency curve whose abscissa is x.

Let $OQ = h$, $QB = a$, $QB' = b$, $OA = a + h$, $OA' = b + h$; then

$$\int_{-\infty}^{+\infty} f(x)\, dx = (a + b + h) h$$

$$\begin{aligned}\int_{-\infty}^{+\infty} & f(x) \cos n(\theta - x)\, dx \\ &= \int_0^a h \cos n(\theta - x)\, dx + \int_a^{a+h} (a + h - x) \cos n(\theta - x)\, dx \\ &+ \int_0^b h \cos n(\theta + x)\, dx + \int_b^{b+h} (b + h - x) \cos n(\theta + x)\, dx \\ &= \frac{4}{n^2} \cos n[\theta - \tfrac{1}{2}(a - b)] \sin \tfrac{1}{2}nh \sin \tfrac{1}{2}n(a + b + h)\end{aligned}$$

The algebraical steps involved in the evaluation of these four integrals and subsequent simplification are omitted.

Hence the result is

$$\frac{\sin \frac{1}{2}nh}{\frac{1}{2}nh} \frac{\sin \frac{1}{2}n(a + b + h)}{\frac{1}{2}n(a + b + h)} \cos n[\theta - \tfrac{1}{2}(a - b)]$$

By reference to the figure it is clear that

$$a + b + h = 15°, \quad h = 24\epsilon, \quad a = 7\tfrac{1}{2}° - 11\tfrac{1}{2}\epsilon, \quad b = 7\tfrac{1}{2}° - 12\tfrac{1}{2}\epsilon, \quad a - b = \epsilon$$

Write, then
$$\mathfrak{F}_n = \frac{12n\epsilon}{\sin 12n\epsilon} \frac{\frac{15}{2}°n}{\sin \frac{15}{2}°n}$$

and we obtain as the mean value of $\cos n\theta$, when found in this way,

$$\frac{1}{\mathfrak{F}_n} \cos n(\theta - \tfrac{1}{2}\epsilon)$$

It is obvious that if we had begun with $\sin n\theta$, the argument in the result and the factor $\mathfrak{F}_n$ would have been the same. Accordingly, a function $R' \cos(n\theta - \zeta')$ would yield the result $\dfrac{R'}{\mathfrak{F}_n} \cos[n(\theta - \tfrac{1}{2}\epsilon) - \zeta']$. If 24 equidistant results of this sort are submitted to harmonic analysis to find A_n, B_n, we shall get

$$A_n = \frac{R'}{\mathfrak{F}_n} \cos(\zeta' + \tfrac{1}{2}n\epsilon) = R \cos \zeta, \text{ suppose}$$

$$B_n = \frac{R'}{\mathfrak{F}_n} \sin(\zeta' + \tfrac{1}{2}n\epsilon) = R \sin \zeta, \text{ suppose}$$

Accordingly $$\mathrm{R} = \frac{\mathrm{R}'}{\mathfrak{F}_n}, \qquad \zeta = \zeta' + \tfrac{1}{2}n\epsilon$$

But it is required to find R′, ζ', so that

$$\mathrm{R}' = \mathfrak{F}_n \mathrm{R}, \qquad \zeta' = \zeta - \tfrac{1}{2}n\epsilon$$

Thus when the 24 observed hourly tide heights on any m.s. day are regrouped so that the observed height at 12^h m.s. time is reputed to appertain to an exact special hour, and each of the previous and subsequent hourly values of that m.s. day are reputed to belong to previous and subsequent exact special hours; and when a long series of m.s. days are treated similarly, and when the mean heights of water at each of the 24 special hours are harmonically analysed, we shall obtain the required result by augmenting R by a factor $\mathfrak{F}_n$, and by subtracting $\frac{1}{2}n\epsilon$ from ζ.

The values of $\mathfrak{F}_n$ and of $\frac{1}{2}n\epsilon$ will be different for each kind of tide, and the following table gives their numerical values.

Table of $\mathfrak{F}_n$ and $\frac{1}{2}n\epsilon$.

Initial of tide	n	log $\mathfrak{F}_n$	$\frac{1}{2}n\epsilon$
M_1	1	0·00212	0°·26
M_2	2	0·00849	0°·53
M_3	3	0·01915	0°·79
M_4	4	0·03416	1°·05
M_6	6	0·07767	1°·57
N	2	0·01361	0°·82
L	2	0·00570	0°·24
ν	2	0·01278	0°·78
O	1	0·00535	0°·57
J	1	0·00225	− 0°·28
Q	1	0·01149	0°·90
μ	2	0·02016	1°·09
2SM	2	0·00805	− 0°·49
MS	4	0·02342	0°·52
λ	2	0·00595	0°·28
2N	2	0·02136	1°·13
OO	1	0·00481	− 0°·53
MK	3	0·01438	0°·50
2MK	3	0·02632	1°·09
MN	4	0·04328	1°·35

As it does not appear worth while to evaluate the tides written below the line, no use will be made of the last six results given in this table.

§ 7. *On the periods over which the means are to be taken in evaluating the tidal constants.*

We have considered in previous sections the treatment of the group of tides which are associated with solar time, when the period of observation is less than a year, and we have now to consider the other tides.

It is important that the means be taken over such a number of days that the perturbation arising from other tides shall be minimised.

The perturbation between semi-diurnal and diurnal tides is always negligible. It is therefore only necessary to consider the action of the tides M_2, S_2 in the case of semi-diurnal tides, and that of K_1 and O for diurnal tides.

It is easy to see that the influence of a disturbing tide is evanescent when the means are taken over a period such that the excess of the argument of the disturbed over that of the disturbing tide has increased through a multiple of 360°. As, however, we are working with integral numbers of days, and as the speeds of tides are incommensurable, this condition cannot be exactly satisfied.

From this consideration it appears that to minimise the perturbation of S_2, 2SM, μ by M_2 (and *vice versâ*) we must stop at an exact multiple of a semi-lunation. To minimise the effect of M_2 on N and L, and of K_1 on J and Q, we must stop at an exact multiple of a lunar anomalistic period. To minimise the effect of M_2 on ν, we must stop at a multiple of the period $2\pi/(\sigma+\varpi-2\eta)$. To minimise the effect of K_1 on O, we must stop at an exact multiple of a semi-lunar period.

For the quater-diurnal tide, MS, it is immaterial where we stop, and so it may as well be taken at a multiple of a semi-lunation.

The following table (p. 243) gives the rules derived from these considerations.

§ 8. *On the tides of long period.*

The annual (Sa) and semi-annual (Ssa) tides are evaluated in the course of the work by which other important tides are found. These are the only two tides of long period which have a practical importance in respect to tidal prediction, but the luni-solar fortnightly (MSf), the lunar fortnightly (Mf), and the lunar monthly (Mm) tides have a theoretical interest.

It will therefore be well to show how they may be found. The process is short, and, although it is less accurate than the laborious plan followed in the Indian reductions, it appears to give fairly good results.

(*Continued at foot of p.* 243)

Number of the last day to be included in the evaluation of the several tides for observations extending over any period up to a year.

For M_2, μ, 2SM, MS. Stop with one of the following days (semi-lunations)	For O. Stop with one of the following days (semi-lunar periods)	For N, L, J, Q. Stop with one of the following days (anom. periods)	For ν. Stop with one of the following days (periods $2\pi/(\sigma+\varpi-2\eta)$)
14	13	27	31
29	26	54	63
43	40	—	—
58	54	74 + 8	74 + 20
73	67	+ 35	+ 52
—	—	+ 63	—
74 + 14	74 + 7	—	148 + 10
+ 28	+ 21	148 + 16	+ 42
+ 43	+ 34	+ 44	—
+ 58	+ 48	+ 71	222 + 0
+ 73	+ 62	—	+ 31
—	—	222 + 25	+ 63
148 + 13	148 + 1	+ 53	—
+ 28	+ 15	—	296 + 21
+ 43	+ 29	296 + 6	+ 53
+ 58	+ 42	+ 34	—
+ 72	+ 56	+ 61	
—	+ 70	—	
222 + 13	—		
+ 28	222 + 9		
+ 43	+ 23		
+ 58	+ 37		
+ 72	+ 50		
—	+ 64		
296 + 13	—		
+ 28	296 + 4		
+ 43	+ 17		
+ 57	+ 32		
+ 72	+ 45		
—	+ 58		
	+ 72		

For the sake of simplicity, let us consider the tide MSf. Its period is about 14 days, and therefore a day does not differ very largely from a twelfth part of the period. Accordingly, if about two days in a fortnight are rejected by proper rules, the mean heights of water on the remaining days may be taken as representatives of twelve equidistant values of water height.

I therefore go through the whole year and reject, according to proper rules, the daily sums of the 24 hourly heights corresponding to certain 69 of the days out of 369. The remaining 300 values are written consecutively into a schedule of 12 columns and 25 rows, of which each corresponds to a half lunation. The 12 columns are summed, and the sums are harmonically analysed for the first pair of harmonic components. These components have to be divided by 24 times 25, or by 600, because the daily mean water height is $\frac{1}{24}$th of the daily sum, and there are 25 semi-lunations.

In the same way the semi-lunar period is about $13\frac{1}{2}$ days, and if we erase by proper rules 45 daily sums out of 369, we are left with 324, which may be written consecutively in a schedule of 12 columns and 27 rows, of which each corresponds to a semi-lunar period. The summing and analysis is the same as in the last case, but the final division is by 24 times 27, or by 648.

In this way we evaluate the luni-solar fortnightly and lunar fortnightly inequalities in the height of the water.

The period of the moon is between 27 and 28 days, and if we erase appropriately about one day in eight we are left with sets of 24 values which may be taken as 24 equidistant values of the daily sums. Accordingly we erase 46 daily sums out of 358, and write the 312 which remain consecutively into a schedule of 24 columns and 13 rows, of which each corresponds to a lunar anomalistic period.

The 24 columns are summed and the sums analysed for the first components. Finally, the components are to be divided by 24 times 13, or by 312. In this way the lunar monthly tide is evaluated.

But the result obtained in this way is, as far as concerns the tide MSf, to some, and it may be to a large, extent fictitious. It represents, in fact, a residuum of the principal lunar tide M_2. That this is the case will now be proved.

Suppose that t_0 is an integral number of days since epoch, being the time of noon on a certain day; then the principal lunar tide M_2 on that day may be written $H_m \cos[2(\gamma-\sigma)(t_0+\tau)-\zeta_m]$, where τ is less than 24 hours. Then the daily sum for that day will be

$$H_m \frac{\sin 24(\gamma-\sigma)}{\sin(\gamma-\sigma)} \cos[2(\gamma-\sigma)t_0 + 23(\gamma-\sigma) - \zeta_m]$$

Now since t_0 is an integral number of days $2(\gamma-\sigma)t_0$ only differs from $-2(\sigma-\eta)t_0$ by an exact multiple of 360°; hence the argument of the cosine may be written $2(\sigma-\eta)t_0 - 23(\gamma-\sigma) + \zeta_m$.

But the true luni-solar fortnightly tide, which we may denote

$$fH \cos [2(\sigma - \eta)(t_0 + \tau) - \zeta]$$

varies so slowly in the course of a day that the daily sum is sensibly equal to

$$24fH \cos [2(\sigma - \eta) t_0 + 23(\sigma - \eta) - \zeta]$$

It thus appears that the residual effect of M_2 is of exactly the same form as that of MSf. It becomes, therefore, necessary to clear the harmonic components, determined as described above, from the effects of M_2.

In order to determine the values of these clearances, I found the values of $\cos 2(\sigma - \eta) t$ and $\sin 2(\sigma - \eta) t$ for every noon in a year of 369 days. I then erased the values selected for the treatment of MSf and analysed the remaining values. In this way it was easy to find the effect of the known M_2 tide.

Suppose that A_1, B_1 are the first harmonic components determined by the treatment of a series of daily sums, and that δA_1, δB_1 are the corrections to be applied to them to eliminate the effects of M_2, then I find that if A_m, B_m are the two components of M_2 as determined by the previous method (§ 6) of analysis,

$$\delta A_1 = +\,0{\cdot}0304 A_m - 0{\cdot}0171 B_m$$

$$\delta B_1 = -\,0{\cdot}0171 A_m - 0{\cdot}0304 B_m$$

$$C = A_1 + \delta A_1, \qquad D = B_1 + \delta B_1$$

$$C - 0{\cdot}047 D = 0{\cdot}992 fH \cos \zeta$$

$$D + 0{\cdot}047 C = 0{\cdot}992 fH \sin \zeta$$

Whence f being known from Baird's *Manual* (being a function of the longitude of moon's node), H and ζ are determinable. We have also

$$\kappa = \zeta + 2(s_0 - \xi - h_0 + \nu) + 11^\circ{\cdot}7$$

In the set of computation forms which I have prepared for use on the present plan, it is shown what days are to be erased for each of the three analyses, and how they are to be entered in schedules, summed, and analysed.

§ 9. *On abridgment in the computations.*

It seemed probable that one decimal of a foot would suffice to express the hourly tide heights. In order to test this, I have taken several individual days of observation at Port Blair, and have found, by harmonic analysis, the time and amplitude of the diurnal and semi-diurnal H.W., first, when the hourly heights are expressed to two decimal places of a foot, and secondly, when they are only entered to the nearest tenth of a foot. I find that the times of H.W. agree within less than a minute of time, and that the amplitudes agree within a fraction of an inch. If this much be true of individual days, the difference of results arising from two or one place of decimals will clearly entirely disappear when a series of days is considered. Hence, by taking as unit the tenth of a foot, or the inch, or even two inches at places with large tides, we may always express all, or nearly all, the heights on which we are to operate by two significant figures. The adoption of this rule not only saves the writing of a large number of figures, but also enormously diminishes the labour of the additions which have to be made.

It also seemed probable that substantial accuracy might be attained from the harmonic analysis of only 12 hourly values instead of 24. In order to test this I took the tidal reductions for Port Blair, Andaman Islands (kindly lent me by the Survey of India), and have compared the results which would have been derived from 12 values with those actually obtained from 24 values by the computers of the Indian Survey. The following tables give the results:

Semi-diurnal tides.

Initial	Results from 12 two-hourly values	Results from 24 hourly values	Error, (12) – (24)
	ft.	ft.	ft.
S A_2 =	+0·6883	+0·6890	−0·0007
S B_2 =	+0·6775	+0·6768	+0·0008
M A_2 =	−1·7032	−1·7005	−0·0027
M B_2 =	+1·0883	+1·0872	+0·0011
K A_2 =	−0·1437	−0·1407	−0·0030
K B_2 =	−0·2527	−0·2515	−0·0012
L A_2 =	+0·0357	+0·0347	+0·0010
L B_2 =	+0·0612	+0·0610	+0·0002
N A_2 =	+0·3422	+0·3486	−0·0064
N B_2 =	+0·2192	+0·2124	+0·0068
ν A_2 =	−0·1217	−0·1165	−0·0052
ν B_2 =	+0·0867	+0·0887	−0·0020
μ A_2 =	+0·0857	+0·0849	−0·0008
μ B_2 =	+0·0383	−0·0388	−0·0005
2SM A_2 =	+0·0055	+0·0037	+0·0018
2SM B_2 =	−0·0198	−0·0200	+0·0002

Diurnal tides.

Initial	Results from 12 two-hourly values	Results from 24 hourly values	Error, (12) − (24)
	ft.	ft.	ft.
S $\begin{cases} A_1 = \\ B_1 = \end{cases}$	+0·0175 +0·0223	+0·0185 +0·0216	−0·0010 +0·0007
M $\begin{cases} A_1 = \\ B_1 = \end{cases}$	+0·0120 −0·0168	+0·0059 −0·0173	+0·0061 +0·0005
K $\begin{cases} A_1 = \\ B_1 = \end{cases}$	+0·3815 +0·1398	+0·3847 +0·1396	−0·0032 +0·0002
O $\begin{cases} A_1 = \\ B_1 = \end{cases}$	−0·0818 +0·1335	−0·0729 +0·1386	−0·0089 −0·0051
P $\begin{cases} A_1 = \\ B_1 = \end{cases}$	−0·0167 −0·1287	−0·0178 −0·1280	+0·0011 −0·0007
J $\begin{cases} A_1 = \\ B_1 = \end{cases}$	−0·0193 +0·0315	−0·0167 +0·0347	−0·0026 −0·0032
Q $\begin{cases} A_1 = \\ B_1 = \end{cases}$	+0·0140 −0·0170	+0·0136 −0·0194	+0·0004 +0·0024

The mean discrepancy in the case of the semi-diurnal tides is 0·0022 ft., and the greatest is +0·0068; in the case of the diurnal tides the mean discrepancy is 0·0026 ft., and the greatest is 0·0089.

In tidal work results derived from different years of observation differ far more than do these two sets of results, and hence the analysis of 12 two-hourly values for diurnal and semi-diurnal tides gives adequate results.

I find that this abbreviation does not give satisfactory results for quater-diurnal tides, and the sixth harmonic is not derivable from 12 values. Therefore, when these tides are to be evaluated the 24 hourly values must be used.

It will still be necessary to write all the 24 hourly heights on each computing strip, but when the strips are put into any one of the arrangements, except where quater-diurnal tides are required, we need only add up the columns 0, 2, 4, ..., 22, and may omit the columns 1, 3, ..., 23.

§ 10. *On a trial of the proposed method of reduction.*

As already mentioned, I have the tidal reductions for one year (beginning April 19, 1880) for Port Blair, Andaman Islands. I am thus able to make a comparison between the results of the old method and of the new. The computation was, in large part, done for me by Mr Wright.

Port Blair. 1880–81.

	I Indian calculation	N New method	I − N (height)	I − N (phase)
	ft.	ft.	ft.	
$A_0 =$	4·792	4·795	−0·003	
Sa { H =	·299	·299	000	
Sa { κ =	163°	162°	...	+ 1°
Ssa { H =	·106	·111	− ·005	
Ssa { κ =	165°	164°	...	+ 1°
T { H =	·099*	·094		
T { κ =	313°*	339°		
R { H =	·020*	·004		
R { κ =	326°*	312°		
S_1 { H =	·028	·026	+ ·002	
S_1 { κ =	49°	53°	...	− 4°
S_2 { H =	·966	·973	− ·007	
S_2 { κ =	316°	315°	...	+ 1°
S_4 { H =	·003	·003	000	
S_4 { κ =	107°	105°	...	+ 2°
K_1 { H =	·403	·401	+ ·002	
K_1 { κ =	326°	326°	...	0
K_2 { H =	·286	·268	+ ·018	
K_2 { κ =	314°	311°	...	+ 3°
P { H =	·130	·139	− ·009	
P { κ =	324°	323°	...	+ 1°
M_1 { H =	·014	·013	+ ·001	
M_1 { κ =	23°	34°	...	−11°
M_2 { H =	2·042	2·043	− ·001	
M_2 { κ =	279°	279°	...	0
M_3 { H =	·004	·004	000	
M_3 { κ =	20°	54°	...	−34°
M_4 { H =	·003	·006	− ·003	
M_4 { κ =	167°	264°	...	−97°
M_6 { H =	·004	·005	− ·001	
M_6 { κ =	342°	315°	...	+27°
(24 values) Q { H =	·023	·023	000	
(24 values) Q { κ =	236°	233°	...	+ 3°
(12 values) Q { H =	·023	·022	+ ·001	
(12 values) Q { κ =	236°	234°	...	+ 2°

* These are derived from 1880–82.

Tides of Long Period.

	I Indian calculation	N New method	I − N (height)	I − N (phase)
	ft.	ft.	ft.	
MSf { H =	·045	·019	+ ·026	
MSf { κ =	163°	168°	...	−5°
Mf { H =	·056	·056	000	
Mf { κ =	356°	356°	...	0
Mm { H =	·016	·020	− ·004	
Mm { κ =	12°	13°	...	−1°

It appeared sufficient to evaluate the tides of the S series and those allied with them, the tides of the M series, and the tide Q; also the tides of long period MSf, Mf, Mm.

The S series test the new process of harmonic analysis of monthly harmonic components for annual and semi-annual inequalities. I chose M because it is the most important tide, and Q because it puts the proposed method of grouping to a severe test, and is very small in amplitude.

In the Q time scale the day is $26^h\ 52^m$ of mean solar time, from which it follows that one of the 24 mean solar hourly observations may fall as much as $2^h\ 0^m$ away from the exact Q hour to which it is reputed to belong. Thus the hourly observations are arranged in wide groups round the Q hours, and the hypothesis involved in the method is put to a severe strain.

Lastly, the results for tides of long period test my proposed abridgment.

It will be seen in the table on p. 248 that the two methods give results in close agreement. There is, however, a sensible discrepancy in the K_2 tide, but in this case I am inclined to accept the new value as better than the old one. This tide is governed by sidereal time, which differs but little from mean solar time. Hence, in the Indian method of grouping, considerable errors of incidence of the S hours in the K time scale prevail for many days together, and the method seems of doubtful propriety. The same is true of the P tide, and here also the two methods give somewhat different results.

The accuracy with which the very small Q tide comes out, whether from 24 values or only from 12, is surprising, and may perhaps be, to some extent, due to accident. It shows, however, that the present method may be safely applied, even when the special time scale differs considerably from mean solar time.

The results for the tides of long period are quite as close to the old values as could be expected.

§ 11. *A comparison of the work involved in the new and old methods of reduction.*

It has been usual in the Indian reductions to use three digits in expressing the height of water, and there have been 15 series, or even more. Now $3 \times 24 \times 365 \times 15$ is 394,000; hence the computer has had to write that number of figures in reducing a year of observation. This does not include the evaluation of the annual and semi-annual tides, so that we may say that there have been about 400,000 figures to write.

I propose to express the heights by two digits, and they only have to be written once. Thus, in the present plan, the number of figures to write is $2 \times 24 \times 365$, or 17,500. Thus the writing of 382,000 figures is saved.

In the old method the computer had to add together all the digits written, say, 394,000 additions of digit to digit.

I propose to use 24 hourly values in three series, viz., S, M, and MS, and 12 two-hourly values in eight others. Therefore, the number of additions will be $3 \times 2 \times 24 \times 365 + 8 \times 2 \times 12 \times 365$ or 123,000. Thus 270,000 additions are saved.

We may say that formerly there were about 800,000 operations (writing and addition), and that in the present method there will be about 140,000. This estimate does not include a saving of several thousands of operations in obtaining the tides of long period. I am therefore within the mark when I claim that the work formerly bestowed on one year of observation will now reduce at least five years.

It has been found that the manufacture of my computing strips of xylonite is rather expensive, but as it formerly cost in England rather more than £20 to reduce a year of observation, the cost of the apparatus will be covered by the saving in the reduction of a single year, and it will serve for any length of time.

§ 12. *On the completion of the record for short gaps and long gaps.*

In any long series of tidal observations there are usually some breaks in the record in consequence of the stoppage of the clock of the tide gauge, or from some other cause. Now the process of elimination by grouping depends essentially on the completeness of the record, and it is therefore necessary to fill in blanks by interpolation.

Such interpolation has not been usual in the operations of the Indian Survey, and it might be thought that the complete omission of the missing entries is the proper course to take; but it is easily shown that this treatment is exactly equivalent to the assumption that the water remained stagnant at mean sea level during the whole time of stoppage of the gauge. It is obvious, therefore, that any conjectural values are better than none.

The process by which it is proposed to interpolate is best shown by an example.

At Port Blair (beginning April 19, 1880) the column of 6^h from days 99 to 112 gives the heights shown in the first column of the table below. I suppose that the tide gauge broke down on day 103, and only came into action again on day 110*. There was really no breakdown, and the actuality during the supposed hiatus is shown in the last column but one.

Now if we look back about a month we find that the water stood about the same height at the same hour of the day (viz., 6^h). Then the "previous record" (which is complete) beginning at 69^d is entered in the next column. Similarly a "subsequent record" is found about a month later, and is entered in a third column. The mean of the previous and subsequent records is then taken as giving the values to be interpolated.

Table of Interpolation.

Defective record		Previous record		Subsequent record		Mean of previous and subsequent		Actuality	Error
Day	6^h	Day	6^h	Day	6^h				
99	2·54	69	2·02	129	2·56	2·29			
100	3·13	70	2·83	130	3·07	2·95			
101	3·86	71	3·70	131	3·70	3·70			
102	4·40	72	4·55	132	4·27	4·41			
103	...	73	5·10	133	4·88	4·99	Interpolation	4·83	+0·16
104	...	74	5·60	134	5·27	5·44		5·24	+0·20
105	...	75	5·72	135	5·35	5·54		5·39	+0·15
106	...	76	5·67	136	5·28	5·48		5·18	+0·30
107	...	77	5·59	137	4·92	5·26		4·90	+0·36
108	...	78	5·04	138	4·44	4·74		4·53	+0·21
109	...	79	4·62	139	3·57	4·10		4·05	+0·05
110	3·32	80	3·81	140	2·95	3·38			
111	2·64	81	3·26	141	2·50	2·38			
112	2·17	82	2·73	142	1·99	2·36			

The last two columns contain a comparison between the interpolation and what in the present case we know to have been actuality. There is a mean error of 0·20 ft. Thus it is clear that a fair record may be interpolated even with so long a break as a week.

In this example I have only shown the interpolation for one column, but of course all the other twenty-three columns would really have to be treated similarly.

* The days are here numbered from 1, instead of from 0. This has been the usage in India hitherto.

I find by trial that the result would be a little improved by a graphical method, but that process is slightly more troublesome than the numerical one.

It may happen that the hiatus is too long for treatment in this way. I do not think it would be safe to treat much more than a fortnight by interpolation.

It has been shown in § 4 how the tides associated with S are to be treated where the record is deficient, and it remains to consider the other tides.

In § 7 are given the days with which we must stop in the analysis of an incomplete year, and this table affords us the means of treating a long hiatus in the observation.

We may in fact omit all the entries between any two of the numbers given in the table without seriously affecting the result.

Let us suppose, as an example, that the tide gauge broke down on day 210 and was only repaired and in operation again on day 226. Now 210 is 148 + 62, and 225 is 222 + 3.

Then we see by the table in § 7 that in finding the means for M, 2SM, MS, when the computing strips are written for the third time, we must remove strips 59, 60, 61 (which have numbers written on them) and may leave the remaining strips of that writing which are blank. When the strips are written for the fourth time strips 0, 1, 2, 3 will be blank, but we must remove strips 4 to 13 inclusive. When all the strips are used in a complete year there are 369, and this is the divisor used in obtaining the harmonic constants, but when there is this supposed hiatus we do not use 15 strips of the third writing and 14 strips of the fourth writing, so that the divisor will be 340.

Again, when we are evaluating O in the third writing, strips 57, 58, 59, 60, 61 must be removed, and in the fourth writing strips 4 to 9 inclusive. In a complete year the divisor is 369, but we now do not use 17 strips of the third writing and 10 of the fourth writing, so that the divisor becomes 342.

Again, in evaluating N, L, J, Q, in the third writing we remove strips 45 to 61 inclusive, and in the fourth writing strips 4 to 25 inclusive. The divisor is reduced from 358 to 303.

Lastly in evaluating ν, in the third writing strips 43 to 61 inclusive are removed, and in the fourth writing strips 4 to 31 inclusive. The divisor is reduced from 350 to 287.

Any hiatus, be it long or short, may be treated in this way, but it is clear that if it be short enough to treat by interpolation, it is best to adopt that method.

INSTRUCTIONS FOR USING THE COMPUTING APPARATUS.

The apparatus for the reduction of tidal observations, together with computation forms, can be purchased from the Cambridge Scientific Instrument Company at a price (as far as can be now foreseen) of about £8*.

In case of any insufficiency in the following instructions recourse must be taken to the preceding paper.

On the degree of accuracy requisite in the hourly heights.

It will usually be sufficient if the heights be measured to within one-tenth of a foot, and the decimal point may, of course, be omitted in computation.

This gives amply sufficient accuracy at a place where the semi-range of the principal lunar tide is 2 ft., and where spring range is from 6 ft. to 7 ft.

At some places with small tides a smaller unit might be necessary, and at others with very large tides a unit of 2 in., or of a fifth of a foot, might suffice.

Whatever unit of length be taken it is important, for the saving of work, and it is sufficient, that all or nearly all the heights should be expressed by two digits.

Completion of record.

If there is an accidental break in the record, it is very important that it should be completed according to the method shown in § 12, or by some other equivalent plan.

The computation forms are drawn up on the supposition that the year of observation is complete, but with proper alterations, which will now be indicated, they may be used in other cases.

In § 4 it is shown how to treat the tides of the S group when the observations have been subject to a long stoppage in the course of the year, and also when the observations extend over any period from six months to a year.

In § 5 it is shown how to treat the S group for a short period of observation.

If the stoppage be a long one, the method explained in § 12 must be adopted for all the other tides. The same section also shows the treatment for observations extending over any period, long or short, less than a year.

* [The price is, I believe, about £10. Each strip has to be carefully made by hand, and the demand for the apparatus is of course very small. (1907).]

Entries and summations.

The computing strips are intended to take writing in *pencil* or *liquid Indian ink,* but not in common ink.

They are to be cleaned with a damp cloth, and a little soda may be put in the water if they become greasy.

Lay the red S sheet on one drawing board and set up the strips with their ends abutting against the corresponding numbers. The strip numbered 60 is also to be put on the board.

Write the hourly heights for each day on the strip bearing the corresponding number, strip 0 for day 0, strip 1 for day 1, and so on up to strip 73 for day 73. The 24 hourly heights are to be written in the 24 divisions of each strip, beginning on the left with 0^h and ending on the right with 23^h.

Remove strip 60.

Sum the 24 columns formed by the divisional marks on consecutive sets of 30 strips. Thus, days 0 to 29 afford 24 sums; days 30 to 59 afford the second set of 24 sums; days 61 to 73 afford 24 sums, which are the beginning of a third 30, to be completed when the second set of 74 days shall have been written on the strips.

The numbers 0, 1, 2,..., 23, 0, 1,..., 23 at the head and foot of the guide sheet indicate the hours corresponding to the columns.

The sums of the columns on the board are to be entered in the corresponding columns of the form "Hourly sums of S series in twelve months."

Lay the red M guide sheet on the other drawing board, and transfer the strips from the first board to the new arrangement shown by the zigzag lines, strip 60 being now reintroduced.

There will now be 48 columns (more exactly 47, since one of the columns will be found to have nothing in it), numbered at top and bottom 0, 1,..., 23, 0, 1,..., 23. Each of the 48 columns is to be summed from bottom to top (not as for S in groups of 30), and the sums are to be entered in the form "Sums of series M." The 24 sums which come from the left half of the board will be entered in the row marked "red left," those from the right in the row marked "red right."

Lay the red N sheet on the other board, and transfer the strips.

In accordance with § 9, it will now usually suffice to sum only the columns appertaining to the even hours 0, 2, 4,..., 22; as these hours are repeated twice, there will now be 24 columns to sum.

The sums are then to be entered on the form "Sums of series N," in the alternate columns. The complete form is provided, so that all the 24 hourly values may be used if it be thought desirable, but this labour seems unnecessary, at least in a long series of observations.

Lay the successive red sheets on the vacant board, transfer, sum, and enter, until all the red sheets are exhausted.

In the case of S, M, MS the sums of all the columns are necessary, but in the other eight arrangements only the sums of the alternate columns, those of the even hours, are usually necessary. For a short series of observations it may be best to use all the columns, but in this case it will certainly not be worth while to attempt the evaluation of ν, J, Q, μ, 2SM, which are all small in amount.

If the tides of long period MSf, Mf, Mm are required, the 24 numbers written on each strip must be added together, and the sum entered in the form "Long period tides—daily sums."

Clean the strips.

In exactly the same way work through the next 74 days, from 74^d to $74^d + 73^d$, with yellow guide sheets. Then clean the strips, and take another 74 days with green guide sheets, and so on with the blue and violet.

In the last (violet) set attention must be paid to the rules as to the places where the analysis is to stop in each arrangement.

If the year of observation is so incomplete that the hiatus cannot be made good by interpolation, or if the series does not run over the complete year, the series must stop with one of the days specified in the table at the end of § 7, and a note must be made of the number of days used in each series.

The strips marked for omission on the violet sheets, or those selected for omission under the rules of § 7, may be hidden by a sheet of paper when the summations are being made.

The additions of S, M, MS may be verified by proving that the grand total of all the numbers (inclusive of omitted strips in S) written in each of the sets of 74 days is the same in whatever way they are arranged. Thus, the sum of the 48 columns should be equal to the sum of the daily sums. An incomplete verification in the other arrangements, when only half the columns are summed, is found by showing that the sum of all the hourly sums of each 74 days is nearly equal to half the grand total of all the numbers written in that period of 74 days.

When the guide sheets become worn with many pin pricks they may easily be patched with adhesive paper. There seems no reason why this patching should not go on almost indefinitely*.

* It is possible that it may be desired to evaluate the tides OO and 2N, for which no guide sheets are provided. I therefore give instructions for the preparation of guide sheets for these cases. They will be understood by any one who has the set of guide sheets before him. With the instructions given below, the computer might indeed set up the strips without a guide sheet.

I describe the staircase as descending from left to right or from right to left, and I define a short step as being one space down and one space to the left or right, as the case may be, and a

Hourly sums and harmonic analysis.

Complete the summations in the forms for hourly sums, and copy into the forms for harmonic analysis. In this copying it will generally suffice if the last figure in the hourly sums be omitted; for example, if the observations are entered to the nearest tenth of a foot the hourly sums will be given in the same unit, and it will suffice if the hourly sums analysed be written to the nearest foot.

There are 12 analyses (one for each month of 30 days) for the hourly sums in S, and one analysis for each of the other 10 arrangements. All the forms are provided with spaces for 24 hourly sums, but in the eight series N, L, ν, O, Q, J, μ, 2SM, where only 12 values will commonly be used, the entries will only be made on the alternate rows of 0^h, 2^h, ..., 22^h. In these cases the divisor 12, which occurs in the penultimate stage of finding the A's and B's, must be replaced by 6.

The large divisors (viz., 369 for M, μ, 2SM, MS; 369 for O; 358 for N, L, J, Q; and 350 for ν) represent the number of days under reduction, and must be altered appropriately (see table, § 7) if there be long gaps in the observations, or if the year be incomplete, or if the series be a short one.

If some one of the monthly analyses of S is deficient the missing 𝔄's and 𝔅's are to be made good by interpolation.

---

long step as one space down and two to the left or right, as the case may be. When I say, for example, that a short follows 2, I mean that 2 to 3 is a short step. The first mark on each sheet is specified by its incidence in the row of hours at the top.

OO: descending from left to right.

The sequence is long several times repeated and then short.

Red; 0 between 0^h and 1^h; shorts follow 2, 7, 13, 19, 24, 30, 36, 41, 47, 53, 58, 64, 69.

Yellow; 0 between 15^h and 16^h; shorts follow 1, 7, 12, 18, 24, 29, 35, 41, 46, 52, 57, 63, 69.

Green; 0 between 6^h and 7^h; shorts follow 0, 6, 12, 17, 23, 29, 34, 40, 45, 51, 57, 62, 68.

Blue; 0 between 21^h and 22^h; shorts follow 0, 5, 11, 17, 22, 28, 33, 39, 45, 50, 56, 62, 67.

Violet; 0 between 11^h and 12^h; shorts follow 5, 10, 16, 21, 27, 33, 38, 44, 50, 55, 61, 67, 72.

The last strip used for a year is 72.

2N: descending from right to left.

The sequence is long, long, short, long, long, short, and at intervals three longs and a short.

Red; 0 between 22^h and 23^h; shorts follow 1, 4, ..., 16; 20, 23, ..., 35; 39, 42, ..., 57; 61, 64, ..., 70.

Yellow; 0 between 18^h and 19^h; shorts follow 2; 6, 9, ..., 21; 25, 28, ..., 40; 44, 47, ..., 59; 63, 66, ..., 72.

Green; 0 between 13^h and 14^h; shorts follow 1, 4; 8, 11, ..., 26; 30, 33, ..., 45; 49, 52, ..., 64; 68, 71.

Blue; 0 between 8^h and 9^h; shorts follow 0, 3, 6, 9; 13, 16, ..., 28; 32, 35, ..., 50; 54, 57, ..., 69; 73.

Violet; 0 between 4^h and 5^h; shorts follow 2, 5, ..., 14; 18, 21, ..., 33; 37, 40, ..., 52; 56, 59, ..., 71.

The last strip used for a year is 61.

It is then necessary to analyse the monthly values of the 𝔄's and 𝔅's derived from the 12 analyses of S. We thus obtain A_0, A_1, B_1, A_2, B_2, C_0, c_0. C_1, D_1, c_1, d_1, E_0, e_0, E_1, F_1, e_1, f_1, E_2, F_2, e_2, f_2. The rules for these analyses when the year is incomplete are given in § 4, and the computation forms only apply to the case of the complete year.

Astronomical data and final reduction.

Determine from the *Nautical Almanac* and Major Baird's *Manual of Tidal Observations** the astronomical data at 0^h local M.T. on day 0, and proceed according to the form to find the initial arguments and factors for reduction. The astronomical data are then to be used in the forms for final reduction.

We have generally $B = R \sin \zeta$, $A = R \cos \zeta$; the forms are arranged so that colog A is to be added to log B to find log tan ζ, and thence ζ. If ζ lies between $-45°$ and $45°$ or between $135°$ and $225°$, log sec ζ is added to log A to find R; if ζ lies between $45°$ and $135°$ or between $225°$ and $315°$, log cosec ζ is added to log B to find R. Accordingly the computation form has log   sec ζ, room being left for the syllable "co" if necessary; underneath this is written log   , and the computer will insert A or B as the case may be.

There is usually required also a numerical factor 1/f or 𝔉, or both; the logarithms of the 1/f are found amongst the astronomical data, and the logarithms of the constant 𝔉's are printed in the forms in their proper places.

The treatment of the 21 harmonic components derived from the harmonic analysis of the five 𝔄's and 𝔅's is shown in the forms.

The tides of long period.

The processes involved in the evaluation of these tides are sufficiently shown in the forms.

* Taylor and Francis, London, 1886, price 7*s*. 6*d*.

7.

ON TIDAL PREDICTION.

[*Philosophical Transactions of the Royal Society of London*, CLXXXII. (1891), A, pp. 159—229.]

TABLE OF CONTENTS.

		PAGE
	Introduction	258
I.	Analysis	261
II.	Computation	290
III.	Comparison and Discussion	323

INTRODUCTION.

AT most places on the North Atlantic the prediction of high and low water is fairly easy, because there is hardly any diurnal tide. This abnormality makes it sufficient to have a table of the mean fortnightly inequality in the height and interval after lunar transit, supplemented by tables of corrections for the declinations and parallaxes of the disturbing bodies. But when there is a large diurnal inequality, as is commonly the case in other seas, the heights and intervals after the upper and lower lunar transits are widely different; the two halves of each lunation differ much in their characters, and the season of the year has great influence. Thus simple tables, such as are applicable in the absence of diurnal tide, are of no avail.

The tidal information supplied by the Admiralty for such places, consists of rough means of the rise and interval at spring and neap, modified by the important warning that the tide is affected by diurnal inequality. Information of this kind affords scarcely any indication of the time and height of high and low water on any given day, and must, I should think, be almost useless.

This is the present state of affairs at many ports of some importance, but at others a specially constructed tide-table for each day of each year is published in advance. A special tide-table is clearly the best sort of information for the sailor, but the heavy expense of prediction and publication is rarely incurred, except at ports of first rate commercial importance.

There is not, to my knowledge, any arithmetical method in use of computing a special tide-table, which does not involve much work and expense. The admirable tide-predicting instrument of the Indian Government renders the prediction comparatively cheap, yet the instrument can hardly be deemed available for the whole world, and the cost of publication is so considerable, that the instrument cannot, or at least will not, be used for many minor ports at remote places. It is not impossible, too, that national pride may deter the naval authorities of other nations from sending to London for their predictions*.

The object then, of the present paper, is to show how a general tide-table, applicable for all time, may be given in such a form that any one, with an elementary knowledge of the *Nautical Almanac*, may, in a few minutes, compute two or three tides for the days on which they are required. The tables will also be such that a special tide-table for any year may be computed with comparatively little trouble.

Any tide-table necessarily depends on the tidal constants of the particular port for which it is designed, and it is here supposed that the constants are given in the harmonic system, and are derived from the reduction of tidal observations. Where the observation has been by tide-gauge, the process of reduction is that explained in the Report to the British Association for 1883, but where the observations were only taken at high and low water a different process becomes necessary. I have given, in a previous paper, a scheme of reduction in these cases†.

At ports not of first rate commercial importance, tidal observation has rarely been by tide-gauge, and thus it is exactly at those ports where the method of this paper may prove most useful, that we are deprived of the ordinary methods of harmonic analysis. On this account I regard my previous paper as preliminary to the present one, although the two are logically independent of one another.

In the harmonic method the complete analytical expression for the height of water at any time consists of the sum of a number of heights, each multiplied by the cosine of an angle. Each of these angles or arguments involves some or all of the mean longitudes of moon, sun, lunar perigee, and solar perigee; there are, also, certain corrections depending on the longitude of the moon's nodes. The variability of height of water depends principally on the mean longitudes of moon and sun, and, to a subordinate degree, on

* The instrument may be used, I believe, on the payment of certain fees.

† "On the Harmonic Analysis of Tidal Observations of High and Low Water," *Roy. Soc. Proc.*, Vol. XLVIII., 1890, p. 278. [Paper 5, in this volume.]

the longitudes of lunar perigee and node—for the solar perigee is sensibly fixed. There are, therefore, two principal variables and two subordinate ones besides the time. This statement suggests the construction of a table of double entry for the variability of water due to the principal variables, and of correctional tables for the subordinate ones; and this is the plan developed below.

It would not be appropriate to retain the mean longitudes of moon and sun as principal variables, nor the mean longitude of lunar perigee as one of the subordinate variables. In the final table the principal variables are the time of moon's transit, and the time of year, or strictly speaking the sun's true longitude. As to the subordinate variables, the moon's parallax is taken as one, whilst the longitude of the node is retained as the other.

Throughout the first part of the analytical development the moon's true longitude, measured in the orbit, is taken as one of the principal variables. Transition is then made to the time of transit of a fictitious satellite whose right ascension is equal to the moon's true longitude, and the final transition is to the transits of the real moon. This transposition of variables has necessitated a complete development of the tide-generating forces from the beginning, and thus the analysis of the paper is almost complete in itself, although reference is necessarily made to the harmonic method.

Such a piece of work as the present can only be deemed complete when an example has been worked out to test the accuracy of the tidal prediction, and when rules have been drawn up for the arithmetical processes, forming a complete code of instructions to the computer. The example below is intended to carry out these requirements.

I chose the port of Aden for the example, because its tides are more complex and apparently irregular than those of any other place, which, as far as I know, has been thoroughly treated. The tidal constants for Aden are well determined, and the annual tide-tables of the Indian Government afford the means of comparison between my predictions and those of the tide-predicting instrument.

The arithmetic of the example has been long, and the plan of marshalling the work has been rearranged many times. An ordinary computer is said to work best when he is ignorant of the meaning of his work, but in this kind of tentative work a satisfactory arrangement cannot be attained without a full comprehension of the reason of the method. I was, therefore, very fortunate in securing the enthusiastic assistance of Mr J. W. F. Allnutt, and I owe him my warm thanks for the laborious computations he has carried out. After computing fully half of the original table, he made a comparison for the whole of 1889 between our predictions and those of the Indian Government.

I had hoped that tables, less elaborate than those exhibited below, might have sufficed. It appeared, however, that the changes during the lunation and with the time of year, which affect the height and interval, are so abrupt and so great, that short tables would give very inaccurate results, unless used with elaborate interpolation. We can clearly never expect sailors to use tables of double entry, in which interpolation to second and third differences is required; and indeed any interpolation is objectionable. It is proposed, therefore, that the tables shall be made so full that interpolation will be unnecessary. This plan, of course, throws the whole of the interpolation on to the computer, and although extensive interpolation—even when done graphically—is tedious, yet it is obviously best to have it done once for all, rather than piecemeal by the user of the tables.

The first part of the paper gives the analytical development, the second contains the numerical example and rules of computation, and in the third I give some account of the comparison with other predictions and with actuality and suggestions for abridgements in cases where less accuracy shall be thought sufficient.

The analytical formulæ of the first part are much scattered, and it may not always be easy to see whither they tend; but the second part virtually contains a summary of the first, and a comparison between these two parts will show both the reasons for the analysis and those for the arithmetical processes.

I have tried to make the instructions to the computer so complete that he need not be troubled with the reason for his operations; but if there is, through inadvertence, any incompleteness, reference must be made to the analytical development for interpretation and instruction.

Part I. Analysis.

§ 1. *Development of the Tide-Generating Potential.*

Let A, B, C (fig. 1) be axes fixed in the earth, AB being the equator and C the north pole.

Let r, ρ be respectively the radius-vectors of M the moon, and of P any other point. Let M_1, M_2, M_3 be the direction cosines of the moon, and ζ_1, ζ_2, ζ_3 those of P, both referred to the axes A, B, C.

Then by the usual theory the potential V, of the second order of harmonics, is

$$V = \tfrac{3}{2}\frac{\mu M}{r^3}\rho^2(\cos^2 \mathrm{PM} - \tfrac{1}{3}) \quad \text{......................(1)}$$

where M is the moon's mass, and μ the attractional constant.

But since $\cos \text{PM} = \zeta_1 M_1 + \zeta_2 M_2 + \zeta_3 M_3$

$$\cos^2 \text{PM} - \tfrac{1}{3} = 2\zeta_1\zeta_2 \,.\, M_1 M_2 + 2 \,.\, \frac{\zeta_1^2 - \zeta_2^2}{2} \,.\, \frac{M_1^2 - M_2^2}{2} + 2\zeta_2\zeta_3 \,.\, M_2 M_3$$
$$+ 2\zeta_1\zeta_3 \,.\, M_1 M_3 + \tfrac{3}{2}(\tfrac{1}{3} - \zeta_3^2)(\tfrac{1}{3} - M_3^2) \quad \text{......(2)}$$

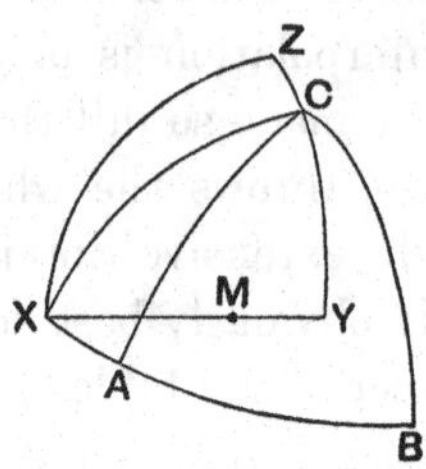

Fig. 1.

Now let X, Y, Z, fig. 1, be a second set of rectangular axes, fixed in space; let M be the projection of the moon in her orbit XY. Let $I = \text{ZC}$, the obliquity of the lunar orbit to the equator; let $\chi = \text{AX} = \text{BCY}$, and $\psi = \text{MX}$, the moon's longitude measured in her orbit from the point X.

From the figure it is clear that

$$\left.\begin{aligned} M_1 &= \cos\psi \cos\chi + \sin\psi \sin\chi \cos I \\ &= \cos^2 \tfrac{1}{2}I \cos(\chi - \psi) + \sin^2 \tfrac{1}{2}I \cos(\chi + \psi) \\ M_2 &= -\cos\psi \sin\chi + \sin\psi \cos\chi \cos I \\ &= -\cos^2 \tfrac{1}{2}I \sin(\chi - \psi) - \sin^2 \tfrac{1}{2}I \sin(\chi + \psi) \\ M_3 &= \sin I \sin\psi \end{aligned}\right\} \quad \text{............(3)}$$

Hence from (3)

$$\left.\begin{aligned} M_1^2 - M_2^2 &= \cos^4 \tfrac{1}{2}I \cos 2(\chi - \psi) \\ &\qquad + \tfrac{1}{2}\sin^2 I \cos 2\chi + \sin^4 \tfrac{1}{2}I \cos 2(\chi + \psi) \\ -2M_1 M_2 &= \cos^4 \tfrac{1}{2}I \sin 2(\chi - \psi) \\ &\qquad + \tfrac{1}{2}\sin^2 I \sin 2\chi + \sin^4 \tfrac{1}{2}I \sin 2(\chi + \psi) \\ M_2 M_3 &= -\sin \tfrac{1}{2}I \cos^3 \tfrac{1}{2}I \cos(\chi - 2\psi) \\ &\qquad + \tfrac{1}{2}\sin I \cos I \cos\chi + \sin^3 \tfrac{1}{2}I \cos \tfrac{1}{2}I \cos(\chi + 2\psi) \\ M_1 M_3 &= -\sin \tfrac{1}{2}I \cos^3 \tfrac{1}{2}I \cos(\chi - 2\psi) \\ &\qquad + \tfrac{1}{2}\sin I \cos I \sin\chi + \sin^3 \tfrac{1}{2}I \cos \tfrac{1}{2}I \sin(\chi + 2\psi) \\ \tfrac{1}{3} - M_3^2 &= \tfrac{1}{3} - \tfrac{1}{2}\sin^2 I + \tfrac{1}{2}\sin^2 I \cos 2\psi \end{aligned}\right\} \quad \text{...(4)}$$

In order to simplify the result of the substitutions from (4) into (2) and (1), the axes fixed in the earth may be taken as follows:

The axis A on the equator in the meridian of the place where the tides are observed, B 90° east of A, and C as already stated at the north pole.

Then if λ be the latitude of the place of observation

$$\zeta_1 = \cos\lambda, \qquad \zeta_2 = 0, \qquad \zeta_3 = \sin\lambda$$

Hence (2) becomes

$$\cos^2 \mathrm{PM} - \tfrac{1}{3} = \tfrac{1}{2}(M_1^2 - M_2^2)\cos^2\lambda + M_1 M_3 \sin 2\lambda + \tfrac{3}{2}(\tfrac{1}{3} - M_3^2)(\tfrac{1}{3} - \sin^2\lambda) \ldots (5)$$

We shall therefore only require the first and the last two of (4), and these may be considerably simplified.

The angle I ranges between 23° 27′ ± 5° 9′; hence the term in $M_1^2 - M_2^2$ which involves $\sin^4 \tfrac{1}{2}I$ is negligible, and similarly that which involves $\sin^3 \tfrac{1}{2}I \cos \tfrac{1}{2}I$ in $M_1 M_3$ may be omitted.

Further, it is only necessary to develop V in as far as it is variable; now as I ranges from 18° 18′ to 28° 36′, and back again in the course of 19 years, $\sin^2 I$ oscillates slightly in value about a mean. The "tides of long period" corresponding to the term $\tfrac{1}{2}\sin^2 I \cos 2\psi$ in $\tfrac{1}{3} - M_3^2$, are, although sensible, so small that we shall take no account of them; *à fortiori* the variability of $\tfrac{1}{3} - \tfrac{1}{2}\sin^2 I$ is negligible. It would be easy, however, to take account of these terms if it were desirable to do so.

We need therefore only pay attention to the first and fourth of (4), and the last term in each may be omitted.

In fig. 2 let M be the moon in her orbit, and let A be the axis A of fig. 1 fixed in the earth.

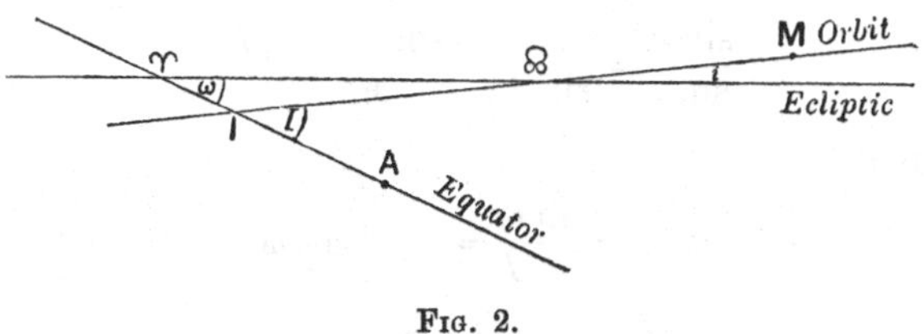

FIG. 2.

Then let ν be the right ascension of the point I, and ξ its longitude measured in the moon's orbit.

Let ω be the obliquity of the ecliptic, i the inclination of the moon's orbit, ☊ the longitude of the moon's node.

Let t be the mean solar hour angle at the place and time of observation, h the sun's mean longitude, and l the moon's longitude measured in the orbit.

Then, by the definition of χ and ψ, we have

$$\chi = \text{IA}, \qquad \psi = \text{IM}$$

Since $\nu = \Upsilon\text{I}$, $\xi = \Upsilon\Omega - \Omega\text{I}$, $\Upsilon\Omega = \Omega$, and $t+h$ the sidereal hour angle,

$$\chi = t + h - \nu, \qquad \psi = l - \xi$$

Thus to the degree of approximation adopted, (4) becomes

$$\left.\begin{aligned} M_1^2 - M_2^2 &= \cos^4 \tfrac{1}{2}I \cos 2(t+h-l-\nu+\xi) + \tfrac{1}{2}\sin^2 I \cos 2(t+h-\nu) \\ M_1M_3 &= \cos^3 \tfrac{1}{2}I \sin \tfrac{1}{2}I \cos (t+h-2l+\tfrac{1}{2}\pi-\nu+2\xi) \\ &\quad + \tfrac{1}{2}\sin I \cos I \cos (t+h-\tfrac{1}{2}\pi-\nu) \\ \tfrac{1}{3} - M_3^2 &= 0 \end{aligned}\right\}\ldots(6)$$

We must now obtain approximate formulæ to express the functions of I, ν, and ξ in terms of ω, i, and Ω. The inclination i may be treated as so small that its square may be neglected.

From fig. 2 we have

$$\cos I = \cos i \cos \omega - \sin i \sin \omega \cos \Omega$$

whence

$$\left.\begin{aligned} \cos I &= \cos\omega - i \sin\omega \cos\Omega \\ \cos^2 \tfrac{1}{2}I &= \cos^2 \tfrac{1}{2}\omega\,(1 - i\tan\tfrac{1}{2}\omega\cos\Omega) \\ \sin^2 \tfrac{1}{2}I &= \sin^2 \tfrac{1}{2}\omega\,(1 + i\cot\tfrac{1}{2}\omega\cos\Omega) \\ \cos^4 \tfrac{1}{2}I &= \cos^4 \tfrac{1}{2}\omega\,(1 - 2i\tan\tfrac{1}{2}\omega\cos\Omega) \\ \sin^2 I &= \sin^2\omega\,(1 + 2i\cot\omega\cos\Omega) \\ \cos^3\tfrac{1}{2}I\sin\tfrac{1}{2}I &= \cos^3\tfrac{1}{2}\omega\sin\tfrac{1}{2}\omega\left(1 + i\frac{2\cos\omega - 1}{\sin\omega}\cos\Omega\right) \\ \sin I\cos I &= \sin\omega\cos\omega\,(1 + 2i\cot 2\omega\cos\Omega) \end{aligned}\right\}\ldots\ldots\ldots(7)$$

Again from the figure

$$\frac{\sin\nu}{\sin i} = \frac{\sin\Omega}{\sin I} = \frac{\sin(\Omega - \xi)}{\sin\omega}$$

Since ξ is small,

$$1 - \xi\cot\Omega = \frac{\sin\omega}{\sin I} = 1 - i\cot\omega\cos\Omega$$

$$\nu = \frac{\sin i \sin\Omega}{\sin I}$$

Therefore

$$\left.\begin{aligned} \nu &= i\frac{\sin\Omega}{\sin\omega} \\ \xi &= i\cot\omega\sin\Omega \\ \nu - \xi &= i\tan\tfrac{1}{2}\omega\sin\Omega \\ 2\xi - \nu &= i\frac{2\cos\omega - 1}{\sin\omega}\sin\Omega \end{aligned}\right\}\ldots\ldots\ldots\ldots\ldots\ldots\ldots(8)$$

Hence from (7) and (8)

$$\left.\begin{aligned}
\cos^4 \tfrac{1}{2}I \cos 2(\nu - \xi) &= \cos^4 \tfrac{1}{2}\omega (1 - 2i \tan \tfrac{1}{2}\omega \cos \Omega)\\
\cos^4 \tfrac{1}{2}I \sin 2(\nu - \xi) &= \cos^4 \tfrac{1}{2}\omega \,.\, 2i \tan \tfrac{1}{2}\omega \sin \Omega\\
\tfrac{1}{2}\sin^2 I \cos 2\nu &= \tfrac{1}{2}\sin^2 \omega (1 + 2i \cot \omega \cos \Omega)\\
\tfrac{1}{2}\sin^2 I \sin 2\nu &= \tfrac{1}{2}\sin^2 \omega \,.\, 2i \operatorname{cosec} \omega \sin \Omega\\
\cos^3 \tfrac{1}{2}I \sin \tfrac{1}{2}I \cos (2\xi - \nu) &= \cos^3 \tfrac{1}{2}\omega \sin \tfrac{1}{2}\omega\, [1 + i(2\cos\omega - 1)\operatorname{cosec}\omega \cos \Omega]\\
\cos^3 \tfrac{1}{2}I \sin \tfrac{1}{2}I \sin (2\xi - \nu) &= \cos^3 \tfrac{1}{2}\omega \sin \tfrac{1}{2}\omega \,.\, i(2\cos\omega - 1)\operatorname{cosec}\omega \sin \Omega\\
\tfrac{1}{2}\sin I \cos I \cos \nu &= \tfrac{1}{2}\sin\omega\cos\omega (1 + 2i \cot 2\omega \cos \Omega)\\
\tfrac{1}{2}\sin I \cos I \sin \nu &= \tfrac{1}{2}\sin\omega\cos\omega \,.\, i \operatorname{cosec}\omega \sin \Omega
\end{aligned}\right\} \quad \text{.........(9)}$$

Now $\omega = 23° \, 27'{\cdot}3$, $i = 5° \, 8'{\cdot}8$; whence

$$2i \tan \tfrac{1}{2}\omega = {\cdot}0372, \quad 2i \cot \omega = \frac{{\cdot}283}{{\cdot}683}, \quad 2i \operatorname{cosec}\omega = \frac{{\cdot}308}{{\cdot}683}, \quad i\,\frac{2\cos\omega - 1}{\sin\omega} = {\cdot}188$$

$$2i \cot 2\omega = \frac{{\cdot}115}{{\cdot}683}, \qquad i \operatorname{cosec}\omega = \frac{{\cdot}154}{{\cdot}683}$$

Four of these are written with ·683 in the denominator for reasons given below. Then from (9) and (6) we have

$$\left.\begin{aligned}
M_1^2 - M_2^2 = {} & \cos^4 \tfrac{1}{2}\omega \cos 2(t + h - l) + \tfrac{1}{2}\sin^2\omega \cos 2(t + h)\\
& + \left\{- \cos^4 \tfrac{1}{2}\omega \,.\, {\cdot}0372 \cos 2(t + h - l) \right.\\
& \qquad \left. + \tfrac{1}{2}\sin^2\omega \,.\, \frac{{\cdot}283}{{\cdot}683}\cos 2(t + h)\right\} \cos \Omega\\
& + \left\{\cos^4 \tfrac{1}{2}\omega \,.\, {\cdot}0372 \sin 2(t + h - l) \right.\\
& \qquad \left. + \tfrac{1}{2}\sin^2\omega \,.\, \frac{{\cdot}308}{{\cdot}683}\sin 2(t + h)\right\} \sin \Omega\\
M_1 M_3 = {} & \cos^3 \tfrac{1}{2}\omega \sin \tfrac{1}{2}\omega \cos (t + h - 2l + \tfrac{1}{2}\pi)\\
& + \tfrac{1}{2}\sin\omega\cos\omega\cos(t + h - \tfrac{1}{2}\pi)\\
& + \left\{\cos^3 \tfrac{1}{2}\omega \sin \tfrac{1}{2}\omega \,.\, {\cdot}188 \cos (t + h - 2l + \tfrac{1}{2}\pi) \right.\\
& \qquad \left. + \tfrac{1}{2}\sin\omega\cos\omega \,.\, \frac{{\cdot}115}{{\cdot}683}\cos(t + h - \tfrac{1}{2}\pi)\right\} \cos \Omega\\
& + \left\{- \cos^3 \tfrac{1}{2}\omega \sin \tfrac{1}{2}\omega \,.\, {\cdot}188 \sin (t + h - 2l + \tfrac{1}{2}\pi) \right.\\
& \qquad \left. + \tfrac{1}{2}\sin\omega\cos\omega \,.\, \frac{{\cdot}154}{{\cdot}683}\sin(t + h - \tfrac{1}{2}\pi)\right\} \sin \Omega
\end{aligned}\right\} \text{...(10)}$$

Now let c be the moon's mean distance, and let $\tau = \tfrac{3}{2}\mu M/c^3$, and

$$1 + P = \frac{c^3}{r^3}$$

Then as far as concerns the moon

$$V = \tau\rho^2(1+P)\{\tfrac{1}{2}\cos^2\lambda\,(M_1^2 - M_2^2) + \sin 2\lambda\,.\,M_1M_3\}\quad\text{........(11)}$$

where $M_1^2 - M_2^2$, M_1M_3 are given by (10).

In writing down the corresponding functions for the sun, we shall write a subscript accent to all the symbols, and accordingly V for the sun is given by

$$V_, = \tau_,\rho_,^2(1+P_,)\{\tfrac{1}{2}\cos^2\lambda\,(M_{,1}^2 - M_{,2}^2) + \sin 2\lambda\,.\,M_{,1}M_{,3}\}\quad\text{......(12)}$$

where, by symmetry with (10)

$$\left.\begin{aligned} M_{,1}^2 - M_{,2}^2 &= \cos^4\tfrac{1}{2}\omega\cos 2(t+h-l_,) + \tfrac{1}{2}\sin^2\omega\cos 2(t+h) \\ M_{,1}M_{,3} &= \cos^3\tfrac{1}{2}\omega\sin\tfrac{1}{2}\omega\cos(t+h-2l_,+\tfrac{1}{2}\pi) \\ &\quad + \tfrac{1}{2}\sin\omega\cos\omega\cos(t+h-\tfrac{1}{2}\pi) \end{aligned}\right\}\text{...(13)}$$

The masses and mean distances of the moon and sun are such that

$$\frac{\tau}{\tau+\tau_,} = \cdot 683, \qquad \frac{\tau_,}{\tau+\tau_,} = \cdot 317\quad\text{..................(14)}$$

Strictly speaking, τ denotes $\tfrac{3}{2}\mu Mc^{-3}(1+\tfrac{3}{2}e^2)$, and $\tau_,$ denotes

$$\tfrac{3}{2}\mu M_, c_,^{-3}(1+\tfrac{3}{2}e_,^2)$$

where e, $e_,$ are the eccentricities of the lunar and solar orbits. A comparison with the Report on Tides to the British Association for 1883 [Paper 1], will show that, for the purpose in hand, it is correct to introduce these factors involving e^2 and $e_,^2$.

The whole potential may be collected from (10), (11), (12), (13), (14).

Since P is small the factor $(1+P)$ may be treated as unity in the small terms which involve $\cos\Omega$ and $\sin\Omega$, and we shall find it convenient to distinguish the several parts of V, denoting by V the principal part, by V_P the part involving P, by V_Ω the part involving Ω, and by $V_{P_,}$ the part involving $P_,$.

For brevity write

$$K = \rho^2\,.\,\tfrac{1}{2}\cos^2\lambda\,.\,\cos^4\tfrac{1}{2}\omega; \qquad K_{,,} = \rho^2\,.\,\tfrac{1}{2}\cos^2\lambda\,.\,\tfrac{1}{2}\sin^2\omega$$

$$K_0 = \rho^2\,.\,\sin 2\lambda\,.\,\cos^3\tfrac{1}{2}\omega\sin\tfrac{1}{2}\omega; \qquad K_, = \rho^2\,.\,\sin 2\lambda\,.\,\tfrac{1}{2}\sin\omega\cos\omega$$

Then remarking that

$$\tau(1+P) + \tau_,(1+P_,) = (\tau+\tau_,)(1+\cdot 683P)(1+\cdot 317P_,)$$

approximately, we have

$$\begin{aligned} V = {} & \tau K\cos 2(t+h-l) + (\tau+\tau_,)K_{,,}\cos 2(t+h) + \tau_, K\cos 2(t+h-l_,) \\ & + \tau K_0\cos(t+h-2l+\tfrac{1}{2}\pi) + (\tau+\tau_,)K_,\cos(t+h-\tfrac{1}{2}\pi) \\ & + \tau_, K_0\cos(t+h-2l_,+\tfrac{1}{2}\pi) \end{aligned}$$

$$V_P = P\{\tau K \cos 2(t+h-l) + \cdot 683(\tau+\tau_{,})K_{,,}\cos 2(t+h) + \tau K_o \cos(t+h-2l+\tfrac{1}{2}\pi) + \cdot 683(\tau+\tau_{,})K_{,}\cos(t+h-\tfrac{1}{2}\pi)\}$$

$$V_{☊} = \cos ☊\{-\tau K \cdot 0372\cos 2(t+h-l) + (\tau+\tau_{,})K_{,,}\cdot 283\cos 2(t+h) + \tau K_o \cdot 188\cos(t+h-2l+\tfrac{1}{2}\pi) + (\tau+\tau_{,})K_{,}\cdot 115\cos(t+h-\tfrac{1}{2}\pi)\}$$
$$+ \sin ☊\{+\tau K\cdot 0372\sin 2(t+h-l) + (\tau+\tau_{,})K_{,,}\cdot 308\sin 2(t+h) - \tau K_o\cdot 188\sin(t+h-2l+\tfrac{1}{2}\pi) + (\tau+\tau_{,})K_{,}\cdot 154\sin(t+h-\tfrac{1}{2}\pi)\}$$

$$V_{P_{,}} = P_{,}\{\cdot 317(\tau+\tau_{,})K_{,,}\cos 2(t+h) + \tau_{,}K\cos 2(t+h-l_{,}) + \cdot 317(\tau+\tau_{,})K_{,}\cos(t+h-\tfrac{1}{2}\pi) + \tau_{,}K_o\cos(t+h-2l_{,}+\tfrac{1}{2}\pi)\} \quad \ldots(15)$$

The whole tide-generating potential is then $V + V_P + V_{☊} + V_{P_{,}}$.

§ 2. *Formula for the Height of Tide.*

If we had been going to consider the complete expression for the tide in terms of a series of simple harmonic functions of the time, it would have been necessary to substitute for $l, l_{,}, P, P_{,}$ their values in terms of the mean longitudes s and h of the two bodies, and of the eccentricities $e, e_{,}$ of their orbits. When the potential is so expanded the principle of forced oscillations allows us to conclude that the oscillations of the sea will be of the same periods and types as the several terms of the potential, but with amplitudes and phases which can only be deduced from observation. The oscillations of the sea will not however be necessarily of the simple harmonic form, and accordingly "over-tides" of double and triple frequency have to be introduced in order to represent the motion according to Fourier's method.

This is the plan pursued in the "harmonic analysis of tidal observations," and each simple harmonic oscillation is known by an arbitrarily chosen initial letter.

It is found in fact, as is suggested by theory, that tides of approximately the same frequency or "speed" have amplitudes approximately proportional to their corresponding terms in the potential, and have their phases retarded by approximately the same amount.

The notation of harmonic analysis will be adopted here, because it is proposed to compute the tide-table from the harmonic constants.

The mean longitudes of the sun, moon, and lunar perigee are denoted by h, s, p, and their hourly changes by σ, η, ϖ (from σελήνη, ἥλιος, perigee); the angular velocity of the earth's rotation is γ (from γῆ), so that $\gamma - \eta$ is 15° per hour.

The eccentricities of the lunar and solar orbits are $e, e_{,}$.

In the harmonic system the tides are denominated by the arbitrarily chosen initials M_2, M_4, S_2, S_4, K_2, K_1, N, L, ν, λ, μ, T, R, O, P, Q, J, Sa, Ssa, and the semi-ranges of these tides will be here denoted by the suffixes m, $2m$, s, $2s$, n, l, ν, λ, μ, t, r, o, p, q, j, sa, ssa to the symbol H, and the retardations of phase by the same suffixes to the symbol κ. But the semi-ranges and retardations of the tides K_2, K_1 are denoted by $H_{,,}$, $\kappa_{,,}$, $H_{,}$, $\kappa_{,}$ instead of conforming to the rest of the notation*.

Then all the H's and κ's are the immediate results of harmonic analysis, and are supposed to be given when the construction of a tide-table is contemplated.

In the development (15) of V, the first term corresponds with the principal lunar tide M_2, its over-tide M_4, parts of the elliptic tides N, L, parts of the evectional tides ν, λ, and part of the variational tide μ.

The first term of V_P contributes the remainder of N, L, ν, λ, μ.

The second term of V corresponds with the luni-solar semi-diurnal tide K_2, and the second term of V_P corresponds with small inequalities in K_2, which are neglected in the harmonic analysis.

The third term of V corresponds with the principal solar tide S_2, its over-tide S_4, and parts of the elliptic tides T, R.

The fourth term of V corresponds with the lunar diurnal tide O, and parts of the elliptic tides M_1 and Q; the third term of V_P corresponds with another part of M_1 and the rest of Q.

The fifth term of V corresponds with the luni-solar diurnal tide K_1, and with part of the elliptic tides M_1 and J; its over-tide fuses with K_2. The fourth term of V_P corresponds with part of the elliptic tide M_1, and the rest of J.

The last term of V corresponds with the solar diurnal tide P, and with small inequalities neglected in the harmonic analysis.

The whole of $V_{☊}$ corresponds with those factors of augmentation and small alterations of phase, which are denoted by f and functions of ν and ξ in harmonic analysis.

$V_{P,}$ corresponds with the remainder of the solar elliptic tides T, R and with other small inequalities usually neglected in the harmonic analysis.

The semi-annual (Ssa) and annual (Sa) tides are due to meteorological causes, and have to be introduced after we have done with astronomical considerations, so that they do not enter through the V's.

Since the terms V, V_P, $V_{☊}$, $V_{P,}$ are approximately simple harmonics, it follows that each term will correspond to an approximately simple harmonic

* I adopted this originally for convenience in writing, and having got used to the notation hall retain it.

tide, with amplitude and phase the same as that given by harmonic analysis, and to its over-tides. If then we write

$$\Delta = 2t + 2h - 2l - \kappa_m \qquad (16)$$

the height of water corresponding to V is

$$\begin{aligned} h = {} & \mathrm{H}_m \cos \Delta + \mathrm{H}_{,,} \cos(\Delta + \kappa_m + 2l - \kappa_{,,}) + \mathrm{H}_s \cos(\Delta + \kappa_m + 2l - 2l_, - \kappa_s) \\ & + \mathrm{H}_o \cos[\tfrac{1}{2}(\Delta + \kappa_m) - l + \tfrac{1}{2}\pi - \kappa_o] + \mathrm{H}_, \cos[\tfrac{1}{2}(\Delta + \kappa_m) + l - \tfrac{1}{2}\pi - \kappa_,] \\ & + \mathrm{H}_p \cos[\tfrac{1}{2}(\Delta + \kappa_m) + l - 2l_, + \tfrac{1}{2}\pi - \kappa_p] + \mathrm{H}_{2m} \cos[2(\Delta + \kappa_m) - \kappa_{2m}] \\ & + \mathrm{H}_{2s} \cos[2(\Delta + \kappa_m) + 4l - 4l_, - \kappa_{2s}] \qquad (17) \end{aligned}$$

We shall, in the developments immediately following, leave $V_{P,}$ out of consideration, and shall resume the effects of the sun's parallax in § 11.

In order to represent, as far as possible, the elliptic tides, it is found to be necessary to introduce certain numerical factors α, β, and certain angles ${}_m\kappa$, ${}_o\kappa$, as shown in the next formula. It is also practically convenient that P should denote the value of c^3/r^3, not at the time corresponding to t, but some hours earlier.

Then the height of water corresponding to V_P is taken as

$$\begin{aligned} h_P = P\{ & \alpha \mathrm{H}_m \cos(\Delta + \kappa_m - {}_m\kappa) + {\cdot}683\mathrm{H}_{,,} \cos(\Delta + \kappa_m + 2l - \kappa_{,,}) \\ & + \beta \mathrm{H}_o \cos[\tfrac{1}{2}(\Delta + \kappa_m) - l + \tfrac{1}{2}\pi - {}_o\kappa] \\ & + {\cdot}683\mathrm{H}_, \cos[\tfrac{1}{2}(\Delta + \kappa_m) + l - \tfrac{1}{2}\pi - \kappa_,]\} \qquad (18) \end{aligned}$$

The height corresponding to V_{Ω} is

$$\begin{aligned} h_{\Omega} = \cos \Omega \, \{ & - \mathrm{H}_m {\cdot}0372 \cos \Delta + \mathrm{H}_{,,} {\cdot}283 \cos(\Delta + \kappa_m + 2l - \kappa_{,,}) \\ & + \mathrm{H}_o {\cdot}188 \cos[\tfrac{1}{2}(\Delta + \kappa_m) - l + \tfrac{1}{2}\pi - \kappa_o] \\ & + \mathrm{H}_, {\cdot}115 \cos[\tfrac{1}{2}(\Delta + \kappa_m) + l - \tfrac{1}{2}\pi - \kappa_,]\} \\ + \sin \Omega \, \{ & \mathrm{H}_m {\cdot}0372 \sin \Delta + \mathrm{H}_{,,} {\cdot}308 \sin(\Delta + \kappa_m + 2l - \kappa_{,,}) \\ & - \mathrm{H}_o {\cdot}188 \sin[\tfrac{1}{2}(\Delta + \kappa_m) - l + \tfrac{1}{2}\pi - \kappa_o] \\ & + \mathrm{H}_, {\cdot}154 \sin[\tfrac{1}{2}(\Delta + \kappa_m) + l - \tfrac{1}{2}\pi - \kappa_,]\} \qquad (19) \end{aligned}$$

§ 3. *The Elliptic Tides.*

The constants α, β, ${}_m\kappa$, ${}_o\kappa$, are insufficient to represent fully the harmonic tides N, L, ν, λ, μ, Q, J, and it would render the proposed method of computing a tide-table too cumbrous for practical use if additional constants were introduced. It is, therefore, necessary to adopt a compromise.

The tides N and L of the harmonic system are given by

$$\mathrm{H}_n \cos[2t + 2h - 2s - (s - p) - \kappa_n] + \mathrm{H}_l \cos[2t + 2h - 2s + (s - p) + \pi - \kappa_l]$$

In the present development the mean lunar tide M_2 and the elliptic tides are to be represented by

$$\mathrm{H}_m \cos[2t + 2h - 2l - \kappa_m] + \alpha \mathrm{H}_m P \cos[2t + 2h - 2l - {}_m\kappa]$$

It has been already remarked that it will be convenient that P should refer to a time earlier than t, and the time referred to is that of the moon's transit. Now H.W. occurs later than moon's transit by an interval whose mean value is $\kappa_m/2(\gamma-\sigma)$, and the mean value of the interval to the preceding L.W. is $(\kappa_m-\pi)/2(\gamma-\sigma)$. If, therefore, we put $\zeta=\frac{\sigma-\varpi}{2(\gamma-\sigma)}\kappa_m$ for H.W., and $\frac{\sigma-\varpi}{2(\gamma-\sigma)}(\kappa_m-\pi)$ for L.W., we have approximately

$$P=\frac{c^3}{r^3}-1=3e\cos[s-p-\zeta]$$

also

$$2l=2s+4e\sin(s-p)$$

Hence the elliptic tides are represented in the present development by

$$2e\mathrm{H}_m\cos(2t+2h-3s+p-\kappa_m)+\tfrac{3}{2}\alpha e\mathrm{H}_m\cos(2t+2h-3s+p-{}_m\kappa+\zeta)$$
$$+2e\mathrm{H}_m\cos(2t+2h-s-p+\pi-\kappa_m)-\tfrac{3}{2}\alpha e\mathrm{H}_m\cos(2t+2h-s-p+\pi-{}_m\kappa-\zeta)$$

In order that the first pair of these terms may give the N tide correctly we must have

$$\mathrm{H}_n=2e\mathrm{H}_m\cos(\kappa_n-\kappa_m)+\tfrac{3}{2}\alpha e\mathrm{H}_m\cos(\kappa_n-{}_m\kappa+\zeta)$$
$$0=2e\mathrm{H}_m\sin(\kappa_n-\kappa_m)+\tfrac{3}{2}\alpha e\mathrm{H}_m\sin(\kappa_n-{}_m\kappa+\zeta)$$

And the condition that the second pair may give the L tide is

$$\mathrm{H}_l=2e\mathrm{H}_m\cos(\kappa_l-\kappa_m)-\tfrac{3}{2}\alpha e\mathrm{H}_m\cos(\kappa_l-{}_m\kappa-\zeta)$$
$$0=2e\mathrm{H}_m\sin(\kappa_l-\kappa_m)-\tfrac{3}{2}\alpha e\mathrm{H}_m\sin(\kappa_l-{}_m\kappa-\zeta)$$

The condition for N may be written

$$\tan(\kappa_n-{}_m\kappa+\zeta)=\frac{2e\mathrm{H}_m\sin(\kappa_m-\kappa_n)}{\mathrm{H}_n-2e\mathrm{H}_m\cos(\kappa_m-\kappa_n)}$$
$$\tfrac{3}{2}\alpha e\mathrm{H}_m=\{\mathrm{H}_n-2e\mathrm{H}_m\cos(\kappa_m-\kappa_n)\}\sec(\kappa_n-{}_m\kappa+\zeta)$$

Then, since both $\kappa_n-\kappa_m$ and $\kappa_n-{}_m\kappa+\zeta$ are small angles, we have approximately

$${}_m\kappa=\kappa_n+\zeta-\frac{2e\mathrm{H}_m(\kappa_m-\kappa_n)}{\mathrm{H}_n-2e\mathrm{H}_m}$$
$$\alpha\mathrm{H}_m=\frac{2}{3e}(\mathrm{H}_n-2e\mathrm{H}_m)$$

Similarly the condition for L gives

$${}_m\kappa=\kappa_l-\zeta-\frac{2e\mathrm{H}_m(\kappa_l-\kappa_m)}{2e\mathrm{H}_m-\mathrm{H}_l}$$
$$\alpha\mathrm{H}_m=\frac{2}{3e}(2e\mathrm{H}_m-\mathrm{H}_l)$$

These conditions cannot be satisfied simultaneously and a compromise must be adopted.

If H_l were zero we should have $\alpha = \frac{4}{3}$. Hence if α be taken as greater than $\frac{4}{3}$ we are virtually making the L tide negative. Now, it is unadvisable to take the L tide as negative even if small, because, if so, we are representing the elliptic tide compounded from N and L as greatest when it should be smallest and *vice versâ*.

I therefore propose to use the condition for the L tide only for the purpose of putting a superior limit to α, and to the constants to be derived from the condition for the more important N tide.

In order to express the result in the form which will ultimately be required I write

$$p_n = \frac{2\gamma - 3\sigma + \varpi}{2\gamma - 2\sigma}, \qquad p_m = \frac{2\gamma - 2\sigma}{2\gamma - 2\sigma} = 1$$

so that p_n is the ratio of the "speed" of the N tide to that of M_2, and p_m is merely introduced for the sake of algebraic symmetry.

Then

$$\zeta = (p_m - p_n)\,\kappa_m \text{ for H.W. and } (p_m - p_n)(\kappa_m - \pi) \text{ for L.W.}$$

We thus have approximately, when referring to H.W.

$$p_m \kappa_m - {}_m\kappa = p_n \kappa_m - \kappa_n + \frac{2eH_m(\kappa_m - \kappa_n)}{H_n - 2eH_m}$$

and αH_m is equal to the smaller of $\frac{2}{3e}(H_n - 2eH_m)$ and $\frac{4}{3}H_m$

When referring to L.W. we have $(\kappa_m - \pi)$ in the above formula in the place of κ_m wherever κ_m occurs multiplied by a p.

A similar argument may be followed with regard to the diurnal tides, save that the smaller tide corresponding to L, has not usually been evaluated in the harmonic analysis. The tide corresponding to N bears the initial Q, and

$$p_q = \frac{\gamma - 3\sigma + \varpi}{2\gamma - 2\sigma}, \qquad p_o = \frac{\gamma - 2\sigma}{2\gamma - 2\sigma}$$

so that p_q and p_o are the ratios of the "speeds" of the Q and O tides to that of M_2. Then

$$\frac{\sigma - \varpi}{2(\gamma - \sigma)} = p_o - p_q$$

and we find for H.W.

$$p_o \kappa_m - {}_o\kappa = p_q \kappa_m - \kappa_q + \frac{2eH_o(\kappa_o - \kappa_q)}{H_q - 2eH_o}$$

and βH_o is equal to the smaller of $\frac{2}{3e}(H_q - 2eH_o)$ and $\frac{4}{3}H_o$

Since $e = {\cdot}0549$, $\frac{2}{3e} = 12{\cdot}144 = 19 \times {\cdot}639$ (the factor being introduced because we shall hereafter have to divide by 19); also $\frac{4}{3}=19\times\frac{4}{57}=19\times{}\cdot070$, and $\frac{1}{2e}=9\cdot11$.

If we put then $\alpha'=\frac{1}{19}\alpha$, $\beta'=\frac{1}{19}\beta$, our formulæ for H.W. become

$$\left.\begin{aligned} p_m\kappa_m-{}_m\kappa &= p_n\kappa_m-\kappa_n+\frac{(\kappa_m-\kappa_n)}{9\cdot11\mathrm{H}_n/\mathrm{H}_m-1} \\ p_o\kappa_m-{}_o\kappa &= p_q\kappa_m-\kappa_q+\frac{(\kappa_o-\kappa_q)}{9\cdot11\mathrm{H}_q/\mathrm{H}_o-1} \end{aligned}\right\}\text{...............(20)}$$

where $p_m=1$, $p_n={}\cdot981$, $p_o={}\cdot481$, $p_q={}\cdot462$, and where for L.W. κ_m is to be replaced by $\kappa_m-\pi$ wherever it is multiplied by a p.

Also, if $36\cdot4\mathrm{H}_n$ is less than $8\mathrm{H}_m$

$$\alpha=12\cdot14\frac{\mathrm{H}_n}{\mathrm{H}_m}-\frac{4}{3}$$

$$\alpha'\mathrm{H}_m={}\cdot639\mathrm{H}_n-{}\cdot070\mathrm{H}_m$$

If $36\cdot4\mathrm{H}_q$ is less than $8\mathrm{H}_o$ ....(21)

$$\beta=12\cdot14\frac{\mathrm{H}_q}{\mathrm{H}_o}-\frac{4}{3}$$

$$\beta'\mathrm{H}_o={}\cdot639\mathrm{H}_q-{}\cdot070\mathrm{H}_o$$

If $36\cdot4\mathrm{H}_n$ is greater than $8\mathrm{H}_m$, $\alpha=\frac{4}{3}$, and $\alpha'\mathrm{H}_m={}\cdot070\mathrm{H}_m$.

If $36\cdot4\mathrm{H}_q$ is greater than $8\mathrm{H}_o$, $\beta=\frac{4}{3}$, and $\beta'\mathrm{H}_o={}\cdot070\mathrm{H}_o$.

Lastly, if the elliptic tides are unknown, $\alpha=\beta=1$, $\alpha'=\beta'=\frac{1}{19}$.

§ 4. *Reference to Transit of Fictitious Moon.*

Suppose that there is a fictitious satellite whose R.A. is equal to the Moon's longitude, and let $l^{(0)}$, $h^{(0)}$ be the R.A. of the fictitious moon and of the mean sun at noon of the day under consideration, and let τ be the mean solar time of the fictitious moon's transit.

Then at transit $$t+h-l=q\pi$$

where q is even for upper and odd for lower transit.

The rate of change of $t+h$ is γ, and if $\dot{l}$ be the rate of change of l, τ is given by

$$\tau=\frac{l^{(0)}-h^{(0)}+q\pi}{\gamma-\dot{l}}$$

Then at time $\tau+T$

$$\Delta=2t+2h-2l-\kappa_m$$

$$=2q\pi+2(\gamma-\dot{l})T-\kappa_m$$

and $$\tfrac{1}{2}\Delta=q\pi+(\gamma-\dot{l})T-\tfrac{1}{2}\kappa_m$$

Now Δ occurs in the arguments of all the semi-diurnal tides, $\frac{1}{2}\Delta$ in those of the diurnal, and 2Δ in those of the quater-diurnal.

Hence we may omit the $2q\pi$ in the expression for Δ, and write

$$\Delta = 2(\gamma - \dot{l})\,T - \kappa_m$$

provided that when q is odd we change the signs of the diurnal terms.

Therefore we may write the diurnal terms with alternative signs $\pm$, and the upper sign will be appropriate when we refer to an upper transit, and the lower sign to a lower transit.

In the application of the formulæ $T + \tau$ will be the time for which the mean solar hour angle is t, and we shall have

$$T = \frac{\kappa_m}{2(\gamma - \dot{l})} + \frac{\Delta}{2(\gamma - \dot{l})} \qquad \text{.......................(22)}$$

The first of these terms is obviously the interval from fictitious transit to the mean lunar H.W., and the second is the interval from lunar H.W. to the time t.

When we wish to discuss low waters it is convenient to put Δ equal to $\delta - \pi$, and in this case we shall have

$$T = \frac{\kappa_m - \pi}{2(\gamma - \dot{l})} + \frac{\delta}{2(\gamma - \dot{l})} \qquad \text{.......................(23)}$$

The first of these terms is the interval from transit to the mean lunar L.W., which precedes the H.W. given by (22), and the second is the interval after lunar L.W. to the time t.

We shall proceed to consider H.W., and deduce therefrom the formulæ for L.W.

§ 5. *Reduction of Longitudes of Moon and Sun to time of Fictitious Transit.*

We have seen in (22) that the interval, from fictitious moon's transit to the time t, is

$$T = \frac{\Delta + \kappa_m}{2(\gamma - \dot{l})}$$

Now the equation of conservation of areas for the moon's motion gives

$$r^2\dot{l} = \sigma c^2 \sqrt{(1 - e^2)}$$

but since $c^3/r^3 = 1 + P$, and since P is small, we have

$$\dot{l} = \sigma(1 + \tfrac{2}{3}P), \text{ very nearly}$$

Therefore $$T = \frac{\Delta + \kappa_m}{2(\gamma - \sigma)}\left(1 + \frac{2}{3}\frac{P\sigma}{\gamma - \sigma}\right)$$

$$\dot{l}T = \frac{(\Delta + \kappa_m)\,\sigma}{2(\gamma - \sigma)} + \tfrac{2}{3}P\frac{\gamma\sigma}{2(\gamma - \sigma)^2}(\Delta + \kappa_m)$$

Let $$p_m = \frac{2\gamma - 2\sigma}{2\gamma - 2\sigma}, \quad p_{,,} = \frac{2\gamma}{2\gamma - 2\sigma}, \quad p_s = \frac{2\gamma - 2\eta}{2\gamma - 2\sigma}$$

$$p_o = \frac{\gamma - 2\sigma}{2\gamma - 2\sigma}, \quad p_{,} = \frac{\gamma}{2\gamma - 2\sigma}, \quad p_p = \frac{\gamma - 2\eta}{2\gamma - 2\sigma}$$

and let $$\epsilon = \tfrac{2}{3} \frac{\gamma\sigma}{2(\gamma - \sigma)^2} = \tfrac{1}{3} p_{,,}(p_{,,} - 1) = \cdot 01311$$

Then if l_o, $l_{o,}$ be the moon's and sun's longitudes at the time of fictitious transit, we have

$$l = l_o + \frac{\sigma}{2\gamma - 2\sigma}(\Delta + \kappa_m) + \epsilon P(\Delta + \kappa_m)$$

$$l_{,} = l_{o,} + \frac{\eta}{2\gamma - 2\sigma}(\Delta + \kappa_m) + \epsilon \frac{\eta}{\gamma} P(\Delta + \kappa_m)$$

Then neglecting the term in P when multiplied by the small fraction η/γ, and introducing the p's, we have

$$\left.\begin{aligned} 2l &= 2l_o + (p_{,,} - 1)(\Delta + \kappa_m) + 2\epsilon P(\Delta + \kappa_m) \\ 2(l - l_{,}) &= 2(l_o - l_{o,}) + (p_s - 1)(\Delta + \kappa_m) + 2\epsilon P(\Delta + \kappa_m) \\ -l &= -l_o + (p_o - \tfrac{1}{2})(\Delta + \kappa_m) - \epsilon P(\Delta + \kappa_m) \\ l &= l_o + (p_{,} - \tfrac{1}{2})(\Delta + \kappa_m) + \epsilon P(\Delta + \kappa_m) \\ l - 2l_{,} &= l_o - 2l_{o,} + (p_p - \tfrac{1}{2})(\Delta + \kappa_m) + \epsilon P(\Delta + \kappa_m) \end{aligned}\right\} \quad \ldots(24)$$

We must now take the several terms of the expressions for the tide in (17) (18) (19), and rearrange them by aid of (24).

(i.) $\cos(\Delta + \kappa_m + 2l - \kappa_{,,}) + \cdot 683 P \cos(\Delta + \kappa_m + 2l - \kappa_{,,})$

$$= \cos[p_{,,}(\Delta + \kappa_m) + 2l_o - \kappa_{,,}] + \cdot 683 P \cos[{}_{,,}p(\Delta + \kappa_m) + 2l_o - \kappa_{,,}]$$

where $${}_{,,}p = p_{,,} + \frac{2\epsilon}{\cdot 683} = p_{,,} + \cdot 0384 = 1\cdot 0379 + \cdot 0384 = 1\cdot 0763$$

(ii.) $\cos(\Delta + \kappa_m + 2(l - l_{,}) - \kappa_s) = \cos[p_s(\Delta + \kappa_m) + 2(l_o - l_{o,}) - \kappa_s]$

$$- 2\epsilon P(\Delta + \kappa_m)\sin[p_s(\Delta + \kappa_m) + 2(l_o - l_{o,}) - \kappa_s]$$

Now the maximum value of P is about $\tfrac{1}{6}$, and $\Delta + \kappa_m$ will not differ largely from κ_m, because Δ will oscillate about the value zero, and κ_m might always be taken as less than 180°, either positively or negatively. Hence the coefficient of this second term cannot exceed $2 \times \cdot 013 \times \tfrac{1}{6} \times \pi$, or about $\cdot 013$.

Thus the second term cannot be more than about an eightieth of the first, and may be neglected. This term may also be safely neglected, even when κ_m is greater than 180°.

(iii.) $\cos[\tfrac{1}{2}(\Delta + \kappa_m) - l + \tfrac{1}{2}\pi - \kappa_o] + P\beta \cos[\tfrac{1}{2}(\Delta + \kappa_m) - l + \tfrac{1}{2}\pi - {}_o\kappa]$

$$= \cos[p_o(\Delta + \kappa_m) - l_o + \tfrac{1}{2}\pi - \kappa_o] + P\beta \cos[{}_op(\Delta + \kappa_m) - l_o + \tfrac{1}{2}\pi - {}_o\kappa]$$

where $${}_op = p_o - \frac{\epsilon}{\beta} = \cdot 4811 - \frac{\cdot 0131}{\beta}$$

The reader who verifies the above formula will perceive the nature of the approximation adopted.

(iv.) $\cos[\frac{1}{2}(\Delta+\kappa_m)+l-\frac{1}{2}\pi-\kappa_{,}]+{\cdot}683P\cos[\frac{1}{2}(\Delta+\kappa_m)+l-\frac{1}{2}\pi-\kappa_{,}]$

$$=\cos[p_{,}(\Delta+\kappa_m)+l_o-\tfrac{1}{2}\pi-\kappa_{,}]+{\cdot}683P\cos[{}_{,}p(\Delta+\kappa_m)+l_o-\tfrac{1}{2}\pi-\kappa_{,}]$$

where $${}_{,}p=p_{,}+\frac{\epsilon}{\cdot 683}=p_{,}+{\cdot}0192={\cdot}5189+{\cdot}0192={\cdot}5381$$

(v.) The P term may be written

$$\cos[p_p(\Delta+\kappa_m)+l_o-2l_{o,}+\tfrac{1}{2}\pi-\kappa_p]$$

with a neglect similar to that involved in (ii.), but, of course, less important.

Before using these transformations the following notation must be introduced

$$\left.\begin{aligned} ☽ &= l_o - l_{o,} \\ \odot &= l_{o,} + 5^\circ \end{aligned}\right\} \quad \ldots\ldots\ldots\ldots\ldots\ldots\ldots\ldots(25)$$

So that $l_o = ☽ + \odot - 5^\circ$, $l_{o,} = \odot - 5^\circ$. We then write

$$\vartheta_m = p_m\kappa_m - \kappa_m$$

$${}_m\vartheta = p_m\kappa_m - {}_m\kappa \text{ (see (20), § 3)}$$

$$\vartheta_{,,} = 2(☽ + \odot - 5^\circ) + p_{,,}\kappa_m - \kappa_{,,}$$

$${}_{,,}\vartheta = \vartheta_{,,} + ({}_{,,}p - p_{,,})\kappa_m = \vartheta_{,,} + {\cdot}0384\kappa_m$$

$$\vartheta_s = 2☽ + p_s\kappa_m - \kappa_s$$

$$\vartheta_o = -(☽ + \odot - 5^\circ) + p_o\kappa_m - \kappa_o + 90^\circ$$

$${}_o\vartheta = -(☽ + \odot - 5^\circ) + p_o\kappa_m - {}_o\kappa - \frac{\cdot 01311}{\beta}\kappa_m + 90^\circ \text{ (see (20), § 3)}$$

$$\vartheta_{,} = (☽ + \odot - 5^\circ) + p_{,}\kappa_m - \kappa_{,} - 90^\circ$$

$${}_{,}\vartheta = \vartheta_{,} + ({}_{,}p - p_{,})\kappa_m = \vartheta_{,} + {\cdot}0192\kappa_m$$

$$\vartheta_p = (☽ - \odot + 5^\circ) + p_p\kappa_m - \kappa_p + 90^\circ \quad \ldots\ldots\ldots\ldots\ldots\ldots\ldots(26)$$

We also require for the quater-diurnal tides

$$\vartheta_{2m} = 2p_m\kappa_m - \kappa_{2m}$$

$$\vartheta_{2s} = 2p_s\kappa_m - \kappa_{2s} \quad \ldots\ldots\ldots\ldots\ldots\ldots\ldots(26)$$

and for the annual and semi-annual tides

$$\vartheta_{sa} = (\odot - 5^\circ) - \kappa_{sa}$$

$$\vartheta_{ssa} = 2(\odot - 5^\circ) - \kappa_{ssa} \quad \ldots\ldots\ldots\ldots\ldots\ldots(26)$$

The parts of the ϑ's which are independent of ☽ and $\odot$ will, in the example, be written K, with appropriate suffix.

The moon's mean parallax is 57′, then when its parallax is 57′ + Π,

$$1 + P = (1 + \tfrac{1}{57}\Pi)^3$$

so that $P = \frac{1}{19}\Pi = {\cdot}0526\Pi$ approximately.

Π will be substituted for P, and the numerical coefficients in h_{P} will be divided by 19 or multiplied by ·0526.

We observe then that $\frac{1}{19} \times {}$·683 = ·0360, and $\frac{1}{19}\alpha = \alpha'$, $\frac{1}{19}\beta = \beta'$.

We can, by aid of the transformations (i.) to (v.) and the formulæ (26), now re-write (17) (18) (19), but some part of what was originally included in h is transferred to h_{P}. The result is as follows:

$$\begin{aligned} h = {} & \text{H}_m \cos(p_m\Delta + \vartheta_m) + \text{H}_{\prime\prime} \cos(p_{\prime\prime}\Delta + \vartheta_{\prime\prime}) + \text{H}_s \cos(p_s\Delta + \vartheta_s) \\ & \pm \text{H}_o \cos(p_o\Delta + \vartheta_o) \pm \text{H}_{\prime} \cos(p_{\prime}\Delta + \vartheta_{\prime}) \pm \text{H}_p \cos(p_p\Delta + \vartheta_p) \\ & + \text{H}_{2m} \cos(2p_m\Delta + \vartheta_{2m}) + \text{H}_{2s} \cos(2p_s\Delta + \vartheta_{2s}) \quad \text{......(27)} \end{aligned}$$

$$\begin{aligned} h_{\text{P}} = \Pi \, \{ & \alpha'\text{H}_m \cos(p_m\Delta + {}_m\vartheta) + {}\cdot0360\text{H}_{\prime\prime} \cos({}_{\prime\prime}p\Delta + {}_{\prime\prime}\vartheta) \\ & \pm \beta'\text{H}_o \cos({}_op\Delta + {}_o\vartheta) \pm {}\cdot0360\text{H}_{\prime} \cos({}_{\prime}p\Delta + {}_{\prime}\vartheta)\} \quad \text{......(28)} \end{aligned}$$

$$\begin{aligned} h_{\Omega} = {} & \cos\Omega \, \{-\text{H}_m \cdot037 \cos(p_m\Delta + \vartheta_m) + \text{H}_{\prime\prime} \cdot283 \cos(p_{\prime\prime}\Delta + \vartheta_{\prime\prime}) \\ & \qquad \pm \text{H}_o \cdot188 \cos(p_o\Delta + \vartheta_o) \pm \text{H}_{\prime} \cdot115 \cos(p_{\prime}\Delta + \vartheta_{\prime})\} \\ & + \sin\Omega \, \{\text{H}_m \cdot037 \sin(p_m\Delta + \vartheta_m) + \text{H}_{\prime\prime} \cdot308 \sin(p_{\prime\prime}\Delta + \vartheta_{\prime\prime}) \\ & \qquad \mp \text{H}_o \cdot188 \sin(p_o\Delta + \vartheta_o) \pm \text{H}_{\prime} \cdot154 \sin(p_{\prime}\Delta + \vartheta_{\prime})\} \quad \text{......(29)} \end{aligned}$$

Annual and semi-annual tides $= \text{H}_{sa} \cos\vartheta_{sa} + \text{H}_{ssa} \cos\vartheta_{ssa}$ ....(30)

Except for the solar parallactic portion to be considered later, this is the final expression for the tide. Apart from the slow variability of Π and Ω the only variable is Δ, for the ϑ's are not continuous variables, but change *per saltum* from one lunar transit to the next. Thus, in finding the maximum or H.W., we may treat Δ as the variable.

§ 6. *The Time and Height of High Water.*

Apart from parallactic and nodal corrections, H.W. occurs when h in (27) is a maximum, and the value of Δ which satisfies that condition will give the time estimated from the epoch when Δ is zero. The value of Δ, when reduced to time in the manner shown below, gives the inequality in the interval after moon's transit, and will be called I. The time from moon's transit until Δ is zero is the mean interval, and will be called i. Then the interval from moon's transit is $i + I$. There is indeed a small parallactic correction to i, which is omitted in this statement, but is included below.

I begin by considering the mean interval.

Δ vanishes when $2(t + h - l) - \kappa_m = 0$, that is to say, at a time $\kappa_m/2(\gamma - \dot{l})$ after moon's transit.

Now, as shown in § 5,

$$\gamma - \dot{l} = (\gamma - \sigma)\left(1 + \tfrac{2}{3}\frac{P\sigma}{\gamma - \sigma}\right)$$

therefore

$$\frac{\kappa_m}{2(\gamma - \dot{l})} = \frac{\kappa_m}{2(\gamma - \sigma)} + \frac{\kappa_m}{2(\gamma - \sigma)} \frac{2}{3} \frac{P\sigma}{\gamma - \sigma}$$

The first of these terms is denoted by $\dot{\iota}$, so that

$$\dot{\iota} = \frac{\kappa_m}{2(\gamma - \sigma)} \quad \text{.............................(31)}$$

The second term is the parallactic correction to the mean interval, and will be considered in § 8, together with the other parallactic corrections.

If we denote by the suffixes 2, 1, 4 the semi-diurnal, diurnal, and quater-diurnal terms, (27) may be written

$$h = \Sigma H_2 \cos(p_2\Delta + \vartheta_2) + \Sigma H_1 \cos(p_1\Delta + \vartheta_1) + \Sigma H_4 \cos(p_4\Delta + \vartheta_4)$$

The condition for maximum is that $dh/d\Delta$ should vanish, and the equation consists of the sum of three pairs of terms of the form

$$\Sigma\,[Hp \cos\vartheta \sin p\Delta + Hp \sin\vartheta \cos p\Delta]$$

equated to zero.

It is well known that

$$\frac{\sin k\alpha}{k \sin\alpha} = 1 + \frac{1-k^2}{3\,!}\sin^2\alpha + \dots$$

$$\frac{\cos k\alpha}{\cos\alpha} = 1 + \frac{1-k^2}{2\,!}\sin^2\alpha + \dots$$

Now all the p_2's are nearly unity, the p_1's nearly $\frac{1}{2}$, the p_4's nearly 2. Hence to a near approximation

$$\sin p_n\Delta = \frac{2p_n}{n}\sin\tfrac{1}{2}n\Delta\left[1 - \tfrac{1}{6}\left(\frac{4p_n^2}{n^2} - 1\right)\sin^2\tfrac{1}{2}n\Delta\right]$$

$$\cos p_n\Delta = \cos\tfrac{1}{2}n\Delta\left[1 - \tfrac{1}{2}\left(\frac{4p_n^2}{n^2} - 1\right)\sin^2\tfrac{1}{2}n\Delta\right]$$

where n is 2, 1, or 4.

The p's are so nearly equal to 1, $\frac{1}{2}$, 2 respectively, that the second terms are small. In the case of the second term of $\sin p_1\Delta$, $\cos p_1\Delta$ we have the factor $\sin^2\Delta$, which is equal to $\frac{1}{2}$ even when Δ is 90°; also the height of the quater-diurnal tide is small. Thus, both for the diurnal and quater-diurnal tides, the second term may be neglected.

In the case of the semi-diurnal terms, indeed, the correction is so small that it may certainly be neglected if Δ be less than 45° or 50°, and without much loss of accuracy even for larger values of Δ. We shall in the first place neglect these second terms entirely.

Now let
$$\left.\begin{array}{l} F = \Sigma Hp \sin\vartheta \\ G = \Sigma Hp^2 \cos\vartheta \end{array}\right\} \quad \text{.............................(32)}$$

with suffixes 2, 1, 4 for semi-diurnal, diurnal, quater-diurnal terms.

The constituent terms of F, G will be written separately below, for example $F_s = H_s p_s \sin\vartheta_s$.

Then the equation $dh/d\Delta = 0$, leads to

$$F_2 \cos \Delta + G_2 \sin \Delta \pm (F_1 \cos \tfrac{1}{2}\Delta + 2G_1 \sin \tfrac{1}{2}\Delta) + F_4 \cos 2\Delta + \tfrac{1}{2}G_4 \sin 2\Delta = 0 \quad \text{.........(33)}$$

With regard to the additional terms referred to above, the values of the semi-diurnal p's are, $p_m = 1$, $p_{,,} = 1{\cdot}038$, $p_s = 1{\cdot}035$, and hence $p_{,,}^2 - 1$ and $p_s^2 - 1$ are both nearly equal to ·072. Therefore if (33) be regarded as the fundamental equation, the additional terms may be taken into account by supposing that there are corrections to F_2 and G_2 given by

$$\left.\begin{aligned} \delta F_2 &= -\,{\cdot}012 \sin^2 \Delta \,.\, 3F_2 \\ \delta G_2 &= -\,{\cdot}012 \sin^2 \Delta\, (G_{,,} + G_s) \end{aligned}\right\} \text{.....................(34)}$$

The equation (33) may be solved thus:

take
$$\tan \Delta_0 = -\frac{F_2 \pm F_1}{G_2 \pm G_1} \quad \text{..........................(35)}$$

Then if we put

$$D_0 = F_2 \sin \Delta_0 - G_2 \cos \Delta_0 \pm \tfrac{1}{2}(F_1 \sin \tfrac{1}{2}\Delta_0 - 2G_1 \cos \tfrac{1}{2}\Delta_0) + 2(F_4 \sin 2\Delta_0 - \tfrac{1}{2}G_4 \cos 2\Delta_0)$$

$$N_0 = F_2 \cos \Delta_0 + G_2 \sin \Delta_0 \pm (F_1 \cos \tfrac{1}{2}\Delta_0 + 2G_1 \sin \tfrac{1}{2}\Delta_0) + (F_4 \cos 2\Delta_0 + \tfrac{1}{2}G_4 \sin 2\Delta_0) \text{...(36)}$$

and
$$\delta\Delta = \frac{N_0}{D_0} \quad \text{.................................(37)}$$

the solution of the equation (33) is $\Delta = \Delta_0 + \delta\Delta$; and if D is the value of D_0 when Δ replaces Δ_0, it is clear that

$$\delta D = D - D_0 = \delta\Delta\,\{(F_2 \cos \Delta_0 + G_2 \sin \Delta_0) \pm \tfrac{1}{4}(F_1 \cos \tfrac{1}{2}\Delta_0 + 2G_1 \sin \tfrac{1}{2}\Delta_0) + 4(F_4 \cos 2\Delta_0 + \tfrac{1}{2}G_4 \sin 2\Delta_0)\} \text{.........(38)}$$

The angle Δ has to be reduced to time by division by $2(\gamma - \dot{l})$. As in the case of the reduction of the mean interval

$$\frac{\Delta}{2(\gamma - \dot{l})} = \frac{\Delta}{2(\gamma - \sigma)}\left(1 + \tfrac{2}{3}\frac{P\sigma}{\gamma - \sigma}\right)$$

Now the greatest value of P is about $\frac{1}{5}$, and $\sigma/(\gamma - \sigma)$ is ·038, hence $\frac{2}{3}P\sigma/(\gamma - \sigma)$ is at greatest ·005; but the inequality in the interval is rarely more than 2^h, and even if $\Delta/2(\gamma - \sigma)$ amounted to 3^h, the second term would only be about 1^m. Therefore I neglect the parallactic inequality in the reduction of Δ to time, and simply divide Δ by $2(\gamma - \sigma)$.

The value of $2(\gamma - \sigma)$ is 28°·98 per hour, and the reciprocal of this is ·0345, or $\frac{1}{30} + \frac{21}{20} \cdot \frac{1}{30^2}$.

Hence, if Δ be given in degrees, we have to multiply by ·0345, or its equivalent, to find its value in hours.

If any correction be given in circular measure the reduction to time is also very simple, for $57^\circ{\cdot}296 \times {\cdot}0345 = 1^h{\cdot}977 = 119^m$, or

$$\frac{1}{2(\gamma - \sigma)} \text{ (circ. meas.)} = 119^m \quad \text{.....................(39)}$$

Thus, we have for H.W.,

$$\left.\begin{aligned} \Delta &= \Delta_o + \delta\Delta \\ D &= D_o + \delta D \\ I &= {\cdot}0345\Delta = \left(\frac{1}{30} + \frac{21}{20}\cdot\frac{1}{30^2}\right)\Delta \end{aligned}\right\} \quad \text{.....................(40)}$$

Turning now to the expression for the height: it is expressible as the sum of three terms of the form

$$\Sigma\,(H \cos\vartheta \cos p\Delta - H \sin\vartheta \sin p\Delta)$$

Let

$$A = \Sigma H \cos\vartheta$$

then

$$h = A_2 \cos\Delta - F_2 \sin\Delta \pm (A_1 \cos\tfrac{1}{2}\Delta - 2F_1 \sin\tfrac{1}{2}\Delta) + A_4 \cos 2\Delta - \tfrac{1}{2}F_4 \sin 2\Delta \quad \text{.........(41)}$$

with a correction to the height corresponding with

$$\left.\begin{aligned} \delta A_2 &= -\,{\cdot}012 \sin^2\Delta\,.\,3\,(A_{\prime\prime} + A_s) \\ \delta F_2 &= -\,{\cdot}012 \sin^2\Delta\,.\,F_2 \end{aligned}\right\} \quad \text{..................(42)}$$

In §§ 8, 9 we shall have to consider the variations of the interval and height due to variations of semi-diurnal and diurnal A, F, G. It is clear from the preceding formulæ and from (39) that

$$\left.\begin{aligned} \delta I &= \frac{119^m}{D}\{\delta F_2 \cos\Delta + \delta G_2 \sin\Delta \pm (\delta F_1 \cos\tfrac{1}{2}\Delta + 2\delta G_1 \sin\tfrac{1}{2}\Delta)\} \\ \delta h &= \delta A_2 \cos\Delta - \delta F_2 \sin\Delta \pm (\delta A_1 \cos\tfrac{1}{2}\Delta - 2\delta F_1 \sin\tfrac{1}{2}\Delta) \end{aligned}\right\} \quad \text{...(43)}$$

A particular case of the application of (43) is to the computation of the corrections referred to in (34) and (42).

§ 7. *On Evanescent Tides.*

At certain parts of the lunation, the diurnal tides sometimes suffice to annul one H.W. and one L.W., so that there is only one tide a day, perhaps for several days running.

If the inferior H.W. be watched as the condition of evanescence approaches, it will be seen to become smaller and smaller, and to occur later and later or earlier and earlier, and the adjacent L.W. undergoes similar changes. In the limit H.W. and L.W. coalesce, and in a tide diagram the coalescence appears as a point of contrary reflexure with horizontal tangent. Beyond this point, the reflexure is still maintained, although the tangent is not horizontal; finally, the tangent again becomes horizontal, and the double H.W. and L.W. again reappear.

Now in the use of the method of this paper, the loss of the double tide is very inconvenient, and I therefore propose to take the point of reflexure as representing both H.W. and L.W. during evanescence.

If N cannot vanish there is evanescence, and the point of reflexure is given by $D=0$. The limit of the H.W. is given by $N=0$, $D=0$ simultaneously, and beyond this only D can vanish. The vanishing of D is taken to represent H.W.

Accordingly, when N cannot vanish, we proceed to make D vanish thus:

$$0 = F_2 \sin \Delta - G_2 \cos \Delta \pm \tfrac{1}{2}(F_1 \sin \tfrac{1}{2}\Delta - 2G_1 \cos \tfrac{1}{2}\Delta) + 2(F_4 \sin 2\Delta - \tfrac{1}{2}G_4 \cos 2\Delta) \quad \text{......(44)}$$

As a first approximation put

$$\tan \Delta_o = \frac{G_2 \pm G_1}{F_2 \pm \frac{1}{4}F_1}$$

Then writing

$$E_o = -\{F_2 \cos \Delta_o + G_2 \sin \Delta_o \pm \tfrac{1}{4}(F_1 \cos \tfrac{1}{2}\Delta_o + 2G_1 \sin \tfrac{1}{2}\Delta_o) + 4(F_4 \cos 2\Delta_o + \tfrac{1}{2}G_4 \sin 2\Delta_o)\} \quad \text{.........(45)}$$

we have

$$\Delta = \Delta_o + \delta\Delta, \quad \text{where} \quad \delta\Delta = \frac{D_o}{E_o} \quad \text{..................(46)}$$

In the rest of the calculation this value of Δ is to be treated exactly as though it had been determined by the former method.

The corrections M, N, P, Q, R, S considered in succeeding sections, however, present a difficulty. In this case, Δ will always be very nearly $\pm 90°$, and I propose to compute P, Q, S (see §§ 8, 9) as though that were the true value of Δ.

But the correctional terms M′, N′, R, defined in §§ 8, 9, become theoretically infinite, and we are therefore compelled not to compute them, and to fill up the hiatus in the manner shown in the example below.

The process here suggested is a makeshift, but it is sufficient for the construction of a trustworthy tide-table, since the real occurrence at these times is a long period of nearly slack water, with or without a small maximum and minimum.

§ 8. *Parallactic Corrections to Time and Height.*

We will first consider the parallactic correction to the mean interval; we saw at the beginning of § 6 that there is a correction to the interval due to the moon's unequal motion in longitude of

$$\frac{\kappa_m}{2(\gamma-\sigma)} \cdot \tfrac{2}{3} \frac{P\sigma}{\gamma-\sigma}$$

Now according to (31) $\kappa_m/2(\gamma-\sigma)$ is i, the mean interval, which we will suppose expressed in hours; also P is $\frac{1}{19}\Pi$ and $\sigma/(\gamma-\sigma)$ is ·03788, and $\frac{2}{3}$ of 60^m is 40^m; hence the correction in minutes is

$$i \,.\, 40^m \,.\, {\cdot}03788 \,.\, \tfrac{1}{19} \,.\, \Pi = i \times 0^m{\cdot}0797\Pi \quad \text{..............(47)}$$

Now turning to the other parallactic terms:

Let us introduce the following notation, which follows the same plan as that used before, viz.:

$$\left.\begin{matrix} {}_m\text{A} \\ {}_m\text{F} \\ {}_m\text{G} \end{matrix}\right\} = \alpha'\text{H}_m \left.\begin{matrix} \cos \\ p_m \sin \\ p_m^2 \cos \end{matrix}\right\} {}_m\vartheta, \qquad \left.\begin{matrix} {}_{\prime\prime}\text{A} \\ {}_{\prime\prime}\text{F} \\ {}_{\prime\prime}\text{G} \end{matrix}\right\} = {\cdot}036\,\text{H}_{\prime\prime} \left.\begin{matrix} \cos \\ {}_{\prime\prime}p \sin \\ {}_{\prime\prime}p^2 \cos \end{matrix}\right\} {}_{\prime\prime}\vartheta$$

$$\left.\begin{matrix} {}_o\text{A} \\ {}_o\text{F} \\ {}_o\text{G} \end{matrix}\right\} = \beta'\text{H}_o \left.\begin{matrix} \cos \\ {}_op \sin \\ {}_op^2 \cos \end{matrix}\right\} {}_o\vartheta, \qquad \left.\begin{matrix} {}_{\prime}\text{A} \\ {}_{\prime}\text{F} \\ {}_{\prime}\text{G} \end{matrix}\right\} = {\cdot}036\,\text{H}_{\prime} \left.\begin{matrix} \cos \\ {}_{\prime}p \sin \\ {}_{\prime}p^2 \cos \end{matrix}\right\} {}_{\prime}\vartheta$$

A comparison of the method of § 6 with (28) § 5 shows that the corrections to the time and height are to be found by applying corrections to the A's, F's, G's, as follows:

$$\delta\text{F}_2 = \Pi\,({}_m\text{F} + {}_{\prime\prime}\text{F}) = \Pi l_2, \qquad \delta\text{G}_2 = \Pi\,({}_m\text{G} + {}_{\prime\prime}\text{G}) = \Pi m_2$$

$$\delta\text{F}_1 = \Pi\,({}_o\text{F} + {}_{\prime}\text{F}) = \Pi l_1, \qquad \delta\text{G}_1 = \Pi\,({}_o\text{G} + {}_{\prime}\text{G}) = \Pi m_1$$

$$\delta\text{A}_2 = \Pi\,({}_m\text{A} + {}_{\prime\prime}\text{A}) = \Pi z_2$$

$$\delta\text{A}_1 = \Pi\,({}_o\text{A} + {}_{\prime}\text{A}) = \Pi z_1 \quad \text{......................(48)}$$

Then comparing (47) (48) with (39) and (43) we see that if

$$\text{R} = 119^m \frac{1}{\text{D}} \{l_2 \cos\Delta + m_2 \sin\Delta \pm (l_1 \cos\tfrac{1}{2}\Delta + 2m_1 \sin\tfrac{1}{2}\Delta)\} + 0^m{\cdot}08\,.\,i$$

$$\text{S} = \qquad \{z_2 \cos\Delta - l_2 \sin\Delta \pm (z_1 \cos\tfrac{1}{2}\Delta - 2l_1 \sin\tfrac{1}{2}\Delta)\} \quad \text{...........(49)}$$

the corrections to the interval and height are

$$\left.\begin{matrix} \delta I_{\text{P}} = \Pi\text{R} \\ \delta h_{\text{P}} = \Pi\text{S} \end{matrix}\right\} \quad \text{..................................(50)}$$

§ 9. *Corrections to Time and Height for Longitude of Moon's Node.*

Here again, we treat the terms as corrections to A, F, G.

Let

$$
\begin{aligned}
c_2 &= \qquad\qquad\qquad\qquad \cdot 283\,\mathrm{H}_{\prime\prime} p_{\prime\prime} \sin \vartheta_{\prime\prime} = \qquad\qquad \cdot 283\,\mathrm{F}_{\prime\prime} \\
d_2 &= -\cdot 037\,\mathrm{H}_m p_m \cos \vartheta_m - \cdot 308\,\mathrm{H}_{\prime\prime} p_{\prime\prime} \cos \vartheta_{\prime\prime} = -\cdot 037\,\mathrm{G}_m - \cdot 296\,\mathrm{G}_{\prime\prime} \\
e_2 &= -\cdot 037\,\mathrm{H}_m p_m^2 \cos \vartheta_m + \cdot 283\,\mathrm{H}_{\prime\prime} p_{\prime\prime}^2 \cos \vartheta_{\prime\prime} = -\cdot 037\,\mathrm{G}_m + \cdot 283\,\mathrm{G}_{\prime\prime} \\
f_2 &= \qquad\qquad\qquad\qquad \cdot 308\,\mathrm{H}_{\prime\prime} p_{\prime\prime}^2 \sin \vartheta_{\prime\prime} = \qquad\qquad \cdot 319\,\mathrm{F}_{\prime\prime} \\
a_2 &= -\cdot 037\,\mathrm{H}_m \cos \vartheta_m + \cdot 283\,\mathrm{H}_{\prime\prime} \cos \vartheta_{\prime\prime} = -\cdot 037\,\mathrm{A}_m + \cdot 283\,\mathrm{A}_{\prime\prime} \\
b_2 &= \qquad\qquad\qquad\qquad \cdot 308\,\mathrm{H}_{\prime\prime} \sin \vartheta_{\prime\prime} = \qquad\qquad \cdot 296\,\mathrm{F}_{\prime\prime} \\
c_1 &= \cdot 188\,\mathrm{H}_o p_o \sin \vartheta_o + \cdot 115\,\mathrm{H}_{\prime} p_{\prime} \sin \vartheta_{\prime} = \cdot 188\,\mathrm{F}_o + \cdot 115\,\mathrm{F}_{\prime} \\
d_1 &= \cdot 188\,\mathrm{H}_o p_o \cos \vartheta_o - \cdot 154\,\mathrm{H}_{\prime} p_{\prime} \cos \vartheta_{\prime} = \cdot 391\,\mathrm{G}_o - \cdot 297\,\mathrm{G}_{\prime} \\
e_1 &= \cdot 188\,\mathrm{H}_o p_o^2 \cos \vartheta_o + \cdot 115\,\mathrm{H}_{\prime} p_{\prime}^2 \cos \vartheta_{\prime} = \cdot 188\,\mathrm{G}_o + \cdot 115\,\mathrm{G}_{\prime} \\
f_1 &= -\cdot 188\,\mathrm{H}_o p_o^2 \sin \vartheta_o + \cdot 154\,\mathrm{H}_{\prime} p_{\prime}^2 \sin \vartheta_{\prime} = -\cdot 0905\,\mathrm{F}_o + \cdot 080\,\mathrm{F}_{\prime} \\
a_1 &= \cdot 188\,\mathrm{H}_o \cos \vartheta_o + \cdot 115\,\mathrm{H}_{\prime} \cos \vartheta_{\prime} = \cdot 188\,\mathrm{A}_o + \cdot 115\,\mathrm{A}_{\prime} \\
b_1 &= -\cdot 188\,\mathrm{H}_o \sin \vartheta_o + \cdot 154\,\mathrm{H}_{\prime} \sin \vartheta_{\prime} = -\cdot 391\,\mathrm{F}_o + \cdot 297\,\mathrm{F}_{\prime}
\end{aligned}
\qquad \text{.........(51)}
$$

A comparison with (29) then shows that

$$
\begin{aligned}
\delta\mathrm{F}_2 &= c_2 \cos \Omega + d_2 \sin \Omega, & \delta\mathrm{F}_1 &= c_1 \cos \Omega + d_1 \sin \Omega \\
\delta\mathrm{G}_2 &= e_2 \cos \Omega + f_2 \sin \Omega, & \delta\mathrm{G}_1 &= e_1 \cos \Omega + f_1 \sin \Omega \\
\delta\mathrm{A}_2 &= a_2 \cos \Omega + b_2 \sin \Omega, & \delta\mathrm{A}_1 &= a_1 \cos \Omega + b_1 \sin \Omega
\end{aligned}
\quad \text{...(52)}
$$

We now put

$$
\begin{aligned}
\mathrm{M}' - \delta\mathrm{M} &= 119^{\mathrm{m}} \frac{1}{\mathrm{D}} \{c_2 \cos \Delta + e_2 \sin \Delta \pm (c_1 \cos \tfrac{1}{2}\Delta + 2e_1 \sin \tfrac{1}{2}\Delta)\} \\
\mathrm{N}' - \delta\mathrm{N} &= 119^{\mathrm{m}} \frac{1}{\mathrm{D}} \{d_2 \cos \Delta + f_2 \sin \Delta \pm (d_1 \cos \tfrac{1}{2}\Delta + 2f_1 \sin \tfrac{1}{2}\Delta)\} \\
\mathrm{P}' &= a_2 \cos \Delta - c_2 \sin \Delta \pm (a_1 \cos \tfrac{1}{2}\Delta - 2c_1 \sin \tfrac{1}{2}\Delta) \\
\mathrm{Q}' &= b_2 \cos \Delta - d_2 \sin \Delta \pm (b_1 \cos \tfrac{1}{2}\Delta - 2d_1 \sin \tfrac{1}{2}\Delta)
\end{aligned}
\qquad \text{.........(53)}
$$

The corrections to interval and height, as far as concerns the investigation up to the present point, are therefore

$$
\begin{aligned}
\delta I_{\Omega} &= (\mathrm{M}' - \delta\mathrm{M}) \cos \Omega + (\mathrm{N}' - \delta\mathrm{N}) \sin \Omega \\
\delta h_{\Omega} &= \mathrm{P}' \cos \Omega + \mathrm{Q}' \sin \Omega \qquad \text{..............................(54)}
\end{aligned}
$$

§ 10. *Reference to the Moon's True Transit.*

The intervals have been referred to the transits of a fictitious satellite whose R.A. is equal to l, and we now require corrections so as to refer to the moon's true transit.

Let T, T′ be the times of true and fictitious transits, and let α be the moon's R.A.

Then if t denotes the mean solar hour angle at time T,

$$t + h - \alpha = q\pi$$

and, dropping the suffix o to l, at time T′

$$t + h - l = q\pi$$

where q is an even integer at upper, and an odd one at lower transit.

If α be the value of the moon's R.A. at time T′, then its value at T is $\alpha + \dot{\alpha}(\mathrm{T} - \mathrm{T}')$; also the $t + h$ of the first equation is equal to the $t + h$ of the second corrected by $\gamma(\mathrm{T} - \mathrm{T}')$.

Hence the two equations may be written

$$t + h + \gamma(\mathrm{T} - \mathrm{T}') - \alpha - \dot{\alpha}(\mathrm{T} - \mathrm{T}') = q\pi$$

$$t + h - l = q\pi$$

Hence $$\mathrm{T} - \mathrm{T}' = \frac{\alpha - l}{\gamma - \dot{\alpha}}$$

It will afford a sufficiently close approximation if we replace $\dot{\alpha}$ by the moon's mean motion σ, so that

$$\mathrm{T}' = \mathrm{T} + \frac{l - \alpha}{\gamma - \sigma}$$

We have, therefore, to find the excess of the moon's longitude above her R.A. at the time of fictitious transit.

It will be seen from fig. 2 that we have to determine the relationship between the R.A. and longitude, both measured from the intersection of the orbit and equator. The formula is exactly analogous with that which gives the reduction of longitudes to the ecliptic, so that

$$l - \xi = \alpha - \nu + \tan^2 \tfrac{1}{2}I \sin 2(l - \xi) - \tfrac{1}{2}\tan^4 \tfrac{1}{2}I \sin 4(l - \xi) + \dots$$

The last of these terms is very small and may be neglected, so that

$$l - \alpha = -(\nu - \xi) + \tan^2 \tfrac{1}{2}I \cos 2\xi \sin 2l - \tan^2 \tfrac{1}{2}I \sin 2\xi \cos 2l$$

By the formula (7)

$$\tan^2 \tfrac{1}{2}I = \tan^2 \tfrac{1}{2}\omega\,[1 + 2i \operatorname{cosec} \omega \cos ☊]$$

$$\sin 2\xi = 2i \cot \omega \sin ☊$$

$$\cos 2\xi = 1$$

$$\nu - \xi = i \tan \tfrac{1}{2}\omega \sin ☊$$

Hence $$l - \alpha = \tan^2 \tfrac{1}{2}\omega \sin 2l + \cos ☊ \left[2i \frac{\tan^2 \frac{1}{2}\omega}{\sin \omega} \sin 2l\right]$$
$$+ \sin ☊ \left[-2i \tan^2 \tfrac{1}{2}\omega \cot \omega \cos 2l - i \tan \tfrac{1}{2}\omega\right]$$

But $2l = 2☽ + 2⊙ - 10° = \Theta$, suppose; and

$$\frac{\tan^2 \frac{1}{2}\omega}{\gamma - \sigma} = 0^{h}{\cdot}172 = 10^{m}{\cdot}32, \qquad \frac{2i \tan^2 \frac{1}{2}\omega}{(\gamma - \sigma) \sin \omega} = 4^{m}{\cdot}60, \qquad \frac{2i \tan^2 \frac{1}{2}\omega \cot \omega}{\gamma - \sigma} = 4^{m}{\cdot}22$$

and $$\frac{i \tan \frac{1}{2}\omega}{\gamma - \sigma} = 4^{m}{\cdot}41$$

Let us put then

$$\left.\begin{aligned} \Theta &= 2☽ + 2⊙ - 10° \\ \delta T &= 0^{h}{\cdot}172 \sin \Theta \\ \delta M &= 4^{m}{\cdot}60 \sin \Theta \\ \delta N &= -\left[4^{m}{\cdot}22 \cos \Theta + 4^{m}{\cdot}41\right] \end{aligned}\right\} \quad \text{..............(55)}$$

and we have $$T' = T + \delta T + \delta M \cos ☊ + \delta N \sin ☊$$

It is now clear that, when the tide is referred to the moon's true transit, δT must be added to I, and δM, δN must be added to $M' - \delta M$, $N' - \delta N$ as given in (53) to find M', N'.

Let $\mathfrak{A}$ be the height of mean sea level above the datum adopted for the tide-table, and let

$$B_o = \mathfrak{A} + H_{ssa} \cos \vartheta_{ssa} \text{.........................(56)}$$

Then i denotes the mean interval from moon's transit to H.W., and the interval is

$$i + I + \delta T + M' \cos ☊ + N' \sin ☊ + \delta I_P$$

and the height is

$$B_o + h + P' \cos ☊ + Q' \sin ☊ + \delta h_P + H_{sa} \cos \vartheta_{sa}$$

In these formulæ all the quantities are functions of ☽; now although ☽ can easily be found from the *Nautical Almanac*, it is not tabulated, and the difficulty of using the tide-table would be considerably augmented if it were necessary to compute ☽ for each tide. We shall proceed, then, to convert our formulæ into others, in which there is direct reference to moon's true transit, although there is some loss of accuracy in the process, and the amount of computation required to form the table is increased.

We have seen that the time T' of fictitious transit is given by

$$T' = T + \delta T + \delta M \cos ☊ + \delta N \sin ☊$$

and that $(\gamma - \eta) T' + h - l = q\pi$ where h and l correspond to the time of fictitious transit.

Now $$l_{,} = h + 2e_{,} \sin (l_{,} - 281°)$$

where $e_{,}$ is the eccentricity of the sun's orbit, and 281° the longitude of perigee.

Also $$l = \text{☽} + l_,$$

Hence $$l - h = \text{☽} + 2e_, \sin(l_, - 281^\circ)$$

But

$$\frac{2e_,}{\gamma - \eta} = 7^{\text{m}}\cdot 69 = 0^{\text{h}}\cdot 128\text{; also } l_, = \odot - 5^\circ\text{, and } -286^\circ = +74^\circ$$

Therefore $$\text{T}' = \frac{q\pi + \text{☽}}{\gamma - \eta} + 7^{\text{m}}\cdot 69 \sin(\odot + 74^\circ),$$

and $$\frac{q\pi + \text{☽}}{\gamma - \eta} = \text{T} + \delta\text{T} + \delta\text{M} \cos ☊ + \delta\text{N} \sin ☊ - 7^{\text{m}}\cdot 69 \sin(\odot + 74^\circ)$$

This formula connects ☽ with T, the time of moon's transit.

Now I is a function of ☽ or of $(q\pi + \text{☽})/(\gamma - \eta)$; hence if $I(\text{T})$ be the value of I for which ☽ $= (\gamma - \eta)\text{T} - q\pi$, and if $I(\text{☽})$ be the true value of I,

$$I(\text{☽}) = I(\text{T}) + \frac{dI}{d\text{T}}\{\delta\text{T} + \delta\text{M} \cos ☊ + \delta\text{N} \sin ☊ - 7^{\text{m}}\cdot 69 \sin(\odot + 74^\circ)\}$$

and similarly

$$h(\text{☽}) = h(\text{T}) + \frac{dh}{d\text{T}}\{\delta\text{T} + \delta\text{M} \cos ☊ + \delta\text{N} \sin ☊ - 0^{\text{h}}\cdot 128 \sin(\odot + 74^\circ)\}$$

These expressions have now to be substituted in those for the height and interval; but in the small terms δT, M′, N′, &c., we may regard ☽ as denoting $(\gamma - \eta)\text{T} - q\pi$.

In carrying out the substitutions preparations will be made for computation.

First put $$\delta I = \frac{dI}{d\text{T}}\delta\text{T} = \frac{dI}{d\text{T}} \times 10^{\text{m}}\cdot 32 \sin\Theta$$

$$= \left(2\frac{dI}{d\text{T}}\right) . (2\cdot 34)(2^{\text{m}}\cdot 2 \sin\Theta) \qquad \text{.....................(57)}$$

Similarly, since $2^{\text{m}}\cdot 2$ is $\frac{11}{300}$ hrs., we put

$$\delta h = \left(2\frac{dh}{d\text{T}}\right)(2\cdot 34)\left(\tfrac{11}{300}\sin\Theta\right) \qquad \text{.....................(58)}$$

Next put $$\text{M}'' = \frac{dI}{d\text{T}}\delta\text{M} = \frac{dI}{d\text{T}} \times 4^{\text{m}}\cdot 60 \sin\Theta$$

$$= \left(2\frac{dI}{d\text{T}}\right)(2^{\text{m}}\cdot 2 \sin\Theta)\left(1 + \tfrac{1}{20}\right) \qquad \text{..................(59)}$$

and $$-\text{N}'' = -\frac{dI}{d\text{T}}\delta\text{N}$$

$$= \left(2\frac{dI}{d\text{T}}\right)(2^{\text{m}}\cdot 2) + \left(2\frac{dI}{d\text{T}}\right)(2^{\text{m}}\cdot 2 \cos\Theta)\left(1 - \tfrac{1}{20}\right) \qquad \text{......(60)}$$

Similarly put

$$P'' = \left(2\frac{dh}{dT}\right)(\tfrac{11}{300}\sin\Theta)(1+\tfrac{1}{20}) \quad \text{..........................(61)}$$

$$-Q'' = \left(2\frac{dh}{dT}\right)(\tfrac{11}{300}) + \left(2\frac{dh}{dT}\right)(\tfrac{11}{300}\cos\Theta)(1-\tfrac{1}{20}) \quad \text{........(62)}$$

Lastly put

$$\mathfrak{i} = -\left(2\frac{dI}{dT}\right)\{3^{m}\cdot 85 \sin(\odot + 74°)\} \quad \text{.................(63)}$$

$$\mathfrak{h}' = -\left(2\frac{dh}{dT}\right)\{0^{h}\cdot 064 \sin(\odot + 74°)\} \quad \text{.................(64)}$$

With this notation

$$I(☽) = I(T) + \delta I + M'' \cos ☊ + N'' \sin ☊ + \mathfrak{i}$$

$$h(☽) = h(T) + \delta h + P'' \cos ☊ + Q'' \sin ☊ + \mathfrak{h}$$

These are now to be substituted in the expressions for interval and height, and in doing so we may drop the (T) after the I and h.

I write

$$\mathfrak{I} = I + \delta T + i + \delta I$$

$$\mathfrak{H} = B_o + h + \delta h$$

$$\mathfrak{h} = \mathfrak{h}' + H_{sa} \cos \vartheta_{sa}$$

$$M = M' + M'', \qquad N = N' + N''$$

$$P^{*} = P' + P'', \qquad Q = Q' + Q'' \quad \text{..................(65)}$$

Our formulæ are designed to serve for any time of year and its opposite; now since $\mathfrak{i}$ and $\mathfrak{h}$ involve the sun's longitude they change their signs in six months, and we write $\pm\mathfrak{i}$, $\pm\mathfrak{h}$, and it is to be understood that the upper sign is to be used for the time of year under computation and the lower for its opposite.

The interval is then

$$\mathfrak{I} \pm \mathfrak{i} + M \cos ☊ + N \sin ☊ + R\Pi \quad \text{....................(66)}$$

and the height is

$$\mathfrak{H} \pm \mathfrak{h} + P \cos ☊ + Q \sin ☊ + S\Pi \quad \text{....................(67)}$$

One part of M, N, P, Q arises from a true change in the tide when the longitude of the moon's node changes, and the remainder (nearly equally large) merely depends on the reference to true instead of fictitious transit. The quantities $\mathfrak{i}$ and $\mathfrak{h}$ depend partly on a portion of the equation of time and partly on the annual tide.

We must now explain the computation of $2dI/dT$ and of $2dh/dT$.

* This P will not be confused with the P defined in § 1, which has been replaced by Π.

If $u_o, u_1, \ldots u_{23}$ are cyclical values of a function, the symmetrical interpolation formula in the neighbourhood of u_m is

$$u_{x+m} = u_m + \tfrac{1}{2}x(\Delta E^{-1} + \Delta) u_m + \frac{x^2}{2!} \Delta^2 E^{-1} u_m$$
$$+ \tfrac{1}{2}\frac{x(x^2-1)}{3!}(\Delta^3 E^{-2} + \Delta^3 E^{-1}) u_m + \ldots$$

where $\qquad E u_m = u_{m+1}$ and $\Delta u_m = u_{m+1} - u_m$

Then when $x = 0$,

$$2\frac{du_m}{dx} = (\Delta E^{-1} + \Delta) u_m - \tfrac{1}{6}(\Delta^3 E^{-2} + \Delta^3 E^{-1}) u_m + \ldots$$

or

$$2\frac{du_m}{dx} = (u_{m+1} - u_{m-1}) - \tfrac{1}{6}\Delta^2 E^{-1}(u_{m+1} - u_{m-1}) + \ldots$$

In the present case the first term will usually suffice, but the second term may be easily computed.

In order then to compute the required differential coefficients we arrange the I's and h's in two columns, the even entries in one column and the odd in another, and take the differences of the two columns independently of one another.

§ 11. *The Correction for Solar Parallax.*

The terms depending on solar parallax arise from the potential $V_{P_\prime}$ in (15), but the correction is so small that I shall omit it from the example below. It is well, however, to show how it may be computed. The only terms of importance are those in H_s, $H_\prime$ corresponding to the tides S_2, K_1.

The variability of $P_\prime$ enters into the calculation in the form of corrections to A_2, G_2, F_2, A_1, G_1, F_1.

The sun's parallax is approximately

$$1 + e_\prime \cos(l_\prime - 281°), \text{ and hence } P_\prime = 3e_\prime \cos(\odot + 74°)$$

Now $\qquad 3e_\prime$ is ·0504, and $119^m \times {}·0504 = 6^m{·}0$

Then, since
$$\left.\begin{matrix}\delta F_2\\ \delta G_2\\ \delta A_2\end{matrix}\right\} = P_\prime \left\{\begin{matrix}F_s\\ G_s\\ A_s\end{matrix}\right\}, \qquad \left.\begin{matrix}\delta F_1\\ \delta G_1\\ \delta A_1\end{matrix}\right\} = {·}317 P_\prime \left\{\begin{matrix}F_\prime\\ G_\prime\\ A_\prime\end{matrix}\right\}$$

it follows that

$$\delta I_{P_\prime} = \frac{6^m{·}0}{D}\cos(\odot + 74°)\{F_s \cos\Delta + G_s \sin\Delta \pm {·}317(F_\prime \cos\tfrac{1}{2}\Delta + 2G_\prime \sin\tfrac{1}{2}\Delta)\}$$

$$\delta h_{P_\prime} = {·}050\cos(\odot + 74°)\{A_s \cos\Delta - F_s \sin\Delta \pm {·}317(A_\prime \cos\tfrac{1}{2}\Delta - 2F_\prime \sin\tfrac{1}{2}\Delta)\}$$

These must be deemed to be corrections to $\mathfrak{i}$ and $\mathfrak{h}$, since they change signs in six months.

§ 12. *The Formulæ for a Low Water Table.*

The formulæ (27), (28), (29), (30), for the height of water would (except in one detail) be applicable to L.W., but they would not be convenient, because Δ oscillates about $-\pi$ at the L.W. which precedes the H.W. for which mean Δ is zero.

Hence put $\Delta = \delta - \pi$.

The L.W. formulæ may be made exactly similar to those for H.W. by making the heights negative, and this condition is satisfied by adding π to all the arguments. Thus the formulæ for the height may be written

$$-\Sigma \mathrm{H} \cos(p\delta + \theta)$$

where $$\theta = \vartheta + (1-p)\pi$$

A similar change may be made in the nodal terms, but the parallactic terms require further consideration. The term $\cos(p_m\Delta + {}_m\vartheta)$ changes, not only because its sign is to be changed and Δ is to be replaced by $\delta - \pi$, but also because, as appears in § 3, ${}_m\vartheta$ changes.

For H.W.

$${}_m\vartheta = p_m\kappa_m - {}_m\kappa = p_n\kappa_m - \kappa_n + \frac{\kappa_m - \kappa_n}{9{\cdot}11\mathrm{H}_n/\mathrm{H}_m - 1}$$

but for L.W. κ_m must be replaced by $\kappa_m - \pi$ wherever it is multiplied by p; hence, for L.W., $p_m\kappa_m - {}_m\kappa$ must have $(p_m - p_n)\pi$ or $(1-p_n)\pi$ added to its previous value.

Hence the term in $\cos(p_m\Delta + {}_m\vartheta)$ for H.W. corresponds in the expression for $-h_p$ for L.W., to a term involving

$$\cos[p_m\delta + (1-p_m)\pi + (1-p_n)\pi + {}_m\vartheta]$$

Since $p_m = 1$, the required term is $\cos(p_m\delta + {}_m\theta)$, where

$${}_m\theta = {}_m\vartheta + (1-p_n)\pi$$

Again the parallactic term for O involves $\cos({}_op\Delta + {}_o\vartheta)$ for H.W., and treating it in the same way, the corresponding term in $-h_p$ for L.W. is found to be

$$\cos({}_op\delta + {}_o\theta)$$

where $${}_o\theta = (1 - {}_op)\pi + (p_o - p_q)\pi + {}_o\vartheta$$

But $${}_op = p_o - \frac{\epsilon}{\beta}$$

Hence $${}_o\theta = {}_o\vartheta + \left(1 + \frac{\epsilon}{\beta} - p_q\right)\pi$$

We have also

$${}_{\prime\prime}\theta = {}_{\prime\prime}\vartheta + (1 - {}_{\prime\prime}p)\pi, \qquad {}_{\prime}\theta = {}_{\prime}\vartheta + (1 - {}_{\prime}p)\pi$$

Then the L.W. formulæ for depths below mean sea level are similar in form to those for H.W. for elevation above mean sea level, with δ in place of Δ, and with θ's in the place of ϑ's.

The connection of θ's with ϑ's is given in the following table.

Initial	Principal and nodal terms	Parallactic terms
M_2	$\theta_m = \vartheta_m = 0$	${}_m\theta = {}_m\vartheta + (1 - p_n)\,\pi = {}_m\vartheta + 3^\circ{\cdot}4$
K_2	$\theta_{\prime\prime} = \vartheta_{\prime\prime} + (1 - p_{\prime\prime})\,\pi$	${}_{\prime\prime}\theta = {}_{\prime\prime}\vartheta + (1 - {}_{\prime\prime}p)\,\pi$
	$= \vartheta_{\prime\prime} - 7^\circ{\cdot}0$	$= {}_{\prime\prime}\vartheta - 13^\circ{\cdot}7$
S_2	$\theta_s = \vartheta_s + (1 - p_s)\,\pi$	
	$= \vartheta_s - 6^\circ{\cdot}4$	(assume $\beta = \frac{4}{3}$)
O	$\theta_o = \vartheta_o + (1 - p_o)\,\pi$	${}_o\theta = {}_o\vartheta + \left(1 + \frac{\epsilon}{\beta} - p_q\right)\pi$
	$= \vartheta_o + \frac{1}{2}\pi + 3^\circ{\cdot}4$	$= {}_o\vartheta + \frac{1}{2}\pi + 8^\circ{\cdot}5$
K_1	$\theta_{\prime} = \vartheta_{\prime} + (1 - p_{\prime})\,\pi$	${}_{\prime}\theta = {}_{\prime}\vartheta + (1 - {}_{\prime}p)\,\pi$
	$= \vartheta_{\prime} + \frac{1}{2}\pi - 3^\circ{\cdot}4$	$= {}_{\prime}\vartheta + \frac{1}{2}\pi - 6^\circ{\cdot}8$
P	$\theta_p = \vartheta_p + (1 - p_p)\,\pi$	
	$= \vartheta_p + \frac{1}{2}\pi - 2^\circ{\cdot}9$	
M_4	$\theta_{2m} = \vartheta_{2m} + (1 - 2p_m)\,\pi$	
	$= \vartheta_{2m} - \pi$	
S_4	$\theta_{2s} = \vartheta_{2s} + (1 - 2p_s)\,\pi$	
	$= \vartheta_{2s} - \pi - 12^\circ{\cdot}8$	

All the θ's differ from the ϑ's by a small angle, or by an angle nearly equal to 90° or 180°. Where the difference is small the A, G, F for L.W. will be nearly equal to those for H.W.; where the difference is nearly 90° the A, G, F for L.W. will be nearly the same as those for H.W., a quarter year earlier or later; and where the difference is nearly 180° the A, G, F for L.W. will be nearly equal and opposite to those for H.W. Hence we may set aside the change of 90° and 180° to be satisfied by a shift of a quarter year, or by change of sign. Suppose then, that $\theta = \vartheta + \alpha$, where α is small, and let [A], [G], [F] denote the values of L.W. A, G, F; then, remembering that

$$A = H\cos\vartheta, \qquad G = Hp^2\cos\vartheta, \qquad F = Hp\sin\vartheta$$

and that [A], [G], [F] are represented by similar formulæ with θ in place of ϑ, we have

$$[A] = A - \frac{\sin\alpha}{p} F$$

$$[G] = G - p\sin\alpha\, F$$

$$[F] = F + p\sin\alpha\, A$$

The values of α are given above, and those of the p's are known; hence it is easy to compute formulæ of transition from L.W. to H.W.

The rules given below in the example are derived from these formulæ, but the coefficients $\sin \alpha/p$ and $p \sin \alpha$ are given in round numbers appropriate for computation, and are sometimes treated as zero.

Part II. Computation.

Remarks on the Computations.

The multiplications are supposed to be done with Crelle's Bremiker's multiplication table*.

The other tables required are tables of squares, natural tangents, circular measure, and a traverse table. Bottomley's tables† are convenient for the purpose, because they give no more than is required. The nautical traverse table, such as that in Inman's tables, or in Chambers' logarithms, is used for finding such quantities as $H \cos \vartheta$ and $H \sin \vartheta$, for if H is "Distance," $H \cos \vartheta$ is "Lat." and $H \sin \vartheta$ is "Dep.," and the position of the decimal point is determined by inspection. For the use of this table it is advantageous to have only angles with a whole number of degrees, so as to avoid cross interpolations‡; the whole calculation is therefore conducted so as to avoid broken degrees. A traverse table is commonly given for "Distances" from 0 to 300, hence, if the "Distance" involves three digits and lies between 300 and 999, an interpolation is required, so as to use the entries between 30 and 99; this interpolation can easily be made by inspection.

All the angles are entered so that the significant part is less than 90°, by treating them as + or −, with π or 180° added where necessary. This facilitates the use of the traverse table.

It is best to determine the signs of the cosines and sines independently from their numerical values, and, accordingly, in the example where ·000 is entered as the value of a sine or cosine, it has a sign attached to it.

I suppose the computer to be able to add up a short column of figures, where some of the entries are + and others −. This is an arithmetical process not much practised, but easily acquired.

The sequences of angles and of cosines and sines, which occur frequently below, appertain (except in the cases of Sa and Ssa) to values of the excess of moon's longitude over sun's (for which the symbol used is ☽) at intervals of 15°, beginning with ☽ = 0°, and ending with ☽ = 345°, 24 values in all. But in the earlier part of the computation, the beginning of the sequence occurs at a different part of the column at different times of the year. Thus,

* *Rechentafeln*, Berlin, Georg Reimer.

† *Four-figure Mathematical Tables*, by J. T. Bottomley, F.R.S. Macmillan.

‡ Inman's Traverse Table is arranged so that the interpolation for a fraction of a degree is not very awkward.

a list of months is written in the margin, to show where we are to begin at any specified time of year. Strictly speaking these months are the times when the sun's longitude + 5° (for which the symbol used is ☉) is equal to a multiple of 30°; thus, when ☉ is 0°, we have March 15th, when ☉ is 30°, April 14th, and so on, as shown in Table VI. If the number of degrees in ☽ be reduced to time, at the rate of 15° per hour, we have, approximately, the time of the moon's transit, and in the later stages of the computation the time of Moon's transit is made to replace ☽. The sequences of angles are found by adding multiples of 30°, adding or subtracting multiples of 15°, or adding multiples of 60° (see Table II.) to certain initial angles (see Table I.). When the sequence has been carried so far that the next addition would reproduce the first angle with π added to it, it is unnecessary to proceed further. In the sequences of cosines and sines of such angles, when we have got to this same point, it is unnecessary to proceed further, since the remainder is the same as the beginning, with the sign changed.

In subsequent stages where a constant has to be added to a sequence, the new sequence will have double as many entries as the old, the first half being formed by addition, and the second half by subtraction; but in repeating the new sequence the signs are *not* to be changed. This follows immediately from what has been said of the signs of sequences.

Before proceeding with the computations I give some tables and rules of general applicability to all ports. It will be best for the computer who is learning the process to pass straight to the example, and to refer back to these tables as they are required; but I give them in the first place, because they will be wanted in the case of any other port.

Tables and Rules applicable to all Ports.

TABLE I. For finding the K's, the initial entries of the several sequences, for H.W. (See (26), § 5.)

Initials	Principal and nodal terms	Parallactic terms
M_2	$K_m = p_m \kappa_m - \kappa_m = 0$	${}_mK = p_n \kappa_m - \kappa_n + \dfrac{\kappa_m - \kappa_n}{9{\cdot}11\, H_n/H_m - 1}$
K_2	$K_{\prime\prime} = p_{\prime\prime} \kappa_m - \kappa_{\prime\prime} - 10^\circ$	${}_{\prime\prime}K = K_{\prime\prime} + ({}_{\prime\prime}p - p_{\prime\prime})\,\kappa_m = K_{\prime\prime} + {\cdot}038\,\kappa_m$
S_2	$K_s = p_s \kappa_m - \kappa_s$	
O	$K_o = p_o \kappa_m - \kappa_o + 95^\circ$	${}_oK = p_q \kappa_m - \kappa_q - \dfrac{{\cdot}013}{\beta}\kappa_m + \dfrac{\kappa_o - \kappa_q}{9{\cdot}11\, H_q/H_o - 1} + 95^\circ$
K_1	$K_\prime = p_\prime \kappa_m - \kappa_\prime - 95^\circ$	${}_\prime K = K_\prime + ({}_\prime p - p_\prime)\,\kappa_m = K_\prime + {\cdot}0192\,\kappa_m$
P	$K_p = p_p \kappa_m - \kappa_p + 95^\circ$	
M_4	$K_{2m} = 2p_m \kappa_m - \kappa_{2m}$	
S_4	$K_{2s} = 2p_s \kappa_m - \kappa_{2s}$	
Sa	$K_{sa} = -\kappa_{sa} - 5^\circ$	
Ssa	$K_{ssa} = -\kappa_{ssa} - 10^\circ$	

TABLE II. For finding the sequences of the ϑ's by putting n successively equal to 0, 1, 2, 3, &c. (See (26), § 5.)

Initials	Principal and nodal terms ϑ	Parallactic terms ϑ
M_2	K_m	$_mK$
K_2	$K_{\prime\prime} +30°n$	$_{\prime\prime}K+30°n$
S_2	$K_s +30°n$	
O	$K_o -15°n$	$_oK-15°n$
K_1	$K_{\prime} +15°n$	$_{\prime}K+15°n$
P	$K_p +15°n$	
M_4	K_{2m}	
S_4	$K_{2s} +60°n$	
Ssa	$K_{ssa}+60°n$	
Sa	$K_{sa} +10°n$	

N.B. The sequence for Sa is required under conditions which differ from those of the other ϑ's.

TABLE III. The numerical values of the p's. (See § 5.)

Initials	Principal and nodal terms	Parallactic terms
M_2	$p_m=1$, $p_m^2=1$	
K_2	$p_{\prime\prime}=1{\cdot}038$, $p_{\prime\prime}^2=1{\cdot}078$	$_{\prime\prime}p=1{\cdot}076$, $_{\prime\prime}p^2=1{\cdot}158$
S_2	$p_s=1{\cdot}035$, $p_s^2=1{\cdot}071$	
N	$p_n={\cdot}981$	
O	$p_o={\cdot}481$, $p_o^2={\cdot}231$	$_op={\cdot}481-{\cdot}013\beta^{-1}$, $_op^2={\cdot}231-{\cdot}012\beta^{-1}$
K_1	$p_{\prime}={\cdot}519$, $p_{\prime}^2={\cdot}269$	$_{\prime}p={\cdot}538$, $_{\prime}p^2={\cdot}289$
P	$p_p={\cdot}516$, $p_p^2={\cdot}266$	
Q	$p_q={\cdot}462$	

TABLE IV. For computing corrections for reference to moon's transit, viz., δT, δM, δN, and the sequence Θ. (See (55) § 10.)

Sequence $(-10^\circ+30^\circ n)$ Θ			
March	-10°	June	$\pi-10^\circ$
	$+20$		$\pi+20$
April	$+50$	July	$\pi+50$
	$+80$		$\pi+80$
May	$\pi-70$	August	-70
	$\pi-40$		-40

Repeat the sequence.

δT $=0^h{\cdot}172\sin\Theta$		δM $=4^m{\cdot}60\sin\Theta$		$-4^m{\cdot}22\cos\Theta$		δN$=$ $-4^m{\cdot}22\cos\Theta-4^m{\cdot}41$	
	h.		m.		m.		m.
March	$-{\cdot}030$	March	$-0{\cdot}80$	March	$-4{\cdot}16$	March	$-8{\cdot}57$
	$+{\cdot}059$		$+1{\cdot}57$		$-3{\cdot}97$		$-8{\cdot}38$
April	$+{\cdot}132$	April	$+3{\cdot}52$	April	$-2{\cdot}71$	April	$-7{\cdot}12$
	$+{\cdot}169$		$+4{\cdot}53$		$-0{\cdot}73$		$-5{\cdot}14$
May	$+{\cdot}162$	May	$+4{\cdot}32$	May	$+1{\cdot}45$	May	$-2{\cdot}96$
	$+{\cdot}111$		$+2{\cdot}96$		$+3{\cdot}23$		$-1{\cdot}18$
June	$+{\cdot}030$	June	$+0{\cdot}80$	Rep. and ch.		June	$-0{\cdot}25$
	$-{\cdot}059$		$-1{\cdot}57$				$-0{\cdot}44$
July	$-{\cdot}132$	July	$-3{\cdot}52$			July	$-1{\cdot}70$
	$-{\cdot}169$		$-4{\cdot}53$				$-3{\cdot}68$
August	$-{\cdot}162$	August	$-4{\cdot}32$			August	$-5{\cdot}86$
	$-{\cdot}111$		$-2{\cdot}96$				$-7{\cdot}64$

The sequences for δT, δM, δN are to be repeated without change of sign. To find the succession of values for any month we begin with the entry opposite to that month, read on down to the bottom, and then begin again at the top. For example, δT for July begins with $-{\cdot}132$, and then, after going on down to $-{\cdot}111$, it begins again at the top with $-{\cdot}030$.

TABLE V. For corrections due to part of the equation of time. (See (63) (64) § 10.)

The following is a table of $-3^{m}\cdot85 \sin(\odot + 74°)$ (which I call c), and of the same when the hour is unit of time (which I call d).

N.B. 286° is the longitude of sun's perigee + 5°, and 74° is its supplement to 360°.

	c $-3^{m}\cdot85 \sin(\odot + 74°)$	d $-0^{h}\cdot064 \sin(\odot + 74°)$
	m.	h.
March . . .	−3·7	−·062
April	−3·7	−·062
May	−2·8	−·046
June	−1·1	−·018
July	+0·9	+·016
August . . .	+2·7	+·045

TABLE VI. Dates and Limits of Applicability of the Tide-Tables.

Heading for tide-table	⊙	Applicability by reference to Sun's long. at Moon's transit Sun's longitude	Heading for tide-table	⊙	Applicability by reference to Sun's long. at Moon's transit Sun's longitude
March 15	0°	from 350° to 0°	Sept. 17	180°	from 170° to 180°
„ 25	10	„ 0 „ 10	„ 28	190	„ 180 „ 190
April 4	20	„ 10 „ 20	Oct. 8	200	„ 190 „ 200
„ 14	30	„ 20 „ 30	„ 18	210	„ 200 „ 210
„ 25	40	„ 30 „ 40	„ 28	220	„ 210 „ 220
May 5	50	„ 40 „ 50	Nov. 7	230	„ 220 „ 230
„ 15	60	„ 50 „ 60	„ 17	240	„ 230 „ 240
„ 26	70	„ 60 „ 70	„ 27	250	„ 240 „ 250
June 5	80	„ 70 „ 80	Dec. 7	260	„ 250 „ 260
„ 16	90	„ 80 „ 90	„ 16	270	„ 260 „ 270
„ 26	100	„ 90 „ 100	„ 26	280	„ 270 „ 280
July 7	110	„ 100 „ 110	Jan. 5	290	„ 280 „ 290
„ 17	120	„ 110 „ 120	„ 15	300	„ 290 „ 300
„ 28	130	„ 120 „ 130	„ 25	310	„ 300 „ 310
Aug. 7	140	„ 130 „ 140	Feb. 4	320	„ 310 „ 320
„ 17	150	„ 140 „ 150	„ 13	330	„ 320 „ 330
„ 28	160	„ 150 „ 160	„ 23	340	„ 330 „ 340
Sept. 7	170	„ 160 „ 170	March 5	350	„ 340 „ 350

This table gives the days of the year on which ⊙, or Sun's longitude + 5°, is nearly equal to a multiple of 10°. These days are used as headings to the several tide-tables. It is intended that the tables shall be used without an interpolation for the time of year, which ought strictly to be made. When the time of a particular moon's transit, with reference to which a tide is to be calculated, falls nearly halfway between any two of the specified days, it becomes uncertain which of the two adjoining tables should be used, and the question can only be decided by reference to the Sun's longitude. A column is therefore given of the limits of applicability of the table.

It would be easy, by means of a table of four columns referring to leap year, to give the Greenwich times at which the Sun's longitude is 0°, 10°, 20°, &c., which would be accurate enough for the present purpose during some twenty-five years.

VII. *The Choice of a Unit of Length.*

In a calculation of this kind it is advantageous to reduce the number of digits as far as possible, consistently with due accuracy, and it is convenient to omit the decimal point when we deal with heights.

The diurnal tides are so various at different places that no general rule can be made to depend on them.

It is required to express as many of the heights as possible by two digits, and it will be best to take such a unit that much of the work shall be conducted with 70's and 80's, but to allow a margin and not to try to bring them into the 90's.

After consideration I think it is best to take such a unit that $\alpha' H_m$ (see below) shall be expressed by 70 or 80. Since α' is usually about $\frac{4}{57}$, or say $\frac{1}{14}$, then when H_m is given in feet and decimals (or any other unit), we are to multiply the heights by a simple factor lying between $14 \times 70 \div H_m$ and $14 \times 80 \div H_m$.

The rule therefore is:—

Multiply the heights by a factor lying between $1000 \div H_m$ *and* $1200 \div H_m$, *and omit decimals.*

In the example below it would have been best to multiply all the heights by 700. We should then have $H_m = 1098$, $H_s = 488$, &c., and this would have made $\alpha' H_m$ equal to 77. The final step in the calculation would then have been to divide all the heights by 700.

In my example I have not followed this plan, and accordingly the decimal point is retained, and an unnecessary number of digits has been written.

VIII. *Rules for the Calculation of a L.W. Table.*

It will be more convenient to state these rules as part of the example for the Port of Aden, although they are of course generally applicable to all ports.

Example of Formation of a Tide-Table.

Table of Constants for the Port of Aden.

		ft.			ft.
M_2	H_m	$= 1{\cdot}568$	O	H_o	$= {\cdot}653$
	κ_m	$= 229^\circ$		κ_o	$= 38^\circ$
S_2	H_s	$= {\cdot}697$	K_1	$H_\prime$	$= 1{\cdot}299$
	κ_s	$= 248^\circ$		$\kappa_\prime$	$= 36^\circ$
K_2	$H_{\prime\prime}$	$= {\cdot}201$	P	H_p	$= {\cdot}388$
	$\kappa_{\prime\prime}$	$= 244^\circ$		κ_p	$= 33^\circ$
N	H_n	$= {\cdot}427$	Q	H_q	$= {\cdot}151$
	κ_n	$= 225^\circ$		κ_q	$= 42^\circ$
L	H_l	$= {\cdot}046$	Sa	H_{sa}	$= {\cdot}390$
	κ_l	$= 230^\circ$		κ_{sa}	$= 357^\circ$
M_4	H_{2m}	$= {\cdot}007$	Ssa	H_{ssa}	$= {\cdot}095$
	κ_{2m}	$= 314^\circ$		κ_{ssa}	$= 126^\circ$
S_4	H_{2s}	$= {\cdot}006$		$\mathfrak{A}$	$= 3{\cdot}859$
	κ_{2s}	$= 271^\circ$			

These are the results of four years of observation, and are the constants from which the tide-table is to be computed.

Formation of Sequences (see Tables I., II., III., and § 6).

K_2	S_2	O	K_1
$p_{\prime\prime}\kappa_m = 238^\circ$ $-\kappa_{\prime\prime} = -244$ $-10^\circ = -10$	$p_s\kappa_m = 237^\circ$ $-\kappa_s = -248$	$p_o\kappa_m = 110^\circ$ $-\kappa_o = -38$ $+95^\circ = 95$ 167	$p_\prime\kappa_m = 119^\circ$ $-\kappa_\prime = -36$ $-95^\circ = -95$
$K_{\prime\prime} = -16$	$K_s = -11$	$K_o = \pi - 13$	$K_\prime = -12$

P	M_4	S_4
$p_p\kappa_m = 118^\circ$ $-\kappa_p = -33$ $+95^\circ = 95$ 180	$2p_m\kappa_m = 458^\circ$ $-\kappa_{2m} = -314$ 144	$2p_s\kappa_m = 474^\circ$ $-\kappa_{2s} = -271$ 203
$K_p = \pi + 0$	$K_{2m} = \pi - 36$	$K_{2s} = \pi + 23$

Sequences of Angles.

M_2 ϑ_m	K_2 $(K_{,,}+30°n)$ $\vartheta_{,,}$	S_2 $(K_s+30°n)$ ϑ_s	O $(K_o-15°n)$ ϑ_o	K_1 $(K_{,}+15°n)$ $\vartheta_{,}$	P $(K_p+15°n)$ ϑ_p	M_4 (K_{2m}) ϑ_{2m}	S_4 $(K_{2s}+60°n)$ ϑ_{2s}
0° (a const.)	Mar. −16°	All months −11°	Mar. π−13°	Mar. −12°	Mar. π+0°	π−36° (const.)	All months π+23°
	+14	+19	π−28	+3	π+15		π+83
	April +44	+49	April π−43	April +18	Feb. π+30		−37
	+74	+79	π−58	+33	π+45		
	May π−76	π−71	May π−73	May +48	Jan. π+60		
	π−46	π−41	π−88	+63	π+75		
			June +77	June +78	Dec. π+90		
			+62	−87	−75		
			July +47	July −72	Nov. −60		
			+32	−57	−45		
			Aug. +17	Aug. −42	Oct. −30		
			+2	−27	−15		

Semi-diurnal: $A = H\cos\vartheta,\ G = Hp^2\cos\vartheta,\ F = Hp\sin\vartheta.$

M_2	S_2			K_2		
	A_s $(H_s=\cdot697)$	G_s $(H_sp_s^2=\cdot746)$	F_s $(H_sp_s=\cdot721)$	$A_{,,}$ $(H_{,,}=\cdot201)$	$G_{,,}$ $(H_{,,}p_{,,}^2=\cdot217)$	$F_{,,}$ $(H_{,,}p_{,,}=\cdot209)$
$H_m=H_mp_m^2=1\cdot568$	All months +·684	+·732	−·137	Mar. +·193	+·209	−·058
$A_m=G_m=1\cdot568$	+·659	+·705	+·234	+·195	+·211	+·051
$F_m=0$	+·457	+·489	+·544	April +·145	+·156	+·145
(constants)	+·133	+·142	+·708	+·055	+·060	+·201
	−·227	−·243	+·682	May −·049	−·053	+·203
	−·526	−·563	+·473	−·140	−·151	+·150

Repeat the sequences for K_2 and S_2 changing the signs.

Add M_2 to S_2.

A_m+A_s	G_m+G_s	F_m+F_s
All months 2·252	2·300	Same as F_s
2·227	2·273	
2·025	2·057	
1·701	1·710	
1·341	1·325	
1·042	1·005	
·884	·836	
·909	·863	
1·111	1·079	
1·435	1·426	
1·795	1·811	
2·094	2·131	

Repeat without change.

Then write out the three K_2 sequences, with the months, *in extenso*, each on a separate strip of paper; place the $A_{\prime\prime}$ strip opposite to the $A_m + A_s$ table, so that the month for which the sequence is required falls in the first place; *e.g.*, for March put + ·193, for April put + ·145, for May put − ·049, for June put − ·193, &c., in the first place opposite 2·252 of $A_m + A_s$. Then add the $A_m + A_s$ and $A_{\prime\prime}$ tables together in a different way for each of the six months from March to August.

Proceed with the $G_{\prime\prime}$, $F_{\prime\prime}$ strips in the same way. The next following table is formed in this way.

Semi-diurnal: $A_2 = A_m + A_{\prime\prime} + A_s$, $G_2 = G_m + G_{\prime\prime} + G_s$, $F_2 = F_m + F_{\prime\prime} + F_s$.

March			April			&c.
A_2	G_2	F_2	A_2	G_2	F_2	Continue as far as August inclusive
2·445	2·509	− ·195	2·397	2·456	+ ·008	
2·422	2·484	+ ·285	2·282	2·333	+ ·435	
2·170	2·213	+ ·689	1·976	2·004	+ ·747	
1·756	1·770	+ ·909	1·561	1·559	+ ·858	
1·292	1·272	+ ·885	1·148	1·116	+ ·740	
·902	·854	+ ·623	·847	·794	+ ·422	
·691	·627	+ ·195	·739	·680	− ·008	
·714	·652	− ·285	·854	·803	− ·435	
·966	·923	− ·689	1·160	1·132	− ·747	
1·380	1·366	− ·909	1·575	1·577	− ·858	
1·844	1·864	− ·885	1·988	2·020	− ·740	
2·234	2·282	− ·623	2·289	2·342	− ·422	

Repeat without change.

Diurnal: $A = H \cos \vartheta$, $G = Hp^2 \cos \vartheta$, $F = Hp \sin \vartheta$.

	O				K_1				P		
	A_o (H_o = ·653)	G_o ($H_o p_o^2$ = ·151)	F_o ($H_o p_o$ = ·314)		$A_{\prime}$ ($H_{\prime}$ = 1·299)	$G_{\prime}$ ($H_{\prime} p_{\prime}^2$ = ·350)	$F_{\prime}$ ($H_{\prime} p_{\prime}$ = ·674)		A_p (H_p = ·388)	G_p ($H_p p_p^2$ = ·103)	F_p ($H_p p_p$ = ·200)
Mar.	− ·636	− ·147	+ ·071	Mar.	+ 1·271	+ ·342	− ·140	Mar.	− ·388	− ·103	·000
	− ·577	− ·133	+ ·148		+ 1·297	+ ·350	+ ·035		− ·375	− ·100	− ·052
April	− ·477	− ·110	+ ·214	April	+ 1·235	+ ·333	+ ·208	Feb.	− ·336	− ·089	− ·100
	− ·346	− ·080	+ ·266		+ 1·089	+ ·294	+ ·367		− ·275	− ·073	− ·141
May	− ·191	− ·044	+ ·300	May	+ ·869	+ ·234	+ ·501	Jan.	− ·194	− ·052	− ·173
	− ·023	− ·005	+ ·314		+ ·590	+ ·159	+ ·601		− ·100	− ·027	− ·193
June	+ ·147	+ ·034	+ ·306	June	+ ·270	+ ·073	+ ·659	Dec.	·000	·000	− ·200
	+ ·307	+ ·071	+ ·278		− ·068	− ·018	+ ·673		+ ·100	+ ·027	− ·193
July	+ ·445	+ ·103	+ ·230	July	− ·402	− ·108	+ ·641	Nov.	+ ·194	+ ·052	− ·173
	+ ·554	+ ·128	+ ·166		− ·708	− ·191	+ ·565		+ ·275	+ ·073	− ·141
Aug.	+ ·625	+ ·144	+ ·092	Aug.	− ·965	− ·260	+ ·451	Oct.	+ ·336	+ ·089	− ·100
	+ ·653	+ ·151	+ ·011		− 1·157	− ·312	+ ·306		+ ·375	+ ·100	− ·052

Repeat changing signs.

Add together the O and K_1 sequences as they stand above:—

$O + K_1$.

	$A_o + A_,$	$G_o + G_,$	$F_o + F_,$
March	+·635	+·195	−·069
	+·720	+·217	+·183
April	+·758	+·223	+·422
	+·743	+·214	+·633
May	+·678	+·190	+·801
	+·567	+·154	+·915
June	+·417	+·107	+·965
	+·239	+·053	+·951
July	+·043	−·005	+·871
	−·154	−·063	+·731
August	−·340	−·116	+·543
	−·504	−·161	+·317

Repeat changing signs.

Write out the P sequences *in extenso* with the months on the margin (24 entries), each on a separate strip of paper; place the A_p strip opposite the $A_o + A_,$ table so that any chosen month in one agrees with that month on the other; add the two tables together, making the first entry in the new sequence that opposite which the chosen month is written. For example the April entry in the A_p sequence (completed) is −·336, and this added to +·758, the April entry of $A_o + A_,$, gives +·422, which is the initial entry in the diurnal sequence for A_1 corresponding to April in the following table. G_p and F_p are operated on in the same way.

The first time the computer does this sort of work he may find it convenient to write out the $O + K_1$ sequences *in extenso*, so as to see exactly how the computation runs, but it will be found with a little practice that this is unnecessary.

Diurnal: $A_1 = A_o + A_, + A_p$, $G_1 = G_o + G_, + G_p$, $F_1 = F_o + F_, + F_p$.

March			April			&c.
A_1	G_1	F_1	A_1	G_1	F_1	Continue up to August, inclusive
+·247	+·092	−·069	+·422	+·134	+·522	
+·345	+·117	+·131	+·368	+·114	+·685	
+·422	+·134	+·322	+·290	+·087	+·801	
+·468	+·141	+·492	+·192	+·054	+·863	
+·484	+·138	+·628	+·081	+·018	+·865	
+·467	+·127	+·722	−·036	−·020	+·810	
+·417	+·107	+·765	−·151	−·057	+·698	
+·339	+·080	+·758	−·254	−·090	+·538	
+·237	+·047	+·698	−·340	−·116	+·343	
+·121	+·010	+·590	−·404	−·134	+·124	
−·004	−·027	+·443	−·441	−·143	−·104	
−·129	−·061	+·265	−·445	−·144	−·324	

Quater-diurnal : $A = H\cos\vartheta$, $G = H(2p)^2\cos\vartheta$, $F = H(2p)\cos\vartheta$.

M_4				S_4		
A_{2m} $(H_{2m} = \cdot007)$	G_{2m} $(4H_m = \cdot028)$	F_{2m} $(2H_m = \cdot014)$		A_{2s} $(H_{2s} = \cdot006)$	G_{2s} $(4H_{2s}p_s^2 = \cdot026)$	F_{2s} $(2H_{2s}p_s = \cdot012)$
$-\cdot006$ const.	$-\cdot023$ const.	$+\cdot008$ const.	All months	$-\cdot006$ $-\cdot001$ $+\cdot005$ Repeat and change	$-\cdot024$ $-\cdot003$ $+\cdot021$ Repeat and change	$-\cdot005$ $-\cdot012$ $-\cdot007$ Repeat and change

Quater-diurnal : $A_4 = A_{2m} + A_{2s}$, $G_4 = G_{2m} + G_{2s}$, $F_4 = F_{2m} + F_{2s}$.

	A_4	G_4	F_4
All months	$-\cdot012$	$-\cdot047$	$+\cdot003$
	$-\cdot007$	$-\cdot026$	$-\cdot004$
	$-\cdot001$	$-\cdot002$	$+\cdot001$
	$\cdot000$	$+\cdot001$	$+\cdot013$
	$-\cdot005$	$-\cdot020$	$+\cdot020$
	$-\cdot011$	$-\cdot044$	$+\cdot015$

Repeat without change of sign.

Semi-annual and Mean Water. (See (56), § 10.)

$$-\kappa_{ssa} = 234^\circ$$
$$-10^\circ = -\ 10$$
$$224$$
$$K_{ssa} = \pi + 44$$

Sequence.

$K_{ssa} + 60^\circ n$ ϑ_{ssa}	$H_{ssa}\cos\vartheta_{ssa}$ $(H_{ssa} = \cdot095)$	$B_0 = \mathfrak{A} + H_{ssa}\cos\vartheta_{ssa}$ $(\mathfrak{A} = 3\cdot859)$
March $\pi + 44^\circ$	$-\cdot068$	March 3·791
April -76	$+\cdot023$	April 3·882
May -17	$+\cdot091$	May 3·950
	Repeat and change	June 3·927
		July 3·836
		Aug. 3·768

Annual.

$$-\kappa_{sa} = -357^\circ$$
$$-5^\circ = -\ \ 5$$
$$\mathrm{K}_{sa} = -\ \ 2$$

Sequence.

$\mathrm{K}_{sa} + 10^\circ n$ ϑ_{sa}	
March -2°	June $+88^\circ$
$+8$	$\pi - 82$
$+18$	$\pi - 72$
April $+28$	July $\pi - 62$
$+38$	$\pi - 52$
$+48$	$\pi - 42$
May $+58$	Aug. $\pi - 32$
$+68$	$\pi - 22$
$+78$	$\pi - 12$

Annual Tide.

$\mathrm{H}_{sa} \cos \vartheta_{sa}$ $(\mathrm{H}_{sa} = \cdot 390)$	
ft.	ft.
March $+\cdot 390$	June $+\cdot 014$
$+\cdot 386$	$-\cdot 054$
$+\cdot 371$	$-\cdot 121$
April $+\cdot 344$	July $-\cdot 183$
$+\cdot 307$	$-\cdot 240$
$+\cdot 261$	$-\cdot 290$
May $+\cdot 207$	Aug. $-\cdot 331$
$+\cdot 146$	$-\cdot 362$
$+\cdot 081$	$-\cdot 381$

Mean Interval i, and Parallactic Correction to i. (See (31), § 6; and (47), § 8.)

$$\kappa_m = 229^\circ$$

$$\frac{1}{30}\kappa_m = 7\cdot 633$$
$$\frac{1}{30^2}\cdot\frac{21}{20}\kappa_m = \cdot 267$$
$$i = 7^{\mathrm{h}}\cdot 900$$

Retaining i in hours, parallactic correction to $i = +\,0^{\mathrm{m}}\cdot 08 \times i$
$= +\,0^{\mathrm{m}}\cdot 63$

N.B. $0^{\mathrm{m}}\cdot 08$ is an absolute constant.

Parallactic Corrections. (See §§ 3, 8.)

$36{\cdot}4\,\mathrm{H}_n$ is greater than $8\,\mathrm{H}_m$; therefore $\alpha=\frac{4}{3}$, $\alpha'\mathrm{H}_m = {\cdot}07\,\mathrm{H}_m = {\cdot}110$

$36{\cdot}4\,\mathrm{H}_q$ is greater than $8\,\mathrm{H}_o$; therefore $\beta=\frac{4}{3}$, $\beta'\mathrm{H}_o = {\cdot}07\,\mathrm{H}_o = {\cdot}046$

(N.B. When either of these inequalities is *less* instead of greater, put

$$\alpha = 12{\cdot}14\,\mathrm{H}_n/\mathrm{H}_m - \tfrac{4}{3},\quad \alpha'\mathrm{H}_m = {\cdot}639\,\mathrm{H}_n - {\cdot}07\,\mathrm{H}_m$$

$$\beta = 12{\cdot}14\,\mathrm{H}_q/\mathrm{H}_o - \tfrac{4}{3},\quad \beta'\mathrm{H}_o = {\cdot}639\,\mathrm{H}_q - {\cdot}07\,\mathrm{H}_o$$

If the N tide is unknown take $\alpha = 1$, $\alpha' = \frac{1}{19}$; if the Q tide is unknown take $\beta = 1$, $\beta' = \frac{1}{19}$.)

(Table III.) $_op = {\cdot}481 - {\cdot}013\beta^{-1} = {\cdot}481 - {\cdot}010 = {\cdot}471$

$_op^2 = {\cdot}231 - {\cdot}012\beta^{-1} = {\cdot}231 - {\cdot}009 = {\cdot}222$

(Table I.) Let

$$\gamma = \frac{\kappa_m - \kappa_n}{9{\cdot}11\,\mathrm{H}_n/\mathrm{H}_m - 1}$$

the denominator is $9{\cdot}11 \times {\cdot}272 - 1 = 1{\cdot}48$; the numerator is $4°$.

Hence $$\gamma = +4° \div 1{\cdot}48 = +3°$$

Let $$\delta = \frac{\kappa_o - \kappa_q}{9{\cdot}11\,\mathrm{H}_q/\mathrm{H}_o - 1} - \frac{{\cdot}013}{\beta}\kappa_m$$

the denominator of the first term is $9{\cdot}11 \times {\cdot}232 - 1 = 1{\cdot}11$; the numerator is $-4°$. Hence the first term is $-4°$. The second term is

$$-{\cdot}010 \times 229° = -2°$$

Hence $$\delta = -6°$$

(See Tables I., II., III.)

M_2	K_2	O	K_1
$p_n\kappa_m = 225°$	${\cdot}0384\kappa_m = 9°$	$p_q\kappa_m = 106°$	${\cdot}0192\kappa_m = 4°$
$-\kappa_n = -225$	$\mathrm{K}_{\prime\prime} = -16$	$-\kappa_q = -42$	$\mathrm{K}_{\prime} = -12$
$+\gamma \quad 3$		$+\delta = -6$	
		$+95° = 95$	
		153	
$_mK = +3$	$_{\prime\prime}K = -7$	$_oK = \pi - 27$	$_{\prime}K = -8$

N.B. My calculations were made on a principle, now abandoned, which led to slightly different values. I therefore now continue the calculation with

$$\alpha'\mathrm{H}_m = {\cdot}122,\quad \beta'\mathrm{H}_o = {\cdot}041$$

$$_m\mathrm{K} = +5°,\quad _{\prime\prime}\mathrm{K} = -8°,\quad _o\mathrm{K} = \pi - 19°,\quad _{\prime}\mathrm{K} = -8°$$

Sequences of Angles.

M_2 $({}_m K)$ ${}_m\vartheta$	K_2 $({}_{\prime\prime}K+30°n)$ ${}_{\prime\prime}\vartheta$	O $({}_o K-15°n)$ ${}_o\vartheta$	K_1 $({}_{\prime}K+15°n)$ ${}_{\prime}\vartheta$
$+5°$ const.	March $-8°$	March $\pi-19°$	March $-8°$
	$+22$	$\pi-34$	$+7$
	April $+52$	April $\pi-49$	April $+22$
	$+82$	$\pi-64$	$+37$
	May $\pi-68$	May $\pi-79$	May $+52$
	$\pi-38$	$+86$	$+67$
		June $+71$	June $+82$
		$+56$	$\pi-83$
		July $+41$	July $\pi-68$
		$+26$	$\pi-53$
		Aug. $+11$	Aug. $\pi-38$
		-4	$\pi-23$

Semi-diurnal.

$${}_m A=\alpha' H_m \cos {}_m\vartheta, \quad {}_m G=\alpha' H_m p_m^2 \cos {}_m\vartheta, \quad {}_m F=\alpha' H_m p_m \sin {}_m\vartheta$$

$${}_{\prime\prime}A=\cdot 036\, H_{\prime\prime} \cos {}_{\prime\prime}\vartheta, \quad {}_{\prime\prime}G=\cdot 036\, H_{\prime\prime}\, {}_{\prime\prime}p^2 \cos {}_{\prime\prime}\vartheta, \quad {}_{\prime\prime}F=\cdot 036\, H_{\prime\prime}\, {}_{\prime\prime}p \sin {}_{\prime\prime}\vartheta$$

$$z_2={}_m A+{}_{\prime\prime}A, \quad m_2={}_m G+{}_{\prime\prime}G, \quad l_2={}_m F+{}_{\prime\prime}F$$

${}_m A$ $(\alpha' H_m=\cdot 122)$	${}_{\prime\prime}A$ $(\cdot 036\, H_{\prime\prime}=\cdot 007)$	${}_m G$ $(\alpha' H_m p_m^2=\cdot 122)$	${}_{\prime\prime}G$ $(\cdot 036\, H_{\prime\prime}\,{}_{\prime\prime}p^2=\cdot 008)$	${}_m F$ $(\alpha' H_m p_m=\cdot 122)$	${}_{\prime\prime}F$ $(\cdot 036\, H_{\prime\prime}\,{}_{\prime\prime}p=\cdot 007)$
$+\cdot 122$ const.	March $+\cdot 007$	$+\cdot 122$ const.	March $+\cdot 008$	$+\cdot 011$ const	March $-\cdot 001$
	$+\cdot 006$		$+\cdot 007$		$+\cdot 003$
	April $+\cdot 004$		April $+\cdot 005$		April $+\cdot 006$
	$+\cdot 001$		$+\cdot 001$		$+\cdot 007$
	May $-\cdot 003$		May $-\cdot 003$		May $+\cdot 006$
	$-\cdot 006$		$-\cdot 006$		$+\cdot 004$

z_2	m_2	l_2
March $+\cdot 129$	March $+\cdot 130$	March $+\cdot 010$
$+\cdot 128$	$+\cdot 129$	$+\cdot 014$
April $+\cdot 126$	April $+\cdot 127$	April $+\cdot 017$
$+\cdot 123$	$+\cdot 123$	$+\cdot 018$
May $+\cdot 119$	May $+\cdot 119$	May $+\cdot 017$
$+\cdot 116$	$+\cdot 116$	$+\cdot 015$
June $+\cdot 115$	June $+\cdot 114$	June $+\cdot 012$
$+\cdot 116$	$+\cdot 115$	$+\cdot 008$
July $+\cdot 118$	July $+\cdot 117$	July $+\cdot 005$
$+\cdot 121$	$+\cdot 121$	$+\cdot 004$
Aug. $+\cdot 125$	Aug. $+\cdot 125$	Aug. $+\cdot 005$
$+\cdot 128$	$+\cdot 128$	$+\cdot 007$

Repeat without change.

It might suffice if the parallactic correction to K_2 were neglected, in which case $z_2=m_2={}_m A$, $l_2={}_m F$. The labour of making the correct table is, however, inconsiderable.

Diurnal: ${}_{o}A = \beta' H_o \cos {}_{o}\vartheta$, ${}_{o}G = \beta' H_o \, {}_{o}p^2 \cos {}_{o}\vartheta$, ${}_{o}F = \beta' H_o \, {}_{o}p \sin {}_{o}\vartheta$

${}_{,}A = \cdot 036 H_{,} \cos {}_{,}\vartheta$, ${}_{,}G = \cdot 036 H_{,} \, {}_{,}p^2 \cos {}_{,}\vartheta$, ${}_{,}F = \cdot 036 H_{,} \, {}_{,}p \sin {}_{,}\vartheta$

$z_1 = {}_{o}A + {}_{,}A$, $m_1 = {}_{o}G + {}_{,}G$, $l_1 = {}_{o}F + {}_{,}F$

	${}_{o}A$ ($\beta' H_o = \cdot 041$)	${}_{,}A$ ($\cdot 036 H_{,} = \cdot 047$)	${}_{o}G$ ($\beta' H_o \, {}_{o}p^2 = \cdot 009$)	${}_{,}G$ ($\cdot 036 H_{,} \, {}_{,}p^2 = \cdot 013$)	${}_{o}F$ ($\beta' H_o \, {}_{o}p = \cdot 019$)	${}_{,}F$ ($\cdot 036 H_{,} \, {}_{,}p = \cdot 025$)
March	− ·039	+ ·047	− ·009	+ ·013	+ ·006	− ·003
	− ·034	+ ·047	− ·007	+ ·013	+ ·011	+ ·003
April	− ·027	+ ·044	− ·006	+ ·012	+ ·014	+ ·009
	− ·018	+ ·038	− ·004	+ ·010	+ ·017	+ ·015
May	− ·008	+ ·029	− ·002	+ ·008	+ ·019	+ ·020
	+ ·003	+ ·018	+ ·001	+ ·005	+ ·019	+ ·023
June	+ ·013	+ ·007	+ ·003	+ ·002	+ ·018	+ ·025
	+ ·023	− ·006	+ ·005	− ·002	+ ·016	+ ·025
July	+ ·031	− ·018	+ ·007	− ·005	+ ·012	+ ·023
	+ ·037	− ·028	+ ·008	− ·008	+ ·008	+ ·020
Aug.	+ ·040	− ·037	+ ·009	− ·010	+ ·004	+ ·015
	+ ·041	− ·043	+ ·009	− ·012	− ·001	+ ·010

	z_1	m_1	l_1
March	+ ·008	+ ·004	+ ·003
	+ ·013	+ ·006	+ ·014
April	+ ·017	+ ·006	+ ·023
	+ ·020	+ ·006	+ ·032
May	+ ·021	+ ·006	+ ·039
	+ ·021	+ ·006	+ ·042
June	+ ·020	+ ·005	+ ·043
	+ ·017	+ ·003	+ ·041
July	+ ·013	+ ·002	+ ·035
	+ ·009	·000	+ ·028
Aug.	+ ·003	− ·001	+ ·019
	− ·002	− ·003	+ ·009

Repeat these sequences, changing the signs.

Nodal Corrections. (See (51) (52) § 9.)

Find by reference to preceding sequences the following nine sequences:—

(i.) $- \cdot 0372 A_m$, (ii.) $+ \cdot 283 A_{,,}$, (iii.) $+ \cdot 296 F_{,,}$, (iv.) $+ \cdot 283 F_{,,}$, (v.) $- \cdot 0372 G_m$, (vi.) $- \cdot 296 G_{,,}$, (vii.) $- \cdot 0372 G_m$, (viii.) $+ \cdot 283 G_{,,}$, (ix.) $+ \cdot 319 F_{,,}$

Then the semi-diurnal sequences are as follows:—

a_2 is (i.) + (ii.); b_2 is (iii.); c_2 is (iv.); d_2 is (v.) + (vi.); e_2 is (vii.) + (viii.); f_2 is (ix.)

For example:—

	a_2.		a_2.
March.........	− ·003	June...........	− ·113
	− ·003		− ·113
April	− ·017	July	− ·099
	− ·042		− ·074
May	− ·072	August.........	− ·044
	− ·098		− ·018

This, and the other semi-diurnal sequences are repeated without change of sign, and in all six of them the months run just as in this example, and denote the places at which to begin reading the sequence for the month in question.

The diurnal sequences are obtained thus:

Find, from preceding sequences, the twelve following—

(i.) $+ \cdot 188 A_o$, (ii.) $+ \cdot 115 A_,$, (iii.) $- \cdot 391 F_o$, (iv.) $+ \cdot 297 F_,$, (v.) $+ \cdot 188 F_o$, (vi.) $+ \cdot 115 F_,$, (vii.) $+ \cdot 391 G_o$, (viii.) $- \cdot 297 G_,$, (ix.) $+ \cdot 188 G_o$, (x.) $+ \cdot 115 G_,$, (xi.) $- \cdot 0905 F_o$, (xii.) $+ \cdot 080 F_,$

Then

a_1 is (i.) + (ii.); b_1 is (iii.) + (iv.); c_1 is (v.) + (vi.); d_1 is (vii.) + (viii.); e_1 is (ix.) + (x.); f_1 is (xi.) + (xii.)

For example:—

	a_1.		a_1.
March.........	+ ·026	June...........	+ ·059
	+ ·041		+ ·050
April	+ ·052	July	+ ·038
	+ ·060		+ ·023
May	+ ·064	August.........	+ ·007
	+ ·064		− ·010

This, and the other diurnal sequences, are repeated with change of sign, and the months in all six of them run just as in this example, and denote the places at which to begin reading the sequence for the month in question.

Calculation of Height, Interval, and Corrections for each Month. (See §§ 7, 8, 9, 10.)

Remarks. Each column in the following computation is arranged exactly like the first, so that it is unnecessary to repeat the letters in the successive columns.

For the month of March, which serves as an example, we refer to the March sequences, and enter the twelve values of G_2 successively, in the top left-hand corners of twelve columns; below these are entered the twelve values of G_1, and the twelve values of G_4, and on the right of the columns are put the twelve values of F_2, F_1, F_4. A similar statement is true of all the other symbols all the way down, and all the sequences are utilised up to twelve entries in each.

The divisions and multiplications may be done by Crelle's table; Δ_0 is found by a table of natural tangents, and $\delta\Delta$ is converted into degrees by a table of circular measure or radians. It is necessary to take as an approximate value of Δ_0 the nearest even number of degrees.

From the places where the values of Δ_0 are found, the left-hand side of each column corresponds to the time of moon's transit written at the head of the column, and the right-hand side to a time of moon's transit 12^h greater than the time specified. But the whole table for any month serves for its opposite (*e.g.*, September opposite to March), by transposing the words right and left in the preceding statement. Thus the whole computation has only to be made for six months (up to August inclusive), instead of for twelve. The diurnal terms with suffix 1, are written in the margin, with alternative signs, and the upper sign is to be used on the left, and the lower on the right of each column.

Thus, in finding, for example, $F_1 \cos \frac{1}{2}\Delta_0$ on the right we deem the F_1 written at the head to have its sign changed. Thus, in the column of 0^h we have $F_1 = -\cdot069$ and on the right-hand, $\Delta_0 = +4°$; then the required entry, on the right-hand, for $F_1 \cos \frac{1}{2}\Delta_0$ is $+\cdot069 \cos(+2°) = +\cdot069$.

The values of δT, δM, δN are extracted from the sequence of those functions in Table IV., and they are the same on each side of the column. The value of B_0 is taken from the sequence of the semi-annual tide and mean water, and changes only with the month.

The parallactic correction to the mean interval, i, is introduced in computing R. This is a constant of the port and is the same in all months.

In computing the height $\mathfrak{H}$, and its corrections, an approximate value of Δ is used, namely, the nearest even number of degrees; this approximate Δ will often be the same as Δ_0.

In this table it appears to me specially important that the signs of the sines and cosines should be determined independently of their numerical values.

Whereas in the right-hand of column 6^h we get, as a result of a second approximation, no value of Δ, the conjectural value $\Delta = +90°$ is adopted for the computation of P, Q, S, and values of M, N, R are not computed.

The table has rows marked δI, δh, M″, N″, P″, Q″; all these are derived from a subsequent table of "*Corrections for reference to the Moon's transit.*" But it appears convenient to finish off the computation on this sheet, although we have to pause in the computation in order to calculate the said table of corrections.

March.

Interval	0° or 0^h		15° or 1^h			90° or 6^h		
G_2 ; F_2	$+2·509$	$-·195$	$+2·484$	$+·285$		$+·627$	$+·195$	
G_1 ; F_1	$+·092$	$-·069$	$+·117$	$+·131$	&c.	$+·107$	$+·765$	
G_4 ; F_4	$-·047$	$+·003$	$-·026$	$-·004$		$-·047$	$+·003$	
G_2+G_1 ; F_2+F_1	$+2·601$	$-·264$	$+2·601$	$+·416$		$+·734$	$+·960$	
G_2-G_1 ; F_2-F_1	$+2·417$	$-·126$	$+2·367$	$+·154$		$+·520$	$-·570$	
$\tan\Delta_o = -\frac{F_2+F_1}{G_2+G_1}$; $-\frac{F_2-F_1}{G_2-G_1}$	$+·102$	$+·052$	$-·160$	$-·065$		$-1·308$	$+1·096$	
Δ_o	$+6°$	$+4°$	$-10°$	$-4°$		$-52°$	$+48°$	
α $F_2\cos\Delta_o$	$-·194$	$-·195$	$+·281$	$+·284$		$+·120$	$+·131$	
β $F_2\sin\Delta_o$	$-·020$	$-·014$	$-·050$	$-·020$		$-·154$	$+·145$	
γ $G_2\cos\Delta_o$	$+2·495$	$+2·503$	$+2·446$	$+2·478$		$+·386$	$+·420$	
δ $G_2\sin\Delta_o$	$+·262$	$+·175$	$-·431$	$-·173$		$-·493$	$+·466$	
ϵ $\pm F_1\cos\frac{1}{2}\Delta_o$	$-·069$	$+·069$	$+·131$	$-·131$		$+·688$	$-·699$	
ζ $\pm F_1\sin\frac{1}{2}\Delta_o$	$-·004$	$+·002$	$-·011$	$+·005$	&c.	$-·336$	$-·311$	
η $\pm 2G_1\cos\frac{1}{2}\Delta_o$	$+·184$	$-·184$	$+·233$	$-·234$		$+·192$	$-·196$	
θ $\pm 2G_1\sin\frac{1}{2}\Delta_o$	$+·010$	$-·006$	$-·020$	$+·008$		$-·094$	$-·087$	
λ $F_4\cos 2\Delta_o$	$+·003$	$+·003$	$-·004$	$-·004$		$-·001$	$·000$	
μ $F_4\sin 2\Delta_o$	$+·001$	$+·000$	$+·001$	$+·001$		$-·003$	$+·003$	
ν $\frac{1}{2}G_4\cos 2\Delta_o$	$-·023$	$-·023$	$-·012$	$-·013$		$+·006$	$+·002$	
ρ $\frac{1}{2}G_4\sin 2\Delta_o$	$-·005$	$-·003$	$+·004$	$+·002$		$+·023$	$-·023$	
$\alpha+\delta$	$+·068$	$-·020$	$-·150$	$+·111$		$-·373$	$+·597$	
$\epsilon+\theta$	$-·059$	$+·063$	$+·111$	$-·123$		$+·594$	$-·786$	
$\lambda+\rho$	$-·002$	$·000$	$·000$	$-·002$		$+·022$	$-·023$	
Sum N_o	$+·007$	$+·043$	$-·039$	$-·014$		$+·243$	$-·212$	
$\zeta-\eta$	$-·188$	$+·186$	$-·244$	$+·239$		$-·528$	$-·115$	
$\frac{1}{2}(\zeta-\eta)$	$-·094$	$+·093$	$-·122$	$+·120$		$-·264$	$-·058$	
$\beta-\gamma$	$-2·515$	$-2·517$	$-2·496$	$-2·498$	&c.	$-·540$	$-·275$	
$2(\mu-\nu)$	$+·048$	$+·046$	$+·026$	$+·028$		$-·018$	$+·002$	
Sum D_o	$-2·561$	$-2·378$	$-2·592$	$-2·350$		$-·822$	$-·331$	
$\delta\Delta = N_o/D_o$	$-·003$	$-·018$	$+·015$	$+·006$		$-·296$	$+·640$	
(In degrees) $\delta\Delta$	$-0°·2$	$-1°·0$	$+0°·9$	$+0°·3$		$-17°·0$	$+36°·7$	
$\Delta = \Delta_o + \delta\Delta$	$+5·8$	$+3·0$	$-9·1$	$-3·7$		$-68·6$*	No H.W.*	
$\frac{1}{30}\Delta$	·193	·100	·303	·123		2·287		
$\frac{1}{30^2}\frac{21}{20}\Delta$	7	4	11	4		80		
(In hours) I	$+·200$	$+·104$	$-·314$	$-·127$	&c.	$-2·367$		
δT	$-·030$	$-·030$	$+·059$	$+·059$		$+·030$		
	$+·170$	$+·074$	$-·255$	$-·068$		$-2·337$		
(Mean int.) i	7·900	7·900	7·900	7·900		7·900		
	8·070	7·974	7·645	7·832		5·563		
	$8^h\ 4^m·2$	$7^h\ 58^m·4$	$7^h\ 38^m·7$	$7^h\ 49^m·9$		$5^h\ 33^m·8$		
(See below) δI	$+0·9$	$+0·4$	$-1·8$	$-0·8$		$+0·6$		
$\mathfrak{I}$	$8^h\ 5^m$	$7^h\ 59^m$	$7^h\ 37^m$	$7^h\ 49^m$		$5^h\ 34^m$		

Take upper sign of diurnal terms on left, lower on right

Continue up to 165° or 11^h, twelve columns in all

* Second approximation (see below) introduced here

MARCH (continued).

Correction of D	0° or 0^h			15° or 1^h			90° or 6^h		
$\alpha+\delta$							$-\cdot531^*$		
$\frac{1}{4}(\epsilon+\theta)$							$+\cdot126^*$		
$4(\lambda+\rho)$							$+\cdot044^*$		
Sum	This correction need only be made when D_o is small and $\delta\Delta$ considerable					&c.	$-\cdot361^*$		&c.
(In circ. meas.) $\delta\Delta$							$+\cdot022^*$		
Product δD							$-\cdot008^*$		
D_o							$-\cdot744^*$		
$D=D_o+\delta D$	$-2\cdot56$		$-2\cdot38$	$-2\cdot59$	$-2\cdot35$		$-\cdot752^*$		
Height									
A_2	2·445	$-F_2$	+ ·195	2·422	− ·285	&c.	·691	− ·195	
A_1	+ ·247	$-F_1$	+ ·069	+ ·345	− ·131		+ ·417	− ·765	
A_4	− ·012	$-F_4$	− ·003	− ·007	+ ·004		− ·012	− ·003	
Approx. Δ	+6°		+4°	−10°	−4°		− 70*	None	
$A_2 \cos\Delta$	2·432	Upper sign on left; lower on right	2·439	2·385	2·416		·236	See below for evanescent tide	
$-F_2 \sin\Delta$	+ ·020		+ ·014	+ ·050	+ ·020		+ ·183		
$\pm A_1 \cos\frac{1}{2}\Delta$	+ ·247		− ·247	+ ·344	− ·345		+ ·342		
$\mp 2F_1 \sin\frac{1}{2}\Delta$	+ ·007		− ·005	+ ·023	− ·009	&c.	+ ·878		
$A_4 \cos 2\Delta$	− ·012		− ·012	− ·007	− ·007		+ ·009		
$-\frac{1}{2}F_4 \sin 2\Delta$	− ·000		− ·000	− ·001	− ·000		+ ·001		
Sum h	2·694		2·189	2·794	2·075		1·649		&c.
B_o	3·791		3·791	3·791	3·791		3·791		
	6·485		5·980	6·585	5·866		5·440		
(See below) δh	− ·004		+ ·001	+ ·004	− ·012		− ·014		
$\mathfrak{h}$	6·48		5·98	6·59	5·85		5·43		
Nodal correction									
Δ	+6°		+4°	−10°	−4°		−70°*	+90*	
$119^m \div D$	-46^m		-50^m	-46^m	-51^m	&c.	-158^{m*}		
c_2	− ·016	e_2	+ ·001	+ ·014	+ ·002		+ ·016	− ·117	
c_1	− ·003	e_1	+ ·011	+ ·032	+ ·015		+ ·134	+ ·014	
$c_2 \cos\Delta$	− ·016	Upper sign on left; lower on right	− ·016	+ ·014	+ ·014		+ ·005		
$e_2 \sin\Delta$	+ 0		+ 0	− 0	− 0		+ ·110		
$\pm c_1 \cos\frac{1}{2}\Delta$	− 3		+ 3	+ 32	− 32		+ ·110		
$\pm 2e_1 \sin\frac{1}{2}\Delta$	+ 1		− 1	− 3	+ 1		− ·16		
Sum	− ·018		− ·014	+ ·043	− ·017	&c.	+ ·209		&c.
Mult. by 119/D	+0·8		+0·7	−2·0	+0·9		− 33·0		
δM	−0·8		−0·8	+1·6	+1·6		+ 0·8		
Sum M′	0·0		−0·1	−0·4	+2·5		− 32·2		
(See below) M″	+0·4		+0·2	−0·8	−0·4		+ 0·3		
Sum M	+0·4		+0·1	−1·2	+2·1		− 31·9	None	

March (continued).

Nodal correction—(continued)	0° or 0^h			15° or 1^h			90° or 6^h		
d_2	− ·120	f_2	− ·019	&c.	&c.	&c.	&c.	&c.	&c.
d_1	− ·159	f_1	− ·017						
$d_2 \cos \Delta$									
$f_2 \sin \Delta$	&c.								
$\pm d_1 \cos \frac{1}{2}\Delta$									
$\pm 2f_1 \sin \frac{1}{2}\Delta$		Compute like M		&c.	&c.	&c.	&c.	&c.	&c.
Sum									
Mult. by 119/D	&c.								
δN									
Sum N′	&c.								
(See below) N″									
Sum N								None	
a_2	− ·003	$-c_2$	+ ·016	− ·003	− ·014	&c.	− ·113	− ·016	&c.
a_1	+ ·026	$-c_1$	+ ·003	+ ·041	− ·032		+ ·059	− ·134	
$a_2 \cos \Delta$	− ·003		− ·003	− ·003	− ·003		− ·039	·000	
$-c_2 \sin \Delta$	+ 2	Upper on left; lower on right	+ 1	+ 2	+ 1		+ 15	− 16	
$\pm a_1 \cos \frac{1}{2}\Delta$	+ 26		− 26	+ 41	− 41		+ 48	− 42	
$\mp 2c_1 \sin \frac{1}{2}\Delta$	+ 0		− 0	+ 6	− 2		+ ·154	+ ·190	
Sum P′	+ ·025		− ·028	+ ·046	− ·045		+ ·178	+ ·132	
(See below) P″	− ·002		·000	+ ·002	− ·005		− ·006	+ ·005	
Sum P	+ ·023		− ·028	+ ·048	− ·050		+ ·172	+ ·137	
b_2	− ·017	$-d_2$	+ ·120	&c.	&c.		&c.	&c.	
b_1	− ·070	$-d_1$	+ ·159						
$b_2 \cos \Delta$									
$-d_2 \sin \Delta$	&c.								
$\pm b_1 \cos \frac{1}{2}\Delta$									
$\mp 2d_1 \sin \frac{1}{2}\Delta$		Compute like P		&c.	&c.	&c.	&c.	&c.	&c.
Sum Q′	&c.								
(See below) Q″									
Sum Q									
Parallactic corrections									
l_2	+ ·010	m_2	+ ·130	+ ·014	+ ·129	&c.	+ ·012	+ ·114	&c.
l_1	+ ·003	m_1	+ ·004	+ ·014	+ ·006		+ ·043	+ ·005	
$l_2 \cos \Delta$	+ ·010		+ ·010	+ ·014	+ ·014		+ ·004		
$m_2 \sin \Delta$	+ 14	Upper sign on left; lower sign on right	+ 9	− 22	− 9		− ·107		
$\pm l_1 \cos \frac{1}{2}\Delta$	+ 3		− 3	+ 14	− 14		+ 35		
$\pm 2m_1 \sin \frac{1}{2}\Delta$	+ 0		− 0	− 1	+ 0	&c.	− 6		&c.
Sum	+ ·027		+ ·016	+ ·005	− ·009		− ·074		
Mult. by 119/D	− 1·2		− 0·8	− 0·2	+ 0·5		+ 11·7		
Par. corr. to i	+ 0·6		+ 0·6	+ 0·6	+ 0·6		+ 0·6		
Sum R	− 0·6		− 0·2	+ 0·4	+ 1·1		+ 12·3	None	

MARCH (continued).

Parallactic corrections —(continued)	0° or 0^h				15° or 1^h			90° or 6^h		
z_2	$+\cdot129$	$-l_2$	$-\cdot010$	&c.	&c.	&c.	&c.	&c.	&c.
z_1	$+\cdot008$	$-l_1$	$-\cdot003$						
$z_2 \cos \Delta$ $-l_2 \sin \Delta$ $\pm z_1 \cos \frac{1}{2}\Delta$ $\mp 2l_1 \sin \frac{1}{2}\Delta$	&c. &c.	Compute like		P and Q		&c.	&c.	&c.	&c.
Sum S	$+\cdot135$		$+\cdot120$	$+\cdot143$	$+\cdot115$		$+\cdot115$	$+\cdot035$	

SECOND approximation.

When the correction $\delta\Delta$ is large, as in the case of column 6^h, this is necessary.

	Column of 90° or 6^h	
Assume Δ_0	$-70°$	$+86°$
α	$+\cdot067$	$+\cdot014$
β	$-\cdot183$	$+\cdot195$
γ	$+\cdot214$	$+\cdot044$
δ	$-\cdot598$	$+\cdot625$
ϵ	$+\cdot627$	$-\cdot560$
ζ	$-\cdot439$	$-\cdot529$
η	$+\cdot175$	$-\cdot157$
θ	$-\cdot123$	$-\cdot146$
λ	$-\cdot002$	$-\cdot003$
μ	$-\cdot002$	$+\cdot000$
ν	$+\cdot018$	$+\cdot023$
ρ	$+\cdot015$	$-\cdot003$
$\alpha+\delta$	$-\cdot531$	$+\cdot639$
$\epsilon+\theta$	$+\cdot504$	$-\cdot706$
$\lambda+\rho$	$+\cdot011$	$-\cdot006$
Sum N_0	$-\cdot016$	$-\cdot073$
$\zeta-\eta$	$-\cdot614$	$-\cdot377$
$\frac{1}{2}(\zeta-\eta)$	$-\cdot307$	$-\cdot189$
$\beta-\gamma$	$-\cdot397$	$+\cdot151$
$2(\mu-\nu)$	$-\cdot040$	$-\cdot046$
Sum D_0	$-\cdot744$	$-\cdot084$
$\delta\Delta = N_0/D_0$	$+\cdot022$	$+\cdot87$
(In degrees) $\delta\Delta$	$+1°\cdot4$	$+50°$
$\Delta = \Delta_0 + \delta\Delta$	$-68\cdot6$	No H. W.

It is concluded that as the correction in the second column is 50° there is no H.W. A conjectural value of $\Delta = +90°$ is used above in computing P, Q, S.*

* There ought in strictness to be further corrections to $\mathfrak{T}$ and $\mathfrak{H}$, but they are of little importance. Thus:—

Further correction to $\mathfrak{T}$ and $\mathfrak{H}$.

When Δ is greater than (say) 50° there are further corrections $[\delta I]$, $[\delta h]$ computed from

$$[\delta I] = -2^h\left(1-\tfrac{1}{85}\right)\frac{\cdot 012}{D}\left[3F_2\sin^2\Delta\cos\Delta + (G_{,,}+G_s)\sin^3\Delta\right]$$

$$[\delta h] = -\cdot 012\left[3(A_{,,}+A_s)\sin^2\Delta\cos\Delta - F_2\sin^3\Delta\right]$$

Thus in the column of 6^h on the left $\Delta = -70°$; then compute thus:—

$$\begin{array}{lll}
F_2 = +\cdot 195, & G_2 = +\ \cdot 627, & A_2 = +\ \cdot 691 \\
3F_2 = +\cdot 585, & -H_m = -1\cdot 568, & -H_m = -1\cdot 568 \\
\hline
& G_2 - H_m = -\ \cdot 941 & A_2 - H_m = -\ \cdot 877 \\
& G_{,,}+G_s = -\ \cdot 941 & 3(A_2 - H_m) = -2\cdot 631 \\
& & 3(A_{,,}+A_s) = -2\cdot 631
\end{array}$$

By successive use of Traverse table,

$$\begin{array}{ll}
(G_{,,}+G_s)\sin\Delta = +\cdot 884, & 3F_2\sin\Delta = -\cdot 550 \\
(G_{,,}+G_s)\sin^2\Delta = -\cdot 831, & 3F_2\sin^2\Delta = +\cdot 517
\end{array}$$

$$\begin{array}{rr}
(G_{,,}+G_s)\sin^3\Delta = & +\ \cdot 781 \\
+3F_2\sin^2\Delta\cos\Delta = & +\ \cdot 177 \\
\hline
& +\ \cdot 958 \\
& \times -\ \cdot 012 \\
\hline
\text{Divide by D or } -\cdot 752 & -\cdot 0115 \\
\hline
& \times 2^h + \cdot 0153 \\
\hline
& +\cdot 0306 \\
-\tfrac{1}{85} & -\quad 4 \\
\hline
[\delta I] = & +\ \cdot 030 \\
= & +\ 1^m\cdot 8 \\
\text{Previous } \mathfrak{T} = & 5^h\ 33^m\cdot 8 \\
\hline
\text{Correct } \mathfrak{T} = & 5^h\ 36^m
\end{array}$$

Again, by successive use of Traverse table,

$$\begin{array}{ll}
3(A_{,,}+A_s)\sin\Delta = +2\cdot 472, & -F_2\sin\Delta = +\cdot 183 \\
3(A_{,,}+A_s)\sin^2\Delta = -2\cdot 323, & -F_2\sin^2\Delta = -\cdot 172
\end{array}$$

$$\begin{array}{rr}
3(A_{,,}+A_s)\sin^2\Delta\cos\Delta = & -\quad \cdot 795 \\
-F_2\sin^3\Delta = & +\quad \cdot 162 \\
\hline
& -\quad \cdot 633 \\
& \times -\ \cdot 012 \\
\hline
[\delta h] = & +\quad \cdot 008 \\
\text{Previous } \mathfrak{H} = & 5\cdot 426 \\
\hline
\text{Correct } \mathfrak{H} = & 5\cdot 43
\end{array}$$

Evanescent Tide. (See § 7.)

The right-hand column of 6^h leads to no H.W., and the tables of 𝕴 and 𝕳 must be completed by other formulæ.

The following calculation is very like the preceding one. The value Δ_o will always be nearly $\pm 90°$, and in our example it is exactly $+90°$.

The computation of M, N, R is to be omitted, and that of P, Q, S has been included in the general calculation with a conjectural $\Delta = +90°$.

MARCH.

Evanescent Tide					Evanescent Tide (continued)		
90° or 6^h					90° or 6^h		
G_2	$+\cdot627$	F_2	$+\cdot195$		$\delta\Delta = \frac{D_o}{E_o}$	,,	$+\cdot110$
G_1	$+\cdot107$	F_1	$+\cdot765$		(In degrees) $\delta\Delta$	,,	$+6°\cdot3$
G_4	$-\cdot047$	F_4	$+\cdot003$		$\Delta = \Delta_o + \delta\Delta$	,,	$+96\cdot3$
		$\frac{1}{4}F_1$	$+\cdot191$		$\frac{1}{30}\Delta$	,,	3·210
G_2+G_1	,,	$F_2+\frac{1}{4}F_1$	,,		$\frac{1}{30^2}\frac{21}{20}\Delta$	,,	·112
G_2-G_1	$+\cdot520$	$F_2-\frac{1}{4}F_1$	$+\cdot004$		I	,,	3·322
$\tan\Delta_o = \frac{G_2+G_1}{F_2+\frac{1}{4}F_1}$	,,	$\frac{G_2-G_1}{F_2-\frac{1}{4}F_1}$	$+130\cdot0$		δT	,,	$+\cdot030$
Δ_o	,,		$+90°$				3·352
a $F_2 \cos\Delta_o$	,,		·000		Mean int. i	,,	7·900
β $F_2 \sin\Delta_o$	,,		$+\cdot195$				11·252
γ $G_2 \cos\Delta_o$	,,		·000				$11^h\ 15^m\cdot1$
δ $G_2 \sin\Delta_o$	,,		$+\cdot627$		(See below) δI	,,	$+\ 2\cdot5$
ϵ $\pm F_1 \cos\frac{1}{2}\Delta_o$	,,		$-\cdot541$		𝕴	,,	$11^h\ 18^m$
ζ $\pm F_1 \sin\frac{1}{2}\Delta_o$	,,		$-\cdot541$		A_2 ·691	F_2	$-\cdot195$
η $\pm 2G_1 \cos\frac{1}{2}\Delta_o$	,,		$-\cdot151$		A_1 $+\cdot417$	F_1	$-\cdot765$
θ $\pm 2G_1 \sin\frac{1}{2}\Delta_o$	,,	Upper sign to left; lower to right	$-\cdot151$		A_4 $-\cdot012$	F_4	$-\cdot003$
λ $F_4 \cos 2\Delta_o$	,,		$-\cdot003$		Δ	,,	$96°$ or $\pi - 84°$
μ $F_4 \sin 2\Delta_o$	,,		·000		$A_2 \cos\Delta$	,,	$-\ \cdot072$
ν $\frac{1}{2}G_4 \cos 2\Delta_o$	,,		$+\cdot024$		$-F_2 \sin\Delta$	,,	$-\ \cdot194$
ρ $\frac{1}{2}G_4 \sin 2\Delta_o$	,,		·000		$\pm A_1 \cos\frac{1}{2}\Delta$	,,	$-\ \cdot279$
$\zeta-\eta$	,,		$-\cdot390$		$\mp 2F_1 \sin\frac{1}{2}\Delta$	,,	$+1\cdot137$
$\frac{1}{2}(\zeta-\eta)$	,,		$-\cdot195$		$A_4 \cos 2\Delta$	,,	$-\ \cdot012$
$\beta-\gamma$	,,		$+\cdot195$		$-\frac{1}{2}F_4 \sin 2\Delta$	,,	$+\ \cdot001$
$2(\mu-\nu)$	,,		$-\cdot048$		Sum h	,,	$+\ \cdot581$
Sum D_o	,,		$-\cdot048$		B_o	,,	3·791
$(\epsilon+\theta)$	,,		$-\cdot692$			Upper sign to left; lower to right	4·372
$\frac{1}{4}(\epsilon+\theta)$	,,		$-\cdot173$		(See below) δh	,,	$+\ \cdot011$
$a+\delta$	,,		$+\cdot627$		𝕳	,,	4·38
$4(\lambda+\rho)$	,,		$-\ \cdot012$				
Sum $-E_o$	,,		$+\cdot442$				
E_o	,,		$-\cdot442$				

Corrections for Reference to Moon's Transit. (See (57)—(64), § 10.)

Of these corrections δI, δh, M″, N″, P″, Q″ have already been used in the preceding calculation, and we have to show how they are to be computed; we also have to compute 𝔦 and 𝔥′.

From March "intervals" and "heights" we extract I and h, and arrange them in double columns—the even entries in one column and the odd in another. The columns 0^h to 11^h afford the 12 values for 0^h to 11^h of I and h by means of their left hand entries, and they afford the 12 values for 12^h to 23^h by means of their right hand entries. The entry for 23^h is repeated at the top and that for 0^h at the bottom, so that each column has 13 entries, and thus each provides 12 first differences. After finding these differences, the distinction of odd and even entries is unnecessary.

The numerical factors 2·2, $\frac{1}{20}$, 2·34, $\frac{11}{300}$, $\frac{1}{20}$, 2·34 in the legends at the top of the columns are absolute constants.

The Θ's are derived from the sequence in Table IV., beginning the sequence with the month treated.

The values of c and d are derived from Table V., for the month named; thus for March c is $-3^m{\cdot}7$ and d is $-0^h{\cdot}062$.

The values of M″, $-$N″, δI are found in columns vii., xi., xii. of the first table, and P″, $-$Q″, δh in vii., xi., xii. in the second table. The entries opposite 0^h to 11^h were used above on the left-hand side of columns 0^h to 11^h, and the entries opposite 12^h to 23^h were used above on the right-hand side of the columns 0^h to 11^h.

The quantity 𝔥′ is not a final result, but after *interpolated* values of 𝔥′ (see below on Interpolation) have been found, we shall add to it *computed* values of the annual tide, so as to form 𝔥.

The arithmetical processes involved in these tables are sufficiently explained by the instructions at the head of each column.

March.—Intervals.

	i. I		$2dI/dT$ ii. Diff. i.	iii. $2\cdot2\times$ ii.	iv. Θ for Mar.	v. iii. sin iv.	vi. $\frac{1}{20}$ v.	M'' vii. v. + vi.	viii. iii. cos iv.	ix. $\frac{1}{20}$ viii.	x. viii. − ix.	$-N''$ xi. iii. + x.	δI xii. $2\cdot34\times$ v.	i xiii. c for Mar. × ii.
h	h	h	h		°			m				m	m	m
23		+0·73												
0	+0·20		−1·04	−2·29	−10	+0·40	+2	+0·4	−2·26	−11	−2·15	−4·4	+0·9	+3·8
1		−0·31	−1·02	−2·24	+20	−0·77	−4	−0·8	−2·11	−11	−2·00	−4·2	−1·8	+3·8
2	−0·82		−1·00	−2·20	+50	−1·69	−8	−1·8	−1·41	−7	−1·34	−3·5	−4·0	+3·7
3		−1·31	−0·96	−2·11	+80	−2·09	−10	−2·2	−0·37	−2	−0·35	−2·5	−4·9	+3·6
&c.	&c.		&c.	&c.	&c.	&c.	&c.	&c.	&c.	&c.	&c.	&c.	&c.	&c.
22	+1·27		−1·09	−2·40	−70	+2·26	+11	+2·4	−0·82	−4	−0·78	−3·2	+5·3	+4·0
23		+0·73	−1·07	−2·35	−40	+1·51	+8	+1·6	−1·80	−9	−1·71	−4·1	+3·5	+4·0
0	+0·20													

March.—Heights.

	i. h		$2dh/dT$ ii. Diff. i.	iii. $\frac{11}{300}\times$ ii.	iv. Θ for Mar.	v. iii. sin iv.	vi. $\frac{1}{20}$ v.	P'' vii. v. + vi.	viii. iii. cos iv.	ix. $\frac{1}{20}$ viii.	x. viii. − ix.	$-Q''$ xi. iii. + x.	δh xii. $2\cdot34\times$ v.	$\mathfrak{h}'$ xiii. d for Mar. × ii.
h	ft	ft	ft.		°			ft				ft	ft	ft
23		2·53												
0	2·69		+·26	+·0095	−10	−·0017	−1	−·002	+·0094	+5	+·0089	+·018	−·004	−·016
1		2·79	+·12	+44	+20	+15	+1	+2	+41	+2	+39	+8	+4	−7
2	2·81		−·08	−29	+50	−22	−1	−2	−19	−1	−18	−5	−5	+5
3		2·71	−·32	−117	+80	−115	−6	−12	−20	−1	−19	−14	−27	+20
&c.	&c.		&c.	&c.	&c.	&c.	&c.	&c.	&c.	&c.	&c.	&c.	&c.	&c.
22	2·28		+·56	+206	−70	−194	−10	−20	+71	+4	+67	+27	−45	−35
23		2·53	+·41	+150	−40	−96	−5	−10	+115	+6	+109	+26	−23	−25
0	2·69													

Additional Values of Intervals and Heights.

Where the intervals change largely between one column and the next, it would add much to the accuracy if additional values were calculated. Thus, in the calculation for March further values between 75° or 5^h and 90° or 6^h, and again, between 90° or 6^h and 105° or 7^h, would be desirable. The like is true for August, where a column between 105° or 7^h and 120° or 8^h would be useful. I choose this last case for my example.

If Inman's traverse table be used, interpolation may be made for $112\frac{1}{2}°$ without much difficulty, but I think it is better to interpolate for an even number of additional degrees, and to compute a column for 113°, found by adding 8° to 105°.

It is proposed then to add a new column between the 8th and 9th.

We begin by interpolating in the sequences of angles. In each sequence we have to find the 8th entry for August; then, if it is semi-diurnal, add 16°; if diurnal add 8° for K_1 and P, and subtract 8° for O; and, if quater-diurnal, add 32°. In the sequence for Θ we add 16°, since it is similar to a semi-diurnal term. The calculation runs thus:—

	K_2	S_2	O	K_1	P	S_4
8th entry . .	$\pi-46°$	$\pi+19°$	$-88°$	$\pi+63°$	$\pi-45°$	$\pi+83°$
Add	$+16$	$+16$	-8	$+8$	$+8$	$+32$
ϑ	$\pi-30$	$\pi+35$	$\pi+84$	$\pi+71$	$\pi-37$	-75

Also $\vartheta_m = 0$, $\vartheta_{2m} = \pi - 36°$, as before.

The 8th entry of Θ is $\pi - 40°$, to which we add 16°, and find $\Theta = \pi - 24°$. With this value of Θ, compute δT, δM, δN. The interpolation amongst the sequences of angles for the parallactic terms is done in the same way. With these new values, and with the former H, Hp^2, Hp we now compute new A's, F's, G's, and are then in a position to compute a new column corresponding to 113° or $7^h\ 32^m$.

In computing δI, δh, M″, N″, P″, Q″, 𝔦, 𝔥′, column ii., for intervals, or $2dI/d\mathrm{T}$, must be put equal to $2(I_{120} - I_{105})$; and similarly, column ii. for heights, or $2dh/d\mathrm{T}$, must be put equal to $2(h_{120} - h_{105})$.

This interpolation would be especially valuable in the case of M, N, P, Q, which change abruptly.

Some interpolation of the kind has been done in my example, but I do not reproduce the work.

The Calculation of a Low Water Table (referred to in general rule VII.).

This may be done almost independently of the H.W. table by replacing the ϑ's by θ's and using the rules given in § 12. The calculation may, however, be materially abridged, and I will now go over the several steps of the calculation noting the mode of transition from one case to the other.

The new A, G, F will be distinguished from the old by enclosing the new ones in square parentheses.

Semi-diurnal $[A_2]$, $[G_2]$, $[F_2]$.

These sequences are derivable directly from the old ones by the rule

$$[A_2] = A_2 + \tfrac{1}{9}F_2, \qquad [G_2] = G_2 + \tfrac{1}{9}F_2, \qquad [F_2] = F_2 - \tfrac{1}{9}(A_{\prime\prime} + A_s)$$

Diurnal $[A_1]$, $[G_1]$, $[F_1]$.

The rule is here more complex:—

$$\begin{array}{lll} [A_o] = -\{A_o - \tfrac{1}{8}F_o\}, & [G_o] = -\{G_o - \tfrac{1}{35}F_o\}, & [F_o] = -\{F_o + \tfrac{1}{35}A_o\} \\ [A_\prime] = A_\prime + \tfrac{1}{9}F_{\prime\prime}, & [G_\prime] = G_\prime + \tfrac{1}{32}F_{\prime\prime}, & [F_\prime] = F_\prime - \tfrac{1}{32}A_\prime \\ [A_p] = A_p + \tfrac{1}{10}F_p, & [G_p] = G_p + \tfrac{1}{38}F_p, & [F_p] = F_p - \tfrac{1}{38}A_p \end{array}$$

and in all these sequences *shift the list of months* in the margin *six places downwards*, so that in the O and K_1 sequences March stands where June stood, and in the P sequence March stands where December stood.

If it be agreed to neglect the terms involving $\frac{1}{8}$, $\frac{1}{35}$, &c., the rule is simply to shift the months and change the signs of the O sequences; but at Aden where the diurnal tide is very large, this would lead to a sensible error.

After the new sequences for O, K_1, P have been found, they are combined to find $[A_1]$, $[G_1]$, $[F_1]$ just as for H.W.

Quater-diurnal

$$[A_4] = -A_4, \qquad [G_4] = -G_4, \qquad [F_4] = -F_4$$

that is to say, simply change signs.

Semi-annual and Mean Water and Annual Tide.

The old calculation serves again.

Mean Interval and its Parallactic Correction.

Here we subtract 180° from κ_m; thus

$$\begin{array}{rr} \kappa_m - \pi = & 49^\circ \\ \tfrac{1}{30}(\kappa_m - \pi) = & 1{\cdot}633 \\ \tfrac{21}{20}\cdot\tfrac{1}{30^2}(\kappa_m - \pi) = & 57 \\ i = & 1{\cdot}690 \end{array}$$

$$\text{Parallactic correction} = +0^m{\cdot}08i = +0^m{\cdot}14$$

It may be well to warn the computer that i may be negative, that is to say L.W. may occur on the average earlier than moon's transit.

Parallactic Corrections.

$\alpha, \alpha', \beta, \beta'$ are unchanged.

The rules are

$$[_mA] = {}_mA - \tfrac{1}{17}\,{}_mF, \qquad [_{\prime\prime}A] = {}_{\prime\prime}A + \tfrac{2}{9}\,{}_{\prime\prime}F$$
$$[_mG] = {}_mG - \tfrac{1}{17}\,{}_mF, \qquad [_{\prime\prime}G] = {}_{\prime\prime}G + \tfrac{1}{4}\,{}_{\prime\prime}F$$
$$[_mF] = {}_mF + \tfrac{1}{17}\,{}_mA, \qquad [_{\prime\prime}F] = {}_{\prime\prime}F - \tfrac{1}{4}\,{}_{\prime\prime}A$$

We then compute $[z_2]$, $[m_2]$, $[l_2]$ by the same rules as before.

In the diurnal terms compute by the following rules:—

$$[_oA] = -\{_oA - \tfrac{1}{3}\,{}_oF\}, \qquad [_{\prime}A] = {}_{\prime}A + \tfrac{2}{9}\,{}_{\prime}F$$
$$[_oG] = -\{_oG - \tfrac{1}{14}\,{}_oF\}, \qquad [_{\prime}G] = {}_{\prime}G + \tfrac{1}{16}\,{}_{\prime}F$$
$$[_oF] = -\{_oF + \tfrac{1}{14}\,{}_oA\}, \qquad [_{\prime}F] = {}_{\prime}F - \tfrac{1}{16}\,{}_{\prime}A$$

the list of months in the margin being shifted six places downwards, so that March stands where June stood. The values of $[z_1]$, $[m_1]$, $[l_1]$ are then computed by the same rule as before.

Nodal Corrections.

We may with sufficient accuracy, take a_2, b_2, c_2, d_2, e_2, f_2 to be unchanged. Referring to the instructions for the computation of a_1, b_1, &c., the new rule may be stated thus:—

$[a_1]$ is (ii.) – (i.), $[b_1]$ is (iv.) – (iii.), $[c_1]$ is (vi.) – (v.)
$[d_1]$ is (viii.) – (vii.), $[e_1]$ is (x.) – (ix.), and $[f_1]$ is (xii.) – (xi.)

and the list of months in the margin is pushed down six places, so that March stands where June stood.

The corrections δT, δM, δN remain unchanged.

When the L.W. sequences have been formed the calculation follows the lines of H.W. calculation precisely, save in three respects—first, in the "heights" h is to be subtracted from B_o, and δh is to be then subtracted from $B_o - h$, instead of the corresponding additions in the H.W. calculation; secondly, the signs of P, Q, S are to be changed as a last step in the calculation of those quantities, in order that the corrections to the heights may be additive instead of subtractive as they would be if we left off exactly as in the case of H.W.; thirdly, after the final table for $\mathfrak{h}'$ has been made, its values must be subtracted from the annual tide.

The reader will easily understand the necessity for these changes when it is remarked that h, δh, $\mathfrak{h}'$ have been estimated as depressions below mean water, whereas B_o and the annual tide are estimated as elevations above the adopted datum; in the result we require, of course, to estimate heights with reference to the datum.

It may be well to warn the computer that $i + I$ may often be negative.

It will be unnecessary to refer henceforth to L.W., since the instructions for H.W. serve also for L.W.

Interpolation.

The sun's longitude increased by 5° is indicated by ⊙, and the months March, April, May, &c., really mean the dates when ⊙ is 0°, 30°, 60°, &c.—that is to say, about the middle of the months. The dates which, on the average, fall the nearest to these times, are given in Table VI.

The 12 columns for any month, headed $0^h, 1^h, 2^h, \ldots 11^h$ contain on the left the 12 values of $\mathfrak{I}$, $\mathfrak{H}$, M, N, &c., corresponding to moon's transit at $0^h, 1^h, 2^h, \ldots 11^h$, and they contain on the right the 12 values for moon's transit at $12^h, 13^h, \ldots 23^h$. This applies to the month named at the head of the table. But these values also appertain to the opposite month (*i.e.*, September opposite to March, October to April, and so on) by reading the right hand entries as appertaining to $0^h, 1^h, \ldots 11^h$, and the left to $12^h, 13^h, \ldots 23^h$. The same is true of $\mathfrak{i}$ and $\mathfrak{h}'$ (see *Corrections for reference to moon's transit*), except that here the values change sign in the opposite month; thus the values of $\mathfrak{i}$ and $\mathfrak{h}'$ which we have computed for March must be taken with the opposite sign when applied for September.

Now it is required to form interpolated tables for every 10° of ⊙, and in all the 18 tables (of which 6 will be originally computed and 12 interpolated) to interpolate for every 20^m of moon's transit.

These interpolations may be done graphically, and I find with millimetre-square paper a convenient scale for ⊙ is 1 mm. to 1°, and for time of moon's transit 15 mm. to 1^h. These will be set out horizontally as abscissæ, and the ordinates will be time in treating $\mathfrak{I}$, and height in treating $\mathfrak{H}$.

A convenient time scale for $\mathfrak{I}$ is 30 mm. to the hour. In the case of $\mathfrak{H}$ the scale must depend on the range of tide at the place—for Aden (with a small range) I have found 50 mm. to the foot convenient.

I will begin with interpolation for ⊙ (*i.e.*, for time of year), and will only refer to $\mathfrak{I}$, since $\mathfrak{H}$ follows the same plan. Write March, April, May, ... January, February, March, at 0, 30, ... 360 mm. along a horizontal line, corresponding to the same number of degrees of ⊙. It may be well to repeat February before the first March, and April after the last March. Set off as an ordinate the left-hand entries from column 0^h for the six months March, April, ... August, and from the right-hand entries of 0^h for the same six months set off ordinates for their opposite months, *e.g.*, the right-hand $\mathfrak{I}$ of 0^h for March, affords $\mathfrak{I}$ of 0^h for September. Through the tops of these ordinates draw a smooth curve of 0^h.* Proceed similarly to form curves of 1^h, 2^h, &c., twelve in all. If the figures get confused we may have two or more, and confusion may often be avoided by drawing parts of the curve with upward or downward shift, so as to make things clear where a number of curves go through nearly the same point.

We now start a fresh figure with time of moon's transit as horizontal line, and $\mathfrak{I}$ as ordinate corresponding to March (or $\odot = 0°$). These 24 (computed) values of $\mathfrak{I}$, joined by a smooth curve, enable us to read off the values of $\mathfrak{I}$ for $\odot = 0°$ for every 20^m of moon's transit, *i.e.*, on the adopted scale, at every 5 mm. of horizontal space.

We now set off from the previous figure the 24 (interpolated) values of $\mathfrak{I}$ corresponding to $\odot = 10°$, of which the first 12 are found at "March + 10 mm." and the last 12 at "September + 10 mm." These 24 values being joined by a curve, give $\mathfrak{I}$ for $\odot = 10°$, and for every 20^m of moon's transit.

We next set off 24 values of $\mathfrak{I}$ corresponding to $\odot = 20°$, of which the first 12 are found in the preceding figure at "March + 20 mm.," and the last 12 at "September + 20 mm." These are treated the same way.

The next in the series are the computed April ($\odot = 30°$) $\mathfrak{I}$'s which are set off like the March ones, joined by a curve and read off to each 20^m of moon's transit.

We then take 24 (interpolated) values of $\mathfrak{I}$ from "April + 10 mm." and "October + 10 mm.," to give $\mathfrak{I}$ for $\odot = 40°$. The 24 from "April + 20 mm." and "October + 20 mm." afford $\mathfrak{I}$ for $\odot = 50°$. The next is the computed May ($\odot = 60°$) series, and we so pass on through six months, the last in the set being derived from "August + 20 mm." and "February + 20 mm." corresponding to $\odot = 170°$.

* The following rule is probably known, but I do not know where it has been stated, except in a note of my own in the *Messenger of Mathematics*. I have found it very useful in drawing good curves.

Rule for Graphical Interpolation half-way between Computed Ordinates.

Draw the polygon (A) joining the tops of a number of equidistant ordinates, and draw the two polygons (B) joining the tops of alternate ordinates. Then every ordinate has marked on it an intercept or sagitta where a side of polygon (B) cuts it. On the half-way ordinates next on each side of a sagitta, set off one-fourth of the sagitta from the points where the two sides of polygon (A) cut those half-way ordinates; the set-off is to be in the direction in which the sagitta would shoot if it were an arrow. When all the quarter sagittas are thus set off, every half-way ordinate (except the first or last of the series) has two points marked on it.

The interpolated curve passes half-way between the pairs of marked points, except in the case of the first and last half-way ordinate, when it passes through the single marked points.

This rule is correct to fourth differences, except in the case of the first and last ordinate, when it is correct only to third differences. In the cases in the text the computed values are cyclical, and there are therefore no first or last.

By means of proportional compasses set to 4, the quarter sagittas may be set off rapidly, and the bisection of the pairs of marked points may be made by eye.

If one person reads the numbers from the figure, whilst another writes, the tabulation may be done very rapidly.

The same process is applied for tabulation of the $\mathfrak{P}$'s. We should, in strictness, do the same by M, N, P, Q, R, S, $\mathfrak{i}$, $\mathfrak{h}'$, but it appears unnecessary to work with so much accuracy.

I have done much of the interpolation by simply writing out the computed values of the quantity to be tabulated in a chess-board table with blanks for the interpolated values. If sixteen squares be considered, a computed value will stand at each corner. Then a great many of the interpolated values may be put in by inspection of the march of the quantity in the two directions. In other parts I make a pencil curve, on millimetre-square paper, of four or five adjacent values, and pass a freehand curve through them to fill in the interpolated values; I rub out the curve when used. It must be remembered that close accuracy in these terms would be mere pedantry.

M, N are computed to the decimal of a minute, and the decimal part may be useful for drawing the pencil interpolation curves, but the result should be tabulated only to the nearest minute. Similarly the third decimal in P and Q may be dropped.

The interpolation of $\mathfrak{i}$ and $\mathfrak{h}'$ follows the same plan, but it must be borne in mind that these functions change sign in opposite months, and this consideration is important when we come to interpolate for $\odot =$ August $+ 10°$, and $+ 20°$.

When there is an evanescent tide (as in the case of March, 18^h) the corrections M, N, R become infinite. As a practical solution this is absurd, and the fact is that there may or may not be a H.W. according to the values of ☊ and Π. Again, in other parts of this and other lunations there may be no H.W., although the tide-table predicts one. In all such cases there is a long period of four or five hours' duration of nearly slack water, and it is accordingly almost a matter of indifference whether or not a small H.W. is predicted. It would necessitate very laborious computations to make correct predictions in these cases, and the result would not be worth the labour. I have adopted, therefore, a makeshift, and have replaced H.W. by the height of water and time when the rate of change of water level is a minimum.

It has been proposed above that P, Q, S shall be computed with a conjectural $\Delta = + 90°$, and this is better than the plan which I actually adopted in my experimental table for Aden, of which a sample is given below*. The practical point to consider in the present instructions is the manner of treatment of M, N, R about the time of evanescence. I propose then that the gap in the values shall be bridged by a conjectural curve, and that the values

* Thus if any one seeks to verify my table, he will not get *exactly* my values for P, Q, S in the neighbourhood of 18^h.

be only given in round numbers. For example, for March we have the following values of M:—$(16^h)+9^{m}{\cdot}3$, $(17^h)+28^{m}{\cdot}5$, (18^h) blank, $(19^h)+45^{m}{\cdot}0$, $(20^h)+17^{m}{\cdot}7$, &c. By drawing a curve I conjecture $+50^m$ for the missing value at 18^h.

A comparison with the corresponding complete curves for February and April helps us in filling the gap.

After the complete table for 𝔥′ is formed we proceed to add to it the values computed in the table of the annual tide for every 10° of ⊙, and so form a table of 𝔥. For example, the first five values of 𝔥′ are −·02*, −·01, −·01, −·01*, ·00 (of which those marked * are computed), and to these we add ·390, the computed annual tide for March, and obtain ·37, ·38, ·38, ·38, ·39.

The final results are then arranged in a table†. If the L.W. were also computed I should propose that the L.W. and H.W. should be given alternately. The following is a sample of the table computed only for H.W. at Aden:

PORT of Aden; High Water Tide-Table.

Times of moon's transit for—		For March 15th (*i.e.*, from sun's long. 350° to 0°), and for September 17th (*i.e.*, from sun's long. 170° to 180°). The upper signs of 𝔦 and 𝔥 apply to March, the lower to September									
March	Sept.	𝔗	𝔦	𝔐	𝔥	M	N	R	P	Q	S
h. m.	h. m.	h. m	m.	ft.	ft.	m.	m.	m.	ft.	ft.	ft.
0 0	12 0	8 5	± 4	6·49	±·37	0	+ 9	− 1·3	+·02	−·08	+·14
20	20	7 55	± 4	·54	±·38	0	+ 9	− 1·0	+·03	−·08	·14
40	40	46	± 4	·57	±·38	− 1	+ 9	− 0·6	+·04	−·09	·14
1 0	13 0	7 37	± 4	6·59	±·38	− 1	+ 9	− 0·2	+·05	−·09	·14
20	20	28	± 4	·60	±·39	− 2	+ 9	+ 0·1	+·06	−·09	·14
40	40	18	± 4	·60	±·39	− 2	+ 9	+ 0·4	+·07	−·09	·15
&c.		&c.		&c.			&c.		&c.		
17 0	5 0	8 29	∓14	4·21	±·37	+28	−11	+ 2·3	−·16	−·01	·09
20	20	9 52	∓15	·21	±·36	+35	− 2	− 7·3	−·10	−·03	·10
40	40	11 2	∓14	·27	±·35	+43	+20	−16	−·05	−·05	·10
18 0	6 0	11 17	∓10	4·38	±·35	+50	+30	−20	·00	−·06	·09
20	20	15	∓ 6	·53	±·34	+54	+25	−30	+·05	−·05	·08
40	40	7	∓ 1	·71	±·33	+50	+20	−30	+·10	−·04	·06
&c.		&c.		&c.			&c.		&c.		

† I have found that it is convenient to cut the constituent tables into strips and paste them together again, so as to save much copying and verification.

If ϖ be the moon's parallax at moon's transit in minutes of arc, and ☊ be the longitude of moon's node, the interval is

$$\mathfrak{I} + \mathfrak{i} + M \cos ☊ + N \sin ☊ + (\varpi - 57') R$$

and the height is

$$\mathfrak{H} + \mathfrak{h} + P \cos ☊ + Q \sin ☊ + (\varpi - 57') S$$

After the table has been completed the computer should test the correctness of the prediction by computing two or three tides in each month, and comparing the results either with the observations from which the harmonic constants were originally derived, or with other known values of high and low water.

Examples of the Use of the Table.

From and after the year 1887, the datum for the tide-tables of the Indian Government for Aden has been 0·37 ft. lower than that used in my table; as I am going to compare my results for 1889 with those of the Indian Government, 0·37 ft. will be added to my heights to make the two comparable.

(A) The moon crossed the meridian at Aden on March 17th, 1889, at $0^h\ 11^m$, Aden M.T. Aden is in $3^h\ 0^m$ E. long., and therefore this is about 21^h, March 16th, G.M.T.; whence from the *Nautical Almanac* we find ϖ the moon's parallax at Aden transit was 58′·2, and $\varpi - 57' = +1·2$. The longitude of moon's node was 108°, and cos ☊ = − ·3, sin ☊ = + 1·0.

Then referring to our table and interpolating between $0^h\ 0^m$ and $0^h\ 20^m$, and taking the upper signs of $\mathfrak{i}$ and $\mathfrak{h}$, we find

$\mathfrak{I} + \mathfrak{i} = 8^h\ 4^m$, $\mathfrak{H} + \mathfrak{h} = 6·89$ ft., $M = 0$, $N = +9^m$, $R = -1^m·2$,
$P = +·03$ ft., $Q = -·08$ ft., $S = +·14$ ft.

Hence $M \cos ☊ + N \sin ☊ + R(\varpi - 57) = +8^m$

and $P \cos ☊ + Q \sin ☊ + S(\varpi - 57) + 0·37 = +0·45$ ft.

Therefore the interval is $8^h\ 12^m$, and the time of H.W. $8^h\ 12^m + 0^h\ 11^m$ or $8^h\ 23^m$ p.m., March 17th; and the height is 7·34 ft. or 7 ft. 4 in.

The Indian tide-table gives as time $8^h\ 12^m$ p.m., and as height 7 ft. 4 in.

(B) On September 17th, 1889, the moon crossed the meridian at $18^h\ 36^m$, ϖ was 54′·2 or $\varpi - 57' = -2·8$, ☊ = 98°, cos ☊ = − ·14, sin ☊ = + 1·0.

Interpolating in our table between $18^h\ 20^m$ and $18^h\ 40^m$, and taking the lower signs of $\mathfrak{i}$ and $\mathfrak{h}$, we find

$\mathfrak{I} - \mathfrak{i} = 11^h\ 22^m$, $\mathfrak{H} - \mathfrak{h} = 4·34$, $M = +50^m$, $N = +20^m$, $R = -30^m$,
$P = +·10$, $Q = -·04$, $S = +·06$

Hence $M \cos ☊ + N \sin ☊ + R(\varpi - 57) = +97^m = +1^h\ 37^m$

and $P \cos ☊ + Q \sin ☊ + S(\varpi - 57) + 0·37 = +0·15$ ft.

Therefore the interval is $12^h\ 59^m$, and the time of H.W. $12^h\ 59^m + 18^h\ 36^m$ or $31^h\ 35^m$ September 17th, *i.e.*, $7^h\ 35^m$ p.m. September 18th, and the height is 4·49 or 4 ft. 6 in.

The Indian tide-table gives "no inferior H.W."

This example shows our table at its worst, for it is clear that a nominally small correction to the time which amounts to an hour and forty minutes must give unsatisfactory results. At this time of year mean water has a height of 3 ft. 10 in., hence our prediction only shows a rise of 8 inches. Although there was probably in reality no maximum (as predicted in the Indian table), I should expect that the water stood at about 4 ft. 6 in. at half-past seven of the evening of September 18th, 1889.

Part III. Comparison and Discussion.

Comparison.

As stated in the Introduction, Mr Allnutt computed a complete H.W. tide-table for 1889 for Aden in order to compare the results with the Indian tide-tables, which are made with the tide-predicting instrument. When this comparison was made our tables had not been brought into exactly the form given above, and Mr Allnutt's work was considerably more laborious than it would have been if undertaken later.

The mechanical predictions were probably made with constants which are the means of the results derived from eight years of observation, whereas the constants used in our tables are derived from only four years. Mr Roberts has supplied me with six weeks of prediction for the year 1887, worked mechanically in duplicate, namely, with the eight-year and the four-year constants. In the latter case the times of H.W. seem to run about 6^m later than in the former, but the difference often rises to 10^m, occasionally to 15^m, and at rare intervals to 20^m. The two sets of constants also give a systematic difference in the heights amounting to about 2 inches, but the difference often rises to 3 inches or falls to 1 inch, and occasionally reaches 4 inches and zero. It follows, therefore, that a sensible part of the discrepancy, or error as it may be called, of our computation is due to the difference of constants. But not nearly all of the "error" can be set down to this cause; it is due to a combination of causes, namely, flaws in the interpolation, imperfect representation of the elliptic tides, and partial inclusion in our method of all the lunar inequalities which are totally neglected in the machine. The principal cause of "error," however, is the imperfect correction for longitude of moon's node and parallax about the time in each lunation when there is partial or total evanescence of the inferior H.W.

Omitting, as we must do, the cases of evanescence, there were 689 H.W. computed by Mr Allnutt, and he finds "the probable errors" in the time and height to be 9^m and 1·2 inches.

In the course of the year there were thirty-two occasions on which the time "error" amounted to 30^m or more. All of these occur in the inferior H.W. at the times of approximate evanescence, where the nodal and parallactic corrections are large. At these times there is always a period of four or five hours of nearly slack water, and the time at which a small maximum occurs is of no importance from a navigational point of view. If these thirty-two cases be taken away the probable error falls to 7^m. A cursory inspection of the table shows also that nearly all the "errors" of 25^m to 30^m fall about the time of evanescence and are therefore unimportant. We are accordingly justified in saying that our predictions do not differ materially from the Indian tables.

It has been already mentioned that I have six weeks of mechanical prediction for 1887, made with the identical constants used in our table. I have, therefore, taken a month, or 58 H.W., out of these six weeks, and compared them with my predictions, made without cross interpolation for date. I find that the errors of time are 27 from 0^m to 5^m, 13 from 5^m to 10^m, 10 from 10^m to 15^m, 4 from 15^m to 20^m, one error of 26^m, and one of 34^m. These give a probable error of 7^m. All the large errors fall on the inferior H.W., at the time when it is very small, and they are, therefore, practically unimportant.

In the heights there are 16 cases of agreement, 22 errors of 1 in., 15 of 2 in., and 5 of 3 in. These give a probable error of 1·0 inch.

The errors of 3 inches all fall about the time when the moon's parallax was small, and I also observe that this was commonly a time when the height errors rose to their greatest in 1889. This is probably due to the imperfect representation of the elliptic tides, which, as shown in § 3, is inherent to our method.

The concordance between the two is good enough, but less perfect in the heights than I expected.

The last comparison is between our predictions and actuality. The observed times and heights for part of 1884 have been furnished me by Colonel Hill, R.E., by the direction of Colonel Strahan, R.E., Deputy Surveyor-General in India. For the purpose of testing the present method, I have computed a H.W. tide-table from 10th March to 9th April, and again from 12th November to 12th December, 1884. In these periods, there are 117 actual and one evanescent H.W. The observation of one H.W. is missing through an accident to the tide-gauge; there are, therefore, 117 H.W. for the purpose of comparison.

The following is a table of errors, regardless of signs:

Time		Height	
Magnitude of error	No. of H.W.	Magnitude of error	No. of H.W.
m. m.		Inches	
0 to 5	35	0	15
5 „ 10	32	1	48
10 „ 15	19	2	28
15 „ 20	19	3	14
20 „ 25	5	4	11
26, 28	2		
33, 36	2		
56, 57	2		
Evanescence	1	Evanescence	1
	117		117

Omitting the case of evanescence and assuming the errors to conform in distribution to the normal exponential law (which is not accurately the case), the probable error is about 9^m in time, and 1·4 inch in height.

When the rise from the higher L.W. to the lower H.W. is *nil* there is evanescence, and when that rise is small there is approximate evanescence. I have accordingly examined the 12 cases in which the error of time is equal to or greater than 20^m, and the following table gives the result.

Rise from L.W. to H.W.	Time errors
Nil	. . .
6^{in} to 8^{in}	22, 26, 28, 56, 57 minutes
13^{in}	36 minutes
17^{in}	22 „
19^{in}	33 „
$2^{ft}\ 10^{in}$	22 „
$3^{ft}\ 9^{in}$	23 „
$3^{ft}\ 11^{in}$	20 „

It appears that where the errors of time were 56^m and 57^m the tide was very nearly evanescent, and that the two other considerable errors of time, viz., 36^m and 33^m, pertain to very small tides.

It has been already pointed out that in such cases a considerable error in the time is of no importance, and it is justifiable, in testing the calculations, to set aside the nine tides in which the rise is less or equal to 19 inches.

There remain 108 H.W., and the greatest error in the times amounts to only 23 minutes. In 58 cases the error is 7 minutes or less, and in 51 cases it is 6 minutes or less; as the half of 108 is 54, it follows that the probable error is a little over 6 minutes.

The Indian predictions maintain their standard of excellence fairly well through the periods of approximate evanescence, but out of 116 tides there are 59 cases with time errors of 10 minutes or less; as the half of 116 is 58, the probable error is about 10 minutes.

Turning now to the heights I find that both mine and the Indian predictions present 63 out of 116 H.W. with zero errors or errors of 1 inch. We may take it then that the probable error for both modes of prediction is about 1 inch. The Indian predictions have, however, the disadvantage that several errors of 5, 6, 7 inches occur. On the other hand, the 11 cases of 4 inches of error which occur in mine have a systematic character; they are all positive (actuality the greater), and all but one affect the higher H.W. about the time when the moon's parallax is small. This defect is doubtless due to the imperfect representation of the elliptic, evectional, and variational tides inherent to my method.

The slight superiority shown over the mechanical prediction must be attributed to the fact that I have used better values of the tidal constants than were available in 1883, when the Indian predictions must have been made.

I learn from Colonel Hill that two independent observers reading the same tide-curve will frequently differ by 5^m and sometimes by 10^m in their estimate of the time, and by 1 and sometimes by 2 inches in the height. Accordingly, predictions which agree with a reading of a tide-curve with probable errors of $6\frac{1}{2}^m$ in time and 1 inch in height may claim to possess a high order of accuracy.

I conclude from the preceding discussion that with good values of the tidal constants the present method leads to excellent predictions, and that they are even better than are required for nautical purposes.

Discussion.

It is probable that methods may be invented by which some abridgement of the computations may be made, but I am, of course, unable to suggest such improvements.

The last-mentioned comparison seemed to show that but little accuracy would be lost if P, Q, were entirely omitted, and if $\mathfrak{h}'$ were treated as zero, so that $\mathfrak{h}$ would consist simply of the annual tide. Indeed, the only advantage gained by the retention of P, Q, $\mathfrak{h}'$ appeared to be the avoidance of a few considerable errors in the inferior H.W. about the time of approximate evanescence. Experience must decide whether the computation and the tables may be lightened by the omission of these quantities.

The advantage gained from M and N is marked, but as these quantities arise almost entirely from the diurnal tides, I am inclined to think that, at places where the diurnal tide is not extremely large, a very fair tide-table might be made without them.

The present method will probably be applied to ports of second-rate importance, where there are not sufficient data for very accurate determination of the tidal constants. In such cases it will be best to omit the computation of P, Q, $\mathfrak{h}'$, and to postpone that of M, N, and perhaps also of R and $\mathfrak{i}$, until the simple tide-table has been tested as to its adequacy for navigational purposes. At most places the annual tide is so large that $\mathfrak{h}$ cannot be omitted, and it is impossible to dispense with the value of S. But it is possible that it might suffice to attribute to S a constant value*, although this would certainly cause very perceptible error in the heights of the lower H.W. and higher L.W. A tide-table which only gave $\mathfrak{I}$, $\mathfrak{H}$, $\mathfrak{h}$, and a constant S would be fairly short, even if computed for every ten days in the year; and this would be a great gain.

The question of how far to go in each case must depend on a variety of circumstances. The most important consideration is, I fear, likely to be the amount of money which can be spent on computation and printing; and after this will come the trustworthiness of the tidal constants and the degree of desirability of an accurate tide-table.

My aim has been to reduce the tables to a simple form, and if, as I imagine, the mathematical capacity of an ordinary ship's captain will suffice for the use of the tables, whether in full or abridged, I have attained the principal object in view.

* I may suppose the elliptic tides unknown, and I should then take $S = \frac{1}{19}H_m + \cdot036H_{,,}$. For Aden this would give $S = \cdot129$, or say $\frac{1}{8}$.

8.

ON THE CORRECTION TO THE EQUILIBRIUM THEORY OF TIDES FOR THE CONTINENTS.

I. By G. H. DARWIN. II. By H. H. TURNER.

[*Proceedings of the Royal Society of London*, XL. (1886), pp. 303—315.]

I.

IN the equilibrium theory of the tides, as worked out by Newton and Bernoulli, it is assumed that the figure of the ocean is at each instant one of equilibrium.

But Sir William Thomson has pointed out that, when portions of the globe are occupied by land, the law of rise and fall of water given in the usual solution cannot be satisfied by a constant volume of water*.

In Part I. of this paper Sir William Thomson's work is placed in a new light, which renders the conclusions more easily intelligible, and Part II. contains the numerical calculations necessary to apply the results to the case of the earth.

If m, r, z be the moon's mass, radius vector, and zenith distance; g mean gravity; ρ the earth's mean density; σ the density of water; a the earth's radius; and $\mathfrak{h}$ the height of tide; then, considering only the lunar influence, the solution of the equilibrium theory for an ocean-covered globe is

$$\frac{\mathfrak{h}}{a} = \frac{3m}{2gr^3} \frac{1}{(1 - \frac{3}{5}\sigma/\rho)} (\cos^2 z - \tfrac{1}{3}) \qquad (1)$$

This equilibrium law would still hold good when the ocean is interrupted by continents, if water were appropriately supplied to or exhausted from the sea as the earth rotates.

* Thomson and Tait's *Natural Philosophy*, 1883, § 808.

Since when water is supplied or exhausted the height of water will rise or fall everywhere to the same extent, it follows that the rise and fall of tide, according to the revised equilibrium theory, must be given by

$$\frac{\mathfrak{h}}{a} = \frac{3ma}{2gr^3} \frac{1}{1 - \frac{3}{5}\sigma/\rho} (\cos^2 z - \tfrac{1}{3}) - \alpha \quad \text{.....................(2)}$$

where α is a constant all over the earth for each position of the moon relatively to the earth, but varies for different positions.

Let Q be the fraction of the earth's surface which is occupied by sea; let λ be the latitude and l the longitude of any point; and let ds stand for $\cos\lambda d\lambda dl$, an element of solid angle. Then we have

$$4\pi Q = \iint ds$$

integrated all over the oceanic area.

The quantity of water which must be subtracted from the sea, so as to depress the sea level everywhere by $a\alpha$, is $4\pi a^3\alpha Q$; and the quantity required to raise it by the variable height $\dfrac{3ma^2}{2gr^3} \dfrac{\cos^2 z - \frac{1}{3}}{1 - \frac{3}{5}\sigma/\rho}$ is the integral of this function, taken all over the ocean. But since the volume of water must be constant, continuity demands that

$$\alpha = \frac{3ma}{2g(1 - \frac{3}{5}\sigma/\rho)r^3} \cdot \frac{1}{4\pi Q} \iint (\cos^2 z - \tfrac{1}{3})\, ds \quad \text{..............(3)}$$

integrated all over the ocean.

On substituting this value of α in (2) we shall obtain the law of rise and fall.

Now if λ, l be the latitude and W. longitude of the place of observation; h the Greenwich westward hour-angle of the moon at the time and place of observation; and δ the moon's declination, it is well known that

$$\cos^2 z - \tfrac{1}{3} = \tfrac{1}{2}\cos^2\lambda \cos^2\delta \cos 2(h - l) + \sin 2\lambda \sin\delta \cos\delta \cos(h - l)$$
$$+ \tfrac{3}{2}(\tfrac{1}{3} - \sin^2\delta)(\tfrac{1}{3} - \sin^2\lambda) \quad \text{.......................(4)}$$

We have next to introduce (4) under the double integral sign of (3), and integrate over the ocean.

To express the result conveniently, let

$$\frac{1}{4\pi Q}\iint \cos^2\lambda \cos 2l\, ds = \cos^2\lambda_2 \cos 2l_2, \quad \frac{1}{4\pi Q}\iint \cos^2\lambda \sin 2l\, ds = \cos^2\lambda_2 \sin 2l_2$$

$$\frac{1}{4\pi Q}\iint \sin 2\lambda \cos l\, ds = \sin 2\lambda_1 \cos l_1, \quad \frac{1}{4\pi Q}\iint \sin 2\lambda \sin l\, ds = \sin 2\lambda_1 \sin l_1$$

$$\frac{1}{4\pi Q}\iint (\tfrac{3}{2}\sin^2\lambda - \tfrac{1}{2})\, ds = \tfrac{3}{2}\sin^2\lambda_0 - \tfrac{1}{2} \quad \text{.................(5)}$$

the integrals being taken over the oceanic area.

These five integrals are called by Sir William Thomson $\mathfrak{A}, \mathfrak{B}, \mathfrak{C}, \mathfrak{D}, \mathfrak{E}$, but by introducing the five auxiliary latitudes and longitudes, $\lambda_2, l_2, \lambda_1, l_1, \lambda_0$, we shall find for the conclusions an easily intelligible physical interpretation.

It may be well to observe that (5) necessarily give real values to the auxiliaries. For consider the first integral as a sample:—

Every element of $\iint \cos^2\lambda \cos 2l ds$ is, whether positive or negative, necessarily numerically less than the corresponding element of $4\pi Q$, and therefore, even if all the elements of the former integral were taken with the same sign, $(4\pi Q)^{-1} \iint \cos^2\lambda \cos 2l ds$ would be numerically less than unity, and *à fortiori* in the actual case it is numerically less than unity.

Now using (5) in obtaining the value of $\iint (\cos^2 z - \frac{1}{3})\, ds$, and substituting in (3), we have

$$\frac{\mathfrak{h}}{a} \div \frac{3ma}{2g(1-\frac{3}{5}\sigma/\rho)r^3} = \tfrac{1}{2}\cos^2\delta\,[\cos^2\lambda\cos 2(h-l) - \cos^2\lambda_2 \cos 2(h-l_2)]$$
$$+ \sin 2\delta\,[\sin\lambda\cos\lambda\cos(h-l) - \sin\lambda_1\cos\lambda_1\cos(h-l_1)]$$
$$+ \tfrac{3}{2}(\tfrac{1}{3} - \sin^2\delta)(\sin^2\lambda_0 - \sin^2\lambda) \quad \text{..........................(6)}$$

The first term of (6) gives the semi-diurnal tide, the second the diurnal, and the third the tide of long period.

The meaning of the result is clear. The latitude and longitude λ_2, l_2 give a certain definite spot on the earth's surface which has reference to the semi-diurnal tide. Similarly λ_1, l_1 give another definite spot which has reference to the diurnal tide; and λ_0 gives a definite parallel of latitude which has reference to the tide of long period.

From inspection we see that at the point λ_2, l_2 the semi-diurnal tide is evanescent, and that at the point $\lambda_2, l_2 + 90°$ there is doubled tide, as compared with the uncorrected equilibrium theory. At the place λ_1, l_1 the diurnal tide is evanescent, and at $-\lambda_1, l_1$ there is doubled diurnal tide.

In the latitude λ_0 the long period tide is evanescent, and in latitude (sometimes imaginary) arc sin $\sqrt{\{\frac{2}{3} - \sin^2\lambda_0\}}$ there is doubled long period tide.

Many or all of these points may fall on continents, so that the evanescence or doubling may only apply to the algebraical expressions, which are, unlike the sea, continuous over the whole globe. But now let us consider more precisely what the points are.

It is obvious that the latitude and longitude λ_2 and l_2, being derived from expressions for $\cos^2\lambda_2 \cos 2l_2$ and $\cos^2\lambda_2 \sin 2l_2$, really correspond with four points whose latitudes and longitudes are

$$\lambda_2, l_2;\ -\lambda_2, l_2;\ \lambda_2, l_2+180°;\ -\lambda_2, l_2+180°$$

Thus there are four points of evanescent semi-diurnal tide, situated on a single great circle or meridian, in equal latitudes N. and S., and antipodal two and two. Corresponding to these four, there are four points of doubled semi-diurnal tide, whose latitudes and longitudes are

$$\lambda_2,\ l_2+90°;\quad -\lambda_2,\ l_2+90°;\quad \lambda_2,\ l_2+270°;\quad -\lambda_2,\ l_2+270°$$

and these also are on a single great circle or meridian, at right angles to the former great circle, and are in the same latitudes N. and S. as are the places of evanescence, and are antipodal two and two.

Passing now to the case of the diurnal tide we see that λ_1, l_1, being derived from expressions for $\sin 2\lambda_1 \cos l_1$ and $\sin 2\lambda_1 \sin l_1$, really correspond with four points whose latitudes and longitudes are

$$\lambda_1,\ l_1;\quad -\lambda_1,\ l_1+180°;\quad 90°-\lambda_1,\ l_1;\quad -90°+\lambda_1,\ l_1+180°$$

Thus there are four points of evanescent diurnal tide, situated on a single great circle or meridian, two of them are in one quadrant in complemental latitudes, and antipodal to them are the two others. Corresponding to these four there are four points of doubled diurnal tide lying in the same great circle or meridian, and situated similarly with regard to the S. pole as are the points of evanescence with regard to the N. pole; their latitudes and longitudes are

$$-\lambda_1,\ l_1;\quad \lambda_1,\ l_1+180°;\quad -90°+\lambda_1,\ l_1;\quad 90°-\lambda_1,\ l_1+180°$$

Lastly, in the case of the long period tide, it is obvious that the latitude λ_0 is either N. or S., and that there are two parallels of latitude of evanescent tide. In case $\sin^2\lambda_0$ is less than $\frac{2}{3}$, or λ_0 less than 54° 44′, there are two parallels of latitude of doubled tide of long period in latitude $\frac{2}{3}$ arc sin $\sqrt{\{\frac{2}{3}-\sin^2\lambda_0\}}$.

From a consideration of the integrals, it appears that as the continents diminish towards vanishing, the four points of evanescent and the four points of doubled semi-diurnal tide close in to the pole, two of each going to the N. pole, and two going to the S. pole; also one of the points of evanescent and one of doubled diurnal tide go to the N. pole, a second pair of points of evanescence and of doubling go to the S. pole, a third pair of points of evanescence and of doubling coalesce on the equator, and a fourth pair coalesce at the antipodes of the third pair; lastly, in the case of the tides of long period the circles of evanescent tide tend to coalesce with the circles of doubled tide, in latitudes 35° 16′ N. and S.

We are now in a position to state the results of Thomson's corrected theory by comparison with Bernoulli's theory.

Consider the semi-diurnal tide on an ocean-covered globe, then at the four points on a single meridian great circle which correspond to the points of evanescence on the partially covered globe, the tide has the same height; and at any point on the partially covered globe the semi-diurnal tide is the excess (interpreted algebraically) of the tide at the corresponding point on the ocean-covered globe above that at the four points.

A similar statement holds good for the diurnal and tides of long period.

By laborious quadratures Mr Turner has evaluated in Part II. the five definite integrals on which the corrections to the equilibrium theory, as applied to the earth, depend.

The values found show that the points of evanescent semi-diurnal tide are only distant about 9° from the N. and S. poles; and that of the four points of evanescent diurnal tide two are close to the equator, one close to the N. pole, and the other close to the S. pole; lastly, that the latitudes of evanescent tide of long period are 34° N. and S., and are thus but little affected by the land.

Thus in all cases the points of evanescence are situated near the places where the tides vanish when there is no land. It follows, therefore, that the correction to the equilibrium theory for land is of no importance.

G. H. D.

II.

For the evaluation of the five definite integrals, called by Sir William Thomson $\mathfrak{A}$, $\mathfrak{B}$, $\mathfrak{C}$, $\mathfrak{D}$, $\mathfrak{E}$, and represented in the present paper by functions of the latitudes and longitudes λ_0, λ_1, λ_2, and l_1, l_2, respectively similar in form to the functions of the "running" latitude and longitude to be integrated, it is necessary to assume some redistribution of the land on the earth's surface, differing as little as possible from the real distribution, and yet with a coast line amenable to mathematical treatment. The integrals are to be taken over the whole ocean, but since the value of any of them taken over the whole sphere is zero, the part of any due to the sea is equal to the part due to the land with its sign changed; and since there is less land than sea, it will be more convenient to integrate over the land, and then change the sign.

Unless specially mentioned, we shall hereafter assume that the integration is taken over the land.

The last of the integrals has already been evaluated by Professor Darwin*, with an approximate coast line, which follows parallels of latitude and longitude alternately.

* Thomson and Tait's *Natural Philosophy*, 1883, § 808 [not reproduced in this volume].

His distribution of land is given in the following table:—

N. lat.	W. long.	E. long.
Lat. 80° to 90°	20° to 50°.	
70 „ 80	22° to 55°: 85° to 115°.	55° to 60°: 90° to 110°.
60 „ 70	35° to 52°: 65° to 80°: 90° to 165°.	10° to 180°.
50 „ 60	0° to 6°: 60° to 78°: 90° to 130°.	10° to 140°: 155° to 160°.
40 „ 50	0° to 5°: 65° to 123°.	0° to 135°.
30 „ 40	0° to 8°: 78° to 120°.	0° to 120°: 135° to 138°.
20 „ 30	0° to 15°: 80° to 82°: 97° to 110°.	0° to 118°.
10 „ 20	0° to 17°: 87° to 95°.	0° to 50°: 75° to 85°: 95° to 108°: 122° to 125°.
0 „ 10	53° to 78°.	0° to 48°: 98° to 105°: 112° to 117°.
S. lat.	**W. long.**	**E. long.**
0° to 10°	37° to 80°	12° to 40°: 110° to 130°.
10 „ 20	37 „ 74	12° to 38°: 45° to 50°: 126° to 144°.
20 „ 30	45 „ 71	15° to 33°: 115° to 151°.
30 „ 40	55 „ 73	20° to 23°: 132° to 140°.
40 „ 50	65 „ 73	170° to 172°.
50 „ 60	67 „ 72	———
60 „ 70	55 „ 65	120° to 130°.
70 „ 80	about 20° of longitude.	
80 „ 90	„ 180° „	

N.B.—*The Mediterranean, being approximately a lake, is treated as land.*

The limits of the 20° and 180° of longitude between S. latitudes 70° and 90° are not specified. For the evaluation of the last integral this is not necessary, for restricting

$$\iint (3\sin^2\lambda - 1)\cos\lambda\, d\lambda\, dl$$

to a representative portion of the land bounded by parallels λ_1 and λ_2, l_1 and l_2, we get $-\frac{1}{4}(l_2 - l_1)\Big[\sin\lambda + \sin 3\lambda\Big]_{\lambda_1}^{\lambda_2} \times \frac{\pi}{180}$; and similarly for Q; so that if t_1 and t_2 be the number of degrees of longitude N. and S. of the equator respectively between latitudes λ_1 and λ_2, the last of the integrals becomes

$$\frac{\Sigma\,\frac{1}{4}(t_1 + t_2)\Big[\sin\lambda + \sin 3\lambda\Big]_{\lambda_1}^{\lambda_2}}{720 - \Sigma\,(t_1 + t_2)\Big[\sin\lambda\Big]_{\lambda_1}^{\lambda_2}}.$$

But for (*e.g.*)

$$\int_{\lambda_1}^{\lambda_2}\int_{l_1}^{l_2} \cos^2\lambda\cos 2l\cos\lambda\, d\lambda\, dl = \frac{1}{24}\Big[9\sin\lambda + \sin 3\lambda\Big]_{\lambda_1}^{\lambda_2}\Big[\sin 2l\Big]_{l_1}^{l_2}$$

the actual limits l_2 and l_1 must be given, and not merely their difference.

It is, however, obvious, on inspection of these integrals, that the land in high latitudes affects them but little; and we shall not lose much by neglecting entirely the Antarctic continent in their evaluation.

This evaluation is reduced by the above process to a series of multiplications, and on performing them the following values of $\mathfrak{A}$, $\mathfrak{B}$, $\mathfrak{C}$, $\mathfrak{D}$, $\mathfrak{E}$, and Q are obtained on the two hypotheses:—

(1) That there is as much Antarctic land as is given in the schedule, which is, however, only taken into account in the last integral $\mathfrak{E}$, and the common denominator $4\pi Q$ of each.

(2) That there is no land between S. latitude 80° and the pole.

The value of Q is given in terms of the whole surface, and represents the fraction of that surface occupied by land; it must be remembered that the Mediterranean Sea is treated as land. Professor Darwin quotes Rigaud's estimate* as 0·266:—

	1st hypothesis	2nd hypothesis
$\mathfrak{A}$	+0·03023	+0·03008
$\mathfrak{B}$	+0·00539	+0·00537
$\mathfrak{C}$	−0·01975	−0·01965
$\mathfrak{D}$	+0·02910	+0·02895
$\mathfrak{E}$	−0·01520	−0·00486
Q	0·283	0·278

These results for $\mathfrak{E}$ and Q have already been given by Professor Darwin in Thomson and Tait's *Natural Philosophy*, and I have found them correct.

We then find for the set of latitudes and longitudes of evanescent tide:—

Nature of tide		1st hypothesis	2nd hypothesis
Long period	lat. λ_0	34° 39′ N.	35° 4′ N.
Diurnal	lat. λ_1	1 0 S.	1 0 S.
	long. l_1	55 50 E.	55 50 E.
Semi-diurnal	lat. λ_2	79 54 N.	79 56 N.
	long. l_2	5 3 W.	5 4 W.

The other points of evanescence are of course easily derivable from these, as shown in the first part of this paper.

As a slightly closer approximation to truth, I have calculated these integrals on another supposition. There are cases where lines satisfying the equations

$$l = \text{const. or } \lambda = \text{const.}$$

* *Transactions of the Cambridge Philosophical Society*, Vol. VI.

diverge somewhat widely from the actual coast line, but a line

$$\pm al \pm b\lambda = \text{const.}$$

(where a and b are small integers) can be found following it more faithfully. An approximate coast line of the land on the earth is defined in the following schedule, west longitudes and north latitudes being considered positive.

Limits of longitude (l)		Equation		Limits of latitude (λ)
+ 20° to + 10°	......	$-\lambda = l - 40$	......	+ 20° to + 30°
——	......	$l = 10$	......	+ 30 „ + 40
+ 10 „ − 23	......	$-\lambda = l - 50$	......	+ 40 „ + 73
− 23 „ + 120	......	$\lambda = 73$	......	——
——	......	$l = 120$	......	+ 73 „ + 80
+ 120 „ + 20	......	$\lambda = 80$	......	——
——	......	$l = 20$	......	+ 80 „ + 70
+ 20 „ + 50	......	$-3\lambda = l - 230$	......	+ 70 „ + 60
+ 50 „ + 70	......	$\lambda = l + 10$	......	+ 60 „ + 80
+ 70 „ + 80	......	$\lambda = 80$	......	——
+ 80 „ + 50	......	$\lambda = l$	......	+ 80 „ + 50
+ 50 „ + 90	......	$-2\lambda = l - 150$	......	+ 50 „ + 30
+ 90 „ + 100	......	$\lambda = 30$	......	——
+ 100 „ + 80	......	$\lambda = l - 70$	......	+ 30 „ + 10
+ 80 „ + 70	......	$\lambda = 10$	......	——
+ 70 „ + 30	......	$2\lambda = l - 50$	......	+ 10 „ − 10
+ 30 „ + 73	......	$-\lambda = l - 20$	......	− 10 „ − 53
——	......	$l = 73$	......	− 53 „ − 14
+ 73 „ + 80	......	$\lambda = 2l - 160$	......	− 14 „ 0
+ 80 „ + 140	......	$\lambda = l - 80$	......	0 „ + 60
+ 140 „ − 150	......	$\lambda = 60$	......	——
− 150 „ − 100	......	$-\lambda = l + 90$	......	+ 60 „ + 10
− 100 „ − 90	......	$+\lambda = l + 110$	......	+ 10 „ + 20
− 90 „ − 80	......	$-\lambda = l + 70$	......	+ 20 „ + 10
− 80 „ − 65	......	$\lambda = l + 90$	......	+ 10 „ + 25
− 65 „ − 40	......	$-\lambda = l + 40$	......	+ 25 „ 0
——	......	$l = -40$	......	0 „ − 20
− 40 „ − 20	......	$-\lambda = l + 60$	......	− 20 „ − 40
− 20 „ − $8\frac{3}{4}$	......	$\lambda = 4l + 40$	......	− 40 „ + 5
− $8\frac{3}{4}$ „ + $12\frac{1}{2}$	......	$\lambda = 5$	......	——
+ $12\frac{1}{2}$ „ + 20	......	$\lambda = 2l - 20$	......	+ 5 „ + 20

New Guinea.

− 130 to − 150	......	$2\lambda = l + 130$	......	0 to − 10
− 150 „ − 140	......	$\lambda = -10$	......	——
− 140 „ − 130	......	$\lambda = l + 130$	......	− 10 „ 0

Australia.

Limits of longitude (l)		Equation		Limits of latitude (λ)
$-140°$ to $-150°$	......	$\lambda = l + 130$	......	$-10°$ to $-20°$
——	......	$l = -150$	......	-20 „ -35
-150 „ -115	......	$\lambda = -35$	......	——
——	......	$l = -115$	......	-35 „ $-22\frac{1}{2}$
-115 „ -140	......	$-2\lambda = l + 160$	......	$-22\frac{1}{2}$ „ -10

It will be seen that it is only rarely necessary to depart from the forms of equation $\pm\lambda = l + x$ and the two original forms $\lambda = \text{const.}$, $l = \text{const.}$ to represent the coast line with considerable accuracy. There are still left one or two outlying portions, of which mention will be made later.

Now supposing we are to find the value of the first integral for the portion of land indicated by the shaded portion of the diagram, EQ being the equator:

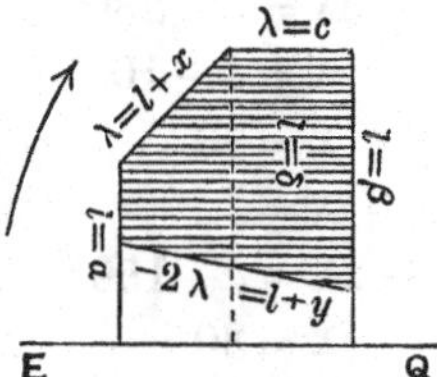

the equations to its boundaries being written at the side of each.

We have

$$\iint \cos^3\lambda d\lambda \cos 2l dl = \frac{1}{12}\int \Big[9\sin\lambda + \sin 3\lambda\Big]_{\lambda_1}^{\lambda_2} \cos 2l dl$$

$$= \frac{1}{12}\int_a^\delta \{9\sin(l+x) + \sin 3(l+x)\}\cos 2l dl$$

$$+ \int_\delta^\beta (9\sin c + \sin 3c)\cos 2l dl$$

$$+ \int_\beta^a - \{9\sin\tfrac{1}{2}(l+y) + \sin\tfrac{3}{2}(l+y)\}\cos 2l dl$$

We may thus simply travel round the boundary omitting the places where $\lambda = \text{constant}$: being careful to go round all the pieces of land in the same direction. If we suppose $l = \alpha$ to be the meridian of Greenwich, and the land to be in the northern hemisphere, the direction indicated above is the wrong one for obtaining the value of the integrals over the land, for the longitudes increase to the left; but by following this direction we shall obtain the values over the sea as is in reality required.

The result of integration has, of course, a different form for each form of the relation between l and λ representing the boundary. In computing the numerical values of the integrals, it is convenient to consider together all the parts of the boundary represented by similar equations.

Below are given as representative the forms which the numerator of the first integral 𝔄 assumes for different forms of the boundary, the quantities within square brackets being taken within limits.

Form		Value of Integral
$\mp \lambda = l + x$	......	$\pm \frac{1}{24} [\frac{1}{5} \cos(5l + 3x) + 3 \cos(3l + x) + \cos(l + 3x) - 9 \cos(-l + x)]$
$\lambda = x$	......	$+ \frac{1}{24} (9 \sin x + \sin 3x) [\sin 2l]$
$l = x$	......	Zero
$\lambda = 2l + x$	......	$- \frac{1}{24} [\frac{9}{4} \cos(4l + x) - 9l \sin x + \frac{1}{8} \cos(8l + 3x) + \frac{1}{4} \cos(4l + 3x)]$
$\lambda = 4l + x$	......	$- \frac{1}{24} [\frac{3}{2} \cos(6l + x) + \frac{9}{2} \cos(2l + x) + \frac{1}{14} \cos(14l + 3x) + \frac{1}{10} \cos(10l + 3x)]$
$\mp 2\lambda = l + x$	......	$\pm \frac{1}{24} [\frac{18}{5} \cos \frac{1}{2}(5l + x) - 6 \cos \frac{1}{2}(-3l + x) + \frac{2}{7} \cos \frac{1}{2}(7l + 3x) - 2 \sin \frac{1}{2}(-l + 3x)]$
$- 3\lambda = l + x$	......	$+ \frac{1}{24} [\frac{27}{7} \cos \frac{1}{3}(7l + x) - \frac{27}{5} \cos \frac{1}{3}(-5l + x) + \frac{1}{3} \cos(3l + x) - \cos(-l + x)]$

Evaluating these integrals on this supposition, we obtain

	1st hypothesis	2nd hypothesis
𝔄	+0·02119	+0·02110
𝔅	+0·00778	+0·00775
ℭ	−0·01890	−0·01882
𝔇	+0·03159	+0·03128
𝔈	−0·04364	−0·03319
Q	0·283	0·278

It will be noticed that the values of Q are exactly the same as before.

From these we deduce

Nature of tide		1st hypothesis	2nd hypothesis
Long period	lat. λ_0	33° 29′ N.	33° 55′ N.
Diurnal	lat. λ_1	1 3 S.	1 3 S.
	long. l_1	59 7 E.	58 58 E.
Semi-diurnal	lat. λ_2	81 22 N.	81 23 N.
	long. l_2	10 5 W.	10 5 W.

The agreement of these values of the quantities with the values calculated on the previous supposition is not quite so close as I anticipated, but it should be remarked that the numerators of the quantities 𝔄, 𝔅, ℭ, 𝔇, 𝔈 are the differences of positive and negative quantities of very much greater magnitude, as becomes obvious on proceeding to the numerical calculation; and thus a comparatively small change in one of the large compensating quantities, due to large tracts of land in different portions of the globe, affects the integrals to a considerable extent.

In this connexion I was led to investigate the effect of counting various small islands and promontories as sea, or small bays and straits as land. For instance, a portion of sea in the neighbourhood of Behring's Straits is included as land, and a corresponding correction must be applied to the integrals. This correction I have *estimated* as follows:—The area of the sea is estimated in square degrees, by drawing lines on a large map corresponding to each degree of latitude and longitude and counting the squares covered by sea, fractions of a square to one decimal place being included, though the tenths have been neglected in the concluded sum. This area has then been multiplied by the value of (say) $\cos^3\lambda \cos 2l$ for the approximate centre of gravity of the portion, to find an approximate value of the integral

$$\iint \cos^3\lambda \cos 2l \, d\lambda \, dl$$

over its surface.

By drawing the assumed coast line on a map, it will become obvious that such corrections may be applied for the following portions, defined by the latitude and longitude of their centres of gravity; remarking that when there is a portion of land which may be fairly considered to compensate a portion of sea in the immediate neighbourhood, no correction has been applied. For instance, it would be seen that part of the Kamschatkan Peninsula is excluded from the coast line, and part of the Sea of Okhotsk is included; but these will produce nearly equal effects on the integrals in opposite directions, and are thus left out of consideration.

Area in square degrees		Longitude		Latitude
+ 160	....	+ 172°	....	+ 64°
+ 240	....	+ 150	....	+ 71½
+ 166	....	+ 85	....	+ 60
+ 80	....	+ 60	....	+ 52
+ 68	....	+ 85	....	+ 9
− 20	....	+ 75	....	+ 21
− 69	....	+ 55	....	+ 4
+ 43	....	+ 34	....	− 11
+ 22	....	− 37	....	− 20
− 48	....	− 47	....	− 19
+ 65	....	− 53	....	+ 18
− 16	....	− 107	....	+ 13
− 34	....	− 102	....	+ 2
− 49	....	− 114	....	+ 1

Area in square degrees		Longitude		Latitude
− 27	....	− 123°	....	+ 12°
− 11	....	− 118	....	− 5
− 43	....	− 138	....	+ 36
− 39	....	− 173	....	− 42

N.B.—*Land-areas are considered positive, sea negative.*

We then find the following corrected values of the integrals :—

	1st hypothesis	2nd hypothesis
𝔄	+0·02237	+0·02247
𝔅	+0·00230	+0·00231
ℭ	−0·01952	−0·01961
𝔇	+0·02665	+0·02676
𝔈	−0·01775	−0·02810
Q	0·279	0·274

and finally the following values of the latitudes and longitudes of evanescent tides :—

Nature of tide		1st hypothesis	2nd hypothesis
Long period	lat. λ_0	34° 33′ N.	34° 7′ N.
Diurnal	lat. λ_1	0 57 S.	0 57 S.
	long. l_1	53 47 E.	53 46 E.
Semi-diurnal	lat. λ_2	81 23 N.	81 21 N.
	long. l_2	2 56 W.	2 56 W.

The estimation of corrections due to these supplementary portions has been checked in two cases by a detailed extension of the method of square blocks of land used previously for evaluation of the whole integrals; that is to say, two of these portions were separately divided into square degrees (instead of squares whose sides were each ten degrees), and the integral evaluated in a similar manner to that previously described. The agreement of the values so calculated with those obtained by the above method of estimation was sufficiently exact to justify a certain confidence in the close agreement of the finally corrected values of the integrals with their theoretically perfect values.

H. H. T.

9.

ATTEMPTED EVALUATION OF THE RIGIDITY OF THE EARTH FROM THE TIDES OF LONG PERIOD.

[*This is* § 848 *of the second edition of Thomson and Tait's* Natural Philosophy (1883). *There have been some changes of notation, so as to make the investigation consistent with the other papers in this volume. Some portions have been omitted, where omissions could be made without any interference with the main result.*]

It appears from Schedule B iii. of the Report to the British Association of 1883 (p. 22 of this volume) that the fortnightly tide Mf is expressed by

$$\tfrac{3}{2}\frac{M}{E}\left(\frac{a}{c}\right)^3 a\,.\left(\tfrac{1}{2}-\tfrac{3}{2}\sin^2\lambda\right)\left(1-\tfrac{5}{2}e^2\right).\tfrac{1}{2}\sin^2 I\cos 2(s-\xi)$$

and the lunar monthly tide Mm by

$$\tfrac{3}{2}\frac{M}{E}\left(\frac{a}{c}\right)^3 a\left(\tfrac{1}{2}-\tfrac{3}{2}\sin^2\lambda\right)e\left(1-\tfrac{3}{2}\sin^2 I\right)\cos(s-p)$$

But in the paper on the correction to the equilibrium theory for the continents (p. 330 of this volume) it is shown that the factor

$$\tfrac{1}{2}-\tfrac{3}{2}\sin^2\lambda=\tfrac{3}{2}\left(\tfrac{1}{3}-\sin^2\lambda\right)=\tfrac{3}{2}\left(\sin^2 35^\circ\,16'-\sin^2\lambda\right)$$

should be replaced by $\tfrac{3}{2}(\sin^2\lambda_0-\sin^2\lambda)$. It appears further from that paper that λ_0 is 34° 39′ according to one hypothesis and 35° 4′ according to another as to the distribution of land on the earth. Hence we may with sufficient accuracy replace this factor by $\tfrac{3}{2}\sin(35^\circ-\lambda)\sin(35^\circ+\lambda)$.

Thus if we write $\tau=\tfrac{3}{2}\dfrac{M}{E}\left(\dfrac{a}{c}\right)^3$, and let ϕ denote the equilibrium fortnightly tide, and μ the equilibrium monthly tide, we have

$$\phi=\tfrac{3}{4}\tau a\left(1-\tfrac{5}{2}e^2\right)\sin^2 I\sin(35^\circ-\lambda)\sin(35^\circ+\lambda)\cos 2(s-\xi)$$

$$\mu=\tfrac{3}{2}\tau ae\left(1-\tfrac{3}{2}\sin^2 I\right)\sin(35^\circ-\lambda)\sin(35^\circ+\lambda)\cos(s-p)$$

Thus the actual fortnightly and monthly tides must be expressed in the forms

$$\left.\begin{aligned}\phi &= \tfrac{3}{4}\tau a\,(1-\tfrac{5}{2}e^2)\sin^2 I\sin(35^\circ-\lambda)\sin(35^\circ+\lambda)\{x\cos 2(s-\xi)+y\sin 2(s-\xi)\}\\ \mu &= \tfrac{3}{2}\tau ae\,(1-\tfrac{3}{2}\sin^2 I)\sin(35^\circ-\lambda)\sin(35^\circ+\lambda)\,\{u\cos(s-p)+v\sin(s-p)\}\end{aligned}\right\} \quad\text{.........(19)}$$

where x, y, u, v are numerical coefficients. If the equilibrium theory be nearly true for the fortnightly and monthly tides, y and v will be small; and x and u will be fractions approaching unity, in proportion as the rigidity of the earth's mass approaches infinity.

If we now put

$$\left.\begin{aligned}\mathrm{a} &= \tfrac{3}{4}\tau a\,(1-\tfrac{5}{2}e^2)\sin^2 I\sin(35^\circ-\lambda)\sin(35^\circ+\lambda)\\ \mathrm{c} &= \tfrac{3}{2}\tau ae\,(1-\tfrac{3}{2}\sin^2 I)\sin(35^\circ-\lambda)\sin(35^\circ+\lambda)\end{aligned}\right\} \quad\text{.........(20)}$$

and for the fortnightly tide write

$$\left.\begin{aligned}\mathrm{a}x &= \mathrm{A}\\ \mathrm{a}y &= \mathrm{B}\end{aligned}\right\} \quad\text{..................................(21)}$$

and for the monthly tide write

$$\left.\begin{aligned}\mathrm{c}u &= \mathrm{C}\\ \mathrm{c}v &= \mathrm{D}\end{aligned}\right\} \quad\text{..................................(22)}$$

we have

$$\begin{aligned}\phi &= \mathrm{A}\cos 2(s-\xi)+\mathrm{B}\sin 2(s-\xi)\\ \mu &= \mathrm{C}\cos(s-p)+\mathrm{D}\sin(s-p)\end{aligned}$$

Every set of tidal observations will give equations for x, y, u, v; and the most probable values of these quantities must be determined by the method of least squares.

For places north of 35° N. lat., or south of 35° S. lat. the coefficients a and c become negative. This would be inconvenient for the arithmetical operations of reduction, and therefore it is convenient to regard the coefficients a and c as being in all cases positive, for we may suppose $(\lambda - 35^\circ)$ to be taken for places in the northern hemisphere North of 35°, and $35^\circ - \lambda$ for places in the same hemisphere to the South of 35°; and similarly for the southern hemisphere.

[This is merely an artifice for avoiding the insertion of many negative signs, and it makes no difference in the result.]

In collecting the results of tidal observation I have to thank Sir William Thomson, General Strachey, and Major Baird* for placing all the materials in my hands, and for giving me every facility. The observations are to be found in the British Association Reports for 1872 and 1876, and in the Tide-tables of the Indian Government.

* [Now Lord Kelvin, Sir Richard Strachey, and Colonel Baird.]

The results of the harmonic analysis of the tidal observations are given altogether for 22 different ports, but of these only 14 are used here. The following are the reasons for rejecting those made at 8 out of the 22 ports.

One of these stations is Cat Island in the Gulf of Mexico; this place, in latitude 30° 14′ N., lies so near to the critical latitude of evanescent fortnightly and monthly tides, that considering the uncertainty in the exact value of that latitude, it is impossible to determine the proper weight which should be assigned to the observation. The result only refers to a single year, viz. 1848, and as its weight must in any case be very small, the omission can exercise scarcely any effect on the result.

Another omitted station is Toulon; this being in the Mediterranean Sea cannot exhibit the true tide of the open ocean.

Another is Hanstal in the Gulf of Cutch. The result is given in an Indian Blue Book. I do not know the latitude, and General Strachey informs me that he believes the observations were only made during a few months for the purpose of determining the mean level of the sea, for the levelling operations of the great survey of India.

The other omitted stations are Diamond Harbour, Fort Gloster and Kidderpore in the Hooghly estuary, and Rangoon, and Moulmein. All these are river stations, and they all exhibit long period tides of such abnormal height as to make it nearly certain that the shallowness of the water has exercised a large influence on the results. The observations higher up the Hooghly seem more abnormal than those lower down. I also learn that the tidal predictions are not found to be satisfactory at these stations.

The following tables exhibit the results for the 14 remaining ports.

. . . .

No attempt has been made to assign weight to each year's observations according to the exact number of months over which the tidal records extend. The data for such weighting are in many cases wanting. In computing the value for a the factor $1-\frac{5}{2}e^2$ was omitted, but it has been introduced finally as explained below.

British and French Ports, North of Latitude 35°.

[*Tidal Reports of Brit. Assoc.* 1872 *and* 1876.]

	Place .... N. Latitude...	Ramsgate 51° 21′	Liverpool 53° 40′				West Hartlepool 54° 41′			Brest 48° 23′
		1	2	3	4	5	6	7	8	9
	Year	1864	1857–8	1858–9	1859–60	1866–7	1858–9	1859–60	1860–1	1875
Fortnightly Tide	A	– ·0112	+ ·0917	+ ·0337	– ·0015	– ·0346	+ ·0497	+ ·0297	+ ·0729	– ·0244
	B	+ ·0311	– ·0156	– ·0153	– ·0240	+ ·0095	+ ·0153	+ ·0438	– ·0027	– ·0959
	a	·0439	·0955	·0946	·0905	·0416	·0996	·0952	·0884	·0684
Monthly Tide	C	– ·0223	– ·0153	– ·1687	+ ·1508	+ ·0127	– ·0688	+ ·1347	– ·0261	– ·0279
	D	– ·0224	+ ·0434	– ·1038	– ·0190	+ ·0708	– ·0299	– ·0100	– ·1365	+ ·0178
	c	·0332	·0302	·0304	·0312	·0392	·0320	·0328	·0339	·0218

Indian Ports.

[*Indian Tide Tables for* 1881.]

	Place ...... Latitude...	Aden 12° 47′		Kurrachee 24° 47′							
		10	11	12	13	14	15	16	17	18	19
	Year	1877–8	1879–80	1868–9	1869–70	1870–1	1873–4	1874–5	1875–6	1876–7	1877–8
Fortnightly Tide	A	+ ·0606	+ ·0597	+ ·0287	+ ·0421	– ·0076	– ·0123	+ ·0429	+ ·0131	+ ·0460	+ ·0508
	B	+ ·0131	+ ·0167	– ·0249	– ·0482	– ·0342	– ·0102	+ ·0328	+ ·0050	– ·0012	+ ·0406
	a	·0818	·0706	·0218	·0248	·0287	·0411	·0440	·0456	·0459	·0448
Monthly Tide	C	+ ·0194	+ ·0329	– ·0288	– ·0429	– ·0140	+ ·0281	+ ·0522	– ·0133	+ ·0631	+ ·0722
	D	– ·0158	+ ·0027	– ·0703	+ ·0035	+ ·0288	+ ·0414	+ ·0229	+ ·0565	+ ·0570	+ ·0831
	c	·0268	·0286	·0185	·0180	·0173	·0153	·0148	·0145	·0145	·0147

INDIAN PORTS (*continued*).

[*Indian Tide Tables for* 1881–2.]

	PLACE...... Latitude..	OKHA POINT AND BEYT HARBOUR 22° 28′	BOMBAY—APOLLO BUNDER 18° 55′			KÁRWÁR 14° 48′		BEYPORE 11° 10′	
		20	21	22	23	24	25	26	27
	YEAR	1874–5	1876–7	1878–9	1879–80	1878–9	1879–80	1878–9	1879–80
Fortnightly Tide	A	+·0529	+·0699	+·0888	+·0653	+·0649	+·0699	+·1040	+·0907
	B	+·0459	+·0121	−·0199	+·0100	−·0168	+·0025	+·0208	+·0281
	a	·0525	·0671	·0617	·0565	·0726	·0665	·0803	·0735
Monthly Tide	C	+·0331	−·0048	+·0372	+·0459	+·0415	+·0562	+·0687	+·0057
	D	−·0375	+·0266	−·0378	−·0036	−·0069	+·0143	+·0072	+·0708
	c	·0177	·0212	·0220	·0229	·0260	·0270	·0287	·0298

	PLACE ...... Latitude ...	PAUMBEN PASS, Island of Ramesweram 9° 16′		VIZAGAPATAM 17° 41′		MADRAS 13° 4′	PORT BLAIR, ROSS ISLAND 11° 40½′
		28	29	30	31	32	33
	YEAR	1878–9	1879–80	1879–80	1880–1	1880–1	1880–1
Fortnightly Tide	A	·0560	·0448	+·0328	+·0524	+·0318	+·0589
	B	−·0010	−·0043	+·0148	−·0167	+·0037	−·0029
	a	·0835	·0764	·0597	·0534	·0626	·0649
Monthly Tide	C	+·0579	+·0279	+·0194	+·0468	+·0304	+·0195
	D	−·0113	+·0439	+·0080	+·0611	+·0260	+·0045
	c	·0298	·0310	·0242	·0253	·0296	·0307

Gauss's notation is adopted for the reductions*. That is to say, [AA] denotes the sum of the squares of the A's, and [Aa] the sum of the products of each A into its corresponding a.

* See Gauss's works, or the Appendix to Chauvenet's *Astronomy*.

In computing the value of a for the fortnightly tide the factor $(1-\frac{5}{2}e^2)$ which occurs therein was treated as being equal to unity; since $\frac{5}{2}e^2 = \cdot 00754$, it follows that the [aa], which would be found from the numbers given in the table, must be multiplied by $(1 - \cdot 01508)$, and the [Aa] and [Ba] by $(1 - \cdot 00754)$. After introducing these correcting factors the following results were found:

$$[\text{aa}] = \cdot 14573,\quad [\text{AA}] = \cdot 09831,\quad [\text{BB}] = \cdot 02576,\quad [\text{Aa}] = \cdot 09836,\quad [\text{Ba}] = \cdot 00291$$
$$[\text{cc}] = \cdot 02253,\quad [\text{CC}] = \cdot 11588,\quad [\text{DD}] = \cdot 07552,\quad [\text{Cc}] = \cdot 01533,\quad [\text{Dc}] = \cdot 00202$$

Then according to the method of least squares, the following are the most probable values of x, y, u, v.

$$x = \frac{[\text{Aa}]}{[\text{aa}]},\qquad y = \frac{[\text{Ba}]}{[\text{aa}]},\qquad u = \frac{[\text{Cc}]}{[\text{cc}]},\qquad v = \frac{[\text{Dc}]}{[\text{cc}]}$$

And if m be the number of observations (which in the present case is 33) the mean errors of x, y, u, v are respectively

$$\frac{1}{[\text{aa}]}\sqrt{\frac{[\text{AA}][\text{aa}] - [\text{Aa}]^2}{m-1}},\qquad \frac{1}{[\text{aa}]}\sqrt{\frac{[\text{BB}][\text{aa}] - [\text{Ba}]^2}{m-1}}$$
$$\frac{1}{[\text{cc}]}\sqrt{\frac{[\text{CC}][\text{cc}] - [\text{Cc}]^2}{m-1}},\qquad \frac{1}{[\text{cc}]}\sqrt{\frac{[\text{DD}][\text{cc}] - [\text{Dc}]^2}{m-1}}$$

The probable errors are found from the mean errors by multiplying by $\cdot 6745$.

I thus find that

$$x = \cdot 675 \pm \cdot 056,\quad y = \cdot 020 \pm \cdot 055,\quad u = \cdot 680 \pm \cdot 258,\quad v = \cdot 090 \pm \cdot 218$$

The smallness of the values of y and v is satisfactory; for, as stated in § 848 (d) of the *Natural Philosophy*, if the equilibrium theory were true for the two tides under discussion, they should vanish. Moreover the signs are in agreement with what they should be, if friction be a sensible cause of tidal retardation. But considering the magnitude of the probable errors, it is of course rather more likely that the non-evanescence of y and v is due to errors of observation*.

If the solid earth does not yield tidally, and if the equilibrium theory is

* Shortly after these computations were completed Professor Adams happened to observe a misprint in the Tidal Report for 1872. This Report gives the method employed in the reduction by harmonic analysis of the tidal observations, and the erroneous formula relates to the reduction of the tides of long period. On inquiring of Mr Roberts, who has superintended the harmonic analysis, it appears that the erroneous formula has been used throughout in the reductions. A discussion of this mistake and of its effects will be found in a paper communicated to the British Association by me in 1882. It appears that the values of the fortnightly tide are not seriously vitiated, but the monthly elliptic tide will have suffered much more. This will probably account for the large probable error which I have found for the value of the monthly tide. If a recomputation of all the long-period tides should be carried out, I think there is good hope that the probable error of the value of the fortnightly tide may also be reduced.

fulfilled, x and u should each be approximately unity, and if it yields tidally they should have equal values. The very close agreement between them is probably somewhat due to chance. From this point of view it seems reasonable to combine all the observations, resulting from 66 years of observation, for both sorts of tides together.

Then writing X and Y for the numerical factors by which the equilibrium values of the two components of either tide are to be multiplied in order to give the actual results, I find

$$X = \cdot 676 \pm \cdot 076, \qquad Y = \cdot 029 \pm \cdot 065$$

These results really seem to present evidence of a tidal yielding of the earth's mass, showing that it has an effective rigidity about equal to that of steel*.

But this result is open to some doubt for the following reason:

Taking only the Indian results (48 years in all), which are much more consistent than the English ones, I find

$$X = \cdot 931 \pm \cdot 056, \qquad Y = \cdot 155 \pm \cdot 068$$

We thus see that the more consistent observations seem to bring out the tides more nearly to their theoretical equilibrium-values with no elastic yielding of the solid.

It is to be observed however that the Indian results being confined within a narrow range of latitude give (especially when we consider the absence of minute accuracy in the evaluation of the critical latitude λ_0) a less searching test for the elastic yielding, than a combination of results from all latitudes.

On the whole we may fairly conclude that, whilst there is some evidence of a tidal yielding of the earth's mass, that yielding is certainly small, and that the effective rigidity is at least as great as that of steel.

[*Postscript.* It is interesting to compare this conclusion with the results obtained by Dr O. Hecker by means of the horizontal pendulum ("Beobachtungen an Horizontalpendeln über die Deformation des Erdkörpers," *K. Preuss. Geodät. Inst.*, Neue Folge, No. 32, 1907), for he finds the deflections of the pendulum to be two-thirds as great as they would be on a rigid earth.]

* It is remarkable that elastic yielding of the upper strata of the earth, in the case where the sea does not cover the whole surface, may lead to an apparent augmentation of oceanic tides at some places, situated on the coasts of continents. This subject is investigated in the Report for 1882 of the Committee of the British Association on "The Lunar Disturbance of Gravity." (Paper 14 in this volume.) It is there, however, erroneously implied that this kind of elastic yielding would cause an apparent augmentation of tide at all stations of observation.

10.

DYNAMICAL THEORY OF THE TIDES.

[*This contains certain sections from the article* "TIDES" *written in* 1906 *for the new edition of the* Encyclopædia Britannica, *being based on the corresponding paragraphs in the original edition and on the article* "TIDES" *in the supplementary volumes. Reproduced by special permission of the Proprietors of the* Enc. Brit.]

§ 12. *Form of Equilibrium.*

Consider the shape assumed by an ocean of density σ on a planet of mass M, density δ and radius a, when acted on by disturbing forces whose potential is a solid spherical harmonic of degree i, the planet not being in rotation.

If S_i denotes a surface spherical harmonic of order i, such a potential is given at the point whose radius vector is ρ by

$$V = \frac{3ma^2}{2r^3}\left(\frac{\rho}{a}\right)^i S_i \quad\text{.............................(1)}$$

In the case considered in an earlier section m is the moon's mass and r her distance, and $i = 2$, while S_i becomes $\cos^2$ (moon's z. d.) $- \frac{1}{3}$; [in the present instance the form of the coefficient is immaterial, save that its dimensions shall be correct].

The theory of harmonic analysis tells us that the form of the ocean, when in equilibrium, must be given by the equation

$$\rho = a + e_i S_i \quad\text{...............................(2)}$$

Our problem is to evaluate e_i.

We know that the external potential of a layer of matter, of depth $e_i S_i$ and density σ, has the value

$$\frac{4\pi\sigma a}{2i+1}\left(\frac{a}{\rho}\right)^{i+1} e_i S_i$$

Hence the whole potential externally to the planet and up to its surface is

$$\frac{M}{\rho}+\frac{3ma^2}{2r^3}\left(\frac{\rho}{a}\right)^i S_i+\frac{4\pi\sigma a}{2i+1}\left(\frac{a}{\rho}\right)^{i+1} e_i S_i \quad\ldots\ldots\ldots\ldots(3)$$

The first and most important term is the potential of the planet, the second that of the disturbing force, and the third that of the departure from sphericity.

Since the surface of the ocean must be a level surface, the expression (3) equated to a constant must be another form of (2). Hence if we put $\rho=a+e_iS_i$ in the first term of (3) and $\rho=a$ in the second and third terms, (3) must be constant; this can only be the case if the coefficient of S_i vanishes.

Hence on effecting these substitutions and equating that coefficient to zero, we find

$$-\frac{M}{a^2}e_i+\frac{3ma^2}{2r^3}+\frac{4\pi\sigma a}{2i+1}e_i=0$$

But by definition of δ and a we have $M=\frac{4}{3}\pi\delta a^3=ga^2$, where g is gravity, and therefore

$$e_i=\frac{\dfrac{3ma^2}{2gr^3}}{1-\dfrac{3\sigma}{(2i+1)\,\delta}} \quad\ldots\ldots\ldots\ldots(4)$$

. . . .

If σ were very small compared with δ the attraction of the water on itself would be very small compared with that of the planet on the water; hence we see that $1\Big/\left(1-\dfrac{3\sigma}{(2i+1)\,\delta}\right)$ is the factor by which the mutual gravitation of the ocean augments the deformation due to the external forces. This factor will occur frequently hereafter, and therefore for brevity we write

$$b_i=1-\frac{3\sigma}{(2i+1)\,\delta} \quad\ldots\ldots\ldots\ldots(5)$$

and (4) may be written

$$e_i=\frac{3ma^2}{2gr^3b_i} \quad\ldots\ldots\ldots\ldots(6)$$

Comparison with (1) then shows that

$$V=gb_i\left(\frac{\rho}{a}\right)^i e_iS_i \quad\ldots\ldots\ldots\ldots(7)$$

is the potential of the disturbing forces under which

$$\rho=a+e_iS_i \quad\ldots\ldots\ldots\ldots(8)$$

is a figure of equilibrium.

We are thus provided with a convenient method of specifying any disturbing force by means of the figure of equilibrium which it is competent to maintain. In considering the dynamical theory of the tides of an ocean-covered planet, we shall specify the disturbing forces in the manner expressed by (7) and (8).

This way of specifying a disturbing force is equally exact whether or not we choose to include the effects of the mutual attraction of the ocean. If the augmentation due to mutual attraction of the water is not included, b_i becomes equal to unity; in this case there is no necessity to use spherical harmonic analysis, and we see that if the equation to the surface of an ocean be

$$\rho = a + S$$

where S is a function of latitude and longitude, it is in equilibrium under forces due to a potential whose value at the surface of the sphere (where $\rho = a$) is gS.

. . . .

§ 14. *Recent Advances in the Dynamical Theory of the Tides.*

The problem of the tidal oscillation of the sea is essentially dynamical. In two papers in the second volume of *Liouville's Journal* (1896), M. Poincaré has considered the mathematical principles involved in the problem, where the ocean is interrupted by land as in actuality. He has not sought to obtain numerical results applicable to any given configuration of land and sea, but he has aimed rather at pointing out methods by which it may some day be possible to obtain such solutions. It would hardly be in place to attempt to follow the details of an investigation of this character, but we may say that it affords a conspicuous example of M. Poincaré's power ot reaching the very heart of a difficult problem.

Even when the ocean is taken as covering the whole earth, the problem presents formidable difficulties, and this is the only case in which it has been solved hitherto*.

Laplace gives the solution in Bks I. and IV. of the *Mécanique Céleste*; but his work is unnecessarily complicated. In the first edition of the present work we gave Laplace's theory without these complications, but the theory is now accessible in Lamb's *Hydrodynamics*, and perhaps in other works of the kind. It will not, therefore, be reproduced here.

In 1897 and 1898 Mr S. S. Hough undertook an important revision of Laplace's theory, and succeeded not only in introducing the effects of the

* Lord Kelvin's (Sir W. Thomson's) paper on the "Gravitational oscillations of rotating water," *Phil. Mag.*, Oct. 1880, bears on this subject. It is the only attempt to obtain numerical results in respect to the effect of the earth's rotation on the oscillations of land-locked seas.

mutual gravitation of the ocean, but also in determining the nature and periods of the free oscillations of the sea*. A dynamical problem of this character cannot be regarded as fully solved unless we are able not only to discuss the "forced" oscillations of the system but also the "free." Hence we regard Mr Hough's work as the most important contribution to the dynamical theory of the tides since the time of Laplace. We shall accordingly present the theory in the form due to Mr Hough, although limitations of space compel us to omit many points of interest.

The analysis is more complex than that of Laplace, where the mutual attraction of the ocean was neglected, but this was perhaps inevitable.

Our first task is to form the equations of motion and continuity, which will be equally applicable to all forms of the theory.

§ 15. *Equations of Motion.*

Let r, θ, ϕ be the radius vector, colatitude, and east longitude of a point with reference to an origin, a polar axis, and a zero-meridian rotating with a uniform angular velocity n from west to east. Then if R, H, Ξ be the radial, colatitudinal, and longitudinal accelerations of the point, we have

$$\left.\begin{aligned} R &= \frac{d^2r}{dt^2} - r\left(\frac{d\theta}{dt}\right)^2 - r\sin^2\theta\left(\frac{d\phi}{dt}+n\right)^2 \\ \Xi &= \frac{1}{r}\frac{d}{dt}\left(r^2\frac{d\theta}{dt}\right) - r\sin\theta\cos\theta\left(\frac{d\phi}{dt}+n\right)^2 \\ H &= \frac{1}{r\sin\theta}\frac{d}{dt}\left[r^2\sin^2\theta\left(\frac{d\phi}{dt}+n\right)\right] \end{aligned}\right\}$$

If the point were at rest with reference to the rotating meridian, we should have $R = -n^2r\sin^2\theta$, $\Xi = -n^2r\sin\theta\cos\theta$, $H = 0$. When these considerations are applied to the motion of an ocean relative to a rotating planet, it is clear that these accelerations, which still remain when the ocean is at rest, are annulled by the permanent oblateness of the ocean. As then they take no part in the oscillations of the ocean, and as we are not considering the figure of the planet, we may omit these terms from R and Ξ. This being so, we must replace $\left(\frac{d\phi}{dt}+n\right)^2$ as it occurs in R and Ξ by $\left(\frac{d\phi}{dt}\right)^2 + 2n\frac{d\phi}{dt}$.

Now suppose that the point whose accelerations are under consideration never moves far from its zero position, and that its displacements ξ, $\eta\sin\theta$ in colatitude and longitude are very large compared with ρ, its radial displacement. Suppose further that the velocities of the point are so small

* *Phil. Trans. Roy. Soc.*, Vol. 189 A, pp. 201—258, and Vol. 191 A, pp. 139—185.

that their squares and products are negligible compared with n^2r^2; then we have

$$\frac{dr}{dt} = \frac{d\rho}{dt}, \text{ a very small quantity}$$

$$r \sin\theta \frac{d\phi}{dt} = \frac{d}{dt}(\eta \sin\theta)$$

$$r\frac{d\theta}{dt} = \frac{d\xi}{dt}$$

Since the radial velocity always remains very small, it is not necessary to concern ourselves further with the value of R, and we only require the two other components which have the approximate forms

$$\left.\begin{aligned} \Xi &= \frac{d^2\xi}{dt^2} - 2n \sin\theta \cos\theta \frac{d\eta}{dt} \\ H &= \sin\theta \frac{d^2\eta}{dt^2} + 2n \cos\theta \frac{d\xi}{dt} \end{aligned}\right\} \quad \ldots\ldots\ldots\ldots\ldots\ldots\ldots\ldots(9)$$

We have now to consider the forces by which an element of the ocean is urged in the direction of colatitude and longitude. These forces are those due to the external disturbing forces, to the pressure of the water surrounding an element of the ocean, and to the attraction of the ocean itself.

If $\mathfrak{e}$ denotes the equilibrium height of the tide, it is a function of co-latitude and longitude, and may be expanded in a series of spherical surface harmonics $\mathfrak{e}_i$. Thus we may write the equation to the equilibrium tide in the form

$$r = a + \mathfrak{e} = a + \Sigma\mathfrak{e}_i$$

Now it appears from (7) and (8) that the value of the potential, at the surface of the sphere where $\rho = a$, under which this is a figure of equilibrium, is

$$V = \Sigma g b_i \mathfrak{e}_i$$

We may use this as specifying the external disturbing force due to the known attractions of the moon and sun, so that $\mathfrak{e}_i$ may be regarded as known.

But in our dynamical problem the ocean is not a figure of equilibrium, and we may denote the elevation of the surface at any moment of time by $\mathfrak{h}$. Then the equation to the surface may be written in the form

$$r = a + \mathfrak{h} = a + \Sigma\mathfrak{h}_i$$

where $\mathfrak{h}_i$ denotes a spherical harmonic just as $\mathfrak{e}_i$ did before.

The surface value of the potential of the forces which would maintain the ocean in equilibrium in the shape it has at any moment is $\Sigma g b_i \mathfrak{h}_i$.

Hence it follows that in the actual case the forces due to fluid pressure and to the attraction of the ocean must be such as to balance the potential just determined. Therefore these forces are those due to a potential $-\Sigma g b_i \mathfrak{h}_i$. If we add to this the potential of the external forces, we have a potential which will include all the forces, the expression for which is $-g\Sigma b_i(\mathfrak{h}_i - \mathfrak{e}_i)$. If further we perform the operations $d/ad\theta$ and $d/a\sin\theta d\phi$ on this potential, we obtain the colatitudinal and longitudinal forces which are equal to the accelerations Ξ and H.

It follows, then, from (9) that the equations of motion are

$$\left.\begin{aligned} \frac{d^2\xi}{dt^2} - 2n\sin\theta\cos\theta\frac{d\eta}{dt} &= -\frac{g}{a}\Sigma b_i\frac{d}{d\theta}(\mathfrak{h}_i - \mathfrak{e}_i) \\ \sin\theta\frac{d^2\eta}{dt^2} + 2n\cos\theta\frac{d\xi}{dt} &= -\frac{g}{a\sin\theta}\Sigma b_i\frac{d}{d\phi}(\mathfrak{h}_i - \mathfrak{e}_i) \end{aligned}\right\} \quad \text{......(10)}$$

It remains to find the equation of continuity. This may be deduced geometrically from the consideration that the volume of an element of the fluid remains constant; but a shorter way is to derive it from the equation of continuity as it occurs in ordinary hydrodynamical investigations. If Φ be a velocity potential, the equation of continuity for incompressible fluid is

$$\delta r\frac{d}{dr}\left(r^2\frac{d\Phi}{dr}\sin\theta\,\delta\theta\,\delta\phi\right) + \delta\theta\frac{d}{d\theta}\left(r\sin\theta\frac{d\Phi}{rd\theta}\delta r\,\delta\phi\right)$$
$$+ \delta\phi\frac{d}{d\phi}\left(r\frac{1}{r\sin\theta}\frac{d\Phi}{d\phi}\delta r\,\delta\theta\right) = 0$$

The element referred to in this equation is defined by r, θ, ϕ, $r+\delta r$, $\theta+\delta\theta$, $\phi+\delta\phi$. The colatitudinal and longitudinal velocities are the same for all the elementary prism defined by θ, ϕ, $\theta+\delta\theta$, $\phi+\delta\phi$, and the sea bottom. Then $\dfrac{d\Phi}{rd\theta} = \dfrac{d\xi}{dt}$, $\dfrac{d\Phi}{r\sin\theta d\phi} = \sin\theta\dfrac{d\eta}{dt}$; and, since the radial velocity is $d\mathfrak{h}/dt$ at the surface of the ocean, where $r = a + \gamma$, and is zero at the sea bottom, where $r = a$, we have $\dfrac{d\Phi}{dr} = \dfrac{r-a}{\gamma}\dfrac{d\mathfrak{h}}{dt}$. Hence, integrating with respect to r from $r = a+\gamma$ to $r = a$, and again with respect to t from the time t to the time when $\mathfrak{h}$, ξ, η all vanish, and treating γ and $\mathfrak{h}$ as small compared with a, we have

$$\mathfrak{h}a\sin\theta + \frac{d}{d\theta}(\gamma\xi\sin\theta) + \frac{d}{d\phi}(\gamma\eta\sin\theta) = 0 \quad \text{...........(11)}$$

This is the equation of continuity, and, together with (10), it forms the system which must be integrated in the general problem of the tides. The difficulties in the way of a solution are so great that none has hitherto been found, except on the supposition that γ, the depth of the ocean, is only a function of latitude. In this case (11) becomes

$$\mathfrak{h}a + \frac{1}{\sin\theta}\frac{d}{d\theta}(\gamma\xi\sin\theta) + \gamma\frac{d\eta}{d\phi} = 0 \quad \text{.................(12)}$$

§ 16. *Adaptation to Forced Oscillations.*

Since we may suppose that the free oscillations are annulled by friction, the solution required is that corresponding to forced oscillations. Now $\mathfrak{e}$, which is proportional to V, has terms of three kinds, the first depending on twice the moon's (or sun's) hour-angle, the second on the hour-angle, and the third independent thereof. The coefficients of the first and second vary slowly, and the whole of the third varies slowly. Hence $\mathfrak{e}$ has a semi-diurnal, a diurnal, and a long-period term. We shall see later that these terms may be expanded in a series of approximately semi-diurnal, diurnal, and slowly varying terms, each of which is a strictly harmonic function of the time.

Thus, according to the usual method of treating oscillating systems, we may make the following assumptions as to the form of the solution

$$\left.\begin{aligned} \mathfrak{e} &= \Sigma\mathfrak{e}_i = \Sigma e_i \cos(2nft + s\phi + \alpha) \\ \mathfrak{h} &= \Sigma\mathfrak{h}_i = \Sigma h_i \cos(2nft + s\phi + \alpha) \\ \xi &= \Sigma b_i x_i \cos(2nft + s\phi + \alpha) \\ \eta &= \Sigma b_i y_i \sin(2nft + s\phi + \alpha) \end{aligned}\right\} \quad \text{..................(13)}$$

where e_i, h_i, x_i, y_i are functions of colatitude only, and e_i, h_i are the associated functions of colatitude corresponding to the harmonic of order i and rank s.

For the semi-diurnal tides $s = 2$, and f approximately unity; for the diurnal tides $s = 1$, and f approximately $\frac{1}{2}$; and for the tides of long period $s = 0$, and f is a small fraction.

Substituting these values in (12), we have

$$\Sigma\left[\frac{1}{\sin\theta}\frac{d}{d\theta}(\gamma b_i x_i \sin\theta) + s\gamma b_i y_i + h_i a\right] = 0 \quad \text{............(14)}$$

Then if we write u_i for $h_i - e_i$, and put $m = n^2 a/g$, substitution from (13) in (14) leads at once to

$$\left.\begin{aligned} f^2 \Sigma b_i x_i + f \sin\theta\cos\theta\, \Sigma b_i y_i &= \frac{1}{4m}\frac{d}{d\theta}\Sigma b_i u_i \\ f^2 \sin\theta\, \Sigma b_i y_i + f\cos\theta\, \Sigma b_i x_i &= -\frac{s}{4m}\frac{\Sigma b_i u_i}{\sin\theta} \end{aligned}\right\} \quad \text{............(15)}$$

Solving (15), we have

$$\left.\begin{aligned} (\Sigma b_i x_i)(f^2 - \cos^2\theta) &= \frac{1}{4m}\left(\frac{d\Sigma b_i u_i}{d\theta} + \frac{s\cos\theta}{f\sin\theta}\Sigma b_i u_i\right) \\ (\Sigma b_i y_i)\sin^2\theta\,(f^2 - \cos^2\theta) &= -\frac{1}{4m}\left(\frac{\cos\theta}{f}\frac{d\Sigma b_i u_i}{d\theta} + \frac{s\Sigma b_i u_i}{\sin\theta}\right) \end{aligned}\right\} \quad \text{......(16)}$$

Then substituting from (16) in (14), we have

$$\frac{1}{\sin\theta}\frac{d}{d\theta}\left[\frac{\gamma\left(\sin\theta\frac{d\Sigma b_i u_i}{d\theta}+\frac{s}{f}\cos\theta\,\Sigma b_i u_i\right)}{f^2-\cos^2\theta}\right]-s\gamma\frac{\frac{\cos\theta}{f}\frac{d\Sigma b_i u_i}{d\theta}+\frac{\Sigma b_i u_i}{\sin\theta}}{\sin\theta\,(f^2-\cos^2\theta)}$$
$$+4ma\Sigma(u_i+e_i)=0\quad\ldots\ldots\ldots\ldots\ldots\ldots(17)$$

This is closely analogous to Laplace's equation for tidal oscillations in an ocean whose depth is only a function of latitude; indeed the only difference is that we have followed Mr Hough in introducing the mutual attraction of the water.

When u_i is found from this equation, its value substituted in (16) will give x_i and y_i.

§ 17. *Zonal Oscillations.*

We might treat the general harmonic oscillations first, and proceed to the zonal oscillations by putting $s=0$. These waves are, however, comparatively simple, and it is well to begin with them. The zonal tides are those which Laplace describes as of the first species, and are now more usually called the tides of long period.

As we shall only consider the case of an ocean of uniform depth, γ the depth of the sea is constant. Then since in this case $s=0$, our equation (17) to be satisfied by u_i or h_i-e_i becomes

$$\frac{d}{d\theta}\left(\frac{\sin\theta\frac{d}{d\theta}\Sigma b_i u_i}{f^2-\cos^2\theta}\right)+\frac{4ma}{\gamma}\Sigma h_i\sin\theta=0$$

This may be written

$$\frac{d}{d\theta}\Sigma b_i u_i+\frac{4ma}{\gamma}\frac{f^2-\cos^2\theta}{\sin\theta}\int^\theta\Sigma h_i\sin\theta d\theta+A=0\quad\ldots\ldots\ldots(18)$$

where A is a constant.

Let us assume $\qquad h_i=C_iP_i,\quad e_i=E_iP_i$

where P_i denotes the ith zonal harmonic of $\cos\theta$. The coefficients C_i are unknown, but the E_i are known because the system oscillates under the action of known forces.

If the term involving the integral in this equation were expressed in terms of differentials of harmonics, we should be able to equate to zero the coefficient of each $dP_i/d\theta$ in the equation, and thus find the conditions for determining the C's.

The task then is to express $\dfrac{f^2-\cos^2\theta}{\sin\theta}\displaystyle\int^\theta P_i\sin\theta d\theta$ in differentials of zonal harmonics.

It is well known that P_i satisfies the differential equation

$$\frac{d}{d\theta}\left(\sin\theta\frac{dP_i}{d\theta}\right)+i(i+1)P_i\sin\theta=0 \quad\text{.................(19)}$$

Therefore $$\int P_i\sin\theta d\theta=-\frac{1}{i(i+1)}\sin\theta\frac{dP_i}{d\theta}$$

and $$\frac{f^2-\cos^2\theta}{\sin\theta}\int P_i\sin\theta d\theta=-\frac{1}{i(i+1)}(f^2-\cos^2\theta)\frac{dP_i}{d\theta}$$
$$=-\frac{1}{i(i+1)}(f^2-1)\frac{dP_i}{d\theta}-\frac{1}{i(i+1)}\sin^2\theta\frac{dP_i}{d\theta}$$

Another well-known property of zonal harmonics is that

$$\sin\theta\frac{dP_i}{d\theta}=\frac{i(i+1)}{2i+1}(P_{i+1}-P_{i-1}) \quad\text{.................(20)}$$

If we differentiate (20) and use (19) we have

$$\frac{i(i+1)}{2i+1}\left(\frac{dP_{i+1}}{d\theta}-\frac{dP_{i-1}}{d\theta}\right)+i(i+1)P_i\sin\theta=0 \quad\text{...........(21)}$$

Multiplying (20) by $\sin\theta$, and using (21) twice over

$$\sin^2\theta\frac{dP_i}{d\theta}=\frac{i(i+1)}{2i+1}\left\{-\frac{1}{2i+3}\left(\frac{dP_{i+2}}{d\theta}-\frac{dP_i}{d\theta}\right)+\frac{1}{2i-1}\left(\frac{dP_i}{d\theta}-\frac{dP_{i-2}}{d\theta}\right)\right\}$$

Therefore

$$\frac{f^2-\cos^2\theta}{\sin\theta}\int P_i\sin\theta d\theta=\frac{1}{(2i-1)(2i+1)}\frac{dP_{i-2}}{d\theta}$$
$$-\left\{\frac{f^2-1}{i(i+1)}+\frac{2}{(2i-1)(2i+3)}\right\}\frac{dP_i}{d\theta}+\frac{1}{(2i+1)(2i+3)}\frac{dP_{i+2}}{d\theta}$$

This expression, when multiplied by $4ma/\gamma$ and by C_i and summed, is the second term of our equation.

The first term is $$\Sigma b_i(C_i-E_i)\frac{dP_i}{d\theta}$$

In order that the equation may be satisfied, the coefficient of each $dP_i/d\theta$ must vanish identically. Accordingly we multiply the whole by $\gamma/4ma$ and equate to zero the coefficient in question, and obtain

$$\frac{b_i\gamma}{4ma}(C_i-E_i)+\frac{C_{i-2}}{(2i-1)(2i-3)}-\left\{\frac{f^2-1}{i(i+1)}+\frac{2}{(2i-1)(2i+3)}\right\}C_i$$
$$+\frac{C_{i+2}}{(2i+3)(2i+5)}=0 \quad\text{..............(22)}$$

This equation (22) is applicable for all values of i from 1 to infinity, provided that we take C_0, E_0, C_{-1}, E_{-1} as being zero.

We shall only consider in detail the case of greatest interest, namely that of the most important of the tides generated by the attraction of the sun and moon. We know that in this case the equilibrium tide is expressed by a zonal harmonic of the second order; and therefore all the E_i, excepting E_2, are zero. Thus the equation (22) will not involve E_i in any case excepting when $i=2$.

If we write for brevity

$$L_i = \frac{f^2-1}{i(i+1)} + \frac{2}{(2i-1)(2i+3)} - \frac{b_i\gamma}{4ma}$$

the equation (22) becomes

$$\frac{C_{i+2}}{(2i+3)(2i+5)} - L_iC_i + \frac{C_{i-2}}{(2i-3)(2i-1)} = 0 \quad \text{..........(23)}$$

save that when $i=2$, the right-hand side is $\dfrac{b_2\gamma}{4ma}E_2$, a known quantity *ex hypothesi*.

The equations naturally separate themselves into two groups, in one of which all the suffixes are even and the other odd. Since our task is to evaluate all the C's in terms of E_2, it is obvious that all the C's with odd suffixes must be zero, and we are left to consider only the cases where $i=2$, 4, 6, &c.

We have said that C_0 must be regarded as being zero; if however we take

$$C_0 = -\frac{3b_2\gamma}{4ma}E_2, \text{ so that } C_0 \text{ is essentially a known quantity}$$

the equation (23) has complete applicability for all even values of i from 2 upwards.

The equations are

$$\frac{C_0}{1.3} - L_2C_2 + \frac{C_4}{7.9} = 0, \quad \frac{C_2}{5.7} - L_4C_4 + \frac{C_6}{11.13} = 0, \text{ \&c.}$$

It would seem at first sight as if these equations would suffice to determine all the C's in terms of C_2, and that C_2 would remain indeterminate; but we shall show that this is not the case.

For very large values of i the general equation of condition (23) tends to assume the form

$$\frac{C_{i+2}+C_{i-2}}{C_i} + \frac{i^2\gamma}{ma} = 0$$

By writing successively $i+2$, $i+4$, $i+6$ in this equation and taking the differences we obtain an equation from which we see that, *unless* C_i/C_{i+2} *tends to become infinitely small*, the equations are satisfied by $C_i = C_{i+2}$ in the limit for very large values of i

Hence, if C_i does not tend to zero, the later portion of the series for h tends to assume the form $C_i(P_i + P_{i+2} + P_{i+4} \ldots)$. All the P's are equal to unity at the pole; hence the hypothesis that C_i does not tend to zero leads to the conclusion that the tide is of infinite height at the pole. The expansion of the height of tide is essentially convergent, and thus this hypothesis is negatived. Thus we are entitled to assume that C_i tends to zero for large values of i.

Now writing for brevity

$$a_i = \frac{1}{(2i+1)(2i+3)^2(2i+5)}$$

we may put (23) into the form

$$\frac{C_{i-2}/C_i}{(2i-3)(2i-1)} = L_i - \cfrac{a_i}{\cfrac{C_i/C_{i+2}}{(2i+1)(2i+3)}}$$

By successive applications of this formula, we may write the right-hand side in the form of a continued fraction.

Let
$$K_i = \frac{a_{i-2}}{L_i -}\;\frac{a_i}{L_{i+2} -}\;\frac{a_{i+2}}{L_{i+4} -}\cdots$$

Then we have
$$\frac{C_{i-2}/C_i}{(2i-3)(2i-1)} = \frac{a_{i-2}}{K_i}$$

or
$$\frac{C_i}{C_{i-2}} = (2i-1)(2i+1)K_i$$

Thus

$$C_2 = 3.5K_2C_0;\quad C_4 = 3.5.7.9K_2K_4C_0;\quad C_6 = 3.5.7.9.11.13K_2K_4K_6C_0 \text{ \&c.}$$

If we assume that any of the higher C's, such as C_{14} or C_{16}, is of negligible smallness, all the continued fractions K_2, K_4, K_6 &c. may be computed; and thus we find all the C's in terms of C_0, which is equal to $-\frac{3b_2\gamma}{4ma}E_2$.

The height of the tide is therefore given by

$$\begin{aligned}\mathfrak{h} &= \Sigma h_i \cos(2nft+\alpha)\\ &= -\frac{3b_2\gamma}{4ma}E_2\{3.5K_2P_2 + 3.5.7.9K_2K_4P_4 + \ldots\}\cos(2nft+\alpha)\end{aligned}$$

It is however more instructive to express $\mathfrak{h}$ as a multiple of the equilibrium tide $\mathfrak{e}$, which is as we know equal to $E_2P_2\cos(2nft+\alpha)$. Whence we find

$$\mathfrak{h} = -\frac{3b_2\gamma}{4ma}\frac{\mathfrak{e}}{P_2}\{3.5K_2P_2 + 3.5.7.9K_2K_4P_4 + 3.5\ldots13K_2K_4K_6P_6 + \ldots\}$$

The number f is a fraction such that its reciprocal is twice the number of sidereal days in the period of the tide. The greatest value of f is that appertaining to the lunar fortnightly tide (Mf in notation of Harmonic Analysis) and in this case f is in round numbers $\frac{1}{28}$, or more exactly $f^2 = {\cdot}00133$.

The ratio of the density σ of sea-water to δ the mean density of the earth is ·18093; which value gives us

$$b_2 = 1 - \frac{3\sigma}{5\delta} = \cdot 89144$$

The quantity m is the ratio of equatorial centrifugal force to gravity, and is equal to $\frac{1}{289}$.

Finally γ/a is the depth of the ocean expressed as a fraction of the earth's radius.

With these numerical values Mr Hough has applied the solution to determine the lunar fortnightly tide for oceans of various depths. Of his results we give two:—

First, when $\gamma = 7260$ ft. $= 1210$ fathoms, which makes $\frac{\gamma}{4ma} = \frac{1}{40}$, he finds

$$\mathfrak{h} = \frac{\mathfrak{C}}{P_2}\{\cdot 2669P_2 - \cdot 1678P_4 + \cdot 0485P_6 - \cdot 0081P_8 + \cdot 0009P_{10} - \cdot 0001P_{12} \ldots\}$$

If the equilibrium theory were true we should have

$$\mathfrak{h} = \frac{\mathfrak{C}}{P_2}\{P_2\}$$

thus we see how widely the dynamical solution differs from the equilibrium value.

Secondly, when $\gamma = 58080$ ft. $= 9680$ fathoms, and $\frac{\gamma}{4ma} = \frac{1}{5}$, he finds

$$\mathfrak{h} = \frac{\mathfrak{C}}{P_2}\{\cdot 7208P_2 - \cdot 0973P_4 + \cdot 0048P_6 - \cdot 0001P_8 \ldots\}$$

From this we see that the equilibrium solution presents some sort of approximation to the dynamical one; and it is clear that the equilibrium solution would be fairly accurate for oceans which are still quite shallow when expressed as fractions of the earth's radius, although far deeper than the actual sea.

The tides of long-period were not investigated by Laplace in this manner, for he was of opinion that a very small amount of friction would suffice to make the ocean assume its form of equilibrium. In the arguments which he adduced in support of this view the friction contemplated was such that the integral effect was proportional to the velocity of the water relatively to the bottom. It is probable the proportionality to the square of the velocity would have been nearer the truth, but the distinction is unimportant.

The most rapid of the oscillations of this class is the lunar fortnightly tide, and the water of the ocean moves northward for a week and then southward for a week. In oscillating systems, where the resistances are proportional to the velocities, it is usual to specify the resistance by a 'modulus of decay,' namely the time in which a velocity is reduced by friction to e^{-1} or 1/2·78 of its initial value. Now in order that the result contemplated by Laplace may be true, the friction must be such that the modulus of decay is short compared with the semi-period of oscillation. It seems practically certain that the friction of the ocean bed would not reduce a slow ocean current to one-third of its primitive value in a day or two. Hence we cannot accept Laplace's discussion as satisfactory, and the investigation which has just been given becomes necessary*.

§ 18. *Tesseral Oscillations.*

The oscillations which we now have to consider are those in which the form of surface is expressible by the tesseral harmonics. The results will be applicable to the diurnal and semidiurnal tides—Laplace's second and third species.

If we write $\sigma = \frac{s}{f}$ the equation (17) becomes

$$\frac{d}{d\theta}\left[\frac{\left(\sin\theta\frac{d}{d\theta}+\sigma\cos\theta\right)\Sigma b_i u_i}{s^2-\sigma^2\cos^2\theta}\right]-\frac{\left(\sigma\cos\theta\frac{d}{d\theta}+s^2\operatorname{cosec}\theta\right)\Sigma b_i u_i}{s^2-\sigma^2\cos^2\theta}$$
$$+\frac{4ma}{\gamma\sigma^2}\Sigma h_i\sin\theta=0\;\ldots\ldots\ldots\ldots\ldots\ldots(24)$$

If D be written for the operation $\sin\theta\frac{d}{d\theta}$, the middle term may be arranged in the form

$$-\frac{\sigma\cot\theta\,(D+\sigma\cos\theta)\,\Sigma b_i u_i}{s^2-\sigma^2\cos^2\theta}-\Sigma b_i u_i\sin\theta$$

Therefore on multiplying (24) by $\sin\theta$ it becomes

$$(D-\sigma\cos\theta)\left[\frac{(D+\sigma\cos\theta)\,\Sigma b_i u_i}{s^2-\sigma^2\cos^2\theta}\right]-\Sigma b_i u_i+\frac{4ma}{\gamma\sigma^2}\Sigma h_i\sin^2\theta=0\;\ldots(25)$$

We now introduce two auxiliary functions, such that

$$\Sigma b_i(h_i-e_i)=\Sigma b_i u_i$$
$$=(D-\sigma\cos\theta)\,\Psi+(s^2-\sigma^2\cos^2\theta)\,\Phi\ldots\ldots\ldots\ldots(26)$$

It is easy to prove that

$$\left.\begin{aligned}(D+\sigma\cos\theta)(D-\sigma\cos\theta)&=D^2-s^2+\sigma\sin^2\theta+(s^2-\sigma^2\cos^2\theta)\\(D-\sigma\cos\theta)(D+\sigma\cos\theta)&=D^2-s^2-\sigma\sin^2\theta+(s^2-\sigma^2\cos^2\theta)\end{aligned}\right\}\ldots(27)$$

Also

$$(D+\sigma\cos\theta)(s^2-\sigma^2\cos^2\theta)\,\Phi=(s^2-\sigma^2\cos^2\theta)(D+\sigma\cos\theta)\,\Phi$$
$$+2\sigma^2\sin^2\theta\cos\theta\,\Phi\;\ldots\ldots\ldots\ldots\ldots\ldots(28)$$

Now perform $D+\sigma\cos\theta$ on (26), and use the first of (27) and (28), and we have

$$(D+\sigma\cos\theta)\,\Sigma b_i u_i=(D^2-s^2+\sigma\sin^2\theta+s^2-\sigma^2\cos^2\theta)\,\Psi$$
$$+(s^2-\sigma^2\cos^2\theta)(D+\sigma\cos\theta)\,\Phi+2\sigma^2\sin^2\theta\cos\theta\,\Phi\;\ldots\ldots(29)$$

* [A fuller discussion of this subject is contained in Paper 11 below.]

The functions Ψ and Φ are as yet indeterminate, and we may impose another condition on them. Let that condition be

$$(D^2 - s^2 + \sigma \sin^2 \theta) \Psi = - 2\sigma^2 \sin^2 \theta \cos \theta \, \Phi \quad \dots\dots\dots\dots (30)$$

Then (29) may be written

$$\frac{(D + \sigma \cos \theta) \Sigma b_i u_i}{s^2 - \sigma^2 \cos^2 \theta} = \Psi + (D + \sigma \cos \theta) \Phi$$

Substituting from this in (25), and using the second of (27), the function Ψ disappears and the equation reduces to

$$(D^2 - s^2 - \sigma \sin^2 \theta) \Phi + \frac{4ma}{\gamma\sigma^2} \Sigma h_i \sin^2 \theta = 0 \quad \dots\dots\dots\dots (31)$$

Since by (30) $-\sigma^2 \cos^2 \theta \, \Phi = \frac{1}{2} \frac{\cos \theta}{\sin^2 \theta} (D^2 - s^2 + \sigma \sin^2 \theta) \Psi$, (26) may be written

$$\Sigma b_i u_i = \left[D - \sigma \cos \theta + \tfrac{1}{2} \frac{\cos \theta}{\sin^2 \theta} (D^2 - s^2 + \sigma \sin^2 \theta) \right] \Psi + s^2 \Phi \quad \dots\dots (32)$$

The equations (30), (31), and (32) define Ψ and Φ, and furnish the equation which must be satisfied.

If we denote $\cos \theta$ by μ the zonal harmonics are defined by

$$P_i = \frac{1}{2^i i!} \left(\frac{d}{d\mu} \right)^i (\mu^2 - 1)^i$$

The following are three well-known properties of zonal harmonics:

$$\frac{d}{d\mu} \left[(1 - \mu^2) \frac{dP_i}{d\mu} \right] + i(i + 1) P_i = 0 \quad \dots\dots\dots\dots (33)$$

$$(i + 1) P_{i+1} - (2i + 1) \mu P_i + i P_{i-1} = 0 \quad \dots\dots\dots\dots (34)$$

$$\frac{dP_{i+1}}{d\mu} - \frac{dP_{i-1}}{d\mu} = (2i + 1) P_i \quad \dots\dots\dots\dots (35)$$

If $P_i^s \frac{\cos}{\sin} s\phi$ are the two tesseral harmonics of order i and rank s, it is also known that

$$P_i^s = (1 - \mu^2)^{\frac{1}{2}s} \frac{d^s P_i}{d\mu^s} \quad \dots\dots\dots\dots (36)$$

Let us now assume

$$h_i = C_i^s P_i^s, \quad e_i = E_i^s P_i^s, \quad \Psi = \Sigma \alpha_i^s P_i^s, \quad \Phi = \Sigma \beta_i^s P_i^s$$

These must now be substituted in our three equations (30), (31), (32), and the result must be expressed by series of the P_i^s functions. It is clear then that we have to transform into P_i^s functions the following functions of P_i^s, namely

$$\frac{1}{\sin^2 \theta} (D^2 - s^2 \pm \sigma \sin^2 \theta) P_i^s, \quad \cos \theta P_i^s$$

$$\left[D - \sigma \cos \theta + \tfrac{1}{2} \frac{\cos \theta}{\sin^2 \theta} (D^2 - s^2 + \sigma \sin^2 \theta) \right] P_i^s$$

If we differentiate (33) s times, and express the result by means of the operator D, we find

$$(D^2 - s^2) P_i^s + i(i+1) P_i^s \sin^2\theta = 0 \quad \text{.................(37)}$$

Again, differentiating (34) s times and using (35), we find

$$(i - s + 1) P_{i+1}^s - (2i+1)\cos\theta P_i^s + (i+s) P_{i-1}^s = 0 \quad \text{........(38)}$$

Lastly, differentiating (36) once and using (33), (35), and (38)

$$DP_i^s = \frac{i(i-s+1)}{2i+1} P_{i+1}^s - \frac{(i+1)(i+s)}{2i+1} P_{i\ 1}^s \quad \text{...........(39)}$$

By means of (37), (38), and (39) we have

$$\frac{1}{\sin^2\theta}(D^2 - s^2 \pm \sigma \sin^2\theta) P_i^s = [-i(i+1) \pm \sigma] P_i^s$$

$$\cos\theta P_i^s = \frac{i+s}{2i+1} P_{i-1}^s + \frac{i-s+1}{2i+1} P_{i+1}^s$$

$$\left[D - \sigma + \tfrac{1}{2}\frac{\cos\theta}{\sin^2\theta}(D^2 - s^2 + \sigma\sin^2\theta)\right] P_i^s$$
$$= -\frac{(i-s+1)[\sigma + i(i-1)]}{2(2i+1)} P_{i+1}^s - \frac{(i+s)[\sigma + (i+1)(i+2)]}{2(2i+1)} P_{i-1}^s$$

Therefore the equations (30), (31), (32) give

$$\Sigma\left[\alpha_i^s\{-i(i+1)+\sigma\} P_i^s + 2\sigma^2\beta_i^s\left\{\frac{i+s}{2i+1} P_{i-1}^s + \frac{i-s+1}{2i+1} P_{i+1}^s\right\}\right] = 0$$

$$\Sigma\left[\beta_i^s\{-i(i+1)-\sigma\} P_i^s + \frac{4ma}{\gamma\sigma^2} C_i^s P_i^s\right] = 0$$

$$\Sigma\left[b_i(C_i^s - E_i^s) P_i^s + \alpha_i^s\left\{\frac{(i-s+1)[\sigma + i(i-1)]}{2(2i+1)} P_{i+1}^s\right.\right.$$
$$\left.\left. + \frac{(i+s)[\sigma + (i+1)(i+2)]}{2(2i+1)} P_{i-1}^s\right\} - \beta_i^s s^2 P_i^s\right] = 0$$

Since these equations must be true identically, the coefficients of P_i^s in each of them must vanish. Therefore

$$\left.\begin{aligned} &\alpha_i^s\{\sigma - i(i+1)\} + 2\sigma^2\left\{\beta_{i+1}^s\frac{i+s+1}{2i+3} + \beta_{i-1}^s\frac{i-s}{2i-1}\right\} = 0 \\ &-\beta_i^s\{\sigma + i(i+1)\} + \frac{4ma}{\gamma\sigma^2} C_i^s = 0 \\ &b_i(C_i^s - E_i^s) + \alpha_{i-1}^s\frac{(i-s)[\sigma + (i-1)(i-2)]}{2(2i-1)} \\ &\quad + \alpha_{i+1}^s\frac{(i+s+1)[\sigma + (i+2)(i+3)]}{2(2i+3)} - \beta_i^s s^2 = 0 \end{aligned}\right\} \quad \text{......(40)}$$

If we eliminate the α's and β's from the third equation (40) by means of the first two, we find

$$\xi^s_{i-2} C^s_{i-2} - L^s_i C^s_i + \eta^s_{i+2} C^s_{i+2} = \frac{\gamma b_i}{4ma} E^s_i \quad \text{...........}\ \ \text{...(41)}$$

where

$$L^s_i = \frac{s^2}{\sigma^2[\sigma + i(i+1)]} + \frac{(i^2 - s^2)[\sigma + (i-1)(i-2)]}{(4i^2-1)[\sigma - (i-1)i][\sigma + i(i+1)]}$$
$$+ \frac{[(i+1)^2 - s^2][\sigma + (i+2)(i+3)]}{[4(i+1)^2 - 1][\sigma - (i+1)(i+2)][\sigma + i(i+1)]} - \frac{\gamma b_i}{4ma}$$

$$\xi^s_{i-2} = \frac{-(i-s)(i-s-1)}{(2i-1)(2i-3)[\sigma - (i-1)i]}$$

$$\eta^s_{i+2} = \frac{-(i+s+1)(i+s+2)}{(2i+3)(2i+5)[\sigma - (i+1)(i+2)]}$$

In the case of the luni-solar semi-diurnal tide (called K_2 in the notation of harmonic analysis) we have $i = 2, s = 2, \sigma = 2$. Hence it would appear that these formulæ for L^s_i, ξ^s_{i-2} fail by becoming indeterminate, but i and s are rigorously integers whereas σ depends on the 'speed' of the tide; accordingly in the case referred to we must regard terms involving $i - s$ as vanishing in the limit when σ approaches to equality with $i(i-1)$. For this particular case then we find

$$L^s_2 = \tfrac{3}{35} - \frac{\gamma b_2}{4ma} \text{ and } \xi^s_0 = 0$$

The equation (41) for the successive C's is available for all values of i provided that C_{-1}, E_{-1}, C_0, E_0 are regarded as being zero.

As in the case of the zonal oscillations, the equations with odd suffixes separate themselves from those with even suffixes, so that the two series may be treated independently of one another. Indeed, as we shall see immediately, the series with odd suffixes are satisfied by putting all the C's with odd suffixes zero for the case of such oscillations as may be generated by the attractions of the moon or sun.

For the semi-diurnal tides $i = 2$, $s = 2$, and f is approximately equal to unity. Hence the equilibrium tide is such that all the E^s_i, excepting E^2_2, are zero.

For the diurnal tides $i = 2$, $s = 1$, and f is approximately equal to $\frac{1}{2}$. Hence all the E^s_i, excepting E^1_2, are zero.

Since in neither case is there any E with an odd suffix, we need only consider those with even suffixes.

In both cases the first equation among the C's is

$$-L^s_2 C^s_2 + \eta^s_4 C^s_4 = \frac{\gamma b_2}{4ma} E^s_2 \quad (s = 2 \text{ or } 1)$$

It follows that if we were to write

$$\xi_0^s C_0^s = -\frac{\gamma b_2}{4ma} E_2^s \ (s=2 \text{ or } 1)$$

the equation of condition amongst the C's would be of general applicability for all even values of i from 2 upwards.

The symbols ξ_0^s, η_2^s do not occur in any of the equations, and therefore we may arbitrarily define them as denoting unity, although the general formulæ of ξ and η would give them other values. Accordingly we shall take

$$\xi_0^s C_0^s = C_0^s = -\frac{\gamma b_2}{4ma} E_2^s \ (s=2 \text{ or } 1)$$

With this definition the equation

$$\xi_{i-2}^s C_{i-2}^s - L_i^s C_i^s + \eta_{i+2}^s C_{i+2}^s = 0 \ (s=2 \text{ or } 1)$$

is applicable for $i = 2, 4, 6$, etc.

It may be proved as in the case of the tides of long period that we may regard C_i^s/C_{i+2}^s as tending to zero.

Then our equation may be written in the form

$$\xi_{i-2}^s \frac{C_{i-2}^s}{C_i^s} = L_i^s - \frac{\xi_i^s \eta_{i+2}^s}{\xi_i^s C_i^s/C_{i+2}^s}$$

and by successive applications the right hand side may be expressed in the form of a continued fraction.

Let us write $$H_i^s = \frac{\xi_{i-2}^s \eta_i^s}{L_i^s -}\;\frac{\xi_i^s \eta_{i+2}^s}{L_{i+2}^s -}\;\frac{\xi_{i+2}^s \eta_{i+4}^s}{L_{i+4}^s -}\cdots$$

Hence our equation may be written

$$\xi_{i-2}^s \frac{C_{i-2}^s}{C_i^s} = \frac{\xi_{i-2}^s \eta_i^s}{H_i^s}$$

Whence $$C_i^s = \frac{H_i^s}{\eta_i^s} C_{i-2}^s$$

It follows that

$$C_2^s = \frac{H_2^s}{\eta_2^s} C_0^s, \qquad C_4^s = \frac{H_2^s H_4^s}{\eta_2^s \eta_4^s} C_0^s, \qquad C_6^s = \frac{H_2^s H_4^s H_6^s}{\eta_2^s \eta_4^s \eta_6^s} C_0^s, \text{ etc.}$$

Then since we have defined

$$\eta_2^s = 1 \quad \text{and} \quad C_0^s = -\frac{\gamma b_2}{4ma} E_2^s$$

all the C's are expressed in terms of known quantities.

Hence the height of tide $\mathfrak{h}$ is given by

$$\mathfrak{h} = \Sigma h_i \cos(2nft + s\phi + \alpha)$$
$$= -\frac{\gamma b_2}{4ma} E_2^s \cos(2nft + s\phi + \alpha) \left[H_2^s P_2^s + \frac{H_2^s H_4^s}{\eta_4^s} P_4^s + \frac{H_2^s H_4^s H_6^s}{\eta_4^s \eta_6^s} P_6^s + \dots \right]$$

But the equilibrium tide $\mathfrak{e}$ is given by

$$\mathfrak{e} = E_2^s P_2^s \cos(2nft + s\phi + \alpha)$$

Hence we may write our result in the following form, which shows the relationship between the true dynamical tide and the equilibrium tide:—

$$\mathfrak{h} = -\frac{\gamma b_2}{4ma}\frac{\mathfrak{e}}{P_2^s}\left\{H_2^s P_2^s + \frac{H_2^s H_4^s}{\eta^s} P_4^s + \frac{H_2^s H_4^s H_6^s}{\eta_4^s \eta_6^s} P_6^s + \dots\right\}$$

From a formula equivalent to this Mr Hough finds for the lunar semi-diurnal tide ($s = 2$), for a sea of 1210 fathoms $\left(\frac{\gamma}{4ma} = \frac{1}{40}\right)$,

$$\mathfrak{h} = \frac{\mathfrak{e}}{P_2^2}\{\cdot 10396 P_2^2 + \cdot 57998 P_4^2 - \cdot 19273 P_6^2 + \cdot 03054 P_8^2 \dots\}$$

This formula shows us that at the equator the tide is 'inverted,' and has 2·4187 times as great a range as the equilibrium tide.

For this same ocean he finds that the solar semi-diurnal tide is 'direct' at the equator, and has a range 7·9548 as great as the equilibrium tide.

Now the lunar equilibrium tide is 2·2 times as great as the solar equilibrium tide, and since 2·2 × 2·4187 is only 5·3 it follows that in such an ocean the solar tides would have a range half as great again as the lunar. Further, since the lunar tides are 'inverted' and the solar 'direct,' spring-tide would occur at quarter-moon and neap-tide at full and change.

We give one more example from amongst those computed by Mr Hough. In an ocean of 9680 fathoms $\left(\frac{\gamma}{4ma} = \frac{1}{5}\right)$, he finds

$$\mathfrak{h} = \frac{\mathfrak{e}}{P_2^2}\{1\cdot 7646 P_2^2 - \cdot 06057 P_4^2 + \cdot 001447 P_6^2 \dots\}$$

At the equator the tides are 'direct' and have a range 1·9225 as great as the equilibrium tide. In this case the tides approximate in type to those of the equilibrium theory, although at the equator, at least, they have nearly twice the range.

Our space will not permit us to give others of the remarkable conclusions reached by Mr Hough.

We do not give any numerical results for the diurnal tides, for reasons which will appear from the following section.

§ 19. *Diurnal tide approximately evanescent.*

The equilibrium diurnal tide is given by

$$\mathfrak{e} = E_2^1 P_2^1 \cos(2nft + \phi + \alpha)$$

where f is approximately $\frac{1}{2}$ and the associated function for $i = 2$, $s = 1$ is

$$P_2^1 = 3 \sin\theta \cos\theta$$

Now the height of tide is given by

$$\mathfrak{h} = \Sigma C_i^s P_i^s \cos(2nft + \phi + \alpha)$$

and the problem is to evaluate the constants C_i^s.

If possible suppose that $\mathfrak{h}$ is also expressed by a single term like that which represents $\mathfrak{e}$, so that

$$\mathfrak{h} = 3C_2^1 \sin\theta \cos\theta \cos(2nft + \phi + \alpha)$$

Then the differential equation (17) to be satisfied becomes

$$\gamma(C_2^1 - E_2^1)\left\{\frac{1}{\sin\theta}\frac{d}{d\theta}\left(\frac{\sin\theta\dfrac{du}{d\theta} + \dfrac{1}{f}u\cos\theta}{f^2 - \cos^2\theta}\right) - \frac{\dfrac{\cos\theta}{f}\dfrac{du}{d\theta} + \dfrac{u}{\sin\theta}}{\sin\theta(f^2 - \cos^2\theta)}\right\} + 4maC_2^1 u = 0$$

where u is written for brevity in place of $\sin\theta\cos\theta$.

Now when f is rigorously equal to $\frac{1}{2}$, it may be proved by actual differentiation that the expression inside { } vanishes identically, and the equation reduces to $C_2^1 = 0$.

We thus find that in this case the differential equation is satisfied by zero oscillation of water level. In other words we reach Laplace's remarkable conclusion that there is no diurnal rise and fall of the tide; there are, it is true, diurnal tidal currents, but they are so arranged that the water level remains unchanged.

In reality f is not rigorously $\frac{1}{2}$ (except for the tide called K_1) and there will be a small diurnal tide. The lunar diurnal tide called O has been evaluated for various depths of ocean by Mr Hough and is found always to be small.

11.

ON THE DYNAMICAL THEORY OF THE TIDES OF LONG PERIOD.

[*Proceedings of the Royal Society of London*, XLI. (1886), pp. 337—342.]

IN the following note an objection is raised against Laplace's method of treating these tides, and a dynamical solution of the problem, founded on a paper by Sir William Thomson, is offered.

Let θ, ϕ be the colatitude and longitude of a point in the ocean, let ξ and $\eta \sin\theta$ be the displacements from its mean position of the water occupying that point at the time t, let $\mathfrak{h}$ be the height of the tide, and let $\mathfrak{e}$ be the height of the tide according to the equilibrium theory; let n be the angular velocity of the earth's rotation, g gravity, a the earth's radius, and γ the depth of the ocean at the point θ, ϕ.

Then Laplace's equations of motion for tidal oscillations are

$$\left.\begin{aligned} \frac{d^2\xi}{dt^2} - 2n\sin\theta\cos\theta\frac{d\eta}{dt} &= -\frac{g}{a}\frac{d}{d\theta}(\mathfrak{h}-\mathfrak{e}) \\ \sin\theta\frac{d^2\eta}{dt^2} + 2n\cos\theta\frac{d\xi}{dt} &= -\frac{g}{a\sin\theta}\frac{d}{d\phi}(\mathfrak{h}-\mathfrak{e}) \end{aligned}\right\} \quad\text{............(1)}$$

And the equation of continuity is

$$\mathfrak{h}a + \frac{1}{\sin\theta}\frac{d}{d\theta}(\gamma\xi\sin\theta) + \gamma\frac{d\eta}{d\phi} = 0 \quad\text{.................(2)}$$

The only case which will be considered here is where the depth of the ocean is constant, and we shall only treat the oscillations of long period in which the displacements are not functions of the longitude.

As the motion to be considered only involves steady oscillation, we assume

$$\left.\begin{aligned} \mathfrak{e} &= e\cos(2nft+\alpha) \\ \mathfrak{h} &= h\cos(2nft+\alpha) \\ \xi &= x\cos(2nft+\alpha) \\ \eta &= y\sin(2nft+\alpha) \\ u &= h-e \end{aligned}\right\} \quad\text{..........................(3)}$$

Hence, by substitution in (1), we have

$$\left.\begin{aligned} xf^2 + yf\sin\theta\cos\theta &= \frac{1}{4m}\frac{du}{d\theta} \\ yf^2\sin\theta + xf\cos\theta &= 0 \end{aligned}\right\}$$

where $$m = \frac{n^2 a}{g}$$

Whence $$x(f^2 - \cos^2\theta) = \frac{1}{4m}\frac{du}{d\theta}$$

$$y\sin\theta(f^2 - \cos^2\theta) = -\frac{1}{4m}\frac{\cos\theta}{f}\frac{du}{d\theta}$$

Then substituting for x and y in (2), which, when γ is constant and η is not a function of ϕ, becomes

$$ha + \frac{\gamma}{\sin\theta}\frac{d}{d\theta}(\xi\sin\theta) = 0$$

we get $$\frac{\gamma}{\sin\theta}\frac{d}{d\theta}\left[\frac{\sin\theta\, du/d\theta}{f^2 - \cos^2\theta}\right] + 4ma(u+e) = 0$$

This is Laplace's equation for tidal oscillations of the first kind*. In these tides f is a small fraction, being about $\frac{1}{28}$ in the case of the fortnightly tide, and e the coefficient in the equilibrium tide is equal to $E(\frac{1}{3} - \cos^2\theta)$, where E is a known function of the elements of the orbit of the tide-generating body, and of the obliquity of the ecliptic.

If now we write $\beta = 4ma/\gamma$, and $\mu = \cos\theta$, our equation becomes

$$\frac{d}{d\mu}\left[\frac{1-\mu^2}{\mu^2 - f^2}\frac{du}{d\mu}\right] = \beta\left[u + E(\tfrac{1}{3} - \mu^2)\right] \quad\text{..................(4)}$$

In treating these oscillations Laplace does not use this equation, but seeks to show that friction suffices to make the ocean assume at each instant its form of equilibrium. His conclusion is no doubt true, but the question remains as to what amount of friction is to be regarded as sufficing to produce the result, and whether oceanic tidal friction can be great enough to have the effect which he supposes it to have.

The friction here contemplated is such that the integral effect is represented by a retarding force proportional to the velocity of the fluid relatively to the bottom. Although proportionality to the square of the velocity would probably be nearer to the truth, yet Laplace's hypothesis suffices for the present discussion. In oscillations of the class under consideration, the water moves for half a period north, and then for half a period south.

Now in systems where the resistances are proportional to velocity, it is usual to specify the resistance by a modulus of decay, namely, that period in

* *Mécanique Céleste.*

which a velocity is reduced by friction to e^{-1} or $1 \div 2{\cdot}783$ of its initial value; and the friction contemplated by Laplace is such that the modulus of decay is short compared with the semi-period of oscillation.

The quickest of the tides of long period is the fortnightly tide, hence for the applicability of Laplace's conclusion, the modulus of decay must be short compared with a week. Now it seems practically certain that the friction of the ocean bed would not much affect the velocity of a slow ocean current in a day or two. Hence we cannot accept Laplace's hypothesis as to the effect of friction.

We now, therefore, proceed to the solution of the equation of motion when friction is entirely neglected.

The solution here offered is indicated in a footnote to a paper by Sir William Thomson (*Phil. Mag.*, Vol. L., 1875, p. 280), but has never been worked out before.

The symmetry of the motion demands that u, when expanded in a series of powers of μ, shall only contain even powers of μ.

Let us assume then

$$\frac{1}{\mu^2 - f^2}\frac{du}{d\mu} = B_1\mu + B_3\mu^3 + \ldots + B_{2i+1}\mu^{2i+1} + \ldots$$

Then

$$\frac{1-\mu^2}{\mu^2 - f^2}\frac{du}{d\mu} = B_1\mu + (B_3 - B_1)\mu^3 + \ldots + (B_{2i+1} - B_{2i-1})\mu^{2i+1} + \ldots$$

$$\frac{d}{d\mu}\left[\frac{1-\mu^2}{\mu^2 - f^2}\frac{du}{d\mu}\right] = B_1 + 3(B_3 - B_1)\mu^2 + \ldots + (2i+1)(B_{2i+1} - B_{2i-1})\mu^{2i} + \ldots \qquad \ldots\ldots\ldots(5)$$

Again

$$\frac{du}{d\mu} = -f^2B_1\mu + (B_1 - f^2B_3)\mu^3 + \ldots + (B_{2i-1} - f^2B_{2i+1})\mu^{2i+1} + \ldots$$

$$u = C - \tfrac{1}{2}f^2B_1^2 + \tfrac{1}{4}(B_1 - f^2B_3)\mu^4 + \ldots + \tfrac{1}{2i}(B_{2i-3} - f^2B_{2i-1})\mu^{2i} + \ldots \qquad \ldots\ldots\ldots(6)$$

where C is a constant.

Then substituting from (5) and (6) in (4), and equating to zero the successive coefficients of the powers of μ, we find

$$\left.\begin{array}{l} C = -\tfrac{1}{3}E + B_1/\beta \\ B_3 - B_1(1 - \tfrac{1}{2.3}f^2\beta) + \tfrac{1}{3}\beta E = 0 \\ \ldots\ldots\ldots\ldots\ldots\ldots\ldots\ldots\ldots\ldots\ldots\ldots \\ B_{2i+1} - B_{2i-1}(1 - \tfrac{1}{2i(2i+1)}f^2\beta) - \tfrac{1}{2i(2i+1)}\beta B_{2i-3} = 0 \\ \ldots\ldots\ldots\ldots\ldots\ldots\ldots\ldots\ldots\ldots\ldots\ldots \end{array}\right\} \qquad \ldots\ldots\ldots(7)$$

Thus the constants C and B_3, B_5, &c., are all expressible in terms of B_1.

We may remark that if

$$-\tfrac{1}{2}\cdot\tfrac{1}{3}\cdot\beta B_{-1}=\tfrac{1}{3}\beta E, \text{ or } B_{-1}=-2E$$

then the general equation of condition in (7) may be held to apply for all values of i from 1 to infinity.

Let us now write it in the form

$$\frac{B_{2i+1}}{B_{2i-1}}=1-\frac{1}{2i(2i+1)}f^2\beta+\frac{1}{2i(2i+1)}\beta\frac{B_{2i-3}}{B_{2i-1}}\quad\text{.................(8)}$$

When i is large, B_{2i+1}/B_{2i-1} either tends to become infinitely small, or it does not do so.

Let us suppose that it does not tend to become infinitely small. Then it is obvious that the successive B's tend to become equal to one another, and so also do the coefficients $B_{2i-1}-f^2B_{2i+1}$ in the expression for $du/d\mu$.

Hence $\dfrac{du}{d\mu}=L+\dfrac{M}{1-\mu^2}$, where L, M are finite, for all values of μ. Hence $\dfrac{du}{d\theta}=-L\sqrt{1-\mu^2}+\dfrac{M}{\sqrt{1-\mu^2}}$, and therefore x is infinite when $\mu=1$ at the pole, and $d\xi/dt$ is infinite there also.

Hence the hypothesis, that B_{2i+1}/B_{2i-1} does not tend to become infinitely small, gives us infinite velocity at the pole. But with a globe covered with water this is impossible, the hypothesis is negatived, and B_{2i+1}/B_{2i-1} tends to become infinitely small.

This being established let us write (8) in the form

$$\frac{B_{2i-1}}{B_{2i-3}}=\frac{-\frac{1}{2i(2i+1)}\beta}{1-\frac{1}{2i(2i+1)}f^2\beta-B_{2i+1}/B_{2i-1}}\quad\text{..................(9)}$$

By repeated applications of (9) we have in the form of a continued fraction

$$\frac{B_{2i-1}}{B_{2i-3}}=\frac{-\frac{1}{2i(2i+1)}\beta}{1-\frac{1}{2i(2i+1)}f^2\beta}\,\overline{\big|+}\,\frac{\frac{1}{(2i+2)(2i+3)}\beta}{1-\frac{1}{(2i+2)(2i+3)}f^2\beta}\,\overline{\big|+}\,\frac{\frac{1}{(2i+4)(2i+5)}\beta}{1-\frac{1}{(2i+4)(2i+5)}f^2\beta}\,\overline{\big|+}\,\&\text{c.}$$

.........(10)

And we know that this is a continuous approximation, which must hold in order to satisfy the condition that the water covers the whole globe.

Let us denote this continued fraction by $-N_i$.

Then, if we remember that $B_{-1}=-2E$, we have

$$B_1=2EN_1,\quad \frac{B_3}{B_1}=-N_2,\quad \frac{B_5}{B_3}=+N_3,\ \&\text{c.}$$

so that

$$B_3 = -2EN_1N_2, \qquad B_5 = -2EN_1N_2N_3, \qquad B_7 = -2EN_1N_2N_3N_4, \text{ \&c.}$$

and
$$C = -\tfrac{1}{3}E + 2\frac{EN_1}{\beta}$$

Then the height of tide $\mathfrak{h}$ is equal to $h\cos(2nft+\alpha)$, the equilibrium tide $\mathfrak{e}$ is equal to $E(\frac{1}{3}-\mu^2)\cos(2nft+\alpha)$, and we have

$$h = u + E(\tfrac{1}{3}-\mu^2)$$
$$= C + \tfrac{1}{3}E - (E+\tfrac{1}{2}f^2B_1)\mu^2 + \tfrac{1}{4}(B_1 - f^2B_3)\mu^4 + \tfrac{1}{6}(B_3 - f^2B_5)\mu^6 + \dots$$
$$\frac{h}{E} = \frac{2N_1}{\beta} - (1+f^2N_1)\mu^2 + \tfrac{1}{2}N_1(1+f^2N_2)\mu^4 - \tfrac{1}{3}N_1N_2(1+f^2N_3)\mu^6 + \dots$$

Now when $\beta = 40$, we have $\gamma = \frac{1}{40}\times 4ma = \frac{1}{2890}a = 7260$ feet; so that $\beta = 40$ gives an ocean of 1200 fathoms.

With this value of β, and with $f = \cdot 0365012$, which is the value for the fortnightly tide, I find

$$N_1 = 3\cdot 040692, \quad N_2 = 1\cdot 20137, \quad N_3 = \cdot 66744, \quad N_4 = \cdot 42819, \quad N_5 = \cdot 29819$$
$$N_6 = \cdot 21950, \quad N_7 = \cdot 16814, \quad N_8 = \cdot 13287, \quad N_9 = \cdot 107, \quad N_{10} = \cdot 1, \text{ \&c.}$$

These values give

$$\frac{2}{\beta}N_1 = \cdot 15203, \qquad 1+f^2N_1 = 1\cdot 0041, \qquad \tfrac{1}{2}N_1(1+f^2N_2) = 1\cdot 5228$$
$$\tfrac{1}{3}N_1N_2(1+f^2N_3) = 1\cdot 2187, \qquad \tfrac{1}{4}N_1N_2N_3(1+f^2N_4) = \cdot 6099$$
$$\tfrac{1}{5}N_1\dots N_4(1+f^2N_5) = \cdot 2089, \qquad \tfrac{1}{6}N_1\dots N_5(1+f^2N_6) = \cdot 0519$$
$$\tfrac{1}{7}N_1\dots N_6(1+f^2N_7) = \cdot 0098, \qquad \tfrac{1}{8}N_1\dots N_7(1+f^2N_8) = \cdot 0014$$
$$\tfrac{1}{9}N_1\dots N_8(1+f^2N_9) = \cdot 00017, \text{ \&c.}$$

So that

$$\frac{h}{E} = \cdot 1520 - 1\cdot 0041\mu^2 + 1\cdot 5228\mu^4 - 1\cdot 2187\mu^6 + \cdot 6099\mu^8 - \cdot 2089\mu^{10}$$
$$+ \cdot 0519\mu^{12} - \cdot 0098\mu^{14} + \cdot 0014\mu^{16} - \cdot 0002\mu^{18} + \dots$$

At the pole, where $\mu = 1$, the equilibrium tide is $-\frac{2}{3}E$; at the equator it is $+\frac{1}{3}E$.

Now at the pole $\quad h = -E\times\cdot 1037 = -\tfrac{2}{3}E\times\cdot 1556$

and at the equator $\quad h = +E\times\cdot 1520 = \tfrac{1}{3}E\times\cdot 4561$

In a second case, namely, with an ocean four times as deep, so that $\beta = 10$, I find

$$\frac{h}{E} = \cdot 2363 - 1\cdot 0016\mu^2 + \cdot 5910\mu^4 - \cdot 1627\mu^6 + \cdot 0258\mu^8 - \cdot 0026\mu^{10} + \cdot 0002\mu^{12} - \dots$$

At the pole $\quad h = -E\times\cdot 3137 = -\tfrac{2}{3}E\times\cdot 471$

at the equator $\quad h = +E\times\cdot 2363 = +\tfrac{1}{3}E\times\cdot 709$

With a deeper ocean we should soon arrive at the equilibrium value for the tide, for N_2, N_3, &c., become very small, and $2N_1/\beta$ becomes equal to $\frac{1}{3}$.

These two cases, $\beta=40$, $\beta=10$, are two of those for which Laplace has given solutions in the cases of the semi-diurnal and diurnal tides. We notice that, with such oceans as we have to deal with, the tide of long period is certainly less than half its equilibrium amount.

In Thomson and Tait's *Natural Philosophy* (edition of 1883) [Paper 9, p. 340 above] I have made a comparison of the observed tides of long period with the equilibrium theory. The probable errors of the results are large, but not such as to render them worthless, and in view of the present investigation it is surprising to find that on the average the tides of long period amount to as much as two-thirds of their equilibrium value.

The investigation in the *Natural Philosophy* was undertaken in the belief of the correctness of Laplace's view as to the tides of long period, and was intended to evaluate the effective rigidity of the earth's mass.

The present result shows us that it is not possible to attain any estimate of the earth's rigidity in this way, but as the tides of long period are distinctly sensible, we may accept the investigation in the *Natural Philosophy* as generally confirmatory of Thomson's view as to the great effective rigidity of the whole earth's mass.

There is one tide, however, of long period of which Laplace's argument from friction must hold true. In consequence of the regression of the nodes of the moon's orbit there is a minute tide with a period of nearly nineteen years, and in this case friction must be far more important than inertia. Unfortunately this tide is very minute, and as I have shown in a Report for 1886 to the British Association on the tides [p. 166, above], it is entirely masked by oscillations of sea level produced by meteorological or other causes.

Thus it does not seem likely that it will ever be possible to evaluate the effective rigidity of the earth's mass by means of tidal observations.

12.

ON THE ANTARCTIC TIDAL OBSERVATIONS OF THE 'DISCOVERY.'

[*To be published hereafter as a contribution to the scientific results of the voyage of the 'Discovery.'*]

THE 'Discovery' wintered in 1902 and in 1903 at the south-eastern extremity of Ross Island, on which Mount Erebus is situated, in south latitude 78° 49′ and east longitude 166° 20′.

The station is near the west coast of a great bay in the Antarctic Continent, and the westerly coast line runs northward from the station for about 9° of latitude. To the eastward of the bay, however, the coast only attains a latitude of about 75° and follows approximately a small circle of latitude. Since the tide-wave comes from the east and travels to the west the station is not sheltered by the coast to the westward, and the continent to the eastward can do but little to impede the full sweep of the tide-wave in the Antarctic Ocean. It is true that Ross Island itself is partially to the east of the anchorage, but it is so small that its influence cannot be important. Of course the westward coast line must exercise an influence on the state of tidal oscillation, for regarding the tide-wave as a free wave coming in from the east, it is clear that it will run up to the end of the bay and then wheel round northward along the westerly coast. It would seem then that the situation is on the whole a good one for such observations. Of course their value would have been much increased if it had been possible to obtain other observations elsewhere.

The following account by Lieutenant Michael Barne, R.N., explains the manner in which the tidal observations were made.

"On our arrival in the vicinity of our winter quarters on Feb. 8th, 1902, a good deal of the previous year's ice remained attached to the land. As there was no foreshore, and pieces of this ice were constantly moving out, it was impossible to erect a tide-pole. With the final departure of the old ice, the temperature fell, and young ice formed continually, only to be quickly broken up by the almost incessant easterly winds.

"As this state of affairs promised to last for a considerable time, an effort was made to obtain records of the tides. A stout graduated pole was erected alongside the ice foot in about 10 feet of water, the lower end being heavily weighted and the upper end securely guyed. Some intermittent observations were secured in this manner, but they are probably of little value, as the ice was continually forming round the pole, which was only with difficulty freed from it. Besides this, communication with the shore, and consequently approach to the tide-pole, was constantly interrupted.

"On the ship being finally frozen in, a tide gauge of the following nature was erected (Fig. 1).

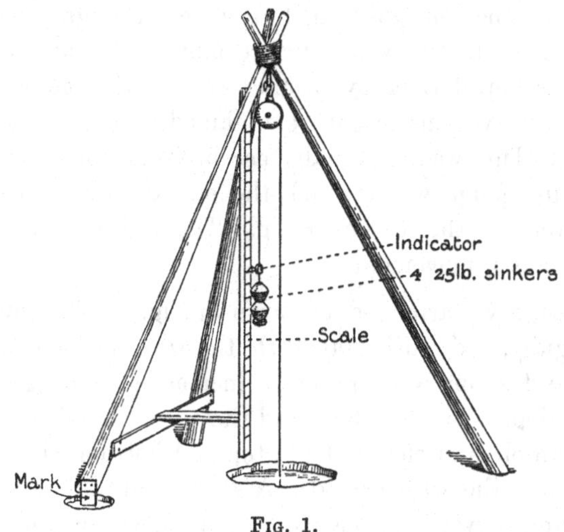

FIG. 1.

"A single length of pianoforte wire (sounding wire) was led through a block, secured to the head of a tripod. One end of this length was attached to eight 25 lb. sinkers, which were lowered to the bottom. Four 25 lb. sinkers were secured to the other end in such a manner as to allow of their free movement, between the ice and the block, as the ice, with the tripod, rose and fell with the tide. An indicator was clamped to the wire, and a suitable scale secured to the tripod.

"It was thought that the motive force supplied by the weight of four sinkers, would be sufficient to draw the smooth surface of the wire through the ice, as the water rose and fell; whilst, in case it should fail to do so, the weight of eight sinkers would not be sufficient to break the wire.

"As it was considered possible that the ice, owing to the proximity of the land, might not maintain a uniform position relative to the surface of the water, a small hole was occasionally opened close to one of the tripod legs, to which was attached a mark, indicating the height to which the water should rise. A few observations shewed that no error from this cause was to be apprehended.

"This gauge was placed about 200 yards from the ship, and two-hourly readings with but slight interruption were continued from April 12th to April 28th.

"Some sluggishness in its movement which was eventually noticed, and its final breakdown was possibly partly due to the thickening of the ice, but principally I think to the fact that too small a block was employed at the tripod head. The scale was, by accident, secured so that the readings increased upwards; consequently they have to be inverted*.

"It was originally intended to place a tide gauge in the ship, owing to the far greater convenience of position, but it was thought that the position of the ship relatively to the water surface might alter and this might lead to errors. It was hoped that, by placing one on the ice as well as one in the ship, check observations might be obtained to determine if this source of error existed. This was eventually accomplished on April 25th, but by the time the ship gauge was erected, the outside gauge had ceased to be entirely satisfactory for the reasons given. The observations however shew a close approximation of movement.

"The ship gauge was arranged as shewn in Fig. 2. The supporting blocks were secured rigidly, and, until May 10th, the wire was led directly through the ice. As the friction was gradually increasing, a suggestion made by Dr Wilson was adopted on that date, and the wire was taken through a tube, filled with paraffin oil, and closed at the top and bottom with a hard wooden plug through which the wire passed. A maximum and minimum arrangement, with balanced weights, was added as shewn in the sketch†. Unfortunately, both on May 10th and May 12th, in re-fitting the gauge, the indicator had to be fixed afresh, and therefore the observations cannot be referred to a common zero.

"A mark was placed on the ship's side to ascertain any vertical movement of the ship relatively to the water surface, and a long plummet was secured in the engine-room, to shew any alteration in her inclination to the vertical.

"On April 6th, 1903, the tide gauge was re-erected, and observations continued; but, owing to the large number of observers employed, the maximum and minimum arrangement was not fitted.

"The height of the mark (*b*) on the ship's side above the water, was ascertained about once a month, in the same manner as during the winter

* This oversight was rectified before May 12. G. H. D.

† No use has been made of this arrangement. G. H. D.

of 1902, i.e. by digging a hole through the ice below the mark, and measuring its height. On these occasions the difference between the heights of the leading blocks was measured in the following manner. A wooden scale (*c*) marked in half-inches was secured to the beam (*a*), in a vertical position,

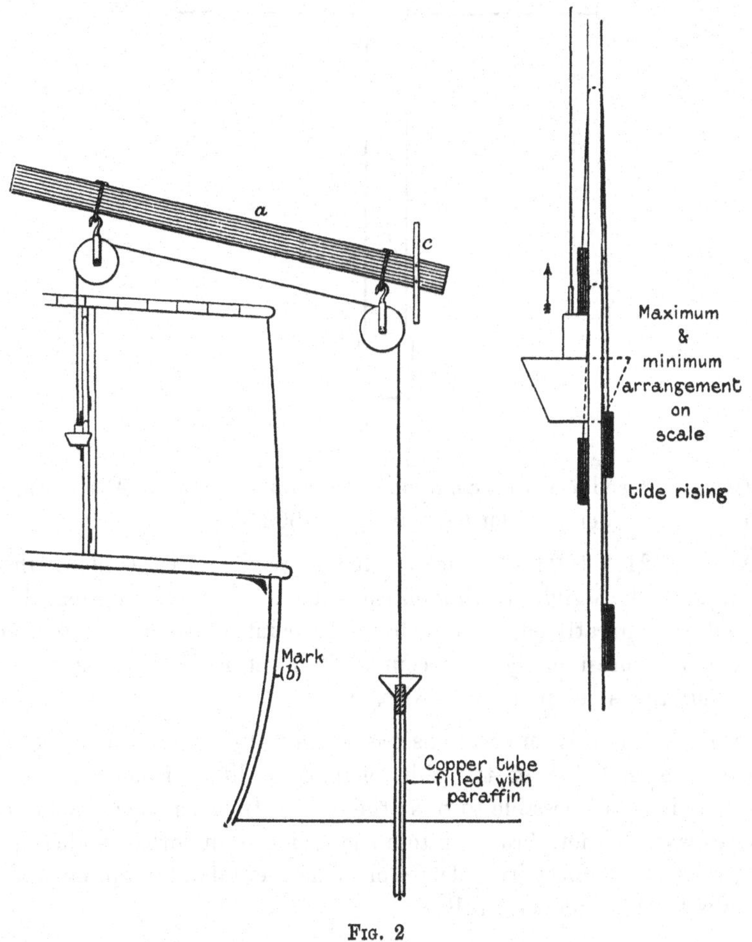

Fig. 2

close to the outer block, with its zero mark on a level with the top of the sheave. A wooden instrument, shaped like the letter T, and having a lead weight attached to its lower end, and a hole *A* (see Fig. 3) in the centre of its upper end, was hung, freely, on a nail, in such a manner that its upper side was horizontal, and on a level with the top of the sheave of the inner block. By bringing the eye on a level with the upper side of this T-piece, and noting the position on the scale at which the upper side, if produced, would cut it, the reading of the scale was obtained which gave the difference in height of the blocks.

"By taking periodical measurements of the height of the mark, and the difference between the heights of the blocks, data were obtained, by which readings could be corrected for alteration of the trim and the list of the ship

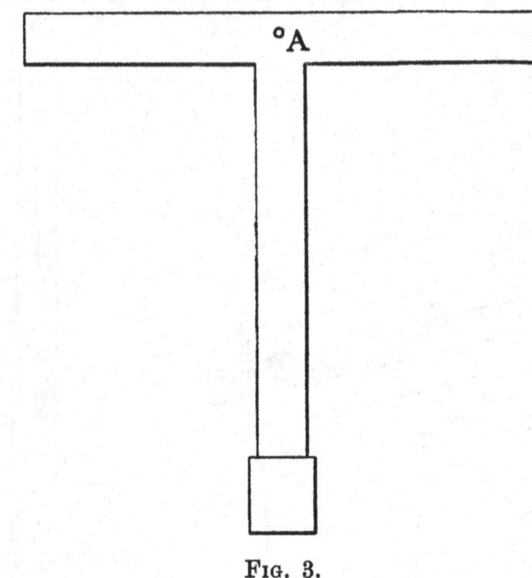

FIG. 3.

respectively and reduced to a common zero, namely that on April 6th, when the tide gauge was erected for the winter of 1903.

"On Sept. 21st, 1903, the wire carried away close to the place where it was secured to the weight resting on the bottom. On examination, the wire was found to be greatly eaten away, from the point of attachment to a height of several feet, presumably on account of an action between the cast-iron sinkers, and the steel pianoforte wire."

The series of hourly observations was occasionally interrupted by accidents, and the trim and list of the ship changed a little from time to time. Accordingly it is not possible to treat the observations as a continuous whole. The series was therefore broken into a succession of months, so chosen as to avoid periods of manifest irregularity or of accidental interruption, and each month was treated independently.

The choice of the method of harmonic analysis to be employed seemed to lie between that explained in the *Admiralty Scientific Manual* and that devised for the use of the tidal Abacus*. The method of the *Manual* is considerably more laborious than the other, and it was highly desirable that the Abacus should be used if it could be trusted for a short series of observations. I therefore asked Mr Wright, who carried out the reductions and was familiar with the use of the Abacus, to reduce the first month in duplicate by the two methods. Curves were drawn through ordinates repre-

* *Manual of Scientific Inquiry*, Article *Tides*; and "On an apparatus for facilitating the reduction of tidal observations," *Proc. Roy. Soc.*, Vol. 52, p. 345. [Papers 4 and 6 in this volume.]

senting the mean height of water at the 24 hours of mean lunar time, as derived in the two ways. Although the whole range of height in the 24 mean lunar hours was only about six inches, the two curves shewed a substantial agreement. The same process was then applied with O-time, when the range was found to be about 15 inches, and the agreement of the two curves was very close. The method of the *Manual* shewed several sharp peaks or irregularities in the curves which were nearly smoothed out by the use of the Abacus. Such peaks would not affect the values of semi-diurnal or of diurnal components to a sensible amount, and as they are clearly accidental, I concluded that the use of the Abacus was quite satisfactory, and accordingly that method was adopted throughout.

In the use of harmonic analysis it is necessary that the month under discussion should differ a little in length according to the tide which is being evaluated. For finding the M_2-tide months of 30 days or of 29 days would be almost equally advantageous, but as 30 days gives us one more day of observation that period was adopted. Similarly 30 days is appropriate for the S_2-tide. For a short period of observation it is necessary to regard this tide as compounded of the S_2 and K_2 tides, and we must also suppose its range to vary with the sun's parallax. The separation of these two tides from one another depends on theoretical considerations, which appear to be well founded.

Similarly in a short series of observations the K_1 and P tides must be treated as fused together in a single tide, and they are separable by theoretical considerations only. For these two tides a month of 27 days is appropriate.

Lastly the analysis for the O-tide demands the use of a month of 28 days*.

I determined, then, to separate the months in such a way that the shortest months (27 days) should follow one another as closely as possible, while the longer months should overlap slightly. Whenever any event occurred, whereby it seemed likely that the observations might be vitiated, the months were chosen so as to omit the time of possible or actual abnormality.

It was clearly desirable that the largest possible number of independent or nearly independent months should be discussed. This consideration led in one case to an overlap of as much as six days; thus the fifth month of 27 days ended on October 19, while the sixth month began on October 13.

In the few cases where hourly observations were missing, the defects were made good by interpolation. Although the observations began in April, 1902, the first satisfactory continuous period began on May 12. It will be well to

* This use of months of various lengths necessitates some small arithmetical changes in the method as explained in the paper on the Apparatus referred to above. [See p. 225 above.]

state the epochs for the succession of twelve months which it was possible to obtain, and to add a few comments on the observations.

First month. This begins with 0^h May 12, 1902. The observations really begin at 2^h, but extrapolated values were used for 0^h and 1^h.

Second month. This begins with 0^h June 5, 1902.

On the afternoon of July 5 the wire attached to the sinker parted and the observations ceased. The apparatus was only reinstalled at 5 p.m. on July 23rd.

Third month. This begins with 0^h July 24, 1902.

Fourth month. This begins with 0^h Aug. 23, 1902.

Fifth month. This begins with 0^h Sept. 23, 1902. The height for 6^h on October 20 was interpolated.

On October 1 it was found that the ship had shifted so as to affect the readings by one inch. The date at which the shift had occurred was unknown, and moreover so small a change could not affect the results sensibly.

Sixth month. This begins with 0^h Oct. 13, 1902. On Nov. 9, the four hourly values, 1^h to 4^h inclusive, were missing and were supplied by interpolation.

As already remarked this month considerably overlaps the one before it. This was necessary if a seventh month was to be secured before the observations ceased for the season, but the choice of the stage at which the overlapping should be made to occur was more or less arbitrary.

Seventh month. This begins with 0^h Nov. 13, 1902. The observation for 22^h of Dec. 9 is missing and was supplied by interpolation. On Dec. 13 the wire parted and the series ended for the year.

In the second winter, that of 1903, somewhat greater care seems to have been taken to note the small shifting of the ship.

Eighth month. This begins with 0^h April 6, 1903. Between April 22 and May 3 the ship shifted so as to make the readings too high by 3 inches, compared with the earlier ones. As an arbitrary correction I deducted one inch from all heights from 0^h April 24 to 0^h April 27; from 1^h April 27 to 0^h April 30 I deducted two inches; and for the rest of the month the full three inches. These arbitrary corrections were submitted to independent harmonic analysis, and it appeared that they afforded corrections so minute as to leave the tidal constants virtually unchanged.

Ninth month. This begins with 0^h May 9, 1903. The ship shifted considerably at some time about June 12, and as it is only possible to obtain one month before that date, there is an unutilized gap of a few days between this month and the one before it.

Tenth month. This begins with 0^h June 15, 1903. On July 10 a sensible shift in the trim and height of the ship was discovered. This necessitates the addition of 4½ inches to all heights, 2½ inches being due to angular movement and 2 inches to vertical movement. As an arbitrary correction I added 2 inches to all heights from 0^h July 8 to 0^h July 9; and afterwards I added the full 4½ inches.

Eleventh month. This begins with 0^h July 14, 1903.

Twelfth month. This begins with 0^h August 14, 1903.

After Sept. 8 the observations were only taken every two hours, and for the remainder of the month the values at the odd hours were interpolated.

The observations stop on Sept. 20, but are not used in the reductions after Sept. 13.

No corrections have been applied for changes in the barometric pressure. As the application of such a correction would have been very laborious and moreover somewhat speculative, I have relied on the automatic elimination of the inequalities produced by taking mean values.

The following are the results of the twelve harmonic analyses, the heights being stated in inches.

		1902							1903				
Month ……		1	2	3	4	5	6	7	8	9	10	11	12
M_2	H (inches)	2·272	2·286	2·180	1·931	1·735	1·562	1·206	1·908	2·195	2·330	2·406	2·177
	κ	0°·93	2°·48	13°·78	22°·37	26°·04	32°·16	30°·50	347°·57	355°·04	3°·83	8°·65	14°·57
S_2	H	·906	1·390	1·051	·928	1·203	1·272	1·196	1·368	1·431	1·009	·829	1·262
	κ	270°·57	277°·69	276°·62	284°·70	275°·40	274°·10	249°·27	268°·61	267°·83	271°·42	280°·66	273°·55
K_2	H	·247	·379	·287	·253	·328	·347	·326	·373	·390	·275	·226	·344
	κ						Same as for S_2						
K_1	H	9·229	9·583	9·759	11·148	8·913	9·465	9·637	9·974	10·119	11·381	10·142	11·561
	κ	13°·05	15°·81	11°·32	18°·92	18°·61	11°·70	16°·78	7°·84	13°·57	9°·25	10°·41	8°·83
P	H	3·076	3·194	3·253	3·716	2·971	3·155	3·212	3·325	3·373	3·794	3·381	3·854
	κ						Same as for K_1						
O	H	8·562	8·251	9·257	9·879	8·456	9·706	9·075	9·353	10·079	9·717	9·460	9·561
	κ	1°·68	354°·55	4°·85	1°·52	2°·62	1°·86	358°·47	357°·90	354°·07	2°·60	2°·58	356°·78
A_0 (inches)		25·46	27·19	34·98	28·99	27·26	27·24	22·94	25·79	25·95	22·64	23·57	15·93

N.B. *The values of* A_0 *represent merely the changes in the position of the ship and have therefore no physical significance; all the heights are stated in inches.*

The values of H and κ are somewhat irregular from month to month, and it is therefore not permissible to adopt the mean values of H and κ as representing the mean tide. I have therefore formed H cos κ and H sin κ for each month and have taken the mean of each as giving the mean values of H cos κ and H sin κ. It is easy to compute from these the proper mean values of H and κ for each tide. The results are given in the following table:—

Mean values of tidal constants.

Semidiurnal tides				Diurnal tides			
M_2	H =	1·966 inches	= 0·164 feet	K_1	H =	9·245 inches	= 0·770 feet
	κ =	9°·9	= 10°		κ =	14°·05	= 14°
S_2	H =	1·142 inches	= 0·095 feet	P	H =	3·082 inches	= 0·257 feet
	κ =	272°·1	= 272°		κ =	14°·05	= 14°
K_2	H =	·311 inches	= 0·026 feet	O	H =	9·264 inches	= 0·772 feet
	κ =	272°·1	= 272°		κ =	359°·5	= 0°

The sum of the semi-ranges of the three diurnal tides is 21·6 inches and of the three semidiurnal tides is only 3·4 inches. This result corresponds with the fact that little trace of the semidiurnal tide is to be discovered from mere inspection of the tide curve.

When tidal observations have been reduced it is always important to verify that the constants found do really represent the tidal oscillation, for in computations of such complexity it is always possible that some gross mistake of principle may have slipped in unnoticed. Such a verification is especially important in a case where the tides are found to be very abnormal, as here, and where the results from month to month are not closely consistent. I accordingly asked Mr Glazebrook to run off curves for two periods with the Indian tide-predicter at the National Physical Laboratory. The constants used were the means for the tides evaluated. It is probable that a better result would have been attained if a number of other tides, with constants assigned by theoretical considerations from analogy with the constants actually evaluated, had also been introduced; but I did not think it was worth while to do so. Evidence will be given hereafter to shew that the smaller elliptic diurnal tides must exercise an appreciable influence.

The periods chosen for the comparison were about three weeks beginning on May 12, 1902, and nearly the same time in November. It does not seem worth while to reproduce the whole of the observed and computed curves for these periods. The observed tide-curve has frequently sharp irregularities presumably produced by weather or by unperceived shifts of the ship, and the maxima are sometimes sharp peaks instead of flowing curves. However, on the whole the computed and observed curves follow one another very well, at least throughout all those portions where the diurnal tide is pronounced. Where the diurnal inequality is nearly evanescent, and the semidiurnal tide becomes perceptible the discordance is sometimes considerable, although even in these cases every rise and fall of the water is traceable in the computed curve. Such discordance was inevitable, for at this part of the curve all those tidal oscillations which have any importance have disappeared, and only those tides remain which are very small; moreover most of these tides are avowedly omitted from the computed curve.

I give two figures. The first (Fig. 4) shews the two curves where the diurnal tide is large, viz. from 0^h May 24th to 0^h May 25th, 1902; it is a rather favourable example of the general agreement referred to above. The second figure (Fig. 5) is selected because it exhibits by far the worst discordance which occurred in the six weeks under comparison.

I conclude that the reductions are quite as good as could be expected from tide-curves which present as much irregularity as these do. It would not be possible to make a very good tide-table from the constants, but no one wants a tide-table for Ross Island. We only need sufficient accuracy to obtain an insight into the nature of the Antarctic tides, and the constants are quite sufficient for that end.

When the mean heights of water at the 24 hours of mean lunar time were plotted in curves for each month, it became obvious that a pure semidiurnal inequality did not represent the facts very closely, and that there remained also a sensible diurnal inequality. Such an inequality is given by the tide M_1, and if we neglect the minute portion of the tide M_1 which depends on the terms in the tide-generating potential which vary as the fourth power of the moon's parallax, such an inequality is found to depend on the composition of two elliptic tides with speeds $\gamma - \sigma - \varpi$ and $\gamma - \sigma + \varpi$. The genesis of this compounded tide is explained in the report to the British Association for 1883 [Paper 1].

I accordingly thought it worth while to evaluate the M_1 tide for each of the twelve months under reduction. The results come out sufficiently discordant to render it impossible to assign any definite value to the tide, yet there appears to be some sort of method in the phases. Thus the phases for the twelve months come out for 1902 9°, −3°, −45°, 6°, −32°, 70°, 12°, and for 1903 6°, −159°, −179°, −42°, −10°.

Two of the phases, those for the 9th and 10th months, are very discordant, but for these months the amplitude of M_1 is small; it is also very small for the 6th month with phase 70°. The mean of all the other phases is such that κ is pretty small, and this agrees with what is to be expected because κ for the tide O is small. It thus appears probable that there has been a sensible disturbance from the M_1 tide of the values of the mean heights of water as arranged in mean lunar time. It should be noted that the whole amplitude of oscillation is so small that it is really surprising that this effect should be traceable at all.

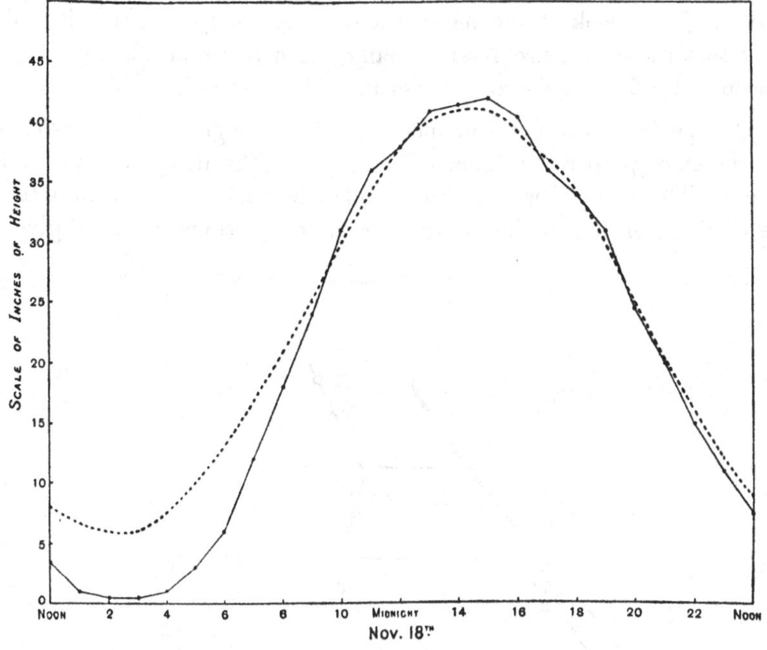
Scale of Inches of Height
45
40
35
30
25
20
15
10
5
0
Noon 2 4 6 8 10 Midnight 14 16 18 20 22 Noon
Nov. 18th

Fig. 4.

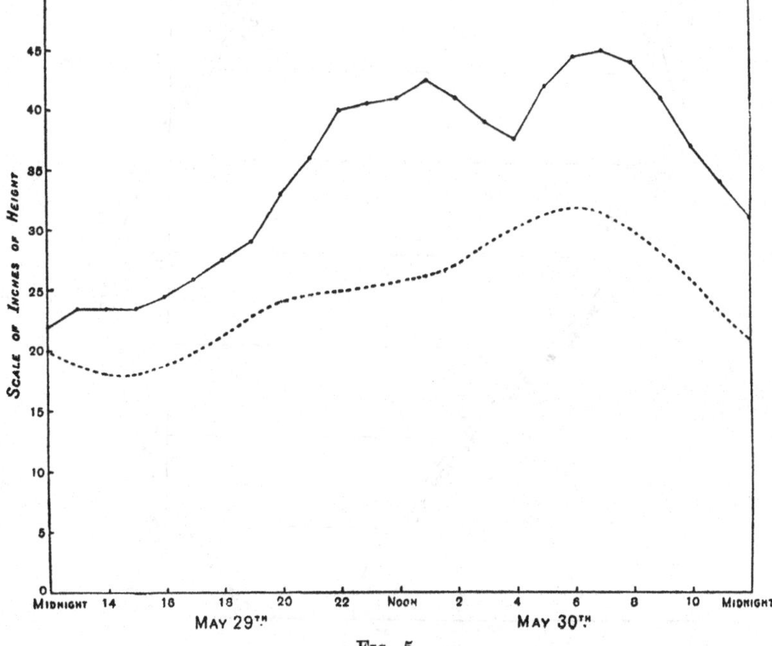
Scale of Inches of Height
45
40
35
30
25
20
15
10
5
0
Midnight 14 16 18 20 22 Noon 2 4 6 8 10 Midnight
May 29th
May 30th

Fig. 5.

There is one feature in the results which is so singular that it is well to refer to it. If we look at the heights and phases of the M_2-tide it will be observed that there is a progressive change both in amplitude and phase as the season of 1902 advances, and this change is repeated in 1903.

Mere inspection does not convince one of the degree of regularity, and I have therefore prepared a figure which exhibits the march of H cos κ and of H sin κ. The values for each month may be taken to appertain to the middle of the month, and the points surrounded by rings in Fig. 6 give the

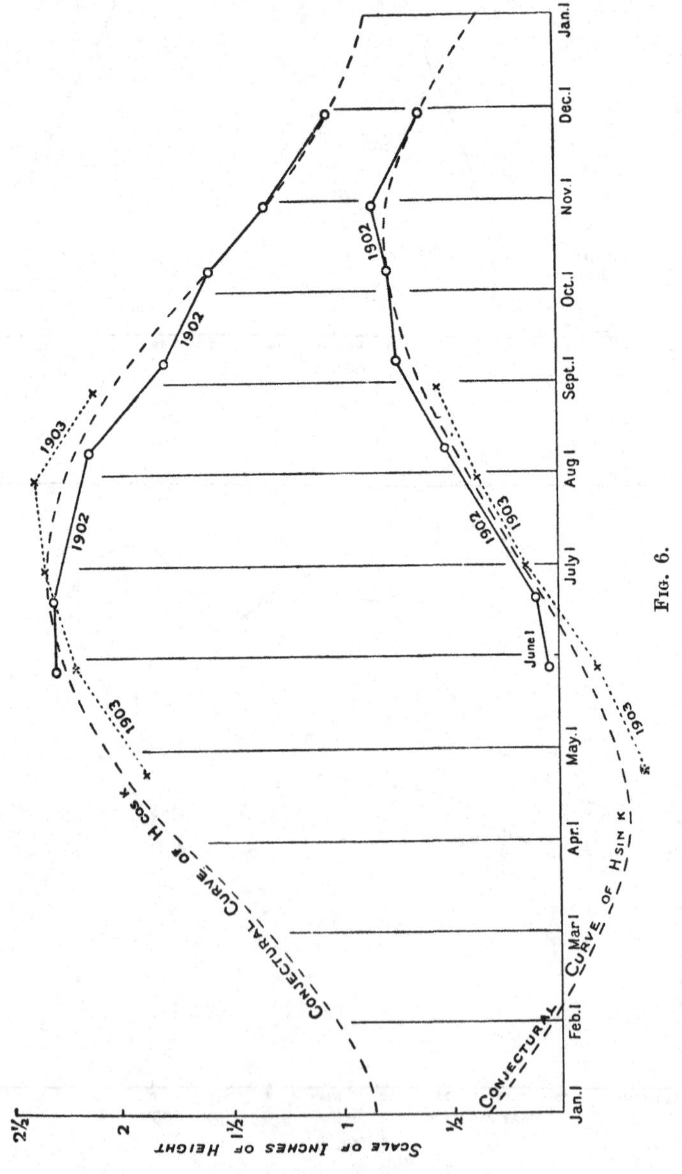

Fig. 6.

values for the season of 1902, while those marked with crosses give the values for 1903. The broken line shews conjectural curves which appear to satisfy the observations. The conjectural curves are such that (in inches)

$$H \cos \kappa = 1{\cdot}65 - {\cdot}75 \cos (\eta t + 2^\circ)$$

$$H \sin \kappa = {\cdot}23 + {\cdot}53 \cos (\eta t + 79^\circ)$$

where η is 360° per annum and t is expressed in months.

There would thus be an annual inequality in $H \cos \kappa$ and $H \sin \kappa$, and their mean values, viz. 1·65 and 0·23 inches, would shew that the mean lunar semidiurnal tide is expressed by $H = 1\frac{2}{3}$ inches, $\kappa = 8^\circ$.

The mean given previously as derived only from the observations was $H = 2$ inches, $\kappa = 10^\circ$.

It will be noticed that the greatest retardation occurs about midsummer, and at the same season there is a considerable decrease of amplitude. It is almost impossible to believe that the thawing of the sea could decrease the amplitude of the tide, although it might possibly increase it.

It would be strange if this result, depending as it does on 12 independent observations, should arise from mere chance. Yet there is no astronomical tide which can give an annual inequality in the lunar semidiurnal tide. I note that if the observations of 1903 were pushed backward one month the whole of the observations would fall into a more perfect curve. Hence an inequality of 13 months would satisfy the conditions more perfectly than one of 12 months. There is theoretically a minute tidal inequality of long period (Laplace's first species) with a period of 14 months due to the variation of latitude; but it is difficult to see how any perturbation of the lunar semidiurnal tide could be produced in this way.

But if we have found a true physical phenomenon, the same kind of effect ought probably to be produced on all the other tides. Yet when the observations for the other tides are plotted out in the same way, the points appear to be arranged almost chaotically. It is true that some slight tendency may be perceived for an increase of amplitude towards mid-winter, but the effect is too uncertain to justify reduction to numbers.

A much longer series of observations would be needed to throw a clear light on the point raised, but the result is so curious that it would not have been right to pass it by in silence.

Tidal observations were made at Ross Island (called Erebus Island on the memorandum) by Dr Wilson from 2^h Jan. 11, 1904 to 8^h Jan. 13. The place of observation was some 40 or 50 miles to the northward of the winter station. As there seemed some reason to suspect a seasonal variability in the tides, it seemed worth while to compare with actuality a tide-curve computed with the constants derived from the winter observations. A curve was therefore run off at the National Physical Laboratory for a few days beginning with 0^h Jan. 11, 1904. Although the sites of the two sets of observations are not identical, comparison with actuality shews a satisfactory agreement. It is unfortunate that these observations were made just after the time when the diurnal inequality has vanished and is beginning to increase again; for at these times the agreement is liable to be imperfect between computed and observed curves. On these grounds no surprise need be felt on account of the fact that the semidiurnal tide is somewhat more clearly marked in the observed tide-curve than in the computed one, and that the whole range of the diurnal tide on Jan. 11 was three inches greater, and on Jan. 12 about six inches (out of 28 inches) greater than appears from the computed curve. The computed and observed times of high and low water agree closely with one another. We may, on the whole, accept these summer observations as proving that our tidal constants are substantially correct.

The semidiurnal tides, although small, exhibit clearly another peculiarity; it is that (κ of S_2) – (κ of M_2) exhibits a seasonal change of roughly the same character in both years.

In all cases 'the age of the tide' is negative and its mean value is about – 4 days; in other words spring-tide occurs four days *before* or ten days *after* full and change of moon.

If the phases of M_2 and S_2 differed by 180° we should have neaps at full and change, and springs at half moon. This case corresponds to 'direct' lunar tide and 'inverted' solar tide. In the actual case

$$(\kappa \text{ of } M_2) - (\kappa \text{ of } S_2) = 370^\circ - 272^\circ = 98^\circ$$

thus the observations shew a result a little nearer to this condition than to the ordinary one where springs coincide with full and change of moon.

The unusual relationship between the M_2 and S_2 tides is such as to make it worth while to examine what would be the condition of affairs in an ocean of uniform depth covering the whole planet. From the few soundings which have been made it would seem that the ocean may be about 600 fathoms in depth, although further north the depth appears to be considerably greater. I have therefore taken the formulæ of Mr Hough (*Phil. Trans.* A, 191 (1878), pp. 177, 180) and evaluated the lunar and solar semidiurnal tides for an ocean of 7260 ft. in latitudes 60°, 65°, 70°, 75° with the following results:—

Lunar semidiurnal tide.

Latitude	60°	65°	70°	75°
H of equilibrium tide	6·052 cm.	4·324 cm.	2·832 cm.	1·622 cm.
Factor of augmentation for dynamical tide	1·932	1·496	1·098	·755
H of dynamical tide for ocean of 7260 ft. (direct tide)	11·69 cm. 4½ inches	6·47 cm. 2½ inches	3·11 cm. 1¼ inches	1·22 cm. ½ inch

Solar semidiurnal tide.

Latitude	60°	65°	70°	75°
H of equilibrium tide	2·816 cm.	2·012 cm.	1·318 cm.	0·755 cm.
Factor of augmentation for dynamical tide	−6·441	−4·390	−2·556	−1·003
H of dynamical tide for ocean of 7260 ft. (inverted tide)	−18·14 cm. −7 inches	−8·83 cm. −3½ inches	−3·37 cm. −1⅓ inches	−0·751 cm. −⅓ inch

We thus find that in these high latitudes the solar tide is more magnified than the lunar, and is inverted. Thus in latitude 60° the solar tide is much larger than the lunar and is inverted, whereas in latitude 70° they are nearly of equal magnitude and the inversion of the solar tide still continues.

For an ocean of twice the depth both the tides are direct, and they are not so much magnified.

Although the Antarctic Ocean runs all round the globe it is of course unjustifiable to apply these results directly to the oscillations of the actual ocean, but they serve to shew that we have no reason to expect considerable semidiurnal tides so near to the pole, and also that the great discrepancy between the phases of M_2 and S_2 is not so surprising a fact as might appear at first sight.

It is useless to carry out a similar investigation for the diurnal tides, because the variations in the depth of ocean exercise so large an influence on the result. We know in fact that for an ocean of uniform depth the K_1 tide vanishes completely, and the O-tide nearly vanishes.

I find that the equilibrium O-tide is 3½ inches in lat. 60° and falls to 2 inches in lat. 75°. Thus the amplitudes of the diurnal tides observed by the 'Discovery' are very much larger than the equilibrium values.

The Scottish Antarctic Expedition passed the winter of 1903 in S. lat. 60° 44′ and W. long. 44° 39′ at the South Orkney Islands; they were thus nearly opposite to the station of the 'Discovery.' Their station was well adapted for determining the general character of the tides in the Antarctic Ocean. The reduction of their observations was made by Mr Selby at the National Physical Laboratory, and gave the following results:—

	M_2	S_2	K_2	K_1	P	O
H	1·522 ft.	·902 ft.	·245 ft.	·496 ft.	·166 ft.	·559 ft.
κ	172°	198°	198°	15°	15°	359°

It will be noticed that these results are quite normal, save that the S_2-tide is rather large compared with M_2; and there is a well-marked diurnal inequality. They acquire a special interest when considered in connection with the 'Discovery's' results. We see that the semidiurnal tides are 'inverted' but have little or no retardation; whereas the M_2-tide of the 'Discovery' is small, but 'direct' also with little retardation. We are thus led to suspect that to the northward of the latitude of the South Orkneys, where the 'Scotia' wintered, the semidiurnal tides are inverted with small retardation; that somewhere between the South Orkneys and near to the Antarctic Continent there is a nodal line for the M_2-tide. There must be also a similar node for the S_2-tide, and we may perhaps suppose that the node of the S_2-tide is nearer to Ross Island than that of the M_2-tide.

When we turn to the diurnal tides we find an entirely different condition, for at both places the phases are virtually identical, and there seems a *primâ facie* case for maintaining that the phase of the diurnal tide throughout the whole Antarctic Ocean is approximately the same as in the equilibrium theory. I cannot venture to offer any theory in explanation of the greater magnitude of the diurnal tide at Ross Island than at the South Orkneys.

PART II

LUNAR DISTURBANCE OF GRAVITY

13.

ON AN INSTRUMENT FOR DETECTING AND MEASURING SMALL CHANGES IN THE DIRECTION OF THE FORCE OF GRAVITY.

By G. H. DARWIN and HORACE DARWIN.

[Report of the Committee, consisting of Mr G. H. DARWIN, Professor Sir WILLIAM THOMSON, Professor TAIT, Professor GRANT, Dr SIEMENS, Professor PURSER, Professor G. FORBES, and Mr HORACE DARWIN, appointed for the Measurement of the Lunar Disturbance of Gravity. This Report is written in the name of G. H. DARWIN merely for the sake of verbal convenience. *British Association Report for* 1881, pp. 93—126.]

I. *Account of the experiments.*

WE feel some difficulty as to the form which this report should take, because we are still carrying on our experiments, and have, as yet, arrived at no final results. As, however, we have done a good deal of work, and have come to conclusions of some interest, we think it better to give at once an account of our operations up to the present time, rather than to defer it to the future.

In November, 1878, Sir William Thomson suggested to me that I should endeavour to investigate experimentally the lunar disturbance of gravity, and the question of the tidal yielding of the solid earth. In May, 1879, we both visited him at Glasgow, and there saw an instrument, which, although roughly put together, he believed to contain the principle by which success might perhaps be attained. The instrument was erected in the Physical Laboratory of the University of Glasgow. We are not in a position to give an accurate description of it, but the following rough details are quite sufficient.

A solid lead cylinder, weighing perhaps a pound or two, was suspended by a fine brass wire, about 5 feet in length, from the centre of the lintel or cross-beam of the solid stone gallows, which is erected there for the purpose of pendulum experiments. A spike projected a little way out of the bottom of the cylindrical weight; a single silk fibre, several inches in length, was cemented to this spike, and the other end of the fibre was cemented to the edge of an ordinary galvanometer-mirror. A second silk fibre, of equal length, was cemented to the edge of the mirror at a point near to the attachment of the former fibre. The other end of this second fibre was then attached to a support, which was connected with the base of the stone gallows. The support was so placed that it stood very near to the spike at the bottom of the pendulum, and the mirror thus hung by the bifilar suspension of two silks, which stood exceedingly near to one another in their upper parts. The instrument was screened from draughts by paper pasted across between the two pillars of the gallows; but at the bottom, on one side, a pane of glass was inserted, through which one could see the pendulum bob and galvanometer-mirror.

It is obvious that a small displacement of the pendulum, in a direction perpendicular to the two silks, will cause the mirror to turn about a vertical axis.

A lamp and slit were arranged, as in a galvanometer, for exhibiting the movement of the pendulum, by means of the beam of light reflected from the mirror.

No systematic observations were made, but we looked at the instrument at various hours of the day and night, and on Sunday also, when the street and railway traffic is very small.

The reflected beam of light was found to be in incessant movement, of so irregular a character that it was hardly possible to localise the mean position of the spot of light on the screen, within 5 or 6 inches. On returning to the instrument after several hours, we frequently found that the light had wandered to quite a different part of the room, and we had sometimes to search through nearly a semicircle before finding it again.

Sir William Thomson showed us that, by standing some 10 feet away from the piers, and swaying from one foot to the other, in time with the free oscillations of the pendulum, quite a large oscillation of the spot of light could be produced. Subsequent experience has taught us that considerable precautions are necessary to avoid effects of this kind, and the stone piers at Glasgow did not seem to be well isolated from the floor, and the top of the gallows was used as a junction for a number of electric connections.

The cause of the extreme irregularity of the movements of the pendulum was obscure; and as Sir William Thomson was of opinion that the instrument was well worthy of careful study, we determined to undertake a series of experiments at the Cavendish Laboratory at Cambridge. We take this opportunity of recording our thanks to Lord Rayleigh* for his kindness in placing rooms at our disposal, and for his constant readiness to help us.

The pressure of other employments on both of us prevented our beginning operations immediately, and the length of time which we have now spent over these experiments is partly referable to this cause, although it is principally due to the number of difficulties to be overcome, and to the quantity of apparatus which has had to be manufactured.

In order to avoid the possibility of disturbance from terrestrial magnetism, we determined that our pendulum should be made of pure copper†. Mr Hussey Vivian kindly gave me an introduction to Messrs Elkington, of Birmingham; and, although it was quite out of their ordinary line of business, they consented to make what we required. Accordingly, they made a pair of electrolytically-deposited solid copper cylinders, $5\frac{1}{2}$ inches long, and $2\frac{3}{4}$ inches in diameter. From their appearance, we presume that the deposition was made on to the inside of copper tubes, and we understand that it occupied six weeks to take place. In November, 1879, they sent us these two heavy masses of copper, and, declining any payment, courteously begged our acceptance of them. Of these two cylinders we have, as yet, only used one; but should our present endeavours lead to results of interest, we shall ultimately require both of them.

Two months before the receipt of our weights, the British Association had reappointed the Committee for the Lunar Disturbance of Gravity, and had added our names thereto. Since that time, with the exception of compulsory intermissions, we have continued to work at this subject. My brother Horace and I have always discussed together the plan on which to proceed; but up to the present time much the larger part of the work has consisted in devising mechanical expedients for overcoming difficulties. In this work he has borne by very far the larger share; and the apparatus has been throughout constructed from his designs, and under his superintendence, by the Cambridge Scientific Instrument Company.

Near the corner of a stone-paved ground-floor room in the Cavendish Laboratory there stands a very solid stone gallows, similar to, but rather more massive than, the one at Glasgow. As it did not appear thoroughly free from rigid connection with the floor, we had the pavement raised all round the piers, and the earth was excavated from round the brick basement to the depth of about 2 feet 6 inches, until we were assured that there was no connection with the floor or walls of the room, excepting

* Professor Maxwell had given us permission to use the 'pendulum room,' but we had not yet begun our operations at the time of his death.

† We now think that this was probably a superfluity of precaution.

through the earth. The ditch, which was left round the piers, was found very useful for enabling us to carry out the somewhat delicate manipulations involved in hanging the mirror by its two silk fibres.

Into the middle of the flat ends of one of our copper weights (which weighed 4797 grammes, with spec. gr. 8·91) were screwed a pair of copper plugs; one plug was square-headed and the other pointed. Into the centre of the square plug was soldered a thin copper wire, just capable of sustaining the weight, and intended to hang the pendulum.

A stout cast-iron tripod was made for the support of the pendulum. Through a hole in the centre of it there ran rather loosely a stout iron rod with a screw cut on it. A nut ran on the screw and prevented the rod from slipping through the hole. The other end of the copper wire was fixed into the end of the rod.

The tripod was placed with its three legs resting near the margin of the circular hole in the centre of the lintel of the gallows. The iron rod was in the centre of the hole, and its lower end appeared about six inches below the lower face of the lintel. The pendulum hung from the rod by a wire of such length as to bring the spiked plug within a few inches of the base of the gallows. This would of course be a very bad way of hanging a pendulum which is intended to swing, but in our case the displacements of the end of the pendulum were only likely to be of a magnitude to be estimated in thousandths or even millionths of an inch, and it is certain that for such small displacements the nut from which the pendulum hung could not possibly rock on its bearings. However, in subsequent experiments we improved the arrangement by giving the nut a flange, from which there projected three small equidistant knobs, on which the nut rested.

The length of the pendulum from the upper juncture with the iron rod down to the tip of the spike in the bob was 148·2 cm.

An iron box was cast with three short legs, two in front and one behind; its interior dimensions were $15 \times 15 \times 17\frac{1}{2}$ cm.; it had a tap at the back; the front face ($15 \times 17\frac{1}{2}$) was left open, with arrangements for fixing a plate-glass face thereon. The top face ($15 \times 17\frac{1}{2}$) was pierced by a large round hole. On to this hole was cemented an ordinary earthenware 4-inch drain pipe, and on to the top of this first pipe there was cemented a second. The box was thus provided with a chimney 144 cm. high. The cubic contents of the box and chimney were about $3\frac{1}{2}$ gallons.

The box was placed standing on the base of the gallows, with the chimney vertically underneath the round hole in the lintel. The top of the chimney nearly reached the lower face of the lintel, and the iron rod of the pendulum extended a few inches down into the chimney. The pendulum wire ran down the middle of the chimney, and the lower half of the pendulum bob was visible through the open face of the iron box. The stone gallows faces towards the S.E., but we placed the box askew on the base, so that its open face was directed towards the S.

The three legs of the box rested on little metal discs, each with a conical hole in it, and these discs rested on three others of a somewhat larger size. When the box was set approximately in position, we could by an arrangement of screws cause the smaller discs to slide a fraction of an inch on the larger ones, and thus exactly adjust the position of the box and chimney.

A small stand, something like a retort stand, about 4 inches high, stood on a leaden base, with a short horizontal arm clamped by a screw on to the thin vertical rod. This was the 'fixed' support for the bifilar suspension of the mirror. The stand was placed to the E. of the pendulum bob, and the horizontal arm reached out until it came very close to the spike of the pendulum.

The suspension and protection from tarnishing of our mirror gave us much trouble, but it is useless to explain the various earlier methods employed, because we have now overcome these difficulties in a manner to be described later. The two cocoon fibres were fixed at a considerable distance apart on the edge of the mirror, and as they were very short they splayed out at nearly a right angle to one another. By means of this arrangement the free period of oscillation of the mirror was made very short, and we were easily able to separate the long free swing of the pendulum from the short oscillations of the mirror.

The mirror was hung so that the upper ends of the silks stood within an eighth of an inch of one another, but the tip of the spike stood $\frac{1}{8}$ or $\frac{1}{16}$ of an inch higher than the fixed support. The plate-glass front of the box was then fixed on with indiarubber packing.

It is obvious that a movement of the box parallel to the front from E. to W. would bring the two fibres nearer together; this operation we shall describe as sensitising the instrument. A movement of the box perpendicular to the front would cause the mirror to show its face parallel to the front of the box; this operation we shall describe as centralising. As sensitising will generally decentralise, both sets of screws had to be worked alternately.

The adjusting screws for moving the box did not work very well; nevertheless, by a little trouble we managed to bring the two silks of the bifilar suspension very close to one another.

After the instrument had been hung as above described, we tried a preliminary sensitisation, and found the pendulum to respond to a slight touch on either pier. The spot of light reflected from the mirror was very unsteady, but not nearly so much so as in the Glasgow experiment; and we were quite unable to produce any perceptible increase of agitation by stamping or swaying to and fro on the stone floor. This showed that the isolation of the pier was far more satisfactory than at Glasgow.

We then filled the box and pipes with water. We had much trouble with slow leakage of the vessel, but the most serious difficulty arose from the air-bubbles which adhered to the pendulum. By using boiled water we obviated this fairly well, but we concluded that it was a great mistake to have a flat bottom to the pendulum. This mistake we have remedied in the final experiment described in the present paper.

The damping effect of the water on the oscillations of the pendulum and of the mirror was very great, and although the incessant dance of the light continued, it was of much smaller amplitude, and comparatively large oscillations of the pendulum, caused by giving the piers a push, died out after two or three swings. A very slight push on the stone piers displaced the mean position of the light, but jumping and stamping on the pavement of the room produced no perceptible effect. If, however, one of us stood on the bare earth in the ditch behind, or before the massive stone pier, a very sensible deflection of the light was caused; this we now know was caused by an elastic depression of the earth, which tilted the whole structure in one or the other direction. A pull of a few ounces, delivered horizontally on the centre of the lintel, produced a clear deflection, and when the pull was 8 lbs., the deflection of the spot of light amounted to 45 cm. We then determined to make some rough systematic experiments.

The room was darkened by shutters over all the windows, and the doors were kept closed. The paraffin lamp stood at three or four feet to the S.E. of the easterly stone pier, but the light was screened from the pier.

We began our readings at 12 noon (March 15, 1880), and took eight between that time and 10.30 P.M. From 12 noon until 4 P.M. the lamp was left burning, but afterwards it was only lighted for about a minute to take each reading. At 12 the reading was 595 mm., and at 4 P.M. it was 936 mm.*; these readings, together with the intermediate ones, showed that the pendulum had been moving northwards with a nearly uniform velocity. After the lamp was put out, the pendulum moved southward, and by 10.30 P.M. was nearly in the same position as at noon.

During the whole of the two following days and a part of the next we took a number of readings from 9 A.M. until 11 P.M. The observations when graphically exhibited showed a fairly regular wave, the pendulum being at the maximum of its northern excursion between 5 and 7 P.M., and probably furthest south between the same hours in the morning. But

* I give the numbers as recorded in the note-book, but the readings would sometimes differ by 2 or 3 mm. within half-a-minute. The light always waves to and fro in an uncertain sort of way, so that it is impossible to assign a mean position with any certainty.

besides this wave motion, the mean position for the day travelled a good deal northward. We think that a part of this diurnal oscillation was due to the warping of the stone columns from changes of temperature. An increase of temperature on the south-east faces of the piers carried the lintel towards the north-west, and of this displacement we observed only the northerly component. The lamp produced a very rapid effect, and the diurnal change lagged some two hours behind the change in the external air. The *difference* between the temperatures of the S.E. and N.W. faces of the pier must have been very slight indeed. At that time, and indeed until quite recently, we attributed the whole of this diurnal oscillation to the warping of the piers, but we now feel nearly certain that it was due in great measure to a real change in the horizon.

We found that warming one of the legs of the iron tripod, even by contact with the finger, produced a marked effect, and we concluded that the mode of suspension was unsatisfactory.

Although we had thus learnt that changes of temperature formed the great obstacle in the way of success, there were a good many things to be learnt from the instrument as it existed at that time.

After the box and pipes had been filled for some days the plate-glass front cracked quite across, and a slow leakage began to take place; we were thus compelled to dismount the whole apparatus and to make a fresh start.

It is obvious that to detect and measure displacements of the pendulum in the N. and S. direction, the azimuth of the silks by which the mirror is suspended must be E. and W., and that although any E. and W. displacement of the pendulum will be invisible, still such displacement will alter the sensitiveness of the instrument for the N. and S. displacements. In order to obviate this we determined to constrain the pendulum to move only in the N. and S. azimuth.

Accordingly we had a T-piece about 4 inches long fixed to the end of the iron rod from which the pendulum hung. The two ends of a fine copper wire were soldered into the ends of the T-piece; a long loop of wire was thus formed. The square-headed plug at the top of the pendulum bob was replaced by another containing a small copper wheel, which could revolve about a horizontal axis. The bearings of the wheel were open on one side.

When the wheel was placed to ride on the bottom of the wire loop, and the pendulum bob hooked on to the axle of the wheel by the open bearings, we had our pendulum hanging by a bifilar suspension. The motion of the pendulum was thus constrained to take place only perpendicular to the plane of the wire loop.

The iron tripod was replaced by a slate slab large enough to entirely cover the hole in the lintel of the gallows. Through the centre of the slab was a round hole, of about one inch in diameter, through which passed the iron rod with the T-piece at the lower end. The iron rod was supported on the slate by means of the flanged nut above referred to. There was also a straight slot, cut quite through the slab, running from the central hole to the margin. The purpose of this slot will be explained presently.

In the preceding experiment we had no means of determining the absolute amount of displacement of the pendulum, although, of course, we knew that it must be very small. There are two methods by which the absolute displacements are determinable; one is to cause known small displacements to the pendulum and to watch the effect on the mirror; and the second is to cause known small horizontal forces to act on the pendulum. We have hitherto only employed the latter method, but we are rather inclined to think that the former may give better results.

The following plan for producing small known horizontal forces was suggested by my brother.

Suppose there be a very large and a very small pendulum hanging by wires of equal length from neighbouring points in the same horizon; and suppose the large and the small pendulum to be joined by a fibre which is a very little shorter than the distance between the points of suspension. Then each pendulum is obviously deflected a little from the vertical, but the deflection of the small pendulum varies as the mass of the larger, and that of the larger as the mass of the smaller. If m be the mass of the small pendulum, and M that of the large one, and if a be the distance between the points of suspension, then it may be easily shown that if a be increased by a small length δa, the increase of the linear deflection of the large pendulum is $m\delta a/(m+M)$. If l be the length of either pendulum, the angular deflection of the larger one is $m\delta a/l\,(m+M)$, and this is the deflection which would be produced by a horizontal force equal to $m\delta a/l\,(m+M)$ of gravity. It is clear, then, that by making the inequality between the two weights m and M very great, and the displacement of the point of suspension very small, we may deflect the large pendulum by as small a quantity as we like. The theory is almost the same if the two pendulums are not of exactly the same length, or if the length of one of them be varied.

Now in our application of this principle we did not actually attach the two pendulums together, but we made the little pendulum lean up against the large one; the theory is obviously just the same.

We call the small pendulum 'the disturber,' because its use is to disturb the large pendulum by known forces. A small copper weight for the disturber weighed ·732 grammes, and the large pendulum bob, with its pulley, weighed 4831·5. Therefore the one was 6600 times as massive as the other. The disturber was hung by a platinum wire about $\frac{1}{1000}$th of an inch in diameter, which is a good deal thinner than a fine human hair.

We must now explain how the disturber was suspended, and the method of moving its point of suspension.

Parallel to the sides of the slot in the slate slab there was riveted a pair of brass rails, one being V-shaped and the other flat; on these rails there slid a little carriage with three legs, one of which slid on one rail, and the other two on the other. A brass rod with an eyelet-hole at the end was fixed to the centre of the carriage, and was directed downwards so that it passed through the centre of the slot. The slot was directed so that it was perpendicular to the T-piece from which the pendulum hung, and the brass rod of the little carriage was bent and of such length, that when the carriage was pushed on its rails until it was as near the centre of the slab as it would go, the eyelet-hole stood just below the T-piece, and half-way between the two wires. A micrometer screw was clamped to the slab and was arranged for making the carriage traverse known lengths on its rails, and as the wires of the pendulum were in the E. and W. plane, the carriage was caused to travel N. and S. by its micrometer screw.

One end of the fine platinum wire was fastened to the eyelet, and the other (as above stated) to the small disturbing weight. The platinum wire was of such length that the disturber just reached the pulley by which the big pendulum hung. We found that by pushing the carriage up to the centre, and very slightly tilting it off one rail, we could cause the disturber weight to rest on either side of the pulley at will. If it was left on the side of the pulley remote from the disturber-carriage, it was in gear, and the traversing of the carriage on its rails would produce a small pressure of the disturber on to the side of the pulley. If it was left on the same side of the pulley as the disturber-carriage, the two pendulums were quite independent and the disturber was out of gear.

On making allowance for the difference in length between the pendulum and the disturber, and for the manner in which the thrust was delivered at the top of the pendulum, but omitting the corrections for the weights of the suspending wires and for the elasticity of the copper wire, we found that one turn of the micrometer screw should displace the spike at the bottom of the pendulum through 0·0001 mm. or $\frac{1}{245000}$th of an inch. The same displacement would be produced by an alteration in the direction of gravity with reference to the earth's surface by $\frac{1}{70}$th of a second of arc.

A rough computation showed that the to and fro motion of the pendulum in the N.S. azimuth, due to lunar attraction, should, if the earth be rigid, be the same as that produced by $2\frac{2}{5}$ turns of the micrometer screw.

We now return to the other arrangements made in re-erecting the instrument.

A new mirror, silvered on the face, was used, and was hung in a slightly different manner.

The fluid in which the pendulum was hung was spirits and water. The physical properties of such a mixture will be referred to later. In order to avoid air-bubbles we boiled 3½ gallons of spirits and water for three hours in vacuo, and the result appeared satisfactory in that respect.

After the mirror was hung, the plate-glass front to the box was fixed and the vessel was filled by the tap in the back of the box. The disturber was not introduced until afterwards, and we then found that the pendulum responded properly to the disturbance.

As the heat of a lamp in the neighbourhood of the piers exercised a large disturbance, we changed the method of observing, and read the reflection of a scale with a telescope. The scale was a levelling staff divided into feet, and tenths and hundredths of a foot, laid horizontally at 15 feet from the piers, with the telescope immediately over it.

Since the amount of fluid through which the light had to pass was considerable, we were forced to place a gas-flame immediately in front of the scale; but the gas was only kept alight long enough to take a reading.

After sensitising the instrument we found that the incessant dance of the image of the scale was markedly less than when the pendulum was hung in water. A touch with the finger on either pier produced deflection by bending the piers, and the instrument responded to the disturber.

The vessel had been filled with fluid for some days, and we had just begun a series of readings, when the plate-glass front again cracked quite across without any previous warning. Thus ended our second attempt.

In the third experiment (July and August, 1880) the arrangements were so nearly the same as those just described that we need not refer to them. The packing for the plate-glass front was formed of red lead, and this proved perfectly successful, whereas the indiarubber packing had twice failed. As we were troubled by invisible leakage and by the evaporation of the fluid, we arranged an inverted bottle, so as always to keep the chimney full. We thought that when the T-piece at the end of the shaft became exposed to the air, the pendulum became much more unsteady, but we now think it at least possible that there was merely a period of real terrestrial disturbance.

From August 10 to 14 we took a series of observations from early morning until late at night. We noted the same sort of diurnal oscillatory motion as before, but the outline of the curve was far less regular. This, we think, may perhaps be explained by the necessity we were under of leaving the doors open a good deal, in order to permit the cord to pass by which Lord Rayleigh was spinning the British Association coil.

Notwithstanding that the weather was sultry the warping of the stone columns must have been very slight, for a thermometer hung close to the pier scarcely showed a degree of change between the day and night, and the *difference* of temperature of the N. and S. faces must have been a very small fraction of a degree. At that time, however, we still thought that the whole of the diurnal oscillation was due to the warping of the columns.

We next tried a series of experiments to test the sensitiveness of the instrument.

As above remarked the image of the scale was continually in motion, and moreover the mean reading was always shifting in either one direction or the other. At any one time it was possible to take a reading to within $\frac{1}{10}$th of a foot with certainty, and to make an estimate of the $\frac{1}{100}$th of a foot, but the numbers given below are necessarily to be regarded as very rough approximations.

As above stated, the gallows faced about to the S.E., and we may describe the two square piers as the E. and W. piers, and the edges of each pier by the points of the compass towards which they are directed.

On August 14, 1880, my brother stood on a plank supported by the pavement of the room close to the S.W. edge of the W. pier, and, lighting a spirit lamp, held the flame for ten seconds within an inch or two of this edge of the pier. The effect was certainly produced of making the pendulum bob move northwards, but as such an effect is fused in the diurnal change then going on, the amount of effect was uncertain. He then stood similarly near the N.E. edge of the E. pier, and held the spirit flame actually licking the edge of the stone during one minute. The effect should now be opposed to the diurnal change, and it was so. Before the exposure to heat was over the reading had decreased ·15 feet, and after the heat was withdrawn the recovery began to take place almost immediately. We concluded afterwards that the effect was equivalent to a change of horizon of about 0″·15.

When the flame was held near but not touching the lintel for thirty seconds, the effect was obvious but scarcely measurable, even in round numbers, on account of the unsteadiness of the image.

When a heated lump of brass was pushed under the iron box no effect whatever was perceived, and even when a spirit flame was held so as to lick one side of the iron box during thirty seconds, we could not be sure that there was any effect. We had expected a violent disturbance, but these experiments seemed to show that convection currents in the fluid produce remarkably little effect.

When a pull of 300 grammes was delivered on to the centre of the lintel in a southward direction, we determined by several trials that the displacement of the reading was about ·30 feet, which may be equal to about 0″·3 change of horizon.

Two-thirds of a watering-can of water was poured into the ditch at the back of the pier. In this experiment the swelling of the ground should have an effect antagonistic to that produced by the cooling of the back face of the pier, and also to the diurnal changes then going on. The swelling of the ground certainly tilted the pier over, so that the reading was altered by ·10 feet. A further dose of water seemed to have the same effect, and it took more than an hour for the piers to regain their former position. As the normal diurnal change was going on simultaneously, we do not know the length of time during which the water continued to produce an effect.

On August 15 we tried a series of experiments with the disturber. When the disturber was displaced on its rails, the pendulum took a very perceptible time to take up its new position, on account of the viscosity of the fluid in which it was immersed.

The diurnal changes which were going on prevented the readings from being very accordant amongst themselves, but we concluded that twenty-five turns of the screw gave between ·4 and ·3 feet alteration in the reading on the scale. From the masses and dimensions of the pendulum and disturber, we concluded that 1 foot of our scale corresponded with about 1″ change in horizon. Taking into account the length of the pendulum, it appeared that 1 foot of our scale corresponded with $\frac{1}{1400}$th of a mm. displacement of the spike at the bottom of the pendulum. Now as a tenth of a foot of alteration of reading could be perceived with certainty, it followed that when the pendulum point moved through $\frac{1}{14000}$th of a mm. we could certainly perceive it.

During the first ten days the mean of the diurnal readings gradually increased, showing that the pendulum was moving northwards, until the reading had actually shifted 8 feet on the scale. It then became necessary to shift the scale. Between August 23 and 25 the reading had changed another foot. We then left Cambridge. On returning in October we found that this change had continued. The mirror had, however, become tarnished, and it was no longer possible to take a reading, although one could just see a gas-flame by reflection from the mirror.

Whilst erecting the pendulum we had to stand on, and in front of, the piers, and to put them under various kinds of stress, and we always found that after such stress some sort of apparently abnormal changes in the piers continued for three or four hours afterwards.

We were at that time at a loss to understand the reason of this long-continued change in the mean position of the pendulum, and were reluctant to believe that it indicated any real change of horizon of the whole soil; but after having read the papers of MM. d'Abbadie and Plantamour, we now believe that such a real change was taking place.

By this course of experiments it appeared that an instrument of the kind described may be brought to almost any degree of sensitiveness. We had seen, however, that a stone support is unfavourable, because the bad conductivity of stone prevents a rapid equalisation of temperature between different parts, and even small inequalities of temperature produce considerable warping of the stone piers. But it now seems probable that we exaggerated the amount of disturbance which may arise from this cause.

A cellar would undoubtedly be the best site for such an experiment, but unfortunately there is no such place available in the Cavendish Laboratory. Lord Rayleigh, however, placed the 'balance room' at our disposal, and this room has a northerly aspect. There are two windows in it, high up on the north wall, and these we keep boarded up.

The arrangements which we now intended to make were that the pendulum and mirror should be hung in a very confined space, and should be immersed in fluid of considerable viscosity. The boundary of that space should be made of a heat-conducting material, which should itself form the support for the pendulum. The whole instrument, including the basement, was to be immersed in water, and the basement itself was to be carefully detached from contact with the building in which it stands. By these means we hoped to damp out the short oscillations due to local tremors, but to allow the longer oscillations free to take place; but above all we desired that changes of temperature in the instrument should take place with great slowness, and should be, as far as possible, equal all round.

We removed the pavement from the centre of the room, and had a circular hole, about 3 feet 6 inches in diameter, excavated in the 'made earth,' until we got down to the undisturbed gravel, at a depth of about 2 feet 6 inches.

We obtained a large cylindrical stone 2 feet 4 inches in diameter and 2 feet 6 inches in height, weighing about three-quarters of a ton. This we had intended to place on the earth in the hole, so that its upper surface should stand flush with the pavement of the room. But the excavation had been carried down a little too deep, and therefore an ordinary flat paving stone was placed on the earth, with a thin bedding of cement underneath it. The cylindrical block was placed to stand upon the paving stone, with a very thin bedding of lime and water between the two stones. The surface of the stone was then flush with the floor. We do not think that any sacrifice of stability has been made by this course.

An annular trench or ditch a little less than a foot across is left round the stone. We have lately had the bottom of the ditch cemented, and the vertical sides lined with brickwork, which is kept clear of any contact with the pavement of the room. On the S. side the ditch is a little wider, and this permits us to stand in it conveniently. The bricked ditch is watertight, and has a small overflow pipe into the drains. The water in the ditch stands slightly higher than the flat top of the cylindrical stone, and thus the whole basement may be kept immersed in water, and it is, presumably, at a very uniform temperature all round.

Before describing the instrument itself we will explain the remaining precautions for equalisation of temperature.

On the flat top of the stone stands a large barrel or tub, 5 feet 6 inches high and 1 foot 10 inches in diameter, open at both ends. The diameter of the stone is about 2 inches greater than the outside measure of the diameter of the tub, and the tub thus nearly covers the whole of the stone. The tub is well payed with pitch inside, and stands on two felt rings soaked in tar. Five large iron weights, weighing altogether nearly three-quarters of a ton, are hooked on to the upper edge of the tub, in order to make the joint between the tub and the stone watertight. Near the bottom is a plate-glass window; when it is in position, the window faces to the S. This tub is filled with water and the instrument stands immersed therein.

We had at first much trouble from the leakage of the tub, and we have to thank Mr Gordon, the assistant at the Laboratory, for his ready help in overcoming this difficulty, as well as others which were perpetually recurring. The mounting of the tub was one of the last things done before the instrument was ready for observation, and we must now return to the description of the instrument itself.

We used the same pendulum bob as before, but we had its shape altered so that the ends both above and below were conical surfaces, whilst the central part was left cylindrical. The upper plug with its pulley is replaced by another plug bearing a short round horizontal rod, with a rounded groove cut in it. The groove stands vertically over the centre of the weight, and is designed for taking the wire of the bifilar suspension of the pendulum; when riding on the wire the pendulum bob hangs vertically.

Part of this upper plug consists of a short thin horizontal arm about an inch long. This arm is perpendicular to the plane of the groove, and when the pendulum is in position, projects northwards. Through the end of the arm is bored a fine vertical hole. This part of the apparatus is for the modified form of disturber, which we are now using.

The support for the pendulum consists of a stout copper tube $2\frac{7}{8}$ inches in diameter inside measure, and it just admits the pendulum bob with $\frac{1}{8}$th inch play all round. The tube is 3 feet 6 inches in height, and is closed at the lower end by a diaphragm, pierced in the centre by a round hole, about $\frac{1}{4}$ inch in diameter. The upper end has a ring of brass soldered on to it, and this ring has a flange to it. The upper part of the brass ring forms a short continuation $\frac{3}{4}$ of an inch in length of the copper tube.

The ring is only introduced as a means of fastening the flange to the copper tube.

The upper edge of the brass continuation has three V notches in it at 120° apart on the circumference of the ring. A brass cap like the lid of a pill-box has an inside measure ¼ inch greater than the outside measure of the brass ring. The brass cap has three rods which project inwards from its circumference, and which are placed at 120° apart thereon. When the cap is placed on the brass continuation of the upper tube, the three rods rest in the three V notches, and the cap is geometrically fixed with respect to the tube. A fine screw works through the centre of the cap, and actuates an apparatus, not easy to explain without drawings, by which the cap can be slightly tilted in one azimuth. The object of tilting the cap is to enable us to sensitise the instrument by bringing the silk fibres attached to the mirror into close proximity.

Into the cap are soldered the two ends of a fine brass wire; the junctures are equidistant from the centre of the cap and on opposite sides of it; they lie on that diameter of the cap which is perpendicular to the axis about which the tilting can be produced.

When the pendulum is hung on the brass wire loop by the groove in the upper plug, the wires just clear the sides of the copper tube.

It is clear that the tilting of the cap is mechanically equivalent to a shortening of one side of the wire loop and the lengthening of the other. Hence the pendulum is susceptible of a small lateral adjustment by means of the screw in the cap.

To the bottom of the tube is soldered a second stout brass ring; this ring bears on it three stout brass legs inclined at 120° to one another, all lying in a plane perpendicular to the copper tube. From the extremity of each leg to the centre of the tube is 8½ inches. The last inch of each leg is hollowed out on its under surface into the form of a radial V groove.

There are three detached short pieces of brass tube, each ending below in a flange with three knobs on it, and at the upper end in a screw with a rounded head. These three serve as feet for the instrument. These three feet are placed on the upper surface of our basement stone at 120° apart, estimated from the centre of the stone. The copper tube with its legs attached is set down so that the inverted V grooves in the legs rest on the rounded screw-head at the tops of the three feet, and each of the feet rests on its three knobs on the stone. The bottom of the copper tube is thus raised 5½ inches above the stone. By this arrangement the copper tube is retained in position with reference to the stone, and it will be observed that no part of the apparatus is under any constraint except such as is just necessary to determine its position geometrically.

The screws with rounded heads which form the three feet are susceptible of small adjustments in height, and one of the three heads is capable of more delicate adjustment, for it is actuated by a fine screw, which is driven by a toothed wheel and pinion. The pinion is turned by a wooden rod, made flexible by the insertion of a Hook's joint, and the wooden rod reaches to the top of the tub, when it is mounted surrounding the instrument.

The adjustable leg is to the N. of the instrument, and as the mirror faces S. we call it the 'back-leg.' When the copper support is mounted on its three legs, a rough adjustment for the verticality of the tube is made with two of the legs, and final adjustment is made by the back-leg.

It is obvious that if the back-leg be raised or depressed the point of the pendulum is carried southwards or northwards, and the mirror turns accordingly. Thus the back-leg with its screw and rod affords the means of centralising the mirror. The arrangements for suspending the mirror must now be described.

The lower plug in the pendulum bob is rounded and has a small horizontal hole through it. When the pendulum is hung this rounded plug just appears through the hole in the diaphragm at the bottom of the copper tube.

A small brass box, shaped like a disk, can be screwed on to the bottom of the copper tube, in such a way that a diameter of the box forms a straight line with the axis of the copper tube. One side of the box is of plate glass, and when it is fastened in position the plate glass faces to the S. This is the mirror-box; it is of such a size as to permit the mirror to swing about 15° in either direction from parallelism with the plate-glass front.

The fixed support for the second fibre for the bifilar suspension of the mirror may be described as a very small inverted retort-stand. The vertical rod projects downwards from the underside of the diaphragm, a little to the E. of the hole in the diaphragm; and a small horizontal arm projects from this rod, and is of such a length that its extremity reaches to near the centre of the hole. This arm has a small eyelet-hole pierced through a projection at its extremity.

The mirror itself is a little larger than a shilling and is of thin plate glass; it has two holes drilled through the edge at about 60° from one another. The mirror was silvered on both sides, and then dipped into melted paraffin; the paraffin and silver were then cleaned off one side. The paraffin protects the silver from tarnishing, and the silver film seen through the glass has been found to remain perfectly bright for months, after having been immersed in fluid during that time. A piece of platinum wire about $\frac{1}{1000}$th of an inch in diameter is threaded twice through each hole in opposite directions, in such a manner that with a continuous piece of wire (formed by tying the two ends together) a pair of short loops are formed at the edge of the mirror, over each of the two holes. When the mirror is hung from a silk fibre passing through both loops, the weight of the mirror is sufficient to pull each loop taut.

A single silk fibre was threaded through the eyelet-hole at the end of the blunt point of the pendulum bob, and tied in such a way that there was no loose end projecting so as to foul the other side of the bifilar suspension. The other end of the silk fibre was knotted to a piece of sewing silk on which a needle was threaded.

The pendulum was then hung from the cap by its wire loop, outside the copper tube, and the silk fibre with the sewing silk and needle attached dangled down at the bottom. The cap, with the pendulum attached thereto, was then hauled up and carefully let down into the copper tube. The sewing silk, fibre, and blunt end came out through the hole in the diaphragm.

We then sewed with the needle through the two loops on the margin of the mirror, and then through the eyelet-hole in the little horizontal arm. The silk was pulled taut, and the end fastened off on to the little vertical rod, from which the horizontal arm projects.

The mirror then hangs with one part of the silk attached to the pendulum bob and the other to the horizontal arm.

The two parts of the silk are inclined to one another at a considerable angle, so that the free period of the mirror is short, but the upper parts of the silk stand very close to one another. The mirror-box encloses the mirror and makes the copper tube watertight.

There is another part of the apparatus which has not yet been explained, namely, the disturber. This part of the instrument was in reality arranged before the mirror was hung.

We shall not give a full account of the disturber, because it does not seem to work very satisfactorily.

In the form of disturber which we now use the variation of horizontal thrust is produced by variation in the length of the disturbing pendulum, instead of by variation of the point of support as in the previous experiment. It is not easy to vary the point of support when the pendulum is hung in a tube which nearly fits it.

The disturber weight is a small lump of copper, and it hangs by fine sewing silk. The silk is threaded through the eyelet in the horizontal arm which forms part of the upper plug of the pendulum; thus the disturber weight is to the N. of the pendulum. The silk after passing between the wires supporting the pendulum has its other end attached to the cap at the top at a point to the S. of the centre of the cap. Thus the silk is slightly inclined to the plane through the wires. The arrangement for varying the length of the disturbing pendulum will not be explained in detail, but it may suffice to say that it is produced by a third weight, which we call the 'guide weight,' which may be hauled up or let down in an approximately vertical line. This guide weight determines by its position how much of the upper part of the silk of the disturber shall be cut off, so as not to form a part of the free cord by which the disturbing weight hangs.

The guide weight may be raised or lowered by cords which pass through the cap. If the apparatus were to work properly a given amount of displacement of the guide weight should produce a calculable horizontal thrust on the pendulum. The whole of the arrangements for the disturber could be made outside the copper tube, so that the pendulum was lowered into the tube with the disturber attached thereto.

After the mirror was hung and the mirror-box screwed on, a brass cap was fixed by screws on to the flange at the top of the copper tube. This cap has a tube or chimney attached to it, the top of which rises five inches above the top of the cap or lid from which the pendulum hangs. From this chimney emerges a rod attached to the screw by which the sensitising apparatus is actuated, and also the silk by which the guide weight is raised or depressed.

The copper tube, with its appendages, was then filled with a boiled mixture of filtered water and spirits of wine by means of a small tap in the back of the mirror-box. The mixture was made by taking equal volumes of the two fluids; the boiling to which it was subjected will of course have somewhat disturbed the proportions. Poiseuille has shown* that a mixture of spirits and water has much greater viscosity than either pure spirits or pure water. When the mixture is by weight in the proportion of about seven of water to nine of spirits, the viscosity is nearly three times as great as that of pure spirits or of pure water. As the specific gravity of spirits is about ·8, it follows that the mixture is to be made by taking equal volumes of the two fluids. It was on account of this remarkable fact that we chose this mixture in which to suspend the pendulum, and we observed that the unsteadiness of the mirror was markedly less than when the fluid used was simply water.

The level of the fluid stood in our tubular support quite up to the top of the chimney, and thus the highest point of the pendulum itself was 5 inches below the surface.

The tub was then let down over the instrument, and the weights hooked on to its edge. The plate-glass window in the tub stood on the S. opposite to the mirror-box. The tub was filled with water up to nearly the top of

* *Poggendorff's Annalen*, 1843, Vol. LVIII., p. 437.

the chimney, and the ditch round the stone basement was also ultimately filled with water. The whole instrument thus stood immersed from top to bottom in water.

Even before the tub was filled we thought that we noticed a diminution of unsteadiness in the image of a slit reflected from the mirror. The filling of the tub exercised quite a striking effect in the increase of steadiness, and the water in the ditch again operated favourably.

We met with much difficulty at first in preventing serious leakage of the tub, and as it is still not absolutely watertight, we have arranged a water-pipe to drip about once a minute into the tub. A small overflow pipe from the tub to the ditch allows a very slow dripping to go into the ditch, and thus both vessels are kept full to a constant level. We had to take this course because we found that a rise of the water in the ditch through half an inch produced a deflection of the pendulum. The ditch, it must be remembered, was a little broader on the S. side than elsewhere.

In May, 1881, we took a series of observations with the light, slit and scale. The scale was about 7 feet from the tub, and in order to read it we found it convenient to kneel behind the scale on the ground. I was one day watching the light for nearly ten minutes, and being tired with kneeling on the pavement, I supported part of my weight on my hands a few inches in front of the scale. The place where my hands came was on the bare earth from which one of the paving stones had been removed. I was surprised to find quite a large change in the reading. After several trials I found that the pressure of a few pounds with one hand only was quite sufficient to produce an effect.

It must be remembered that this is not a case of a small pressure delivered on the bare earth at say 7 feet distance, but it is the difference of effect produced by this pressure at 7 feet and 8 feet; for of course the change only consisted in the change of distribution in the weight of a small portion of my body.

We have, however, since shown that even this degree of sensitiveness may be exceeded.

We had thought all along that it would ultimately be necessary to take our observations from outside the room, but this observation impressed it on us more than ever; for it would be impossible for an observer always to stand in exactly the same position for taking readings, and my brother and I could not take a set of readings together on account of the difference between our weights.

In making preliminary arrangements for reading from outside the room we found the most convenient way of bringing the reflected image into the field of view of the telescope was by shifting a weight about the room. My brother stood in the room and changed his position until the image was in the field of view, and afterwards placed a heavy weight where he had been standing; after he had left the room the image was in the field of view.

On the S.W. wall of the room there is a trap-door or window which opens into another room, and we determined to read from this.

In order to read with a telescope the light has to undergo two reflections and twelve refractions, besides those in the telescope; it has also to pass twice through layers of water and of the fluid mixture. In consequence of the loss of light we found it impossible to read the image of an illuminated scale, and we had to make the scale self-luminous.

On the pavement to the S. of the instrument is placed a flat board on to which are fixed a pair of rails; a carriage with three legs slides on these rails, and can be driven to and fro by a screw of ten threads to the inch. Backlash in the nut which drives the carriage is avoided by means of a spiral spring. A small gas-flame is attached to the carriage; in front of it is a piece of red glass, the vertical edge of which is very distinctly visible in the telescope after reflection from the mirror. The red glass was introduced to avoid prismatic effects, which had been troublesome before. The edge of the glass was found to be a more convenient object than a line which had been engraved on the glass as a fiducial mark.

The gas-flame is caused to traverse by pullies driven by cords. The cords come to the observing window, and can be worked from there. A second telescope is erected at the window, for reading certain scales attached to the traversing gear of the carriage, and we find that we can read the position of the gas-flame to within a tenth of an inch, or even less, with certainty.

From the gas the ray of light enters the tub and mirror-box, is reflected by the mirror, and emerges by the same route; it then meets a looking-glass which reflects it nearly at right angles and a little upwards, and finally enters the object-glass of the reading telescope, fixed to the sill of the observing window.

When the carriage is at the right part of the scale the edge of the red glass coincides with the cross wire of the reading telescope, and the reading is taken by means of the scale telescope.

Arrangements had also to be made for working the sensitiser, centraliser, and disturber from outside the room.

A scaffolding was erected over the tub, but free of contact therewith, and this supported a system of worm-wheels, tangent-screws, and pullies by which the three requisite movements could be given. The junctures with the sensitising and centralising rods were purposely made loose, because it was found at first that a slight shake to the scaffolding disturbed the pendulum.

The pullies on the scaffolding are driven by cords which pass to the observing window.

On the window-sill we now have two telescopes, four pullies, an arrangement, with a scale attached, for raising and depressing the guide weight, and a gas tap for governing the flame in the room.

After the arrangements which have been described were completed we sensitised the instrument from outside the room. The arrangements worked so admirably that we could produce a quite extraordinary degree of sensitiveness by the alternate working of the sensitising and centralising wheels, without ever causing the image of the lamp to disappear from the field of view. This is a great improvement on the old arrangement with the stone gallows.

We now found that if one of us was in the room and stood at about 16 feet to the S. of the instrument with his feet about a foot apart, and slowly shifted his weight from one foot to the other, then a distinct change was produced in the position of the mirror. This is the most remarkable proof of sensitiveness which we have yet seen, for the instrument can detect the difference between the distortion of the soil caused by a weight of 140 lbs. placed at 16 feet and at 17 feet. We have not as yet taken any great pains to make the instrument as sensitive as possible, and we have little doubt but that we might exceed the present degree of delicacy, if it were desirable to do so.

The sensitiveness now attained is, we think, only apparently greater than it was with the stone gallows, and depends on the improved optical arrangements, and the increase of steadiness due to the elimination of changes of temperature in the support.

From July 21 to July 25 we took a series of readings. There was evidence of a distinct diurnal period with a maximum about noon, when the pendulum stood furthest northwards; in the experiment with the stone gallows in 1880 the maximum northern excursion took place between 5 and 7 P.M.

The path of the pendulum was interrupted by many minor zigzags, and it would sometimes reverse its motion for nearly an hour together. During the first four days the mean position of the pendulum travelled southward, and the image went off the scale three times, so that we had to recentralise it. In the night between the 24th and 25th it took an abrupt turn northward, and the reading was found in the morning of the 25th at nearly the opposite end of the scale.

On the 25th the dance of the image was greater than we had seen it at any time with the new instrument, so that we went into the room to see whether the water had fallen in the tub and had left the top of the copper tube exposed; for on a previous occasion this had appeared to produce much unsteadiness. There was, however, no change in the state of affairs. A few days later the image was quite remarkable for its steadiness.

On July 25, and again on the 27th, we tried a series of observations with the disturber, in order to determine the absolute value of the scale.

The guide weight being at a known altitude in the copper tube, we took a series of six readings at intervals of a minute, and then shifting the guide weight to another known altitude, took six more in a similar manner; and so on backwards and forwards for an hour.

The first movement of the guide weight produced a considerable disturbance of an irregular character, and the first set of readings were rejected. Afterwards there was more or less concordance between the results, but it was to be noticed there was a systematic difference between the change from 'up' to 'down' and 'down' to 'up.' This may perhaps be attributed to friction between certain parts of the apparatus. We believe that on another occasion we might erect the disturber under much more favourable conditions, but we do not feel sure that it could ever be made to operate very satisfactorily.

The series of readings before and after the change of the guide weight were taken in order to determine the path of the pendulum at the critical moment, but the behaviour of the pendulum is often so irregular, even within a few minutes, that the discrepancy between the several results and the apparent systematic error may be largely due to unknown changes, which took place during the minute which necessarily elapsed between the last of one set of readings and the first of the next. The image took up its new position deliberately, and it was necessary to wait until it had come to its normal position.

Between the first and second sets of observations with the disturber, it had been necessary to enter the room and to recentralise the image. We do not know whether something may not have disturbed the degree of sensitiveness, but at any rate the results of the two sets of observations are very discordant*.

The first set showed that one inch of movement of the gas-flame, which formed the scale, corresponds with $\frac{1}{13}$th of a second of arc of change of horizon; the second gave $\frac{1}{8}$th of a second to the inch.

As we can see a twentieth of an inch in the scale, it follows that a change of horizon of about 0″·005 should be distinctly visible. In this case the

* See, however, the postscript at the end of this part.

point of the pendulum moves through $\frac{1}{40000}$th of a millimetre. At present we do not think that the disturber gives more than the order of the changes of horizon which we note, but our estimate receives a general confirmation from another circumstance.

From the delicacy of the gearing connected with the back-leg, we estimate that it is by no means difficult to raise the back-leg by a millionth of an inch. The looseness in the gearing was purposely kept so great that it requires a turn or two of the external pulley on the window-sill before the backlash is absorbed, but after this a very small fraction of a turn is sufficient to move the image in the field.

We are now inclined to look to this process with the back-leg to enable us to determine the actual value of our scale, but this will require a certain amount of new apparatus, which we have not yet had time to arrange. In erecting the instrument we omitted to take certain measurements which it now appears will be necessary for the use of the back-leg as a means of determining the absolute value of our scale, but we know these measurements approximately from the working drawings of the instrument. Now it appears that one complete revolution of a certain tangent-screw by which the back-leg is raised should tilt the pendulum-stand through almost exactly half a second of arc, and therefore this should produce a relative displacement of the pendulum of the same amount. We have no doubt but that a tenth of the turn of the tangent-screw produces quite a large deflection of the image, and probably a hundredth of a turn would produce a sensible deflection. Therefore, from mere consideration of the effect of the back-leg we do not doubt but that a deflection of the pendulum through a $\frac{1}{200}$th of a second of arc is distinctly visible. This affords a kind of confirmation of the somewhat unsatisfactory deductions which we draw from the operation of the disturber.

Postscript.—The account of our more recent experiments was written during absence from Cambridge from July 29 to August 9. In this period the gradual southerly progression of the pendulum bob, which was observed up to July 28, seems to have continued; for on August 9 the pendulum was much too far S. to permit the image of the gas-flame to come into the field of view of the telescope. On August 9 the image was recentralised, and on the 9th and 10th the southerly change continued; on the 11th, however, a reversal northwards again occurred. During these days the unsteadiness of the image was much greater than we have seen it at any time with the new instrument. There was some heavy rain and a good deal of wind at that time. We intend to arrange a scale for giving a numerical value to the degree of unsteadiness, but at present it is merely a matter of judgment.

It seems possible that earthquakes were the cause of unsteadiness on

August 9, 10, and 11, and we shall no doubt hear whether any earthquakes have taken place on those days.

After August 11 we were both again absent from Cambridge. On August 16 my brother returned, and found that the southerly progression of the pendulum bob had reasserted itself, so that the image was again far out of the field of view. After recentralising he found the image to be unusually steady.

This appeared a good opportunity of trying the effect of purely local tremors.

One observer therefore went into the room and, standing near the instrument, delivered some smart blows on the brickwork coping round the ditch, the stone pavement, the tub, and the large stone basement underneath the water. Little or no effect was produced by this. Very small movements of the body, such as leaning forward while sitting in a chair, or a shift of part of the weight from heels to toes, produced a sensible deflection, and it was not very easy for the experimenter to avoid this kind of change whilst delivering the blows. To show the sensitiveness of the instrument to steady pressure we may mention that a pressure of three fingers on the brick coping of the ditch produces a marked deflection.

On August 17 I returned to Cambridge, and noted, with my brother, that the image had never been nearly so steady before. The abnormal steadiness continued on the 18th. There was much rain during those days.

On the afternoon of the 19th there was a high wind, and although the abnormal steadiness had ceased, still the agitation of the image was rather less than we usually observe it.

The image being so steady on the 17th, we thought that a good opportunity was afforded for testing the disturber. At 6.15 P.M. of that day we began the readings. The changes from 'up' to 'down' were made as quickly as we could, and in a quarter of an hour we secured five readings when the guide weight was 'up,' and four when it was 'down.'

When a curve was drawn, with the time as abscissa, and the readings as ordinates, through the 'up's,' and similarly through the 'down's,' the curves presented similar features. This seems to show that movement of the disturber does not cause irregularities or changes, except such as it is designed to produce.

The displacement of the guide weight was through 5 cm. on each occasion.

The four changes from 'up' to 'down' showed that an inch of scale corresponded with 0″·0897, with a mean error of 0″·0021; the four from 'down' to 'up' gave 0″·0909 to the inch, with a mean error of 0″·0042. Thus the systematic error on the previous occasions was probably only apparent.

Including all the eight changes together, we find that the value of an inch is 0″·0903 with a mean error of 0″·0030.

A change in the scale reading amounting to a tenth of an inch is visible without any doubt, and even less is probably visible. Now it will give an idea of the delicacy of the instrument when we say that a tenth of an inch of our scale corresponds to a change of horizon* through an angle equal to that subtended by an inch at 384 miles.

II. *On the work of previous observers.*

In the following section we propose to give an account of the various experiments which have been made in order to detect small variations of horizon, as far as they are known to us; but it is probable that other papers of a similar kind may have escaped our notice.

In a report of this kind it is useful to have references collected together, and therefore, besides giving an account of the papers which we have consulted, we shall requote the references contained in these papers.

In *Poggendorff's Annalen* for 1873 there are papers by Prof. F. Zöllner, which had been previously read before the Royal Saxon Society, and which are entitled "Ueber eine neue Methode zur Messung anziehender und abstossender Kräfte," Vol. 150, p. 131, "Beschreibung und Anwendung des Horizontalpendels," Vol. 150, p. 134. A part of the second of these papers is translated, and the figure is reproduced in the supplementary number of the *Philosophical Magazine* for 1872, p. 491, in a paper "On the Origin of the Earth's Magnetism."

The horizontal pendulum was independently invented by Prof. Zöllner, and, notwithstanding assertions to the contrary, was probably for the first time actually realised by him; it appears, however, that it had been twice invented before. The history of the instrument contains a curious piece of scientific fraud, of which we shall give an account below.

The instrument underwent some modifications under the hands of Professor Zöllner, and the two forms are described in the above papers.

The principle employed is as follows:—There is a very stout vertical stand, supported on three legs. At the top and bottom of the vertical shaft are fixed two projections. Attached to each projection is a fine straight steel clock spring; the springs are parallel to the vertical shaft of the stand, the one attached to the lower projection running upwards, and

* We use the expression 'change of horizon' to denote relative movement of the earth, at the place of observation, and the plumb-line. Such changes may arise either from alteration in the shape of the earth, or from displacement of the plumb-line; our experiments do not determine which of these two really takes place.

that attached to the upper one running downwards. The springs are of equal length, each being equal to half the distance between their points of attachment on the projections.

The springs terminate in a pair of rings, which stand exactly opposite to one another, so that a rod may be thrust through both.

A glass rod has a heavy weight attached to one end of it, and the other end is thrust through the two rings. The rings are a little separated from one another, and the glass rod stands out horizontally, with its weight at the end, and is supported by the tension of the two springs. It is obvious that if the point of attachment of the upper spring were vertically over that of the lower spring, and if the springs had no torsional elasticity, then the glass rod would be in neutral equilibrium, and would stand equally well in any azimuth.

The springs being thin have but little torsional elasticity, and Professor Zöllner arranges the instrument so that the one support is very nearly over the other. In consequence of this the rod and weight have but a small predilection for one azimuth more than another. The free oscillations of the horizontal pendulum could thus be made extraordinarily slow; and even a complete period of one minute could be easily attained.

A very small horizontal force of course produces a large deflection of the pendulum, and a small deflection of the force of gravitation with reference to the instrument must produce a like result. He considers that by this instrument he could, in the first form of the instrument, detect a displacement of the horizon through 0″·00035; in the second his estimate is 0″·001.

The observation was made by means of a mirror attached to the weight, and scale and telescope.

The maximum change of level due to the moon's attraction is at St Petersburg 0″·0174, and from the sun 0″·0080 [C. A. F. Peters, *Bull. Acad. Imp. St Pétersbourg*, 1844, Vol. III., No. 14]; and thus the instrument was amply sensitive enough to detect the lunar and solar disturbances of gravity*.

* We are of opinion that M. Zöllner has made a mistake in using at Leipzig Peters' results for St Petersburg. Besides this he considers the changes of the vertical to be 0″·0174 on *each* side of a mean position, and thus says the change is 0″·0348 altogether. Now a rough computation which I have made for Cambridge shows that the maximum meridional horizontal component of gravitation, as due to lunar attraction, is $4{\cdot}12 \times 10^{-8}$ of pure gravity. This force will produce a deflection of the plumb-line of 0″·0085, and the total amplitude of meridional oscillation will be 0″·0170. The maximum deflection of the plumb-line occurs when the moon's hour-angle is $\pm 45°$ and $\pm 135°$ at the place of observation. The change at Cambridge when the moon is S.E. and N.W. is 0″·0216. The deflection of the plumb-line varies as the cosine of the latitude, and is therefore greater at Cambridge than at St Petersburg. Multiplying ·0216 by sec 51° 43′ cos 60° we get ·0174, and thus my calculation agrees with that of Peters.

Professor Zöllner found, as we have done, that the readings were never the same for two successive instants. The passing of trains on the railway at a mile distant produced oscillations of the equilibrium position. He seems to have failed to detect the laws governing the longer and wider oscillations performed. Notwithstanding that he took a number of precautions against the effects of changes of temperature, he remarks that "the external circumstances under which the above experiments were carried out must be characterised as extremely unfavourable for this object (measuring the lunar attraction), so that the sensitiveness might be much increased in pits in the ground, provided the reaction of the glowing molten interior against the solid crust do not generate inequalities of the same order."

Further on he says that if the displacements of the pendulum should be found not to agree in phase with the theoretical phase as given by the sun's position, then it might be concluded that gravitation must take a finite time to come from the sun.

It appears to me that such a result would afford strong grounds for presuming the existence of frictional tides in the solid earth, and that Professor Zöllner's conclusion would be quite unjustifiable.

Earlier in the paper he states that he preferred to construct his instrument on a large scale, in order to avoid the disturbing effects of convection currents. We cannot but think, from our own experience, that by this course Professor Zöllner lost more than he gained, for the larger the instrument the more it would necessarily be exposed in its various parts to regions of different temperature, and we have found that the warping of supports by inequalities of temperature is a most serious cause of disturbance.

The instrument of which we have given a short account appears to us very interesting from its ingenuity, and the account of the attempts to use it is well worthy of attention, but we cannot think that it can ever be made to give such good results as those which may perhaps be attained by our plan or by others. The variation in the torsional elasticity of the suspending springs, due to changes of temperature, would seem likely to produce serious variations in the value of the displacements of the pendulum, and it does not seem easy to suspend such an instrument in fluid in such a manner as to kill out the effects of purely local tremors.

Moreover, the whole instrument is kept permanently in a condition of great stress, and one would be inclined to suppose that the vertical stand would be slightly warped by the variation of direction in which the tensions of the springs are applied, when the pendulum bob varies its position.

In a further paper in the same volume, p. 140, "Zur Geschichte des Horizontalpendels," Zöllner gives the priority of invention to M. Perrot, who had described a similar instrument on March 31, 1862 (*Comptes Rendus*, Vol. 54, p. 728), but as far as he knows M. Perrot did not actually construct it.

He also quotes an account of an "Astronomische Pendelwage," by Lorenz Hengler, published in 1832, in Vol. 43 of *Dingler's Polytechn. Journ.*, pp. 81—92. In this paper it appears that Hengler gives the most astonishing and vague accounts of the manner in which he detected the lunar attraction with a horizontal pendulum, the points of support being the ceiling and floor of a room 16 feet high. The terrestrial rotation was also detected with a still more marvellous instrument.

Zöllner obviously discredits these experiments, but hesitates to characterise them, as they deserve, as mere fraud and invention.

The university authorities at Munich state that in the years 1830–1 there was a candidate in philosophy and theology named Lorenz Hengler, of Reichenhofen, "der weder früher noch später zu finden ist."

At p. 150 of the same volume Professor Šafařik contributes a "Beitrag zur Geschichte des Horizontalpendels." He says that the instrument takes its origin from Professor Gruithuisen, of Munich, whose name has "keinen guten Klang" in the exact sciences.

This strange person, amongst other eccentricities, proposed to dig a hole quite through the earth, and proposes a catachthonic observatory. Gruithuisen says, in his *Neuen Analekten für Erd- und Himmelskunde* (Munich, 1832), Vol. I., Part I.: "I believe that the oscillating balance (Schwung-wage) of a pupil of mine (named Hengeller), when constructed on a large scale, will do the best service."

Some of the most interesting observations which have been made are those of M. d'Abbadie. He gave an account of his experiments in a paper, entitled "Études sur la verticale," *Association Française pour l'avancement des Sciences, Congrès de Bordeaux*, 1872, p. 159. As this work is not very easily accessible to English readers, and as the paper itself has much interest, we give a somewhat full abstract of it. He has also published two short notes with reference to M. Plantamour's observations (noticed below), in Vol. 86, p. 1528 (1878), and Vol. 89, p. 1016 (1879), of the *Comptes Rendus*. We shall incorporate the substance of his remarks in these notes in our account of the original paper.

When at Olinda, in Brazil, in 1837, M. d'Abbadie noticed the variations of a delicate level which took place from day to day. At the end of the two months of his stay there the changes in the E. and W. azimuth had compensated themselves, and the level was in the same condition as at first; but the change in the meridian was still progressing when he had to leave.

In 1842, at Gondar, in Ethiopia, and at Saqa, he noticed a similar thing. In 1852 he gave an account to the French Academy (*Comptes Rendus*, May, p. 712) of these observations, as well as of others, by means of levels, which were carried out in a cellar in the old castle of Audaux, Basses Pyrénées.

Leverrier, he says, speaks of sudden changes taking place in the level of astronomical instruments, apparently without cause. Airy has proved that the azimuth of an instrument may change, and Hough notes, in America, capricious changes of the Nadir.

Henry has collected a series of levellings and azimuths observed at Greenwich during ten years, and during eight of the same years at Cambridge (*Monthly Notices R. A. S.*, Vol. VIII., p. 134). The results with respect to these two places present a general agreement, and show that from March to September the western Y of the transit instrument falls through 2″·5, whilst it deviates at the same time 2″ towards the north. Ellis has made a comparison of curves applying to Greenwich, during eight years, for level and azimuth. He shows that there is a general correspondence with the curves of the external temperature (*Memoirs of the R. Ast. Soc.*, Vol. XXIX., pp. 45—57).

In the later papers M. d'Abbadie says that M. Bouquet de la Grye has observed similar disturbances of the vertical at Campbell Island, S. lat. 52° 34′. M. Bouquet used a heavy pendulum governing a vertical lever, by which the angle was multiplied*. He found that the great breakers on the shore at a distance of two miles caused a deviation of the vertical of 1″·1. On one occasion the vertical seems to have varied through 3″·2 in 3¼ hours.

M. d'Abbadie also quotes Elkin, Yvon Villarceau, and Airy as having found, from astronomical observations, notable variations in latitude, amounting to from 7″ to 8″.

As M. d'Abbadie did not consider levels to afford a satisfactory method of observation of the presumed changes of horizon, he determined to proceed in a different manner.

The site of his experiments was Abbadia, in Subernoa, near Hendaye. The Atlantic was 400 metres distant, and the sea-level 62 metres below the place of observation. The subsoil was loamy rock (*roche marneuse*), belonging to cretaceous deposits of the south of France. Notwithstanding the steep slope of the soil, water was found at about 5 metres below the surface.

* I do not find a reference to M. Bouquet in the R.S. catalogue of scientific papers. It appears from what M. d'Abbadie says that certain observations have been made with pendulums in Italy, but that it does not distinctly appear that the variations of level are simultaneous over wide areas. No reference is given as to the observers.

In this situation he had built, in 1863, a steep concrete cone, of which the external slope was ten in one (une inclinaison d'une dixième). The concrete cone is truncated, and the flat surface at the top is 2 metres in diameter. It is pierced down the centre by a vertical hole or well 1 metre in diameter. This well extends to within half a metre of the top, at which point the concrete closes in, leaving only a hole of 12 centimetres up to the flat upper surface.

From the top of the concrete down to the rock is 8 metres, and the well is continued into the rock to a further depth of 2 metres: thus from top to bottom is 10 metres.

A tunnel is made to the bottom of the well in order to drain away the water, and access of the observer to the bottom is permitted by means of an underground staircase. Access can also be obtained to a point half-way between the top and bottom by means of a hole through the concrete. At this point there is a diaphragm across the well, pierced by a hole 21 centimetres in diameter. The diaphragm seems to have been originally made in order to support a lens, but the mode of observation was afterwards changed. The diaphragm is still useful, however, for allowing the observer to stand there and sweep away cobwebs.

The cone is enclosed in an external building, from the roof of which, as I understand, there hangs a platform on which the observer may stand without touching the cone; and the two staircases leading up to the top are also isolated*.

On the hole through the top of the cone is riveted a disk of brass pierced through its centre by a circular hole 21 mm. in diameter. The hole in the disk is traversed across two perpendicular diameters by fine platinum wires; at first there were only two wires, but afterwards there were four, which were arranged so as to present the outline of a right-angled cross. The parallel wires were very close together, so that the four wires enclosed in the centre a very small square space.

At the bottom of the well is put a pool of mercury. The mercury was at first in an iron basin, but the agitation of the mercury was found sometimes to be so great that no reflection was visible for an hour together. At the suggestion of Leverrier the iron basin was replaced by a shallow wooden tray with a corrugated bottom, and a good reflection was then generally obtainable. Immediately over the mercury pool there stood a lens of 10 cm. diameter and 10 metres focal length, and over the brass disk there stood a microscope with moveable micrometer wires in the eye-piece, and a position circle. The platinum wires were illuminated, and on looking through the microscope the observer saw the wires both directly and

* This passage appears to me a little obscure, and I cannot quite understand the arrangement.

by reflection. The observations were taken by measuring the azimuth and displacement of the image of the central square relatively to the real square enclosed by the wires.

One division of the micrometer screw indicated a displacement of vertical of 0″·03, so that the observations were susceptible of considerable refinement.

The whole of the masonry was finished in 1863, and M. d'Abbadie then allowed the structure five years to settle before he began taking observations. The arrangements for observing above described were made in 1868 and 1869.

In the course of a year he secured 2000 observations, and the results appear to be very strange and capricious.

Throughout March, 1869, the perturbations of the mercury were so incessant that observations (taken at that time with the iron basin) were nearly impossible; on the 29th he waited nearly an hour in vain in trying to catch the image of the wires. Two days later the mercury was perfectly tranquil. On April 6 it was much agitated, although the air and sea were calm. A tranquil surface was a rare exception.

In 1870 the corrugated trough was substituted for the iron basin; and M. d'Abbadie says:—

"Cependant, ni le fond inégal du bain rainé ni sa forme ne m'ont empêché d'observer, ce que j'appelle des *ombres fuyantes.* Ce sont des bandes sombres et parallèles qui traversent le champ du microscope avec plus ou moins de vitesse, et qu'on explique en attribuant au mercure des ondes très ténues, causées par une oscillation du sol dans un seul sens. Le plus souvent ces *ombres* semblent courir du S.E. au N.O., approximativement selon l'axe de la chaîne des Pyrénées; mais je les ai observées, le 15 Mars 1872, allant vers le S.O. À cette époque, le mercure était depuis le 29 février, dans une agitation continuelle, comme mon aide l'avait constaté en 1869, aussi dans le mois de Mars*."

He observed also, from time to time, certain oscillations of the mercury too rapid to be counted, which he calls 'tremoussements.' There were also sudden jumpings of the image from one point to another, or 'frétillements,' indicating a sudden change of vertical through 0″·49 to 0″·65.

He observed many microscopic earthquakes, and in some cases the image was carried quite out of the field of view.

He also detected the difference of vertical according to the state of the tide in the neighbouring sea; but the change of level due to this cause was often masked by others occurring contemporaneously.

* M. d'Abbadie writes to me that this phenomenon was ultimately found to result from air currents (Nov. 5, 1881).

From observations during the years 1867 to 1872 (with the exception of 1870) he finds that in every year but one the plumb-line deviated northwards during the latter months of the year, but in 1872 it deviated to the south.

He does not give any theoretical views as to the causes of these phenomena, but remarks that his observations tend to prove that the causes of change are sometimes neither astronomical nor thermometrical.

The most sudden change which he noted was on October 27, 1872, when the vertical changed by 2″·4 in six hours and a quarter. Between January 30 and March 26 of the same year the plumb-line deviated 4″·5 towards the south.

We now come to the valuable observations of M. Plantamour, which we believe are still being prosecuted by him. His papers are "Sur le déplacement de la bulle des niveaux à bulle d'air," *Comptes Rendus*, June 24, 1878, Vol. 86, p. 1522, and "Des mouvements périodiques du sol accusés par des niveaux à bulle d'air," *Comptes Rendus*, December 1, 1879, Vol. 89, p. 937.

The observations were made at Sécheron, near Geneva, at first at the Observatory, and afterwards at M. Plantamour's house. After some preliminary observations, he obtained a very sensitive level and laid it on the concrete floor of a room in which the variations of temperature were very small. The azimuth of the level was E. and W., and the observations were made every hour from 9 A.M. until midnight. Figures are given of the displacement of the bubble during April 24, 25, and 26, 1878. The results indicate a diurnal oscillation of level, the E. end of the level being highest towards 5.30 P.M.; the amplitudes of the oscillations were 8″·4, 11″·2, 15″·75 during these three days. It also appeared that there was a gradual rising of the mean diurnal position of the E. end during the same time.

The level was then transported to a cellar in M. Plantamour's house, when the temperature only varied by half a degree centigrade. The bubble of the level often ran quite up to one end. A new and larger level was obtained, together with the great 'chevalet de fer,' which is used by the manufacturers in testing levels. Both levels were placed E. and W., at about two metres apart. During May 3 and 4, 1878, the bubble travelled eastward without much return, and it is interesting to learn that simultaneous observations by M. Turretini, at the Level Factory, three kilometres distant, at Plainpalais, showed a similar change.

Between May 3 and 6 the level actually changed through 17″. Up to the 19th the level still showed the eastward change.

M. Plantamour remarks that the eastern pier of a transit instrument is known to rise during a part of the year, but not by an amount comparable with that observed by him, and that the diurnal variations are unknown.

After further observations of a similar kind, one of the levels was arranged in the N. and S. azimuth.

The same sort of diurnal oscillations, although more irregular, were observed, but the hours of maximum were not the same in the two levels. During the four days, May 24 to 28, the maximum rising of the north generally took place about noon. This is exactly the converse of what we have recently observed.

In the second paper he remarks:

"Dans le sens du méridien, les mouvements diurnes sont très rares irréguliers et toujours très faibles, le niveau en accuse parfois, quand il n'y en a point de l'est à l'ouest, et inversement, quand ces derniers sont très prononcés, on n'en aperçoit que très rarement du sud au nord."

In our experiment of March 15 to 18, 1880, we found that the pendulum stood furthest north about 6 P.M., so that at that time the S. was most elevated; and in the short series of observations during the present summer the maximum elevation of the S. took place about noon.

On October 1, 1878, M. Plantamour began a new series of observations, which lasted until September 30, 1879. The levels were arranged in the two azimuths as before, and the observations were taken five times a day, namely, at 9 A.M., noon, 3, 6, and 9 P.M. The mean of these five readings he takes as the diurnal value.

During October and November the eastern end of the level fell, which is exactly the converse of what happened during the spring of the same year; he concludes that the eastern end falls when the external temperature falls.

When a curve of the external temperature was placed parallel with that for the level, it appeared that there was a parallelism between the two, but the curve for the level lagged behind that for temperature by a period of from one to four days.

This parallelism was maintained until the end of June, 1879, when it became disturbed. From then until the beginning of September the E. rose, but in a much greater proportion than the rise of mean temperature. It must be noted that July was a cold and wet month.

Although the external temperature began to fall on August 5, the E. end continued to rise until September 8. This he attributes to an accumulation of heat in the soil. The total amplitude of the annual oscillation from E. to W. amounted to 28″·08.

There was also a diurnal oscillation in this azimuth which amounted to 3″·2 on September 5. The east end appeared to be highest between 6 and 7.45 P.M., and lowest at the similar hour in the morning*.

The meridional oscillations were much smaller, the total annual amplitude being only 4″·89. From December 23, 1878, until the end of April, 1879, there was a correspondence between the external temperature curve and that for N. and S. level. We have already quoted the remark on the diurnal meridional oscillations.

M. Plantamour tells us that in 1856 Admiral Mouchez detected no movement of the soil by means of the levels attached to astronomical instruments. On the other hand, M. Hirsch established, by several years of observation at Neuchâtel, that there was an annual oscillation of a transit instrument from E. to W., with an amplitude of 23″, and an azimuthal oscillation of 75″. Similar observations with the transit instrument were made at the observatory at Berne in the summer of 1879.

It is to be regretted that M. Plantamour does not give us more information concerning the manner in which the iron support for the levels was protected from small changes of temperature, nor with regard to the effect of the observer's weight on the floor of the room. We have concluded that both these sources of disturbance should be carefully eliminated.

Some interesting observations were made at Pulkova on a subject cognate to that on which we are writing. M. Magnus Nyrèn contributed, on February 28, 1878, an interesting note to the Imperial Academy of St Petersburg, entitled "Erderschütterung beobachtet an einem feinem Niveau 1877 Mai 10†." On May 10 (April 28), 1877, at 4.16 A.M., a striking disturbance of the level on the axis of the transit was observed by M. Nyrèn in the observatory at Pulkova. The oscillations were watched by him for three minutes; their complete period was about 20 seconds, and their amplitude between 1″·5 and 2″. At 4.35 A.M. there was no longer any disturbance. He draws attention to the fact that it afterwards appeared that one hour and fourteen minutes earlier there had been a great earthquake at Iquique. The distance from Iquique to Pulkova is 10,600 kilometres in a straight line, and 12,540 kilometres along the arc of a great circle. He does not positively connect the two phenomena together; but he observes that if the wave came through the earth from Iquique to Pulkova it must have travelled at the rate of about

* It seems that M. Plantamour sent a figure to the French Academy with the paper, but no figure is given. This figure would doubtless have explained the meaning of some passages which are somewhat obscure. Thus he speaks of the *minimum* occurring between 6 and 7.45, but it is not clear whether minimum means E. highest or E. lowest. I interpret the passage as above, because this was the state of things in the observations recorded in the first of the two papers. There is a similar difficulty about the meridional oscillations.

† *Bull. Acad. St Petersb.*, Vol. XXIV., p. 567.

2·4 kilometres per second. This is the speed of transmission through platinum or silver.

M. Nyrèn thinks the wave-motion could not have been so regular as it was, if the transmission had been through the solid, and suggests that the transmission was through the fluid interior of the earth.

It appears to us that this argument is hardly sound, and that it would be more just to conclude that the interior of the earth was a sensibly perfectly elastic solid; because oscillations in molten rock would surely be more quickly killed out by internal friction than those in a solid. However, M. Nyrèn does not lay much stress on this argument. He also draws attention to the fact that on September 20 (8), 1867, M. Wagner observed at Pulkova an oscillation of the level, with an amplitude of 3″, and that seven minutes before the disturbance there had been an earthquake at Malta. On April 4 (March 23), 1868, M. Gromadzki observed an agitation of the level, and it was afterwards found that there had been an earthquake in Turkestan five minutes before.

Similar observations of disturbances had been made twice before, once by M. Wagner and once by M. Romberg; but they had not been connected with any earthquakes—at least with certainty.

Dr C. W. Siemens has invented an instrument of extraordinary delicacy, which he calls an "Attraction-meter." An account of the instrument is given in an addendum to his paper "On determining the depth of the sea without the use of the sounding-line" (*Phil. Trans.*, 1876, p. 659). We shall not give any account of this instrument, because Dr Siemens is a member of our committee, and will doubtless bring any observations he may make with it before the British Association at some future time.

III. *Remarks on the present state of the subject.*

Although our experiments are not yet concluded, it may be well to make a few remarks on the present aspects of the question, and to state shortly our intentions as to future operations.

Our experiments, as far as they go, confirm the results of MM. d'Abbadie and Plantamour, and we think that there can remain little doubt that the surface of the earth is in incessant movement, with oscillations of periods extending from a fraction of a second to a year.

Whether it be a purely superficial phenomenon or not, this consideration should be of importance to astronomical observers, for their instruments are necessarily placed at the surface of the earth. M. Plantamour and others have shown that there is an intimate connection between the changes of level and those of the temperature of the air; whence it follows that the principal part of the changes must be superficial. On the other hand, M. d'Abbadie has shown that it is impossible to explain all the changes by means of changes of temperature. It would be interesting to determine whether changes of a similar kind penetrate to the bottom of mines, and Gruithuisen's suggestion of a catachthonic observatory seems worthy of attention, although he perhaps went rather far in the proposition that the observatory should be ten or fifteen miles below the earth's surface.

It may appear not improbable that the surface of the soil becomes wrinkled all over, when it is swollen by increase of temperature and by rainfall. If this, however, were the case, then we should expect that instruments erected at a short distance apart would show discordant results. M. Plantamour, however, found that, at least during three days, there was a nearly perfect accordance between the behaviour of two sets of levels at three kilometres apart; and during eight years there appeared to be general agreement between the changes of level of the astronomical instruments at Greenwich and Cambridge. It would be a matter of much interest to determine how far this concordance would be maintained if the instrument of observation had been as delicate as that used by M. d'Abbadie or as our pendulum.

M. Plantamour speaks as though it were generally recognised that one pier of a transit circle rises during one part of the year and falls at another*. But if this be so throughout Europe, we must suppose that there is a kind of tide in the solid earth, produced by climatic changes; the rise and fall of the central parts of continents must then amount to something considerable in vertical height, and the changes of level on the easterly and westerly coasts of a continent must be exactly opposite to one another. We are not aware that any comparison of this kind has been undertaken. The idea seems of course exceedingly improbable, but we understand it to be alleged that it is the eastern pier of transit instruments in Europe which rises during the warmer part of the year. Now if this be generally true for Europe, which has no easterly coast, it is not easy to see how the change can be brought about except by a swelling of the whole continent.

We suggest that in the future it will be thought necessary to erect at each station a delicate instrument for the continuous observation of changes of level. Perhaps M. d'Abbadie's pool of mercury might be best for the longer inequalities, and something like our pendulum for the shorter ones;

* "Dans l'opération au moyen de laquelle on vérifie l'horizontalité de l'axe d'une lunette méridienne, il parait qu'on remarque bien un léger mouvement d'exhaussement de l'est pendant une partie de l'année, mais il n'est pas aussi considérable que celui qu'accuse mon niveau, et l'on n'a jamais remarqué, que je sache, une oscillation diurne comme celle qu'a indiquée le niveau dans le pavillon." *Comptes Rendus*, June 24, 1878, Vol. LXXXVI., p. 1525.

or possibly the pendulum, when used in a manner which we intend to try, might suffice for all the inequalities.

At present the errors introduced by unknown inequalities of level are probably nearly eliminated by the number of observations taken; but it could not fail to diminish the probable error of each observation if a correction were applied for this cause of disturbance from hour to hour, or even from minute to minute. If the changes noted by M. Plantamour are not entirely abnormal in amount, such corrections are certainly sufficient to merit attention.

In our first set of experiments we found that stone piers are exceedingly sensitive to changes of temperature and to small stresses. Might it not be worth while to plate the piers of astronomical instruments with copper, and to swathe them with flannel? We are not aware as to the extent to which care is taken as to the drainage of the soil round the piers, or as to the effect of the weight of the observer's body; but we draw attention to the effect produced by the percolation of water round the basement, and to the impossibility we have found of taking our observations in the same room with the instrument.

In connection with this subject we may notice an experiment which was begun 3½ years ago by my brother Horace. The experiment was undertaken in connection with my father's investigation of the geological activity of earthworms, and the object was to determine the rate at which stones are being buried in the ground in consequence of the excavations of worms.

The experiment is going on at Down, in Kent. The soil is stiff red clay, containing many flints lying over the chalk. There are two stout metal rods, one of iron and the other of copper. The ends were sharpened and they were hammered down vertically into the soil of an old grass field, and they are in contact with one another, or nearly so. When they had penetrated 8 feet 6 inches it was found very difficult to force them deeper, and it is probable that the ends are resting on a flint. The ends were then cut off about three inches above the ground.

A stone was obtained like a small grindstone, with a circular hole in the middle. This stone was laid on the ground with the two metal rods appearing through the hole. Three brass V grooves are leaded into the upper surface of the stone, and a moveable tripod-stand with three rounded legs can be placed on the stone, and is, of course, geometrically fixed by the nature of its contact with the V's. An arrangement with a micrometer screw enables the observer to take contact measurements of the position of the upper surface of the stone with regard to the rods. The stone has always continued to fall, but during the first few months the rate of fall was probably influenced by the decaying of the grass underneath it. The general falling of the stone can only be gathered from observations taken at many months apart, for it is found to be in a state of continual vertical oscillation.

The measurements are so delicate that the raising of the stone produced by one or two cans full of water poured on the ground can easily be perceived. Between September 7 and 19, 1880, there was heavy rain, and the stone stood 1·91 mm. higher at the latter date than at the former. The effect of frost and the wet season combined is still more marked, for on January 23, 1881, the stone was 4·12 mm. higher than it had been on September 7, 1880.

The prolonged drought of the present summer has had a great effect, for between May 8 and June 29 the stone sank through 5·79 mm. The opposite effects of drought and frost are well shown by the fact that on January 23 the stone stood 8·62 mm. higher than on June 29, 1881. The observations are uncorrected for the effect of temperature on the metal rods, but the fact that the readings from the two rods of different metals always agree very closely *inter se*, shows that such a correction would amount to very little.

The changes produced in the height of the stone are, of course, entirely due to superficial causes; but the amounts of the oscillations are certainly surprising, and although the basements of astronomical instruments may be very deep, they cannot entirely escape from similar oscillations*.

In his address to the mathematical section at the meeting of the British Association at Glasgow in 1876, Sir William Thomson tells us† that Peters, Maxwell, Nyrèn, and Newcomb‡ have examined the observations at Pulkova, Greenwich, and Washington, in order to discover whether there is not an inequality in the latitude of the observatories having a period of about 306 days. Such an inequality must exist on account of the motion in that period of the instantaneous axis of rotation of the earth round the axis of maximum moment of inertia. The inequality was detected in the results, but the probable error was very large, and the epochs deduced by the several investigators do not agree *inter se.* It remains, therefore, quite uncertain whether the detection of the inequality is a reality or not. But now we ask whether it is not an essential first step in such an enquiry to make an elaborate investigation by a very delicate instrument of the systematic changes of vertical at each station of observation?

We will next attempt to analyse the merits and demerits of the various methods which have been employed for detecting small changes in the vertical.

* [An account of this experiment is given in *Proc. Roy. Soc.*, Vol. LXVIII., 1901, pp. 253—261.]

† *B. A. Report* for 1876, p. 10. For "Nysen" read "Nyrèn."

‡ Peters' paper is in *Bull. St Pet. Acad.*, 1844, p. 305, and *Ast. Nach.*, Vol. XXII., 1845, pp. 71, 103, 119. Nyrèn's paper is in *Mém. St Pet. Acad.*, Vol. XIX., 1873, No. 13. With regard to Maxwell, see Thomson and Tait's *Nat. Phil.*, 2nd edit., Part I., Vol. I. An interesting letter from Newcomb is quoted in Sir W. Thomson's address.

The most sensitive instrument is probably the horizontal pendulum of Professor Zöllner, and its refinement might be almost indefinitely increased by the addition of the bifilar suspension of a mirror as a means of exhibiting the displacements of the pendulum bob. If this were done it might be possible to construct the instrument on a very small scale and yet to retain a very high degree of sensitiveness. We are inclined to think, however, that the variation of the torsional elasticity of the suspending springs under varying temperature presents an objection to the instrument which it would be very difficult to remove. The state of stress under which the instrument is of necessity permanently retained seems likely to be prejudicial.

Next in order of sensitiveness is probably our own pendulum, embodying the suggestion of Sir William Thomson. We are scarcely in a position as yet to feel sure as to its merits, but it certainly seems to be capable of all the requisite refinement. We shall give below the ideas which our experience, up to the present time, suggest as to improvements and future observations.

Although we know none of the details of M. Bouquet de la Grye's pendulum actuating a lever, it may be presumed to be susceptible of considerable delicacy, and it would be likely to possess the enormous advantage of giving an automatic record of its behaviour. On the other hand the lever must introduce a very unfavourable element in the friction between solids.

M. d'Abbadie's method of observation by means of the pool of mercury seems on the whole to be the best which has been employed hitherto. But it has faults which leave ample fields for the use of other instruments. The construction of a well of the requisite depth must necessarily be very expensive, and when the structure is made of a sufficient size to give the required degree of accuracy, it is difficult to ensure the relative immobility of the cross-wires and the bottom of the well.

Levels are exceedingly good from the point of view of cheapness and transportability, but the observations must always be open to some doubt on account of the possibility of the sticking of the bubble from the effects of capillarity. The justice of this criticism is confirmed by the fact that M. Plantamour found that two levels only two metres apart did not give perfectly accordant results. Levels are moreover, perhaps, scarcely sensitive enough for an examination of the smaller oscillations of level. Dr Siemens' form of level possesses ample sensibility, but is probably open to the same objections on the score of capillarity.

In the case of our own experiments we think that the immersion of the whole instrument in water from top to bottom has proved an excellent precaution against the effects of change of temperature, and our experience leads us to think that much of the agitation of the pendulum in the earlier set of experiments was due to small variations of temperature against which we are now guarded.

The sensitiveness of the instrument leaves nothing to be desired, and were such a thing as a firm foundation attainable, we could measure the horizontal component of the lunar attraction to a considerable degree of accuracy. We believe that this is the first instrument in which the viscosity of fluids has been used as a means of eliminating the effects of local tremors. In this respect we have been successful, for we find that jumping or stamping in the room itself produces no agitation of the pendulum, or at least none of which we can feel quite sure. We are inclined to try the effect of fluids of greater viscosity, such as glycerine, syrup of sugar, or paraffin oil. But along with such fluids we shall almost inevitably introduce air-bubbles, which it may be hard to get rid of. If a fluid of great viscosity were used, we should then only observe the oscillations of level of periods extending over perhaps a quarter to half a minute. The oscillations of shorter periods are, however, so inextricably mixed up with those produced by carriages and railway trains, that nothing would be lost by this.

In connection with this point Mr Christie writes to me, that "In the old times of Greenwich Fair, some twenty years ago, when crowds of people used to run down the hill, I find the observers could not take reflection observations for two or three hours after the crowd had been turned out.......We do not have anything like such crowds now, even on Bank holidays, and I have not heard lately of any interference with the observations." If the observers attributed the agitation of the mercury to the true cause, the elasticity of the soil must be far more perfect than is generally supposed. It would be surprising to find a mass of glass or steel continuing to vibrate for as long as two hours after the disturbance was removed. May it not be suspected that times of agitation, such as those noted by M. d'Abbadie, happened to coincide on two or three occasions with Greenwich Fair?

As the sensitiveness of our present instrument is very great, although the sensitising process has never been pushed as far as possible, we think that it will be advantageous to construct an instrument on half, or even less than half, the present scale. The heavy weights which we now have to employ will thus be reduced to one-eighth of the present amount. The erection of the instrument may thus be made an easy matter, and an easily portable and inexpensive instrument may be obtained.

Our present form of instrument has several serious flaws. The image is continually travelling off the scale, the gearing both internal and external to the room for observing is necessarily complex and troublesome to erect, and lastly we have not yet succeeded in an accurate determination of the value of the scale.

We are in hopes of being able to overcome all these objections. We propose to have a fixed light, which may be cast into the room from the outside. This will free us from the obviously objectionable plan of having a gas-flame in the room, and at the same time will abolish the gearing for traversing the lamp on the scale. We should then abolish the disturbing pendulum and thus greatly simplify the instrument. The readings would be taken by the elevation or depression of the back-leg, until the image of the fixed light was brought to the cross-wire of the observing telescope.

The ease with which the image may be governed with our present arrangements leads us to be hopeful of the proposed plan. The use of the back-leg will, of course, give all the displacements in absolute measure.

The only gearings which it will be necessary to bring outside the room will be those for sensitising and for working the back-leg. The sensitising gearing, when once in order, will not have to be touched again.

The objections to this plan are, that it is necessary to bring one of the supports of the instrument under very slight stresses, and that it will not be possible to take readings at small intervals of time, especially if a more viscous fluid be used*.

Our intention is to proceed with our observations with the present instrument for some time longer, and to note whether the general behaviour of the pendulum has any intimate connection with the meteorological conditions. We intend to observe whether there is a connection between the degree of agitation of the pendulum and the occurrence of magnetic storms. M. Zöllner has thrown out a suggestion for this sort of observation, but we find no notice of his having acted on it†.

We shall also test how far the operation by means of the back-leg may be made to satisfy our expectations.

We have no hope of being able to observe the lunar attraction in the present site of observation, but we think it possible that we may devise a portable instrument, which shall be amply sensitive enough for such a purpose, if the bottom of a deep mine should be found to give a sufficiently invariable support for the instrument.

The reader will understand that it is not easy to do justice to an incomplete apparatus, or to give a very satisfactory account of experiments still in progress; but as it is now two years since the Committee was appointed, we have thought it best to give to the British Association such an account as we can of our progress.

* [Mr Horace Darwin has designed a new form of bifilar pendulum, in which the mirror itself is the bob of the pendulum. Such an instrument, with continuous photographic record, has been used at Birmingham and at Edinburgh. See *Committee on Earth Tremors, B.A. Reports for* 1893 *and* 1894.]

† *Phil. Mag.*, Dec. 1872, p. 497.

14.

THE LUNAR DISTURBANCE OF GRAVITY; VARIATIONS IN THE VERTICAL DUE TO ELASTICITY OF THE EARTH'S SURFACE.

[Second Report of the Committee, consisting of Mr G. H. DARWIN, Professor Sir WILLIAM THOMSON, Professor TAIT, Professor GRANT, Dr SIEMENS, Professor PURSER, Professor G. FORBES, and Mr HORACE DARWIN, appointed for the Measurement of the Lunar Disturbance of Gravity. Written by Mr G. H. DARWIN. *British Association Report for* 1882, pp. 95—119.]

SHORTLY after the meeting of the British Association last year (1881), the instrument with which my brother and I were experimenting at the Cavendish Laboratory, at Cambridge, broke down, through the snapping of the wire which supported the pendulum. A succession of unforeseen circumstances have prevented us, up to the present time, from resuming our experiments.

The body of the present Report, therefore, will merely contain an account of such observations by other observers as have come to our knowledge within the past year, and it must be taken as supplementary to the second part of the Report for 1881. The Appendix, however, contains certain theoretical investigations, which appear to me to throw doubt on the utility of very minute gravitational observations.

The readers of the Report for 1881 will remember that, in the course of our experiments, we were led away from the primary object of the Committee, namely, the measurement of the Lunar Disturbance of Gravity, and found ourselves compelled to investigate the slower oscillations of the soil.

It would be beyond the scope of the present Report to enter on the literature of seismology. But, the slower changes in the vertical having been found to be intimately connected with earthquakes, it would not have been possible, even if desirable, to eliminate all reference to seismology from the present Report.

The papers which are quoted below present evidence of a very miscellaneous character, and therefore this Report must necessarily be rather disjointed. It has seemed best in our account of work done rather to classify together the observers than the subjects. This rule will, however, be occasionally departed from, when it may seem desirable to do so.

The interesting researches in this field made during the last ten years by the Italians, are, I believe, but little known in this country, and as the accounts of their investigations are not easily accessible (there being, for example, no copy of the *Bulletino*, referred to below, at Cambridge), it will be well to give a tolerably full account of the results attained. I have myself only seen the Transactions for four years.

The great extension which these investigations have attained in Italy has been no doubt due to the fact of the presence of active volcanos and of frequent sensible earthquakes in that country. But it is probable that many of the same phenomena occur in all countries.

In 1874 the publication of the *Bulletino del Vulcanismo Italiano* was commenced at Rome under the editorship of Professor S. M. de Rossi, of Rome*. As the title of this publication shows, it is principally occupied with accounts of earthquakes, but the extracts made will refer almost entirely to the slower oscillations of level.

I learn from the *Bulletino* that in 1873 Professor Timoteo Bertelli, of Florence, had published an historical account of small spontaneous movements of the pendulum, observed since the seventeenth century up to that time†.

In 1874 (Anno 1 of the *Bulletino*) Rossi draws attention to the fact that there are periods lasting from a few days to a week or more, in which the soil is in incessant movement, followed by a comparative cessation of such movement. This he calls a 'seismic period.' In the midst or at the end of a seismic period there is frequently a sensible earthquake.

At page 51 he remarks, in a review of some observations of Professor Pietro Monte (Director of the Observatory of Leghorn), that he was led to suspect that the crust of the earth is in continuous and slow movement

* I am compelled to make this abstract from manuscript notes, but my papers having become somewhat disarranged, I am not absolutely certain in one or two places of the year to which the observations refer.

† *Bulletino Boncampagni*, t. VI., Gennaio, 1873. Reprinted Via Lata, No. 2114, Rome.

during the seismic period, and that this movement is influenced by variations of barometric pressure. This suspicion was, he says, confirmed by finding, in his observations of a pendulum at Rocca di Papa (of which we shall speak again below), that during the seismic period the excursions of the pendulum were mostly in the S.W. and N.E. azimuth. This is perpendicular to the volcanic fracture, which runs towards the Alban lake and the sea. The lips of the fracture rise and fall, and there result two sets of waves along and perpendicular to the fracture. In an earthquake these waves are propagated with great velocity (the phenomenon being in fact dynamical), but during the seismic period the same class of changes takes place slowly. This view accords with observations at Velletri made by Professor D. G. Galli.

With regard to the influence of barometric pressure Rossi elsewhere quotes M. Poey (October 15, 1857 ?) as having attributed the deviations of the vertical to this cause, and remarks :—

"Although he (Poey) gave too much weight to the baro-seismic action of large variations of atmospheric pressure, yet after very numerous observations made by me in these last three years (I suppose 1871–4), I can affirm that no marked barometric depression has occurred without having been immediately preceded, accompanied, or followed by marked micro-seismic movements; but besides these there are other irregular, often considerable and instantaneous movements, which occur under high pressure. To distinguish them, I have called the first *baro-seismic*, and the second *vulcano-seismic*, movements." The reader will find a theoretical investigation on this subject in the Appendix to the present Report.

Rossi states (page 118, Anno 1 ?) that whilst Etna was in a condition of activity his pendulums at Rocca di Papa were extraordinarily agitated at the beginning of each barometric storm.

At page 90 of the second year are given graphical illustrations of the simultaneous deflections of pendulums at Rome, Rocca di Papa, Florence, Leghorn, and Bologna. There is some appearance of concordance between them, and this shows that the agitations sometimes affect considerable tracts of land, but that the minor deflections are purely local phenomena.

M. d'Abbadie, in presenting a memoir on micro-seismic movements by Father Bertelli to the French Academy, relates (*Comptes Rendus*, 1875, Vol. 81, p. 297) the following experiment made by Count Malvasia, as proving the independence of the disturbances of the pendulum from the tremors produced by traffic. Two batteries of artillery were marching through Bologna, and it was arranged that at 30 metres from the Palazzo Malvasia they should break into a trot. The pendulum, situated only 6 metres from the street, was observed to be unaffected by this, and continued its oscillations in the E.W. azimuth. A pool of mercury was violently agitated, and it was concluded that the motion communicated to the ground by the artillery was exclusively vertical.

At page 5 of the *Bulletino* for 1876 (January to May), Rossi writes a "Guida pratica per le osservazioni sismiche." This article contains a description of the instruments which have been used by the Italian observers.

Bertelli used a pendulum protected from the air, with a microscope and micrometer for evaluating the oscillations. The upper part of the support of the pendulum consisted of a spiral spring, so that vertical movements of the ground could be recorded. This instrument he calls a tromo-seismometer.

Professor Egidi, of Anagni, proposed to use the reflection from mercury. The object observed was to be a mark fixed on a wall, and the reflected image of the mark was to be observed with a telescope. The deviation of the vertical was to be evaluated by noting the amount of movement required to bring the cross-wires of the telescope on to the mark. This instrument has not, I think, the advantages of M. d'Abbadie's, because the light was incident at about 45° on the mercury, and thus the mark and telescope were remote from one another; whereas in the arrangement of M. d'Abbadie the mark and microscope are close together, and only a micrometer wire in the microscope is movable.

Cavalleri used ten pendulums of graduated length, and found that sometimes one of the pendulums was agitated and sometimes another. Rossi observed the same with his pendulums at Rocca di Papa. It thus appears that the free period of oscillation of the pendulum is a disturbing element.

In order to obviate the discrepancies which must arise in the use of various kinds of pendulums for simultaneous observations in different places, Bertelli and Rossi propose a normal 'tromometer,' of which a drawing is given. The length of the pendulum is 1½ metres, the weight 100 grammes, and it makes forty-nine free oscillations in a minute. To the bottom of the pendulum is attached a horizontal disk, on the underside of which are engraved two fine lines at right-angles to one another. These lines are observed, after total internal reflection in a glass prism placed immediately below the disk, by a horizontal microscope, furnished with a micrometer. The azimuth of the deflection of the vertical is observed by a position-circle.

This paper also contains a description of the author's observatory at Rocca di Papa. It is established in a cave at 700 metres above the sea, on the external slope of the extinct Latian volcano. There is a large central pendulum hanging from the roof, and there are four others with different weights and lengths hanging in tubes cut in the native rock. Only the ends of these pendulums are visible, and they are protected by glass at the visible parts. A great part of this paper is occupied with descriptions of seismometers, and this is outside the scope of the present Report.

In presenting a pamphlet by Father Bertelli, entitled "Riassunto delle osservazioni microsismiche, &c.," to the French Academy (*Comptes Rendus*, 1877, Vol. 84, p. 465), M. d'Abbadie summarises Bertelli's conclusions somewhat as follows:—

The oscillation of the pendulum is generally parallel to valleys or chains of mountains in the neighbourhood. The oscillations are independent of local tremors, velocity and direction of wind, rain, change of temperature and atmospheric electricity.

Pendulums of different lengths betray the movements of the soil in different manners, according to the agreement or disagreement of their free-periods with the period of the terrestrial vibrations.

The disturbances are not strictly simultaneous in the different towns of Italy, but succeed one another at short intervals.

After earthquakes the 'tromometric' or microseismic movements are especially apt to be in a vertical direction. They are always so when the earthquake is local, but the vertical movements are sometimes absent when the shock occurs elsewhere. Sometimes there is no movement at all, even when the shock occurs quite close at hand.

The positions of the sun and moon appear to have some influence on the movements of the pendulum, but the disturbances are especially frequent when the barometer is low.

The curves of 'the monthly means of the tromometric movement' exhibit the same forms in the various towns of Italy, even those which are distant from one another.

The maximum of disturbance occurs near the winter solstice and the minimum near the summer solstice; this agrees with Mallet's results about earthquakes.

At Florence a period of earthquakes is presaged by the magnitude and frequency of pendulous movements in a vertical direction. These movements are observable at intervals and during several hours after each shock.

At page 103 of the first part of the *Bulletino* for 1878?, there is a review of a work by Giulio Grablovitz, "Dell' attrazione luni-solare in relazione coi fenomeni mareo-sismici," Milano, Tipografia degli Ingegneri, 1877.

In this work it appears that M. Grablovitz attributes a considerable part of the deviations of the vertical to bodily tides in the earth, but as he apparently enters into no computations to show the competency of this cause to produce the observed effects, it does not seem necessary to make any further comment on his views.

At page 99 of the volume for September—December, 1878, Rossi writes on the use of the microphone for the purpose of observing earthquakes ("Il microfono nella meteorologia endogena"). He begins by giving an account of a correspondence, beginning in 1875, between himself and Count Giovanni Mocenigo*, of Vicenza, who seems to have been very near to the discovery of the microphone. When the invention of the microphone was announced, Mocenigo and Armellini adopted it for their experiments, and came to the conclusion that the mysterious noises which they heard arose from minute earthquakes or microsisms.

Rossi then determined to undertake observations in his cavern at Rocca di Papa, with a microphone, made of silver instead of carbon, mounted on a stone beam. The sensitiveness of the instrument could be regulated, and he found that it was not much influenced by external noises.

The instrument was placed 20 metres underground, and remote from houses and carriage-roads. It was protected against insects, and was wrapped up in wool. Carpet was spread on the floor of the cave to deaden the noise from particles of stone which might possibly fall. Having established his microphone, he waited till night and then heard noises which he says revealed 'natural telluric phenomena.' The sounds which he heard he describes as 'roarings, explosions occurring isolated or in volleys, and metallic or bell-like sounds' [fremiti, scopii isolati o di moschetteria, e suoni metallici o di campana]. They all occurred mixed indiscriminately, and rose to maxima at irregular intervals. By artificial means he was able to cause noises which he calls 'rumbling (?) or crackling' [rullo o crepito]. The roaring [fremito] was the only noise which he could reproduce artificially, and then only for a moment. It was done by rubbing together the conducting wires, in the same manner as the rocks must rub against one another when there is an earthquake.

A mine having been exploded in a quarry at some distance, the tremors in the earth were audible in the microphone for some seconds subsequently.

There was some degree of coincidence between the agitation of the pendulum-seismograph and the noises heard with the microphone.

At a time when Vesuvius became active, Rocca di Papa was agitated by microsisms, and the shocks were found to be accompanied by the very same microphonic noises as before. The noises sometimes became 'intolerably loud'; on one occasion in the middle of the night, half an hour before a sensible earthquake. The agitation of the microphone corresponded exactly with the activity of Vesuvius.

Rossi then transported his microphone to Palmieri's Vesuvian observatory,

* Count Mocenigo has recently published at Vicenza a book on his observations. It is reviewed in *Nature* for July 6, 1882.

and worked in conjunction with him. He there found that each class of shock had its corresponding noise. The sussultorial shocks, in which I conceive the movement of the ground is vertically up and down, gave the volleys of musketry [i colpi di moschetteria], and the undulatory shocks gave the roarings [i fremiti]. The two classes of noises were sometimes mixed up together.

Rossi makes the following remarks: "On Vesuvius I was put in the way of discovering that the simple fall and rise in the ticking which occurs with the microphone [battito del orologio unito al microfono] (a phenomenon observed by all, and remaining inexplicable to all) is a consequence of the vibration of the ground." This passage alone might perhaps lead one to suppose that clockwork was included in the circuit; but that this was not the case, and that 'ticking' is merely a mode of representing a natural noise, is proved by the fact that he subsequently says that he considers the ticking to be 'a telluric phenomenon.'

Rossi then took the microphone to the Solfatara of Pozzuoli, and here, although no sensible tremors were felt, the noises were so loud as to be heard simultaneously by all the people in the room. The ticking was quite masked by other natural noises. The noises at the Solfatara were imitated by placing the microphone on a vessel of boiling water. Other seismic noises were then imitated by placing the microphone on a marble slab, and scratching and tapping the under surface of it.

The observations on Vesuvius led him to the conclusion that the earthquake oscillations have sometimes fixed nodes and loops, for there were places on the mountain where no effects were observed. Hence, as he remarks, although there may sometimes be considerable agitation in an earthquake, the true centre of disturbance may be very distant.

In conclusion Rossi gives a description of a good method of making a microphone. A common nail has a short piece of copper wire wound round it, and the other end of the wire is wound round a fixed metallic support The nail thus stands at the end of a weak horizontal spring; but the nail is arranged so that it stands inclined to the horizon, instead of being vertical. The point of the nail is then put to rest on the middle of the back of a silver watch, which lies flat on a slab. The two electrodes are the handle of the watch and the metallic support. He says that this is as good as any instrument. The telephone is a seismological instrument, and therefore, strictly speaking, beyond the scope of this Report; but as some details of its use have already been given, I will here quote portions of an interesting letter by Mr John Milne, of the Imperial Engineering College of Tokio, which appeared in *Nature* for June 8, 1882. Mr Milne writes:—

"In order to determine the presence of these earth-tremors, at the end of 1879 I commenced a series of experiments with a variety of apparatus, amongst which were microphones and sets of pendulum apparatus, very similar in general arrangement, but, unfortunately, not in refinement of construction, to the arrangements now being used in the Cavendish Laboratory.

"The microphones were screwed on to the heads of stakes driven in the ground, at the bottom of boxed-in pits. In order to be certain that the records which these microphones gave were not due to local actions, such as birds or insects, two distinct sets of apparatus were used, one being in the middle of the lawn in the front of my house, and the other in a pit at the back of the house. The sensitiveness of these may be learnt from the fact that if a small pebble was dropped on the grass within six feet of the pit, a distinct sound was heard in the telephone, and a swing produced in the needle of the galvanometer placed in connection with these microphones. A person running or walking in the neighbourhood of the pits, had each of his steps so definitely recorded, that a Japanese neighbour, Mr Masato, who assisted me in the experiments, caused the swinging needle of his galvanometer to close an electric circuit and ring a bell, which, it is needless to say, would alarm a household. In the contrivance we have a hint as to how earth-tremors may be employed as thief-detectors.

"The pendulum apparatus, one of which consisted of a 20-lb. bob of lead at the end of 20 feet of pianoforte wire provided with small galvanometer mirrors, and bifilar suspensions were also used in pairs. With this apparatus a motion of the bob relatively to the earth was magnified 1000 times, that is to say, if the spot of light which was reflected from the mirror moved a distance equal to the thickness of a sixpence, this indicated there had been a relative motion of the bob to the extent of 1000th part that amount.

"The great evil which everyone has to contend with in Japan when working with delicate apparatus is the actual earthquakes, which stop or alter the rate of ordinary clocks.

"Another evil which had to be contended with was the wind, which shook the house in which my pendulums were supported, and I imagine the ground by the motion of some neighbouring trees. A shower of rain also was not without its effects upon the microphones. After many months of tiresome observation, and eliminating all motions which by any possibility have been produced by local influence, the general result obtained was that there were movements to be detected every day and sometimes many times per day....

"A great assistance to the interpretation of the various records which an earthquake gives us on our seismographs is what I may call a barricade of post-cards. At the present moment Yedo is barricaded, all the towns around for a distance of 100 miles being provided with post-cards. Everyone of them is posted with a statement of the shocks which have been felt.

"For the months of October and November it was found from the records of the post-cards that nearly all the shocks came from the north and passed Yedo to the south-west. When coming in contact with a high range of mountains, they were suddenly stopped, as was inferred from the fact that the towns beyond this range did not perceive that an earthquake had occurred. This fact having been obtained, the barricade of post-cards has been extended to towns lying still farther north. The result of this has been that several earthquake origins have, so to speak, been surrounded or corralled, whilst others have been traced as far as the seashore. For the latter shocks, earthquake hunting with post-cards has had to cease, and we have solely to rely upon our instruments. Having obtained our earthquake centres, at one or more of these our tremor instruments might be erected, and it would soon be known whether an observation of earth-tremors would tell us about the coming of an earthquake as the cracklings of a bending do about its approaching breakage. To render these experiments more complete, and to determine the existence of a terrain tide, a gravitimeter might be established. I mention this because if terrain tides exist, and they are sufficiently great from a geological point of view, it would seem that they might be more pronounced and therefore easier to measure in a country like Japan, resting in a heated and perhaps plastic bed, than in a country like England, where volcanic activity has so long ceased, and the rocks are, comparatively speaking, cold and rigid, if an instrument, sufficiently delicate to detect differences in the force of gravity, in consequence of our being lifted farther from the centre of the earth every time by the terrain tide as it passed between (*sic*) our feet, could be established in conjunction with the experiments on earth-tremors."

The only account which I have been able to find of M. Bouquet de la Grye's observations (mentioned in the last report) is contained in the *Comptes Rendus* for March 22, 1875, page 725. M. Bouquet writes:—

"... The observation of the levels of our meridian telescopes put us on the track of a curious fact. Not only is Campbell Island subject to earthquakes, but it also exhibits movements when the great swell falls in breakers on the coast. I thought that it would be interesting to study this new phenomenon. The instrument, which was quickly put together, consisted of a steel wire supporting a weight, to which was soldered a needle; the movements of the weight were amplified 240 times by means of a lever; by passing an electric current through this multiplying pendulum, which was terminated at the bottom by a small cup of amalgamated tin, regular oscillations of $\frac{1}{1000}$th of a mm. could be registered. I propose to repeat these observations with a pendulum of much larger amplifying power, so as to try to register the variations of the plumb-line."

In a letter to me, M. d'Abbadie mentions an attempt by Brunner to improve M. Bouquet de la Grye's apparatus, but considers that the attempt was a failure.

He also tells me that Delaunay directed M. Wolf to devise an apparatus for detecting small deviations of the vertical, and that the latter, without M. d'Abbadie's knowledge, adopted his rejected idea of a pendulum, about 30 metres long, bearing a prism at the end by reflection from which a scale was to be read by means of a distant small refractor. The pendulum was actually set up, but the wire went on twisting and untwisting until Delaunay's death, and no observations were made with it. Our own experience is enough to show that nothing could have been made of such an instrument.

M. d'Abbadie gives further explanations of a passage in his own paper about the arrangement of the staircases for access to and observation with his Nadirane. In writing the Report of 1881 I had found the description of the arrangements difficult to understand.

The woodcut below is a copy of the rough diagram that he sent me.

There were three staircases:—

T cut in the rock; CB to ascend from the cellar-flags CD; and, lastly, AS to mount from the boarded ground floor, AB, to the small floor SN, which was hung from the roof. The two upper staircases did not touch the truncated cone of concrete anywhere.

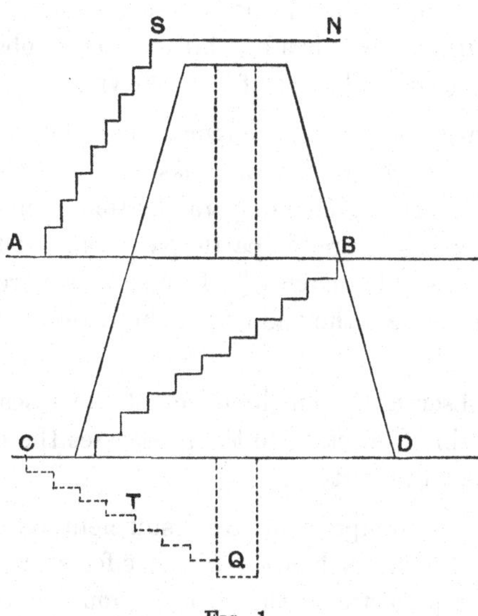

Fig. 1.

Judging from this figure, I imagine that the concrete cone has an external slope of ten in one; the French expression was 'une inclinaison d'une dixième.'

M. d'Abbadie informs me that the apparently curious phenomenon of the 'ombres fuyantes,' which were observed in the reflection from the pool of mercury, to which we drew attention last year, was of no significance. It arose from the currents of air caused by a candle left standing on the staircase T cut in the rock. The light was required for pouring out the mercury, and it was left burning whilst the observation was being taken; but now that this operation is done entirely from above, the phenomenon has disappeared.

In a paper entitled "Recherches sur la Verticale" (*Ann. de la Soc. Scient. de Bruxelles*, 1881), M. d'Abbadie continues the account of his observations with his instrument, called by him a Nadirane. It was described in the last year's Report, and some further details have been given above. A portion of this paper refers to his old observations, and gives further important details as to the exact method of making observations, and of various modifications which have been introduced.

Each complete observation consists of the following processes:—measurement of the distance between the cross-wires and their image, (1) in the meridian, (2) in the prime vertical, (3) in the N.W. azimuth, (4) observation of barometer, (5) of thermometer, (6) of direction and force of the wind, (7) condition and movement of the image estimated with the micrometer, (8) condition of the heavens, (9) of the breakers called 'les Criquets,' which can be observed from the neighbouring room.

This last is to determine whether it is possible to have a rough sea with a calm image; a condition which has not hitherto been observed. This statement seems somewhat contradictory of the following:—

"Aucunes des variations dans les circonstances concomitantes n'a paru se rattacher à l'état de l'image qui, pendant des journées entières, paraît tantôt belle, tantôt faible, et parfois même disparaît entièrement, bien que ce dernier inconvénient ait été évité en grande partie par l'usage d'un récipient en bois à fond rainé pour contenir le mercure." I presume we are to understand that the roughness of the sea and the badness of the image is the only congruence hitherto observed.

M. d'Abbadie's observations on the effect of the tides will be referred to in the Appendix to this Report. He then discusses the various causes which may perhaps influence the vertical.

The variations of air temperature are insufficient, because the vertical has been seen to vary $2''\cdot4$ in six hours. If the effects are to be attributed to variations in the temperature of the rock, it would be necessary to suppose that that temperature varies discontinuously, which it is difficult to admit.

If it be supposed that the changes take place in the instrument itself, the like must be true of astronomical instruments. And there is no reason to admit the reality of such strange variations.

Another cause, more convenient because more vague, is variation of a chemical or mechanical nature in the crust of the earth. But if this be so, why does the vertical ever return to its primitive position? Another cause may be variation in the position of the earth's axis of rotation.

The azimuthal variations in astronomical instruments, referred to by M. d'Abbadie (see a paper by Mr Henry, Vol. VIII. p. 134, *Month. Not. R.A.S.*), are difficult to explain without having recourse to such variation in the axis of rotation.

He also tells us that Ellis (Vol. XXIX. 1861, p. 45, *Mem. R.A.S.*) has discussed the Greenwich observations from 1851 to 1858. A comparison of the results obtained from two neighbouring meridian instruments seemed to show that the azimuthal variations are partly purely instrumental.

M. d'Abbadie's paper contains diagrams illustrating the variations of the vertical observed with the Nadirane during nearly two years. He sums up the results as follows:—

"En résumé le maximum d'écart du sud au nord entre le fil et son image a été égal à 49‵‵·2 (this is 15″·94; it seems as though this should be twice the deviation of the vertical) le 30 Novembre à 8 h. 43 m. du matin. Ce même jour, à 7 h. 28 m., on a lu 40‵‵·1, chiffre porté ici au tableau, et 37‵‵·6 seulement à 1 h. 32 m. du soir. Dans l'espace de six heures la verticale a donc varié de 2‵‵·5 ou 0″·81 (as this is the deviation of the image, should not the deviation of the vertical be half as much?). Le minimum de l'année, ou 3·06, fut atteint le 19 Janvier à 3 h. 3 m. du matin, ainsi que le 21 du même mois à midi, bien qu'on eût observé 3·44 et 3·30 dans les matinées de ces deux jours, ainsi qu'on le voit au tableau ci-après .... Pendant l'année entière la verticale, considérée selon le plan du méridien, a donc varié d'un angle de 12‵‵·45 ou 4″·034 .... On aura .... 8‵‵·3 ou 2″·7 pour la plus grande variation dans le sens Est-Ouest où l'on nivelle les tourillons des lunettes méridiennes."

Towards the end M. d'Abbadie makes the excellent remark, that in discussing latitudes and declinations of stars, account should be taken of the instantaneous position of the vertical at the moment of taking the observation.

In the *Archives des Sciences*, 1881, Vol. V. p. 97, M. P. Plantamour continues the account of his observations on oscillations of the soil at Sécheron, near Geneva. The account of the earlier observations, which we quoted from the *Comptes Rendus* in our previous Report, is also contained in Vol. II. of the *Archives*, p. 641. The paper to which we are now referring contains a graphical reproduction of the previous series of observations, as far as concerns the daily means.

The new series extends from October 1, 1879, to December 31, 1880, the disposition of the levels being the same as was described in our last Report. The observations were taken at 9 A.M. and 6 P.M., which hours are respectively a little before the diurnal minimum and maximum. The meanings of the terms maximum and minimum were somewhat obscure in the *Comptes Rendus*, but I now find that the right interpretation was placed on M. Plantamour's words, for maximum means for the two levels E. end highest and S. end highest.

The N.S. level seems to have behaved very similarly in the two years of observation; the total annual amplitudes in the two years being 4″·89 and 4″·56 respectively. In both years this level followed, with some retardation, the curve of external temperature, except between April and October, when the curves appear to be inverted. The E.W. level behaved very differently in the two years. In 1879 the E. end began to fall rapidly at the end of November, and continued to fall until December 26, when the reading was – 88″·71; it rose a little early in January and then fell again, so that on January 28, 1880, the reading was – 89″·95. The amplitude of the total fall (viz. from October 4, 1879, to January 28, 1880) was 95″·80. In the preceding year the amplitude was only 28″·08. The E. end has never recovered its primitive position, and remains nearly 80″ below its point of departure.

It is difficult to believe that so enormous a variation of level is normal, and one is tempted to suspect that there is some systematic error in his mode of observation. If such oscillations as these were to take place in an astronomical observatory, accurate astronomical observations would be almost impossible.

I have seen nothing which shows that M. Plantamour takes any special precaution with regard to the weight of the observer's body, nor is it expressly stated that the observer always stands in exactly the same position, although, of course, it is probable that this is the case. It would be interesting, also, to learn whether any precautions have been taken for equalising the temperature of the level itself. To hold the hand in the neighbourhood of a delicate level is sufficient to quite alter the reading. In one of his letters to me M. d'Abbadie also remarks on the slow molecular changes in glass, which render levels untrustworthy for comparisons at considerable intervals of time. Although we must admire M. Plantamour's indomitable perseverance, it is to be regretted that his mode of observation is by means of levels; and we are compelled to regard, at least provisionally, these enormous changes of level either as a local phenomenon, or as due to systematic error in his mode of observation.

In the Report for 1881 we referred to some observations by Admiral Mouchez, made in 1856, on changes of level. A short paper by Admiral Mouchez on these observations will be found in the *Comptes Rendus* for 1878, Vol. 87, p. 665. I now find that the observations were, in fact, discussed by M. Gaillot, in a paper entitled "Sur la direction de la verticale à l'observatoire de Paris," at p. 684 of the same volume. The paper consists of the examination of 1077 determinations of latitude, made between 1856 and 1861, with the Gambey circle.

M. Gaillot concludes that the variation from year to year is accidental, and that the variation of latitude in the course of the year is represented by

$$\delta\lambda = +0''{\cdot}20 \sin\left[\frac{360^\circ (t-95)}{365{\cdot}25}\right]$$

where t is the number of days since January 1.

By a comparison of day and night observations he concludes that there is no trace of a diurnal variation. On this we may remark that, if the maximum and minimum occur at 6 P.M. and 6 A.M. (which is, roughly speaking, what we found to be the case), then the diurnal oscillation must necessarily disappear by this method of treatment.

Individual observations ranged from $2''{\cdot}48$ above to $3''{\cdot}17$ below the mean. On this he remarks:—

"Ceux qui savent combien l'observation du nadir présente parfois de difficulté dans un observatoire situé au milieu d'une grande ville, ... ceux-là ne trouveront pas ces écarts exagérés, et ne croiront nullement avoir besoin de faire intervenir une déviation de la verticale pour les expliquer."

M. Gaillot concludes by remarks adverse to any sensible deviations of the vertical.

It seems to me, however, that in the passage about the influence of the traffic of a great town, M. Gaillot begs the whole question by setting down to that disturbing influence all remarkable deviations of the vertical. Our observations, and those of many others, are entirely adverse to such a conclusion.

M. d'Abbadie, in a letter to me, also expresses himself as to the inconclusiveness of M. Gaillot's discussion.

He also tells me that M. Tisserand, in his observations of latitude in Japan, found variations amounting to nearly 7″; and when asked "How he could be so much in error," answered "That he was sure of his observations and calculations, but could not explain the cause of such variations."

The following further references may perhaps be useful:—Maxwell's paper on the 306-day inequality in the earth's rotation, which was mentioned in the Report of last year, is in the *Trans. Roy. Soc. of Edinburgh*, 1857, Vol. XXI. pp. 559–70. See also Bessel's *Abhandlungen*, Vol. II. p. 42, Vol. III. p. 304. In *Nature* for January 12, 1882 (p. 250), there is an account of the work of the Swiss Seismological Commission. The original sources appear to be a text-book on Seismology by Professor Heim, of Bern, the *Annuaire* of the Physical Society of Bern, and the *Archives des Sciences* of Geneva. I learn from M. d'Abbadie that Colonel Orff has been making systematic observations twice a day with levels at the Observatory at Munich, and that Colonel Goulier has been doing the same at Paris, with levels filled with bisulphide of carbon.

APPENDIX.

On Variations in the Vertical due to Elasticity of the Earth's Surface.

By G. H. Darwin.

1. *On the Mechanical Effects of Barometric Pressure on the Earth's Surface.*

The remarks of Signore de Rossi, on the observed connection between barometric storms and the disturbance of the vertical, have led me to make the following investigation of the mechanical effects which are caused by variations of pressure acting on an elastic surface. The results seem to show that the direct measurement of the lunar disturbance of gravity must for ever remain impossible*.

The practical question is to estimate the amount of distortion to which the upper strata of the earth's mass are subjected, when a wave of barometric depression or elevation passes over the surface. The solution of the following problem should give us such an estimate.

Let an elastic solid be infinite in one direction, and be bounded in the other direction by an infinite plane. Let the surface of the plane be everywhere acted on by normal pressures and tractions, which are expressible as a simple harmonic function of distances measured in some fixed direction along the plane. It is required to find the form assumed by the surface, and generally the condition of internal strain.

This is clearly equivalent to the problem of finding the distortion of the earth's surface produced by parallel undulations of barometric elevation and depression. It is but a slight objection to the correctness of a rough estimate of the kind required, that barometric disturbances do not actually occur in parallel bands, but rather in circles. And when we consider the magnitude of actual terrestrial storms, it is obvious that the curvature of the earth's surface may be safely neglected.

This problem is mathematically identical with that of finding the state of stress produced in the earth by the weight of a series of parallel mountains. The solution of this problem has recently been published in a paper by me in the *Philosophical Transactions* (Part II. 1882, pp. 187—230†), and the solution there found may be adapted to the present case in a few lines.

The problem only involves two dimensions. If the origin be taken in the mean horizontal surface, which equally divides the mountains and valleys, and if the axis of z be horizontal and perpendicular to the mountain chains,

* [This prevision has now been falsified; see p. 346.]

† [To be reproduced in Vol. II. of these collected papers.]

and if the axis of x be drawn vertically downwards, then the equation to the mountains and valleys is supposed to be

$$x = -h\cos\frac{z}{b}$$

so that the wave-length from crest to crest of the mountain ranges is $2\pi b$.

The solution may easily be found from the analysis of section 7 of the paper referred to. It is as follows:—

Let α, γ be the displacements at the point x, z vertically downwards and horizontally (α has here the opposite sign to the α of (44)). Let w be the density of the rocks of which the mountains are composed; g gravity; υ modulus of rigidity, then

$$\left.\begin{aligned} \alpha &= \frac{1}{2\upsilon} b \left[x\frac{dW}{dx} - W\right] \\ \gamma &= \frac{1}{2\upsilon} bx\frac{dW}{dz} \\ \text{where}\qquad W &= -gwh e^{-x/b}\cos\frac{z}{b} \end{aligned}\right\} \quad\text{.........................(1)}$$

From these we have at once

$$\left.\begin{aligned} \alpha &= \frac{gwh}{2\upsilon} b\left(1+\frac{x}{b}\right)e^{-x/b}\cos\frac{z}{b} \\ \gamma &= \frac{gwh}{2\upsilon} xe^{-x/b}\sin\frac{z}{b} \\ \frac{d\alpha}{dz} &= -\frac{gwh}{2\upsilon}\left(1+\frac{x}{b}\right)e^{-x/b}\sin\frac{z}{b} \end{aligned}\right\} \quad\text{....................(2)*}$$

The first of these gives the vertical displacement, the second the horizontal, and the third the inclination to the horizon of strata primitively plane.

$$\left.\begin{aligned} \text{At the surface}\qquad \alpha &= \frac{gwh}{2\upsilon} b\cos\frac{z}{b}, \quad \gamma = 0 \\ \frac{d\alpha}{dz} &= -\frac{gwh}{2\upsilon}\sin\frac{z}{b} \end{aligned}\right\} \quad\text{.......................(3)}$$

* It is easy to verify that these values of α and γ, together with the value $p = gwh\,e^{-x/b}\cos z/b$ for the hydrostatic pressure, satisfy all the conditions of the problem, by giving normal pressure $gwh\cos z/b$ at the free surface of the infinite plane, and satisfying the equations of internal equilibrium throughout the solid. I take this opportunity of remarking that the paper from which this investigation is taken contains an error, inasmuch as the hydrostatic pressure is erroneously determined in section 1. The term $-W$ should be added to the pressure as determined in (3). This adds W to the normal stresses P, Q, R throughout the paper, but leaves the difference of stresses (which was the thing to be determined) unaffected. If the reader should compare the stresses, as determined from the values of α, γ in the text above, and from the value of p given in this note, with (38) of the paper referred to, he is warned to remember the missing term W.

[The mistake, referred to, will be corrected in the reproduction of the paper.]

Hence the maximum vertical displacement of the surface is $\pm gwhb/2v$, and the maximum inclination of the surface to the horizon is

$$\pm \operatorname{cosec} 1'' \times gwh/2v \text{ seconds of arc}$$

Before proceeding further I shall prove a very remarkable relation between the slope of the surface of an elastic horizontal plane and the deflection of the plumb-line caused by the direct attraction of the weight producing that slope. This relation was pointed out to me by Sir William Thomson, when I told him of the investigation on which I was engaged; but I am alone responsible for the proof as here given. He writes that he finds that it is not confined simply to the case where the solid is incompressible, but in this paper it will only be proved for that case.

Let there be positive and negative matter distributed over the horizontal plane according to the law $wh \cos(z/b)$; this forms, in fact, harmonic mountains and valleys on the infinite plane. We require to find the potential and attraction of such a distribution of matter.

Now the potential of an infinite straight line, of line-density ρ, at a point distant d from it, is well-known to be $-2\mu\rho \log d$, where μ is the attraction between unit masses at unit distance apart. Hence the potential V of the supposed distribution of matter at the point x, z, is given by

$$V = -2\mu wh \int_{-\infty}^{+\infty} \cos\frac{\zeta}{b} \log \sqrt{\{x^2 + (\zeta - z)^2\}}\, d\zeta$$

$$= -\mu whb \left\{\left[\sin\frac{\zeta}{b} \log\{x^2 + (\zeta - z)^2\}\right]_{-\infty}^{+\infty} - 2\int_{-\infty}^{+\infty} \frac{(\zeta - z)\sin(\zeta/b)}{x^2 + (\zeta - z)^2}\, d\zeta\right\}$$

It is not hard to show that the first term vanishes when taken between the limits.

Now put $t = \dfrac{\zeta - z}{x}$, so that $\sin\dfrac{\zeta}{b} = \sin\dfrac{tx}{b}\cos\dfrac{z}{b} + \cos\dfrac{tx}{b}\sin\dfrac{z}{b}$, and we have

$$V = 2\mu whb \int_{-\infty}^{+\infty} \left(\sin\frac{tx}{b}\cos\frac{z}{b} + \cos\frac{tx}{b}\sin\frac{z}{b}\right)\frac{t\,dt}{1+t^2}$$

But it is known* that

$$\int_{-\infty}^{+\infty} \frac{t \sin ct\, dt}{1+t^2} = \pi e^{-c}, \quad \int_{-\infty}^{+\infty} \frac{t\cos ct}{1+t^2}\, dt = 0$$

Therefore $$V = 2\pi\mu whbe^{-x/b}\cos\frac{z}{b}$$

If g be gravity, a earth's radius, and δ earth's mean density, $2\pi\mu = \dfrac{3g}{2a\delta}$.

And

$$V = \frac{3gwh}{2a\delta} be^{-x/b}\cos\frac{z}{b} \quad \text{.........................(4)}$$

* See Todhunter's *Int. Calc.*; Chapter on "Definite Integrals."

The deflection of the plumb-line at any point on the surface denoted by $x=0$, and z, is clearly dV/gdz, when $x=0$. Therefore,

$$\text{the deflection} = -\frac{1}{g} \times \frac{3gwh}{2a\delta} \sin\frac{z}{b} \text{..............................} (5)$$

But from (2) the slope $\left(\text{or } \frac{d\alpha}{dz}, \text{ when } z \text{ is zero}\right)$, is $-\frac{gwh}{2v}\sin\frac{z}{b}$.

Therefore deflection bears to slope the same ratio as v/g to $\frac{1}{3}a\delta$. This ratio is independent of the wave-length $2\pi b$ of the undulating surface, of the position of the origin, and of the azimuth in the plane of the line normal to the ridges and valleys. Therefore the proposition is true of any combination whatever of harmonic undulations, and as any inequality may be built up of harmonic undulations, it is generally true of inequalities of any shape whatever.

Now $a = 6{\cdot}37 \times 10^8$ cm., $\delta = 5\frac{2}{3}$; and $\frac{1}{3}a\delta = 12{\cdot}03 \times 10^8$ grammes per square centimetre. The rigidity of glass in gravitation units ranges from $1{\cdot}5 \times 10^8$ to $2{\cdot}4 \times 10^8$. Therefore the slope of a very thick slab of the rigidity of glass, due to a weight placed on its surface, ranges from 8 to 5 times as much as the deflection of the plumb-line due to the attraction of that weight. Even with rigidity as great as steel (viz., about 8×10^8), the slope is $1\frac{1}{2}$ times as great as the deflection.

A practical conclusion from this is that in observations with an artificial horizon the disturbance due to the weight of the observer's body is very far greater than that due to the attraction of his mass. This is in perfect accordance with the observations made by my brother and me with our pendulum in 1881, when we concluded that the warping of the soil by our weight when standing in the observing room was a very serious disturbance, whilst we were unable to assert positively that the attraction of weights placed near the pendulum was perceptible. It also gives emphasis to the criticisms we have made on M. Plantamour's observations—namely, that he does not appear to take special precautions against the disturbance due to the weight of the observer's body.

We must now consider the probable numerical values of the quantities involved in the barometric problem, and the mode of transition from the problem of the mountains to that of barometric inequalities.

The modulus of rigidity in gravitation units (say grammes weight per square centimetre) is v/g. In the problem of the mountains, wh is the mass of a column of rock of one square centimetre in section and of length equal to the height of the crests of the mountains above the mean horizontal plane. In the barometric problem, wh must be taken as the mass of a column of mercury of a square centimetre in section and equal in height to a half of the maximum range of the barometer.

This maximum range is, I believe, nearly two inches, or, let us say, 5 cm.

The specific gravity of mercury is 13·6, and therefore $wh = 34$ grammes.

The rigidity of glass is from 150 to 240 million grammes per square centimetre; that of copper 540, and of steel 843 millions.

I will take $v/g = 3 \times 10^8$, so that the superficial layers of the earth are assumed to be more rigid than the most rigid glass. It will be easy to adjust the results afterwards to any other assumed rigidity.

With these data we have $\dfrac{gwh}{2v} = \dfrac{5{\cdot}67}{10^8}$; also $\dfrac{648{,}000}{\pi} \times \dfrac{5{\cdot}67}{10^8} = 0''{\cdot}0117$.

It seems not unreasonable to suppose that 1500 miles ($2{\cdot}4 \times 10^8$ cm.) is the distance from the place where the barometer is high (the centre of the anti-cyclone) to that where it is low (the centre of the cyclone). Accordingly the wave-length of the barometric undulation is $4{\cdot}8 \times 10^8$ cm., and $b = 4{\cdot}8 \times 10^8 \div 6{\cdot}28$ cm., or, say, $b = {\cdot}8 \times 10^8$ cm.

Thus, with these data, $\dfrac{gwh}{2v} b = 4{\cdot}5$ cm.

We thus see that the ground is 9 cm. higher under the barometric depression than under the elevation.

If the sea had time to attain its equilibrium slope, it would stand $5 \times 13{\cdot}6$, or 68 cm. lower under the high pressure than under the low. But as the land is itself depressed 9 cm., the sea would apparently only be depressed 59 cm. under the high barometer.

It is probable that, in reality, the larger barometric inequalities do not linger quite long enough over particular areas to permit the sea to attain everywhere its due slope, and therefore the full difference of water-level can only be attained occasionally.

On the other hand, the elastic compression of the ground must take place without any sensible delay. Thus it seems probable that the elastic compression of the ground must exercise a very sensible effect in modifying the apparent depression or elevation of the sea under high and low barometer.

It does not appear absolutely chimerical that, at some future time when both tidal and barometric observations have attained to great accuracy, an estimate might thus be made of the average modulus of rigidity of the upper 500 miles of the earth's mass.

Even in the present condition of barometric and tidal information, it might be interesting to make a comparison between the computed height of tide and the observed height, in connection with the distribution of barometric pressure. It is probable that India would be the best field for such an attempt, because the knowledge of Indian tides is more complete than that for any other part of the world. On the other hand we shall see in the following section that tidal observations on coast-lines of continents are liable to disturbance, so that an oceanic island would be a more favourable site.

It has already been shown that the maximum apparent deflection of the plumb-line, consequent on the elastic compression of the earth, amounts to 0″·0117, and this is augmented to 0″·0146, when we include the true deflection due to the attraction of the air. It is worthy of remark that this result is independent of the wave-length of the barometric inequality, and thus we get rid of one of the conjectural data.

Thus if we consider the two cases of high pressure to right and low to left, and of low pressure to right and high to left, we see that there will be a difference in the position of the plumb-line relatively to the earth's surface of 0″·0292. Even if the rigidity of the upper strata of the earth were as great as that of steel, there would still be a change of 0″·011.

A deflection of magnitude such as 0″·03 or 0″·01 would have been easily observable with our instrument of last year, for we concluded that a change of $\frac{1}{200}$th of a second could be detected, when the change occurred rapidly.

It was stated in our previous Report that at Cambridge the calculated amplitude of oscillation of the plumb-line due directly to lunar disturbance of gravity amounts to 0″·0216. Now as this is less than the amplitude due jointly to elastic compression and attraction, with the assumed rigidity (300 millions) of the earth's strata, and only twice the result if the rigidity be as great as that of steel, it follows almost certainly that from this cause alone the measurement of the lunar disturbance of gravity must be impossible with any instrument on the earth's surface.

Moreover the removal of the instrument to the bottom of the deepest known mine would scarcely sensibly affect the result, because the flexure of the strata at a depth so small, compared with the wave-length of barometric inequalities, is scarcely different from the flexure of the surface.

The diurnal and periodic oscillations of the vertical observed by us were many times as great as those which have just been computed, and therefore it must not be supposed that more than a fraction, say perhaps a tenth, of those oscillations was due to elastic compression of the earth.

The Italian observers could scarcely, with their instruments, detect deflections amounting to $\frac{1}{100}$th of a second, so that the observed connection between barometric oscillation and seismic disturbance must be of a different kind.

It is not surprising that in a volcanic region the equalisation of pressure, between imprisoned fluids and the external atmosphere, should lead to earthquakes.

If there is any place on the earth's surface free from seismic forces, it might be possible (if the effect of tides as computed in the following section could be eliminated) with some such instrument as ours, placed in a deep mine, to detect the existence of barometric disturbance many hundreds of miles away. It would of course for this purpose be necessary to note the positions of the sun and moon at the times of observation, and to allow for their attraction.

2. *On the Disturbance of the Vertical near the Coasts of Continents due to the Rise and Fall of the Tide.*

Consider the following problem:—

On an infinite horizontal plane, which bounds in one direction an infinite incompressible elastic solid, let there be drawn a series of parallel straight lines, distant l apart. Let one of these be the axis of y, let the axis of z be drawn in the plane perpendicular to the parallel lines, and let the axis of x be drawn vertically downwards through the solid.

At every point of the surface of the solid, from $z=0$ to l, let a normal pressure $gwh(1-2z/l)$ be applied; and from $z=0$ to $-l$ let the surface be free from forces. Let the same distribution of force be repeated over all the pairs of strips into which the surface is divided by the system of parallel straight lines. It is required to determine the strains caused by these forces.

Taking the average over the whole surface there is neither pressure nor traction, since the total traction on the half-strips subject to traction is equal to the total pressure on the half-strips subject to pressure.

The following is the analogy of this system with that which we wish to discuss: the strips subject to no pressure are the continents, the alternate ones are the oceans, g is gravity, w the density of water, and h the height of tide above mean water on the coast-line.

We require to find the slope of the surface at every point, and the vertical displacement.

It is now necessary to bring this problem within the range of the results used in the last section. In the first place, it is convenient to consider the pressures and tractions as caused by mountains and valleys whose outline is given by $x=-h(1-2z/l)$ from $z=0$ to l, and $x=0$ from $z=0$ to $-l$. To utilise the analysis of the last section, it is necessary that the mountains and valleys should present a simple-harmonic outline. Hence the discontinuous function must be expanded by Fourier's method. Known results of that method render it unnecessary to have recourse to the theorem itself. It is known that

$$\pm\tfrac{1}{2}\pi-\tfrac{1}{2}\theta=\sin\theta+\tfrac{1}{2}\sin 2\theta+\tfrac{1}{3}\sin 3\theta+\dots$$

$$-\tfrac{1}{2}\theta=-\sin\theta+\tfrac{1}{2}\sin 2\theta-\tfrac{1}{3}\sin 3\theta+\dots$$

$$\tfrac{1}{2}\pi\mp\theta=\frac{4}{\pi}\left\{\cos\theta+\frac{1}{3^2}\cos 3\theta+\frac{1}{5^2}\cos 5\theta+\dots\right\}$$

The upper sign being taken for values of θ between the infinitely small positive and $+\pi$, and the lower for values between the infinitely small negative and $-\pi$.

Adding these three series together we have

$$2\{\tfrac{1}{2}\sin 2\theta + \tfrac{1}{4}\sin 4\theta + \ldots\} + \frac{4}{\pi}\left\{\cos\theta + \frac{1}{3^2}\cos 3\theta + \frac{1}{5^2}\cos 5\theta + \ldots\right\}$$

equal to $\pi - 2\theta$ from $\theta = 0$ to $+\pi$, and equal to zero from $\theta = 0$ to $-\pi$. Hence the required expansion of the discontinuous function is

$$-\frac{2h}{\pi}\{\tfrac{1}{2}\sin 2\theta + \tfrac{1}{4}\sin 4\theta + \ldots\}$$
$$-\frac{4h}{\pi^2}\left\{\cos\theta + \frac{1}{3^2}\cos 3\theta + \frac{1}{5^2}\cos 5\theta + \ldots\right\} \quad \ldots\ldots\ldots\ldots\ldots(6)$$

where $$\theta = \frac{\pi z}{l} \quad \ldots\ldots\ldots\ldots\ldots\ldots\ldots\ldots\ldots\ldots\ldots(7)$$

For it vanishes from $z = -l$ to 0, and is equal to $-h(1 - 2z/l)$ from $z = 0$ to $+l$.

Now looking back to the analysis of the preceding section we see that if the equation to the mountains and valleys had been $x = -h\sin(z/b)$, α would have had the same form as in (2) but of course with sine for cosine, and γ would have changed its sign and a cosine would have stood for the sine. Applying then the solution (2) to each term of our expansion separately, and only writing down the solution for the surface at which $x = 0$, we have at once that $\gamma = 0$, and

$$\alpha = \frac{gwh}{\pi v}\frac{l}{\pi}\left\{\frac{1}{2^2}\sin 2\theta + \frac{1}{4^2}\sin 4\theta + \frac{1}{6^2}\sin 6\theta + \ldots\right\}$$
$$+\frac{gwh}{\pi v}\cdot\frac{2l}{\pi^2}\left\{\cos\theta + \frac{1}{3^3}\cos 3\theta + \frac{1}{5^3}\cos 5\theta + \ldots\right\}\ldots\ldots\ldots\ldots\ldots(8)$$

The slope of the surface is $\frac{d\alpha}{dz}$ or $\frac{\pi}{l}\frac{d\alpha}{d\theta}$; thus

$$\frac{d\alpha}{dz} = \frac{gwh}{\pi v}\{\tfrac{1}{2}\cos 2\theta + \tfrac{1}{4}\cos 4\theta + \tfrac{1}{6}\cos 6\theta + \ldots\}$$
$$-\frac{gwh}{\pi v}\cdot\frac{2}{\pi}\left\{\sin\theta + \frac{1}{3^2}\sin 3\theta + \frac{1}{5^2}\sin 5\theta + \ldots\right\} \quad \ldots\ldots\ldots\ldots(9)$$

The formulæ (8) and (9) are the required expressions for the vertical depression of the surface and for the slope.

It is interesting to determine the form of surface denoted by these equations. Let us suppose then that the units are so chosen that $gwhl/\pi^2 v$ may be equal to one. Then (8) becomes

$$\alpha = \frac{1}{2^2}\sin 2\theta + \frac{1}{4^2}\sin 4\theta + \ldots + \frac{2}{\pi}\left\{\frac{1}{1^3}\cos\theta + \frac{1}{3^3}\cos 3\theta + \ldots\right\} \ldots(10)$$

$$\frac{d\alpha}{d\theta}=\tfrac{1}{2}\cos 2\theta+\tfrac{1}{4}\cos 4\theta+\ldots-\frac{2}{\pi}\left\{\frac{1}{1^2}\sin\theta+\frac{1}{3^2}\sin 3\theta+\ldots\right\}\ldots(11)$$

When θ is zero or $\pm\pi$, $d\alpha/d\theta$ becomes infinite, which denotes that the tangent to the warped horizontal surface is vertical at these points. The verticality of these tangents will have no place in reality, because actual shores shelve, and there is not a vertical wall of water when the tide rises, as is supposed to be the case in the ideal problem. We shall, however, see that in practical numerical application, the strip of sea-shore along which the solution shows a slope of more than 1″ is only a small fraction of a millimetre. Thus this departure from reality is of no importance whatever.

When $\theta=0$ or $\pm\pi$,

$$\alpha=\frac{2}{\pi}\left\{\frac{1}{1^3}+\frac{1}{3^3}+\frac{1}{5^3}+\ldots\right\}=\frac{2}{\pi}\times 1{\cdot}052={\cdot}670$$

being + when $\theta=0$, and − when $\theta=\pm\pi$.

When $\theta=\pm\tfrac{1}{2}\pi$, α vanishes, and therefore midway in the ocean and on the land there are nodal lines, which always remain in the undisturbed surface, when the tide rises and falls. At these nodal lines, defined by $\theta=\pm\tfrac{1}{2}\pi$,

$$\frac{d\alpha}{d\theta}=-\tfrac{1}{2}\log_e 2\mp\frac{2}{\pi}\left\{\frac{1}{1^3}-\frac{1}{3^3}+\frac{1}{5^3}-\ldots\right\}$$

$$=-{\cdot}3466\mp{\cdot}6168=-{\cdot}9634 \text{ and } +{\cdot}2702$$

Thus the slope is greater at mid-ocean than at mid-land. By assuming θ successively as $\tfrac{1}{6}\pi$, $\tfrac{1}{4}\pi$, $\tfrac{1}{3}\pi$, and summing arithmetically the strange series which arise, we can, on paying attention to the manner in which the signs of the series occur, obtain the values of α corresponding to 0, $\pm\tfrac{1}{6}\pi$, $\pm\tfrac{1}{4}\pi$, $\pm\tfrac{2}{6}\pi$, $\pm\tfrac{3}{6}\pi$, $\pm\tfrac{4}{6}\pi$, $\pm\tfrac{5}{6}\pi$, $\pm\tfrac{6}{6}\pi$. The resulting values, together with the slopes as obtained above, are amply sufficient for drawing a figure, as shown annexed.

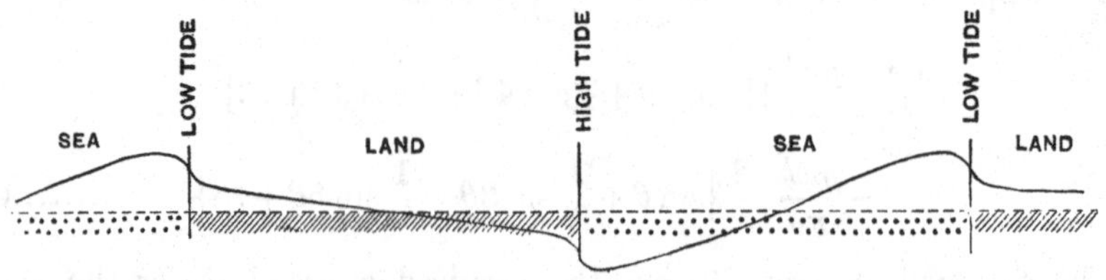

The straight line is a section of the undisturbed level, the shaded part being land, and the dotted sea. The curve shows the distortion, when warped by high and low-tide as indicated.

The scale of the figure is a quarter of an inch to $\tfrac{1}{6}\pi$ for the abscissas, and a quarter of an inch to unity for the ordinates; it is of course an enormous exaggeration of the flexure actually possibly due to tides.

It is interesting to note that the land regions remain very nearly flat, rotating about the nodal line, but with slight curvature near the coasts. It is this curvature, scarcely perceptible in the figure, which is of most interest for practical application.

The series (8) and (9) are not convenient for practical calculation in the neighbourhood of the coast, and they must be reduced to other forms. It is easy, by writing the cosines in their exponential form, to show that

$$\cos\theta + \tfrac{1}{2}\cos 2\theta + \tfrac{1}{3}\cos 3\theta + \dots = -\log_e(\pm 2\sin\tfrac{1}{2}\theta) \quad\text{.........(13)}$$

$$\cos\theta - \tfrac{1}{2}\cos 2\theta + \tfrac{1}{3}\cos 3\theta + \dots = \log_e(2\cos\tfrac{1}{2}\theta) \quad\text{............(14)}$$

Where the upper sign in (13) is to be taken for positive values of θ and the lower for negative.

For the small values of θ, for which alone we are at present concerned, the series (13) becomes $-\log_e(\pm\theta)$ and the lower $\log_e 2$.

Taking half the difference and half the sum of the two series we have

$$\tfrac{1}{2}\cos 2\theta + \tfrac{1}{4}\cos 4\theta + \dots = -\tfrac{1}{2}\log(\pm\theta) - \tfrac{1}{2}\log 2 \quad\text{.........(15)}$$

$$\cos\theta + \tfrac{1}{3}\cos 3\theta + \tfrac{1}{5}\cos 5\theta + \dots = -\tfrac{1}{2}\log(\pm\theta) + \tfrac{1}{2}\log 2 \quad\text{.........(16)}$$

Integrating (16) with regard to θ, and observing that the constant introduced on integration is zero, we have

$$\sin\theta + \frac{1}{3^2}\sin 3\theta + \frac{1}{5^2}\sin 5\theta = -\tfrac{1}{2}\theta\left[\log(\pm\theta) - 1\right] + \tfrac{1}{2}\theta\log 2 \quad\text{...(17)}$$

Then from (15) and (17)

$$\tfrac{1}{2}\cos 2\theta + \tfrac{1}{4}\cos 4\theta + \dots - \frac{2}{\pi}\left\{\sin\theta + \frac{1}{3^2}\sin 3\theta + \dots\right\}$$

$$= -\tfrac{1}{2}\left(1 - \frac{2\theta}{\pi}\right)\log(\pm\theta) - \tfrac{1}{2}\left(1 + \frac{2\theta}{\pi}\right)\log 2 - \frac{\theta}{\pi} \quad\text{...(18)}$$

Integrating (15), and observing that the constant is zero, we have

$$\frac{1}{2^2}\sin 2\theta + \frac{1}{4^2}\sin 4\theta + \dots = -\tfrac{1}{2}\theta\left[\log(\pm\theta) - 1\right] - \tfrac{1}{2}\theta\log 2 \quad\text{......(19)}$$

Integrating (17) and putting in the proper constant to make the left-hand side vanish when $\theta = 0$, we have

$$\frac{1}{1^3} + \frac{1}{3^3} + \frac{1}{5^3} + \dots - \left(\frac{1}{1^3}\cos\theta + \frac{1}{3^3}\cos 3\theta + \dots\right)$$

$$= -\tfrac{1}{4}\theta^2\log(\pm\theta) + \tfrac{1}{4}\theta^2\left(\tfrac{3}{2} + \log 2\right) \quad\text{......(20)}$$

For purposes of practical calculation θ may be taken as so small that the right-hand side of (18) reduces to $-\tfrac{1}{2}\log(\pm 2\theta)$, and the right-hand sides of (19) and (20) to zero.

Hence by (8) and (9), we have in the neighbourhood of the coast

$$\left.\begin{aligned} \alpha &= \frac{gwh}{\pi v} \times \frac{2l}{\pi^2}\left[\frac{1}{1^3}+\frac{1}{3^3}+\frac{1}{5^3}+\dots\right] \\ &= \frac{gwh}{\pi v} \times \frac{l}{\pi^2} \times 2{\cdot}1037 \\ \frac{d\alpha}{dz} &= -\frac{gwh}{2\pi v}\log_e 10 \log_{10}\frac{2\pi z}{l} \end{aligned}\right\} \qquad \text{(21)}$$

I shall now proceed to compute from the formulæ (21) the depression of the surface and the slope, corresponding to such numerical data as seem most appropriate to the terrestrial oceans and continents.

Considering that the tides are undoubtedly augmented by kinetic action, we shall be within the mark in taking h as the semi-range of equilibrium tide. At the equator the lunar tide has a range of about 53 cm., and the solar tide is very nearly half as much. Therefore at the spring-tides we may take $h = 40$ cm. It must be noticed that the highness of the tides, say 15 or 20 feet, near the coast is due to the shallowing of the water, and it would not be just to take such values as representing the tides over large areas; w, the density of the water is, of course, unity.

If we suppose it is the Atlantic Ocean and the shores of Europe with Africa, and of North and South America, which are under consideration, it is not unreasonable to take l as 3,900 miles or $6{\cdot}28 \times 10^8$ cm. Then

$$2\pi z/l = z \times 10^{-8}$$

Taking v/g as 3×10^8, that is to say, assuming a rigidity greater than that of glass, we have for the slope in seconds of arc, at a distance z from the sea-shore

$$\left.\begin{aligned} &\operatorname{cosec} 1'' \times \frac{40}{2\pi \times 3 \times 10^8} \times \log_e 10 \times (8 - \log_{10} z) \\ &\quad = 0''{\cdot}01008\,(8 - \log_{10} z) \end{aligned}\right\} \qquad \text{(22)}$$

From this the following table may be computed by simple multiplication:—

Distance from mean water-mark		Slope
1 cm.	= 1 cm.	0″·0806
10 cm.	= 10 cm.	·0706
10^2 cm.	= 1 metre	·0605
10^3 cm.	= 10 m.	·0504
10^4 cm.	= 100 m.	·0403
10^5 cm.	= 1 kilom.	·0302
10^6 cm.	= 10 kilom.	·0202
2×10^6 cm.	= 20 kilom.	·0170
5×10^6 cm.	= 50 kilom.	·0131
10^7 cm.	= 100 kilom.	·0101

On considering the formula (22) it appears that z must be a very small fraction of a millimetre before the slope becomes even as great as 1′. This proves that the rounded nick in the surface, which arises from the discontinuity of pressure at our ideal mean water-mark, is excessively small, and the vertical displacement of the surface is sensibly the same, when measured in centimetres, on each side of the nick, in accordance with the first of (21).

The result (5) of section 1 shows that, with rigidity 3×10^8, the true deflection of plumb-line due to attraction of the water is a quarter of the slope. Hence an observer in a gravitational observatory at distance z from mean water-mark, would note deflections from the mean position of the vertical $1\frac{1}{4}$ times as great as those computed above. And as high water changes to low, there would be oscillations of the vertical $2\frac{1}{2}$ times as great. We thus get the practical results in the following table:—

Distance of observatory from mean water-mark	Amplitude of apparent oscillation of the vertical
10 metres	0″·126
100 m.	·101
1 kilom.	·076
10 kilom.	·050
20 kilom.	·042
50 kilom.	·035
100 kilom.	·025

It follows, from the calculations made for tracing the curve, that halfway across the continent (that is to say, 3,142 kilometres from either coast) the slope is $\frac{648{,}000}{\pi} \times \frac{gwh}{\pi v} \times \cdot 2703$ seconds of arc, $= 0''\cdot 00237$; and the range of apparent oscillation is 0″·006.

In these calculations the width of the sea is taken as 6,283 kilometres. If the sea be narrower, then to obtain the same deflections of the plumb-line, the observatory must be moved nearer the sea in the same proportion as the sea is narrowed. If, for example, the sea were 3,142 kilometres wide, then at 10 kilometres from the coast the apparent amplitude of deflection is 0″·042. If the range of tide is greater than that here assumed (viz., 80 cm.), the results must be augmented in the same proportion. And, lastly, if the rigidity of the rock be greater or less than the assumed value (viz., 3×10^8) the part of the apparent deflection depending on slope must be diminished or increased in the inverse proportion to the change in rigidity.

I think there can be little doubt that in narrow seas the tides are generally much greater than those here assumed; and it is probable that at a gravitational observatory actually on the sea-shore on the south-coast of England, apart from seismic changes, perceptible oscillations of the vertical would be noted.

Sir William Thomson has made an entirely independent estimate of the probable deflection of the plumb-line at a sea-side gravitational observatory*. He estimates the attraction of a slab of water, 10 feet thick (the range of tide), 50 miles broad perpendicular to the coast, and 100 miles long parallel with the coast, on a plummet 100 yards from the low water mark, and opposite the middle of the 100 miles of length. He thinks this estimate would very roughly represent the state of things say at St Alban's Head. He finds then that the deflection of the plumb-line as high-tide changes to low would be $\frac{1}{4000000}$th of the unit angle or 0''·050. The general theorem proved above, as to the proportionality of slope to attraction, shows that, with rigidity 3×10^8 for the rocks of which the earth is formed, the apparent deflection of the plumb-line would amount to 0''·25.

It is just possible that a way may in this manner be opened for determining the modulus of rigidity of the upper 100 or 200 miles of the earth's surface, although the process would be excessively laborious. The tides of the English Channel are pretty well known, and therefore it would be possible by very laborious quadratures to determine the deflection of the plumb-line due to the attraction of the tide at any time at a chosen station. If then the deflection of the plumb-line could be observed at that station (with corrections applied for the positions of the sun and moon), the ratio of the calculated to the observed and corrected deflection, together with the known value of the earth's radius and mean density, form the materials for computing the rigidity. But such a scheme would be probably rendered abortive by just such comparatively large and capricious oscillations of the vertical, as we, M. d'Abbadie and others, have observed.

It is interesting to draw attention to some observations of M. d'Abbadie on the deflections of the vertical due to tides. His observatory (of which an account was given in the Report for 1881) is near Hendaye in the Pyrenees, and stands 72 metres above, and 400 metres distant from, the sea. He writes†:—

"J'ai réuni 359 comparaisons d'observations spéciales faites lors du maximum du flot et du jusant; 243 seulement sont favorables à la théorie de l'attraction exercée par la masse des eaux, et l'ensemble des résultats pour une différence moyenne de marées égale à 2·9 mètres donne un résultat moyen de 0''·56 ou 0''·18 pour le double de l'attraction angulaire vers le Nord-Ouest. Ceci est conforme à la théorie, car les différences observées doivent être partagées par moitié, selon la loi de la réflexion; mais comme il y a toujours de l'inattendu dans les expériences nouvelles, on doit ajouter que sur les 116 comparaisons restantes il y en a eu 57 où le flot semble repousser le mercure au lieu de l'attirer. Mes résultats ont été confirmés pendant

* Thomson and Tait's *Nat. Phil.* § 818.

† "Recherches sur la Verticale." *Ann. de la Soc. Scient. de Bruxelles*, 1881.

l'hiver dernier par M. l'abbé Artus, qui a eu la patience de comparer ainsi 71 flots et 73 jusants consécutifs, de janvier à mars 1880. Lui aussi a trouvé un tiers environ de cas défavorables à nos théories admises. On est donc en droit d'affirmer que si la mer haute attire le plus souvent le pied du fil à plomb, il y a une, et peut-être plusieurs, autres forces en jeu pour faire varier sa position."

We must now consider the vertical displacement of the land near the coast. In (21) it is shown to be $\alpha_0 = \frac{gwh}{\pi v} \times \frac{l}{\pi^2} \times 2{\cdot}1037$, where α_0 indicates the displacement corresponding to $z = 0$.

With the assumed values, $h = 40$, $v = 3 \times 10^8$, $l = 6{\cdot}28 \times 10^8$, I find $\alpha_0 = 5{\cdot}684$ cm. Hence the amplitude of vertical displacement is 11·37 cm. As long as hl remains constant this vertical displacement remains the same; hence the high tides of 10 or 15 feet which are actually observed on the coasts must probably produce vertical oscillations of quite the same order as that computed.

If the land falls the tide of course rises higher on the coast-line than it would do otherwise; hence the apparent height of tide would be $h + \alpha_0$. But this shows there is more water resting on the earth than according to the estimated value h; hence the depression of the soil is greater in the proportion $1 + \alpha_0/h$ to unity; this again causes more tide, which reacts and causes more depression, and so on. Thus on the whole the augmentation of tide due to elastic yielding is in the ratio of

$$1 + \frac{\alpha_0}{h} + \left(\frac{\alpha_0}{h}\right)^2 + \left(\frac{\alpha_0}{h}\right)^3 + \ldots \text{ or } \frac{1}{1 - \alpha_0/h} \text{ to unity}$$

This investigation is conducted on the equilibrium theory, and it neglects the curvature of the sea bed, assuming that there is a uniform slope from mid-ocean to the sea-coast. The figure shows that this is not rigorously the case, but it is quite near enough for a rough approximation. The phenomena of the short period tides are so essentially kinetic that the value of this augmentation must remain quite uncertain, but for the long-period tides (the fortnightly and monthly elliptic) the augmentation must correspond approximately with the ratio

$$1 : \left(1 - \frac{gwl}{\pi^3 v} \times 2{\cdot}1037\right)$$

The augmentation in narrow seas will be small, but in the Atlantic Ocean the augmenting factor must agree pretty well with that which I now compute.

With the previous numerical values we have α_0/h (which is independent of h) equal to ·1421, and $1 - \alpha_0/h = {\cdot}8579 = \frac{6}{7}$ very nearly.

Thus the long-period tides may probably undergo an augmentation at the coasts of the Atlantic in some such ratio as 6 to 7.

The influence of this kind of elastic yielding is antagonistic to that reduction of apparent tide, which must result from an elastic yielding of the earth's mass as a whole.

The reader will probably find it difficult to estimate what degree of probability of correctness there is in the conjectural value of the rigidity, which has been used in making the numerical calculations in this paper. The rigidity has not been experimentally determined for many substances, but a great number of experiments have been made to find Young's modulus. Now, in the stretching of a bar or wire the compressibility plays a much less important part than the rigidity, and the formula for Young's modulus shows that for an incompressible elastic solid the modulus is equal to three times the rigidity*. Hence a third of Young's modulus will form a good standard of comparison with the assumed rigidity, namely, 3×10^8 grammes weigh. per square centimetre. The following are a few values of a third of Young's modulus and of rigidity, taken from the tables in Sir William Thomson's article on Elasticity† in the *Encyclopædia Britannica.*

Material	A third of Young's mod. and rigidity in terms of 10^8 grammes weight per sq. cm.
Stone	About 1·2
Slate	About 3 to 4
Glass	Rigidity 1·5 to 2·4
Ice	4·7
Copper	4 and rigidity 4·6 to 5·4
Steel	7 to 10 and rigidity 8·4

It will be observed that the assumed rigidity 3 is probably a pretty high estimate in comparison with that of the materials of which we know the superficial strata to be formed.

It is shown, in another paper read before the Association at this meeting, that the rigidity of the earth as a whole is probably as great as that of steel. That result is not at all inconsistent with the probability of the assumption that the upper strata have only a rigidity a little greater than that of glass‡.

3. *On Gravitational Observatories.*

In the preceding sections estimates have been made of the amount of distortion which the upper strata of the earth probably undergo, from the shifting weights corresponding to barometric and tidal oscillations. These results appear to me to have an important bearing on the probable utility of gravitational observatories.

* Thomson and Tait's *Nat. Phil.* § 683.

† Also published separately by Black, Edinburgh.

‡ [See Paper 9, p. 340 above.]

It is not probable, at least for many years to come, that the state of tidal and barometric pressure, for a radius of 500 miles round any spot on the earth's surface, will be known with sufficient accuracy to make even a rough approximation to the slope of the surface a possibility. And were these data known, the heterogeneity of geological strata would form a serious obstacle to the possibility of carrying out such a computation. It would do little in relieving us from these difficulties to place the observatory at the bottom of a mine.

Accordingly the prospect of determining experimentally the lunar disturbance of gravity appears exceedingly remote, and I am compelled reluctantly to conclude that continuous observations with gravitational instruments of very great delicacy are not likely to lead to results of any great interest. It appears likely that such an instrument, even in the most favourable site, would record incessant variations of which no satisfactory account could be given. Although I do not regard it as probable that such a delicate instrument should be adopted for regular continuous observations, yet, by choosing a site where the flexure of the earth's surface is likely to be great, it is conceivable that a rough estimate might be made of the average modulus of elasticity of the upper strata of the earth for one or two hundred miles from the surface.

These conclusions, which I express with much diffidence, are by no means adverse to the utility of a coarser gravitational instrument, capable, let us say, of recording variations of level amounting to 1″ or 2″. If barometric pressure, tidal pressure, and the direct action of the sun and moon, combined together to make apparent slope in one direction, then at an observatory remote from the sea-shore, that slope might perhaps amount to a quarter of a second of arc. Such a disturbance of level would not be important compared with the minimum deviations which could be recorded by the supposed instrument.

It would then be of much value to obtain continuous systematic observations, after the manner of the Italians, of the seismic and slower quasi-seismic variations of level.

I venture to predict that at some future time practical astronomers will no longer be content to eliminate variations of level merely by taking means of results, but will regard corrections derived from a special instrument as necessary to each astronomical observation.

INDEX TO VOLUME I.

A

Abacus, tidal—an apparatus for facilitating harmonic analysis, 216

Abbadie, A. d'—deflection of the vertical, 416, 420, 434, 439, 440

Adams, J. C.—on notation for harmonic analysis, 2; confirms numerical values of coefficients, 26; fusion of lunar and solar K_1, K_2 tides, 43; equivalent multipliers for tides of long period, 66; B. A. Reports, 70, 97; correction to B.A. Report of 1872, 345

Admiralty Scientific Manual—paper on Tides from, 117

Airy, Sir G. B.—Tides and Waves, 156; changes in azimuth of transit, 417

Analysis—*see* Harmonic

Antarctic tidal observations—reduction of, 372

Armellini—seismological observations, 435

B

Baird, Col. A. W.—in charge of Indian Tidal Survey, 2; auxiliary tables for harmonic analysis, 68; Manual for tidal observations, 156

Barne, Lieut. M., R.N.—method of observing tides on the 'Discovery,' 372

Barometric pressure—apparent deflection of vertical due to, 444

Bertelli, Prof.—seismological observations, 431, 434

Bouquet de la Grye—deflection of vertical at Campbell Island, 417, 438; form of seismometer, 438

British Association Reports on harmonic analysis, 1, 2, 3, 71, 341; on lunar disturbance of gravity, 389, 430

Burdwood—directions for reducing tidal observations, 121

C

Christie, Sir W. H. M.—vibrations of the ground at Greenwich, 428

D

d'Abbadie, *see* Abbadie

Darwin, Horace—instrument for measuring deflections of the vertical, 389; experiment on variations of surface soil, 425

Deflection of vertical—d'Abbadie, Plantamour, 416, 417; Nyrèn, 422, 456

'Discovery'—antarctic tidal observations, 372

Dynamical theory of tides, 347, 366

E

Earth, rigidity of—estimated from tides of long period, 340; Hecker on, 346

Earth's surface—changes of level due to elasticity of, 444

Eccles, J.—on treatment of M_1 tide, 39

Egidi, Prof.—seismological observations, 433

Elasticity of earth's surface—changes of level due to, 444

Encyclopædia Britannica, extract from—on the dynamical theory of the tides, 347

Equilibrium theory of tides—correction for continents, 328

F

Ferrel, Prof.—tidal results of U. S. Coast Survey, 70

G

Gaillot—changes of level at Paris, 443

Galli, D. G.—seismological observations, 432

Glazebrook, R. T.—antarctic tide-curves for verification of antarctic reductions, 381, 386

Grablovitz, G.—seismological observations, 434

Gruithuisen, Prof.—on horizontal pendulum, 416

H

Harmonic analysis—*see* table of contents Paper 1, 1; Indian computation forms, 4; choice of periods for analysis, 71; comparison with method of hour-angles and parallaxes, 78; treatment of a short series of observations, 98, 122; of high and low-water observations, 157; an instrument for facilitating, 216

Harmonic analysing instrument, 49

Hecker, Dr O.—rigidity of earth determined by horizontal pendulum, 346

Hengler—fraudulent account of horizontal pendulum, 416

Henry—changes in transit at Greenwich, 417

Hough, S. S.—introduction of mutual gravitation of water in dynamical theory of tides, 349

I

Indian tidal operations—Colonel Baird in charge of, 2

K

Kelvin, Lord—use of *astres fictifs* in equilibrium theory of tides, 3; on tidal instruments, 48; harmonic analysing instrument, 49; correction of equilibrium theory for land, 328; oscillations of rotating water, 349; tides of long period, 366; instrument for measuring deflections of vertical, 389; computation of deflection of vertical due to the tide, 456; values of modulus of rigidity of rock, 458

L

Laplace—*astres fictifs* in equilibrium theory, 3; theory of tides in *Mécanique Céleste*, 349, 367

Latitude—variations of, 422, 426, 443

Levels—Plantamour's observations with, 420

Levels, Datum—for tidal observations, 97

Long period tides—equivalent multipliers in reduction of, 66; dynamical theory of, 366; rigidity of earth estimated from, 340, 371

Lunar deflection of vertical—Hecker measurement of, 346; attempted measurement of, 389; theoretical amount of, 414

M

Malvasia, Count—observation as to local tremors, 432

Microphone as seismological instrument, 435

Milne, Prof. J.—seismological observations, 437

Mocenigo, Count—seismological observations, 435

Monte, Pietro—seismological observations, 431

N

Nadirane—d'Abbadie's instrument, 418, 439—441, 456

National Physical Laboratory—tide-curves for verification of antarctic tidal reductions, 381, 386

Nyrèn—distant earthquakes observed by, 422

O

Observations, tidal—harmonic analysis, 1; of short series, 98, 122; of high and low water, 120; Whewell's treatment of, 120

P

Pendulum, horizontal—Zöllner on, 413; suggestion by Perrot, 416; Šafařik on, 416; Hengler's fraudulent account of, 416; Gruithuisen on, 416

Pendulum, form of—devised by Bouquet de la Grye for observing deflection of vertical, 417, 427, 438

Perrot—suggestion of horizontal pendulum, 416

Peters—amount of lunar deflection of gravity, 414

Plantamour—variations of level, 416; observations with levels, 420

Poincaré, H.—dynamical theory of tides in ocean interrupted by land, 349

Poiseüille—viscosity of alcohol and water, 406

Prediction of tides—Kelvin on instrument for, 48; from approximate tidal constants, 104, 141; method of, 258; comparison with actuality, 325

R

Rigidity of earth—estimated by tides of long period, 340; by horizontal pendulum, 346

Roberts, E.—in charge of Indian tidal prediction, 1, 36; computation forms, 49; correction of the same, 345

Rossi, S. M. de—seismological observations, 431, 435

S

Šafařik—on horizontal pendulum, 416
Scottish antarctic tidal observations, 387
Sea-level, datum—for tidal observations, 97
Sea-level—slow oscillations of, 116
Seismological observations—Italian, 431; Japanese, 437
Selby—reduction of Scottish antarctic tidal observations, 387
Strachey, Sir R.—rules for harmonic analysis, 101, 125, 132

T

Thomson, James—harmonic analyser, 49
Thomson, Sir William—also Thomson and Tait, *see* Kelvin, Lord
Tides—*see* several other titles
Tides—distortion of coast due to weight of, 450
Tide-table—*see* Prediction
Transit—variation of piers of, 417, 422, 443
Turner, Prof. H. H.—correction of equilibrium theory for continents, 328

U

United States Coast Survey—tidal results, 70

V

Vertical, deflection of—an instrument for measuring, 389

W

Whewell, Dr—treatment of tidal observations, 120
Wilson, Dr—antarctic observations in summer, 385

Z

Zöllner—on the horizontal pendulum, 413

CAMBRIDGE: PRINTED BY JOHN CLAY, M.A. AT THE UNIVERSITY PRESS.

Printed in the United States
By Bookmasters

Printed in the United States
By Bookmasters